ANNUAL REVIEW OF
EARTH AND
PLANETARY SCIENCES

EDITORIAL COMMITTEE (1978)

ANNUAL REVIEW OF EARTH AND PLANETARY SCIENCES

FRED A. DONATH, *Editor*
University of Illinois – Urbana

FRANCIS G. STEHLI, *Associate Editor*
Case Western Reserve University

GEORGE W. WETHERILL, *Associate Editor*
Carnegie Institution of Washington

VOLUME 6

1978

ANNUAL REVIEWS INC.   4139 EL CAMINO WAY   PALO ALTO, CALIFORNIA 94306

ANNUAL REVIEWS INC.
Palo Alto, California, USA

REPRINTS   The conspicuous number aligned in the margin with the title of
each article in this volume is a key for use in ordering reprints. Available reprints
are priced at the uniform rate of $1.00 each postpaid. The minimum acceptable
reprint order is 10 reprints and/or $10.00 prepaid. A quantity discount is available.

International Standard Serial Number: 0084-6597
International Standard Book Number: 8243-2006-9
Library of Congress Catalog Card Number: 72-82137

Annual Reviews Inc. and the Editors of its publications assume no responsibility
for the statements expressed by the contributors of this Review.

FILMSET BY TYPESETTING SERVICES LTD, GLASGOW, SCOTLAND
PRINTED AND BOUND IN THE UNITED STATES OF AMERICA

PREFACE

With the appearance of the sixth volume of the *Annual Review of Earth and Planetary Sciences*, we think it worthwhile to explain to readers and future contributors what we are trying to accomplish in these volumes.

As we see it, the central purpose of this series is to make it easier for earth and planetary scientists to cope with the rising volume of research papers in every branch of our science. We attempt to accomplish this by inviting creative reviews of work on topics of high current interest from those persons best qualified to prepare such reviews. We expect that every author who accepts our invitation will provide a good review on schedule. Of course, some articles fail to appear for various reasons, but every default is a disappointment to us.

By a "good review" we mean more than an interesting, informed, and important paper on a timely subject. At the editorial committee meetings at which these volumes are planned, it frequently happens that "hot topics" are suggested, topics associated with the names of particular scientists. Upon further discussion it often becomes apparent that this association arose because essentially all the new and exciting work on the subject had come from the proposed author's own work or that of his research group. In spite of their great scientific interest, such suggested topics are usually not accepted; we believe that the regular scientific journals are the proper place for these topics to be presented. For the *Annual Review of Earth and Planetary Sciences*, we seek to cover topics of similar importance for which the journals contain such a large number of specialized, conflicting, and often non-interacting papers that readers would have to devote an inordinate amount of time to bring order into the apparent chaos. We hope to help readers to understand how intelligent, conscientious workers can come to opposing conclusions. To write the most useful kinds of reviews, an author must stand back and view his subject in perspective, strive to rise above the natural tendency to emphasize his own contributions, and occasionally exercise an unusual degree of generosity in the treatment of the work of his "competitors."

We require "critical reviews," but these must do much more than criticize. They should extract that which is useful even from flawed research, and synthesize this into a product of value to a broad class of readers that includes specialists, non-specialists, and students trying to find their way into a subject. Although we, like everyone else, are sometimes unsuccessful, these remain our goals.

THE EDITORS AND THE EDITORIAL COMMITTEE

SOME RELATED ARTICLES APPEARING IN OTHER ANNUAL REVIEWS

From the *Annual Review of Astronomy and Astrophysics*, Volume 15 (1977)
  The Origin of Solar Activity, E. N. Parker
  Mercury, D. E. Gault, J. A. Burns, P. Cassen, and R. G. Strom
  The Interaction of Supernovae with the Interstellar Medium, Roger A. Chevalier
  Mass and Energy Flow in the Solar Chromosphere and Corona, George L.
    Withbroe and Robert W. Noyes
  Jupiter's Magnetosphere, C. F. Kennel and F. V. Coroniti

From the *Annual Review of Biochemistry*, Volume 46 (1977)
  A Long View of Nitrogen Metabolism, Sarah Ratner
  Biochemical Evolution, Allan C. Wilson, Steven S. Carlson, and Thomas J. White

From the *Annual Review of Ecology and Systematics*, Volume 8 (1977)
  The Evolution of Viruses, Andre J. Nahmias and Darryl C. Reanney
  Carbon Balance in Terrestrial Detritus, William H. Schlesinger
  Control of Forest Growth and Distribution on Wet Tropical Mountains,
    P. J. Grubb
  The Evolution of Life History Traits, Stephen C. Stearns
  The Ecology of Fish Migrations, William C. Leggett
  A History of Savanna Vertebrates in the New World (Part I), S. David Webb
  Circannual Rhythms in Bird Migration, Eberhard Gwinner
  Relative Brain Size and Behavior in Archosaurian Reptiles, James A. Hopson

From the *Annual Review of Fluid Mechanics*, Volume 10 (1978)
  Objective Methods for Weather Prediction, C. E. Leith
  River Meandering, R. A. Callander
  Turbulence and Mixing in Stably Stratified Waters, Frederick S. Sherman, Jorg
    Imberger, and Gilles M. Corcos
  Numerical Methods in Water-Wave Diffraction and Radiation, Chiang C. Mei
  Magnetohydrodynamics of the Earth's Dynamo, F. H. Busse

From the *Annual Review of Materials Science*, Volume 7 (1977)
  Defect Chemistry in Crystalline Solids, F. A. Kröger
  Acoustic Emission in Brittle Materials, A. G. Evans and M. Linzer

From the *Annual Review of Microbiology*, Volume 31 (1977)
  Phototrophic Prokaryotes: the Cyanobacteria, R. Y. Stanier and G. Cohen-Bazire
  Phototrophic Green and Purple Bacteria: A Comparative, Systematic Survey,
    Norbert Pfennig
  Biogenesis of Methane, R. A. Mah, D. M. Ward, L. Baresi, and T. L. Glass
  Oil Tankers and Pollution: A Microbiological Approach, David L. Gutnick and
    Eugene Rosenberg

From the *Annual Review of Nuclear Science*, Volume 27 (1977)
  Element Production in the Early Universe, David N. Schramm and
    Robert V. Wagoner

Annual Review of Earth and Planetary Sciences
Volume 6, 1978

CONTENTS

ANNUAL REVIEWS INC. is a nonprofit corporation established to promote the advancement of the sciences. Beginning in 1932 with the *Annual Review of Biochemistry,* the Company has pursued as its principal function the publication of high quality, reasonably priced Annual Review volumes. The volumes are organized by Editors and Editorial Committees who invite qualified authors to contribute critical articles reviewing significant developments within each major discipline.

Annual Reviews Inc. is administered by a Board of Directors whose members serve without compensation.

Annual Reviews are published in the following sciences: Anthropology, Astronomy and Astrophysics, Biochemistry, Biophysics and Bioengineering, Earth and Planetary Sciences, Ecology and Systematics, Energy, Entomology, Fluid Mechanics, Genetics, Materials Science, Medicine, Microbiology, Neuroscience, Nuclear Science, Pharmacology and Toxicology, Physical Chemistry, Physiology, Phytopathology, Plant Physiology, Psychology, and Sociology. In addition, two special volumes have been published by Annual Reviews Inc.: *History of Entomology* (1973) and *The Excitement and Fascination of Science* (1965).

Fred L. Whipple

Ann. Rev. Earth Planet. Sci. 1978. 6:1–8

THE EARTH AS PART OF THE UNIVERSE ⋊10083

Fred L. Whipple

Center for Astrophysics, Cambridge, Massachusetts 02138

My active scientific life now spans half a century, a half century of breathtaking progress in understanding our Earth and the universe, a half century of constantly accelerating science, technology, and their applications. In this time we have learned more about the interior of the Moon than we knew about the interior of the Earth when I was born. Probably the same is true for Mars, Mercury, and perhaps Venus. Comparative planetology has become a viable subject for concentrated study. Day by day a sounder picture forms of the evolutionary processes that brought these solid planets into existence. Even so, the Earth still stands unique in the known universe. We have yet to observe liquid water elsewhere, or for that matter any liquid, although Mars exhibits convincing evidence for past episodes of torrential flooding.

This article—highly personal, following the policy of the editors—portrays the development of one astronomer's view of the Earth. I was raised on an Iowa farm until the age of 15, with no formal training or parental background involving the intellectual or scientific world. I did have a flair for arithmetic and mechanical devices, accompanied by an inborn fascination with the infinities of time and space and possessed by a stubborn curiosity. Dating some memories: by World War I before the age of ten I was very conscious of the immensity of the Earth and its global features, and impressed by events such as the sinking of the *Titanic* and *Lusitania* and the unparalleled ballistic range of the German Big Berthas, a fantastic 40 miles. The mystery of glaciation and the nature of its processes were brought to my attention in an elementary geology course at Occidental College, my only formal geological training. But my weak long-term memory for dissociated facts such as rock identification kept me from a direct career in Earth processes. At UCLA my mathematics major veered me through physics and finally focused on astronomy where time, space, mathematics, and physics had a common meeting ground.

AGES

The Earth has aged dramatically during the three centuries of the scientific era. The estimated age doubled on the average every 16 years since the middle of the

0084-6597/78/0515-0001$01.00

17th century when Bishop Ussher concluded that the Earth was made in 4004 BC. I have long admired geologist James Hutton for his astuteness and daring, a little more than a century later, to voice the opinion that the geological record set no upper limit on the possible age of the Earth. Early in this century few would hazard an age above 10^9 years, or an aeon by Harold Urey's definition.

Edwin Hubble first extended the age determination to include the known universe. By the 1920s astronomers definitely realized that in stars the atomic packing fractions and Einstein's $E = mc^2$ must permit energy generation by atomic mutations to account for the long life of the Sun. Uncertainty about the detailed processes, however, prevented the calculation of meaningful ages. Hubble's concept about the red shift of galaxies, taken only as a red shift proportional to the distance of the galaxies, provided a link that led to a measure of time. His evaluation of the constant, $H = 500\,\mathrm{km\,sec^{-1}\,10^{-6}\,pc^{-1}}$ (1 pc $= 3.1 \times 10^{18}$ cm), is an inverse time, or an "age" of 1.95 aeons. Note that Hubble carefully avoided calling the red shift a velocity-expansion effect. His determination in 1929, coupled with Claire Patterson's lead-lead age determination of 4.55 aeons in 1955, still doubled the age in 21 years, but the whole universe had become involved. In the late 1950s, however, the Earth and the solar system suddenly reached maturity. Their aging stopped abruptly. The unbelievable precision of laboratory measurements by the strontium-rubidium and other isotopic methods placed the age of the Earth definitively at 4.6 aeons. Indeed the meteoritic studies and the great Apollo return of samples from Moon now leave an uncertainty of only a decimal of an aeon for the time since the Earth, Moon, and meteorites came into existence. There can be little question that this also represents closely the age of the other planets, the Sun and the comets as well.

Oddly enough the universe keeps on aging at much the same rate that the Earth did earlier. Current values of the Hubble constant concentrate in the range 50–$60\,\mathrm{km\,sec^{-1}\,10^{-6}\,pc^{-1}}$ with a few estimates as high as 75 and rare ones as low as 40. The value of $H = 60$, probably the most favored value today, gives 16 aeons as the age of the universe, or a doubling in each 16 years since Hubble's first determination, exactly the rate from Ussher to Hubble. Our galaxy, however, remains a bit vague about her age; perhaps she is about 10 aeons old, having mothered the solar system in middle life as measured to date.

The interpretation of the inverse Hubble constant as the age of the universe is, of course, subject to question. But few astronomers doubt that the red shift is a measure of distance and time in the universe, leading us back more than 10 aeons to the most remote and earliest phenomena directly observable. The next logical step and perhaps the most popular view concludes that we live in an expanding universe that does not contain enough matter to stop the expansion gravitationally. I consider this point of view premature. We still can only speculate generally about unobservable or dark material in the universe. And is it 99.9% certain that the red shift is a velocity measure? But perhaps we really do live in a "black hole" and should talk only about *our* universe, not *the* universe, any number of universes being possible, though isolated observationally.

On the other hand, the great red shifts observed for distant galaxies and quasars combined with the other astronomical measures over the wavelength range from gamma rays to long radio waves give one a sense of confidence about the physical constants. The spectroscopic constants appear to be unaltered in ancient sources (see e.g. Pagel 1977), and the local velocity of light is extraordinarily constant to the order of 1 part in 10^{11} over a corresponding frequency range (pulsars). Thus I should be extremely surprised if any of the physical constants used in geophysics should have changed significantly during the history of the solar system. The Brans-Dicke scalar tensor theory based on Mach's principle predicts that the gravitation constant decreases with time and that various scalar quantities should change. Observational support for the theory, however, has not materialized. As a result we have no substantive evidence that any physical constants depend either on time or position in the universe.

My only direct activity in the age determinations was a continuous monitoring of the work by F. A. Paneth, who measured the helium content of meteorites for many years. He calculated ages from the expected helium production by radioactive atoms such as uranium and thorium. When these ages began to exceed the accepted ages by large factors, Carl A. Bauer, a graduate student, looked into the size distribution of meteorites versus Paneth's helium contents. We thought that helium might be leaking from meteorites, less from large ones than small ones. Bauer's results were quite surprising in that large meteorites showed consistently low helium content, whereas small ones showed a great range from small to large. The interpretation was clear: the helium must be induced by cosmic rays. A large meteorite on surviving passage through the atmosphere loses a thick layer in ablation, a layer that had protected the interior from cosmic-ray exposure. A small meteorite loses much less material by ablation in the atmosphere, and hence receives a much greater exposure to cosmic rays. Some small ones, of course, are pieces from larger bodies. It finally turned out that the residence times in interplanetary space from breakup to encounter with the Earth range up to 0.5 or more aeons for irons but only up to 50×10^6 years for stones. Hence stony meteorites must be broken off from Earth-crossing asteroidal bodies whereas the irons survive long enough to be broken off from the asteroids somewhere in the asteroid belt, finally to be perturbed into Earth-crossing orbits, the half-life for which is $\sim 50 \times 10^6$ years.

THE UPPER ATMOSPHERE

The atmosphere from altitudes of 50–110 km became of interest to me as I began the serious study of meteors. This study came about largely through the theory of Ernst J. Öpik that we are daily encountering meteoritic bodies from interstellar space, so-called hyperbolic meteors. I was eager to prove this theory because it meant that we might actually recover material coming from beyond the solar system. Öpik had shown that clouds of meteoritic and cometary material could be stable about stars for aeons without serious depredation by the gravitational effects of passing stars. Hence, for example, we might encounter meteors belonging to Sirius or other large nearby stars. To find the answer we set up at Harvard a double-station meteor photographic program with rotating shutters powered by the newly developed lightweight synchronous motors. Finally the program gave the answer. We are observing, alas, at least 99% solar-system meteors with no proven interlopers from outer space.

As partial compensation for this disappointment I discovered that the logarithmic gradient of atmospheric density in the meteor region dominates the light curves of meteors. The temperature of the atmosphere actually plays an inverse role. Where the temperature is low, the scale height is small and the density gradient large, so that meteors ablate more rapidly and become brighter more rapidly than in regions of higher temperature. The meteor, of course, depending upon its velocity, chooses a suitable atmospheric density to first become observable. Thus the study of the decelerations and light curves of doubly photographed meteors provided measures of the upper-atmospheric density gradients and, quite poorly, the atmospheric density itself. In the 1930s only indirect methods were available for measuring the density of the atmosphere above the relatively low altitudes attainable by large balloons. The best data came from anomalous sound propagation measures, mostly from the blowing up of post World War I ammunition dumps. Blast waves in the atmosphere can be turned back to the surface when passing through regions of positive temperature gradient. The upper part of the wave front moves faster than the lower, turning the wave front. F. J. W. Whipple, the well-known British meteorologist, collected the records and plotted on maps the locations where the explosions had been heard. He found that a temperature maximum occurs in the 40–60-km altitude region.

For some years the photographic meteors provided an improved measurement of atmospheric densities in the range of 50–110-km altitude. The low-temperature zone at the noctilucent-cloud region near 82-km altitude was clearly delineated, and the rapid rise in temperature above was evident from the meteor studies.

When the V-2 rockets produced by Germany for World War II were "liberated" for upper-atmospheric studies at the White Sands Proving Grounds of the U.S. Army, the situation quickly reversed. When it became possible to measure upper-atmospheric densities directly by rockets, these measures could then be combined with the meteoric data for determinations of meteoroid masses and densities. The result was surprising: meteoroids are essentially fragile low-density bodies averaging only ~ 0.4 g cm^{-3}. Some meteor showers exhibit densities less than 0.01 g cm^{-3}. The explanation came first from the meteor orbits about the Sun. Meteors are clearly cometary in origin. Secondly, my "dirty snowball" theory of comets confirmed the idea that meteoroids from comets should be friable and of low density. Indeed they are so weak structurally that none have been identified on the ground. Only in recent years has Donald E. Brownlee collected and identified tiny cometary particles in the high atmosphere.

A little publicized fact of the rocket studies from 1946 until the beginning of the space era in 1957 concerned the Upper Atmosphere Rocket Research Panel, which later became the Rocket and Satellite Research Panel. The first meeting occurred at Princeton University on February 27, 1946, one of the few meetings that I ever missed during the 12-year active life of the Panel. Originally membership consisted of representatives from the Naval Research Laboratory, the General Electric

Company, Princeton University, Harvard University, the University of Michigan, the U.S. Army Signal Corps, and Johns Hopkins University. Also present were representatives of the National Committee for Aeronautics (NACA), Wright Field, AAF, and the Signal Corps Electronic Laboratory. The UARRP had no official status, no budget, and no vested authority. But, for more than a decade, it controlled the policies and details of upper-atmospheric research because its members represented the scientists and the organizations involved in the research effort by rockets. Its final act before its effective dissolution in 1958 was to provide a considerable part of concept and language in the U.S. Senate Bill concerning the establishment of the National Aeronautics and Space Administration, the metamorphosed NACA. One member of the original committee became a scientific immortal— James A. Van Allen.

From today's vantage point of bureaucratic complexities and suffocating safeguards it is a nostalgic treat to recall the "free-wheeling" days when the military agencies, universities, and research laboratories, both industrial and private, could work together cooperatively and highly effectively on research and engineering problems involved in studying the newly exposed upper atmosphere. That post-war decade marked the transition period from miniscule to massive support of science by the Federal Government. The formation of the National Science Foundation, the great International Geophysical Year, and the opening of the Space Age were progressive steps in public-policy recognition that a bright future for society must depend on the judicious cultivation of science and technology.

The power of the scientific method has produced a wonderous technology whose complexity is beyond the scope of any single mind. Society today is completely dependent on the reasonably smooth operation of this gigantic complex. Science has always depended upon technology, which predated science. Now the two are entwined in a symbiotic explosion of knowledge and human action potential. Such are the dimensions of this explosion that our social and governing institutions can do no better than stagger under the load of utilizing the potential, on the one hand, and guiding its growth on the other. In periods of accelerated change such as this both individual and national goals become clouded. Fortunate are the scientists and those developing new technologies, because they have—in their professional roles at least—visible new and vast frontiers to conquer, an exciting and rewarding goal. The future always beckons as great concepts, such as the IGY and the Space Program, can be dreamed of and implemented.

SATELLITE GEODESY

As an astronomer I was appalled, in the early 1950s, at the dreadful inaccuracy of geodetic positions on a global scale determined by astronomical methods. For short distances radio-electronic methods were beginning to rival optical methods, but relative geodetic positions were at best good to ± 100 m between the continents and were much worse for islands, even as bad as a kilometer in error. I had long been an enthusiast for space travel. The V-2 rocket proved that it could be done, that we could indeed conquer space. Although my major long-term interest was the exploration of the Moon, Planets (and Comets), I saw the potential in artificial satellites for improving geodesy. All we needed were markers at high altitude for triangulation from ground stations. Artificial satellites were just right for the purpose, allowing for the continued progress in computer technology to match the demands of orbit calculation.

When dreams of artificial satellites began to solidify as real plans, in preparation for the International Geophysical Year, I realized that the techniques I had developed for photographing meteors could be modified for tracking satellites. The accuracy in global geodesy could clearly be increased by an order of magnitude, to ± 10 m. Fortunately there was support for the optical satellite tracking program. I could move the Smithsonian Astrophysical Observatory from Washington, D.C. to the Harvard College Observatory in Cambridge, Massachusetts so as to provide a scientific base for the operation. Twelve photographic camera stations distributed around the world with enormous international cooperation were the basic observing units. Communications, computing, and analysis were centered in Cambridge.

As an adjunct to the photographic program we anticipated the need for a world-wide net of visual observers to locate errant satellites and to track some on re-entry paths that could not be followed well enough for good positional predictions. The response of the Amateur Astronomers to band together in the *Moonwatch* program was exhilarating. In 1959 there were more than 200 teams distributed over the entire world and more than 100 were still active when the program ended in 1975.

I stress the Moonwatch activity here because there exists in modern societies a large contingent of highly competent nonprofessional people who are interested in contributing to science. They will respond to any phase of science where there is a real need and a proper leadership for some specific activity that lies within their power, which incidentally is difficult to overestimate. I am convinced that this valuable potential can be developed in many areas of science where widespread regions of the world are involved and where the scientific requirements are clear-cut. The Chinese earthquake program is an excellent current example. In Moonwatch the individual ingenuity, precision in execution, and devotion to the program were a joy to behold. Not only were important scientific results obtained from more than 100,000 observations, but the rapport between the professional and the nonprofessional helped to erase public misunderstandings about the meaning and significance of science in the modern world.

As for the satellite geodesy program, some ten years were required, rather longer than the 1.5-year IGY, to improve geodetic positions and the gravitational potential terms by an order of magnitude. A second decade improved the accuracy by another order of magnitude and before the third decade the laser and retroflectors on satellites will define the global system to an accuracy of a very few centimeters. Then we can observe continental drifts directly over all land masses.

The similar laser observations of retroreflectors planted on the Moon in the Apollo Program are now accurate to some 40 cm. The results have proven that any self energy carried in the gravitational field of the Earth has contributed equally to its inertial mass and to its gravitational mass (to 3% accuracy, Williams et al 1976). So many results of practical as well as scientific value result from the space program that one can only marvel!

SPACE EXPLORATION

The almost incredible potential of space exploration is now beginning to be realized. Economic forces may well accelerate the program in view of highly successful space ventures such as the communication satellites and the Earth Resources Program. The old science fiction scenarios of asteroid mining are now being considered seriously. Thus space exploration should not have to depend solely on the purely scientific and idealistic goals of exploration for its own sake, justifiable as are these goals fundamentally.

When asteroid mining becomes of interest to Wall Street the concomitant geological exploration will make the Earth-crossing asteroids as available for study as are the Rocky Mountains today. The detailed story of asteroid and planetary evolution can then be read along with and as a part of prospecting and subsequent mining operations. If some or many of these asteroids are old comet nuclei, as many suspect, they may or may not be good mineral deposits, but they certainly will provide basic knowledge of early solar-system history and evolution. Perhaps the volatile elements of comets will be of value. Water, at least at this moment, would be the most valuable mineral to be found on the Moon for human settlement there.

But a benefit of an entirely different kind can stem from asteroid exploitation: a knowledge of potentially hazardous asteroids that might strike the Earth. The great Arizona Meteor crater barely predates the recent Wisconsin glaciation. The Tunguska fall in Siberia this century (1908) devastated 1600 km^2 of forest. Its estimated explosive energy of $\sim 10^{25}$ ergs was equivalent in damage to that of a 100 megaton nuclear warhead. We were fortunate that its random target was isolated from human habitation. Such areas are becoming scarce on the land masses. Meteoric impacts in the oceans could produce catastrophic damage in coastal areas, not to mention their hazards to ships at sea. How many ships, indeed, may have been lost by such impacts, unrecordable until recently? There used to be a wide-spread fantasy among nonscientists that the astronomers knew of such a potential catastrophe but were keeping the knowledge a secret to prevent general tension or panic. Would that we had access to such information! Access *is* possible, however, from space, both by optics and radar. Once a potential small asteroid were known to have an orbit that could lead to a collision with the Earth, a relatively small propulsion unit set on the body could divert its path and prevent a collision for thousands of years in the future.

But scientifically, comets present the most exciting promise. We now have the competence to send rendezvous missions to comets and to asteroids. Comets, like the planetesimals of the terrestrial planets, were the building blocks for the outer planets, Uranus and Neptune. The difference was only one of temperature, lower farther out in the solar system where ices could collect along with the rock forming elements. But the comets may well have contributed the volatiles to the Earth and terrestrial planets if, as seems likely, these planets lost their volatiles during formation. Whether or not the atoms of our bodies come largely from comets, we can study in comets the most primitive material left over from the construction of our Solar System. This material, along with the lunar samples under study and samples from other planets to come, should clarify the details of formation of the only·planet in the universe we know to have developed intelligent life. How long, indeed, will we remain unique, alone in the universe?

Literature Cited

Pagel, B. E. P. 1977. On the limits to past variability of the proton-electron mass ratio set by quasar absorption line red-shifts. *MNRAS* 179:81P–85P

Williams, J. G. et al. 1976. New test of the equivalence principle from lunar laser ranging. *Phys. Rev. Let.* 36:551

Ann. Rev. Earth Planet. Sci. 1978. 6: 9–19

MANTLE CONVECTION MODELS   ×10084

Frank M. Richter

Department of the Geophysical Sciences, The University of Chicago, Chicago,
Illinois 60637

INTRODUCTION

The term thermal convection implies a flow driven by thermal density variations,
these variations being in turn maintained against the action of diffusion by the
advection of heat by the flow itself. The energetics is a balance between viscous
dissipation and the release of potential energy by the rising and sinking of warm
and cold material. Convection connotes a buoyancy-driven flow in a *fluid*, but it
can apply equally well to a solid medium so long as the solid can be continuously
deformed. There is ample evidence that earth-forming materials at temperatures
above about 1000°C are readily deformed on geologic time scales, and therefore
there is no obvious inconsistency in applying concepts derived from an understanding
of convection to flows in the earth's interior. In fact, there seems to be little alternative
to convection as the source of the large-scale motions described by plate tectonics.
The simplest argument to this effect is that the release of potential energy by a convective process is the only recognized energy source large enough to balance the
dissipation associated with plate motions. Potential energy can be continuously
released only if the system is in some way being heated, and in the earth, this is
accomplished in large part by radiogenic heating.

Energy arguments suggest that plate tectonics is a surface expression of convection in the earth but gives no indication as to the thermal and flow structure
within the system. In the absence of direct measurements, such specific information
requires explicit models. It is evident from the discussion below that our present
understanding of plate tectonics as a convective process is not based on any single
model or simulation, but instead depends on a synthesis of ideas derived from a
variety of models, each treating a specific aspect of the larger problem. The purpose
of this review then is to discuss the different types of models that contribute to this
synthesis. The reader should realize that in the absence of a definitive model, the
choice of topics and discussion that follows is perforce somewhat subjective.

PLATE TECTONICS AND RAYLEIGH-BENARD CONVECTION

Many of the difficulties inherent in modelling plate tectonics as a convective flow
become apparent when one compares specific properties of the plate tectonic system

9

0084-6597/78/0515-0009$01.00

to the equivalent property of convection in a fluid layer heated from below. Figure 1 shows the major plates and their absolute velocity with respect to the "hot spot" frame of reference (Minster et al 1974). A striking feature is the variety of horizontal scales, which in the Pacific basin alone range in size from about 1000 km in the case of the Cocos plate to about 10,000 km for the Pacific plate. Figure 2 represents a section through the earth from the South Pole to the equator along the 110° east meridian. The figure has been drawn without scale distortion and the 650-km seismic discontinuity is indicated because it is often assumed that flows in the mantle associated with the motion of the plates do not penetrate much below this discontinuity. This common assumption is not universally accepted, but if valid, implies that the horizontal scales represented by plate tectonics are not only varied but also large compared to the depth extent. Figure 2 shows schematically another important property: The spreading at ridges is generally symmetric but the downwelling at convergent boundaries is always one-sided. Returning to Figure 1, we can see that the absolute plate velocities are not correlated with plate size. Instead the plate velocities form a bimodal distribution in which subducting plates move relatively fast (~ 9 cm yr^{-1}) compared to the slower (~ 2 cm yr^{-1}) nonsubducting plates (Forsyth & Uyeda 1975). Minster et al (1974) prefer the interpretation that the bimodal distribution is due to the large proportion of continental area on the slower moving plates. The correlation is equally good, but the explanation in terms of subduction seems more likely since the subducted lithosphere represents the major localized release of potential energy and thus must be an important component of the force balance maintaining plate motions. In brief, plate tectonics suggests a convecting system composed of cells of diverse size and amplitude in

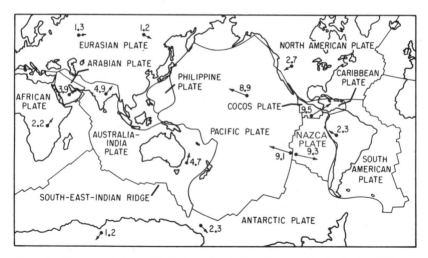

Figure 1   The major plates and their absolute velocity with respect to the "hot spot" frame of reference. This reference frame results in essentially no net rotation of the plate system as a whole.

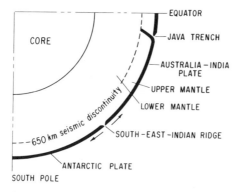

Figure 2   A schematic section through the earth taken along the 110° east meridian and drawn without scale distortion. The plates, shown by the dark shading, are about 100 km thick and move without significant internal deformation except where they bend in the one-sided subduction process. The 650-km seismic discontinuity is a region of rapid increase with depth of the density and seismic velocity, and has been included in the diagram to give some idea of the aspect ratio of the circulation involving the plates under the common assumption that the flow does not penetrate much below this discontinuity.

which the upwellings are symmetric, the downwellings are one-sided, and the velocity of each cell depends not on its size but on its relation to the unsymmetric subduction process.

The properties of convection in the earth suggested by plate tectonics are quite different from those observed in a convecting layer of uniform viscosity fluid heated from below (Rayleigh-Benard convection). This is particularly true of the spatial structure of the two flows. In contrast to plate tectonics, Rayleigh-Benard convection is characterized by cells of relatively uniform horizontal scale approximately equal to the layer depth (Figure 6a). Both upwelling and downwelling are symmetric (Figure 3a). The question of whether cells of different sizes would have similar amplitudes does not arise in connection with experiments since they exhibit a single typical scale, but theoretically it seems very likely that large cells would have relatively small amplitudes. All the properties mentioned above indicate that Rayleigh-Benard convection is not a particularly good analogue for plate tectonics. This is not surprising given that the natural system involves the deformation of internally heated rocks while the experiments use simple fluids. In the next section models incorporating more sophisticated material properties are discussed.

NUMERICAL CONVECTION MODELS

Numerical convection models have been studied extensively in an effort to determine the effect of such properties as temperature-dependent viscosity (see Torrance & Turcotte 1971), non-Newtonian deformation mechanisms (Parmentier, Turcotte & Torrance 1976), internal heating (see McKenzie, Roberts & Weiss 1974), and mineralogical phase transformations (see Schubert, Yuen & Turcotte 1975).

All of these are thought to be important in the mantle. The numerical approach has the advantage of being able to treat otherwise unresolvable material properties, but on the other hand, it has the disadvantage of generally being restricted to two dimensions. The numerical models will find two-dimensional solutions even in those cases in which the actual solution is known from laboratory experiments to be three-dimensional. I therefore regard these numerical models as experiments that use an abstract two-dimensional reference flow to study the effect of particular material properties, and the results must be combined with laboratory observations when applied to real systems in which the flow is most likely to be three-dimensional.

The material properties that have received the greatest attention are variable viscosity and internal heating. Figure 3 illustrates their effect. Case *a* is taken from a numerical convection calculation using a uniform viscosity fluid heated entirely from below. This case can be regarded as the control experiment showing that the

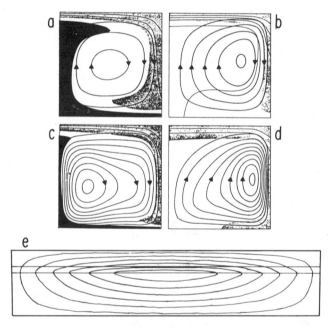

Figure 3 Results from numerical convection calculations showing the streamlines (the bolder lines that close on themselves) and isotherms (the thinner lines that intersect the boundaries) for different types of viscosity and modes of heating. The thermal boundary layers and plumes have been shaded to show better the different thermal structures. The calculations assume that the vertical velocity and temperature field have reflection symmetry across vertical boundaries. The results are as follows: (*a*) uniform viscosity and heated from below; (*b*) uniform viscosity and heated from within; (*c*) variable viscosity and heated from below; (*d*) variable viscosity and heated from within; (*e*) streamlines for the most unstable mode of convection when the fluid is heated from below and the viscosity between the two horizontal lines is three orders of magnitude lower than elsewhere in the layer.

streamlines and thermal field have almost perfect rotational symmetry. Case *b* is heated uniformly from within, instead of from below, and there is no longer a thermal boundary layer at the bottom of the cell or an associated rising warm plume. The flow is dominated by the relatively concentrated sinking of cold material from the upper thermal boundary layer. Case *c* illustrates the effect of temperature-dependent viscosity (warm fluid being less viscous) when the heating is from below. The upwelling region now becomes somewhat concentrated. Parmentier, Turcotte & Torrance (1976) find that convection with non-Newtonian rheology that simulates the actual deformation of rocks is not much different from the computationally simpler Newtonian variable viscosity case. When both variable viscosity and internal heating are included in the calculation (case *d*), the effect of internal heating dominates. Case *e* shows the most unstable cell size for convection in a layer having a thin low viscosity zone within it (Richter & Daly 1977). The cell size under these particular circumstances becomes large compared with the depth, a property that is desirable when modelling plate tectonics, particularly if the flow associated with the plates cannot penetrate much below the 650 km discontinuity.

Models such as those discussed above have many other legitimate uses besides studying the effect of specific material properties. Perhaps the most important result is that they show that a material with an apparent viscosity of about 10^{22} poise [the value estimated by analyzing glacial rebound (Cathles 1975)] and heated at a rate consistent with the observed terrestrial heat flow will have convective velocities similar to those observed in plate tectonics. The ability to reconcile those independent observations is an essential source of confidence in the initial assumption that plate tectonics is the result of convective dynamics. Numerical models are also used to test or illustrate general concepts. An example is how the average temperature in a convective planetary interior is principally controlled by the variation in viscosity with temperature and not by the rate of heat production (Tozer 1972, McKenzie & Weiss 1975). The role of viscosity as a thermoregulator is a consequence of the large viscosity variations that accompany small changes in temperature. The numerical models serve another purpose in providing the only quantitative link between convection and geophysical observables such as gravity anomalies (see McKenzie 1977). The connection between convection and geophysical observables, other than the motion of the plates themselves, is especially important in that the very existence of the plates will make it difficult to detect convective motions not directly associated with the plate tectonic process. The possibility of such flows is discussed in a later section.

The fact remains that the numerical models do not yet come to terms with the principal difficulty in modelling plate tectonics as a convective flow. The difficulty is not one of the minor details of the flow or thermal structure, but instead is caused by the regular size, amplitude, and symmetry of convection cells in the analogue systems as opposed to the observed properties of plate tectonics. At this point one can only speculate as to what processes are important in giving the plate system its distinctive properties. Clearly, the cold surface plates, which are undeformed except when subducted, must have an important effect, but even this assertion is based more on intuition than on specific model calculations.

SIMPLE PLATE MODELS

The numerical models of the preceding section fail to reproduce many obviously important features of the plate tectonic system, and as a consequence, they are not particularly useful for determining in any detail the balance of forces maintaining plate motions. An alternative approach uses simple plate models such as the one shown in Figure 4, which assumes at the outset many of the observed spatial properties of plate tectonics (Richter 1973a, 1977). A further assumption is that the plates overlie and penetrate a material whose deformation is adequately represented by a Newtonian rheology. Under these circumstances the flow under the plates and the resulting resistance to plate motions can be calculated. The resistive forces in the model are of two types: shear stress acting on the base of the plates and sides of the downgoing slab, and a nonhydrostatic pressure acting on the leading edge of the subducted lithosphere. This nonhydrostatic pressure, which drives the return flow from trenches to ridges, was overlooked in earlier discussions of the driving mechanism (Richter 1973a), but recent models (Richter 1977) suggest that it may be the single most important term resisting the motion of subducting plates. Another force resisting plate motions is due to stresses on the thrust faults forming the boundary between converging plates. The magnitude of these stresses is uncertain, but a rough estimate can be made using the observed stress drop associated with the large earthquakes along the thrust planes. The combined effect of these resisting forces is balanced by driving forces resulting from horizontal density variations under trenches and ridges. The negative buoyancy of the downgoing slab can be estimated using models of the thermal structure of the subducted lithosphere (McKenzie 1969, Minear & Toksoz 1970). The density variations under ridges are reflected in the shallow bathymetry there and result in a net force pushing the plates away from the ridge crest (see Richter & McKenzie 1977). An overall balance of forces is formed by combining the results of the simple plate models, the estimate of stresses on the thrust faults, and the auxiliary calculations for the buoyancy forces.

Given the many assumptions already used in constructing even the simplest plate model, the only real remaining unknown connected with such a simulation is the viscosity and its variation in the material under the plates. The viscosity is found by

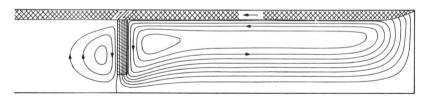

Figure 4   A simple plate model used to discuss the driving mechanism of plate tectonics. The plate on the right is assumed to be moving and subducting at a uniform rate. The nonsubducting plate on the left is in this case stationary. The resulting flow under the plates is given by the streamlines, which show the return flow under the moving plate and a local circulation, induced by the downgoing slab, under the nonsubducting plate.

requiring that models having different horizontal extent of surface plate and depth of penetration by the downgoing slab reproduce in at least a general way observations on the plates they simulate. The principal observations that are useful are the velocity of the plates, the state of stress of the downgoing slabs as inferred from earthquake mechanism studies (Isacks & Molnar 1971), and the lack of obvious gravity or bathymetry anomaly associated with the pressure gradient driving the return flow. Some of the principal conclusions from the simple plate models are given below [see also Richter (1977) and Richter & McKenzie (1977)].

1. The negative buoyancy of the downgoing slab is the principal driving force, and it accounts for the observed differences in velocity between subducting and nonsubducting plates.

2. The lack of correlation between plate velocity and plate size implies that the shear stresses on the base of plates are not an important part of the overall force balance and requires a low viscosity zone at the base of the plates.

3. The negative buoyancy of the downgoing slab is balanced locally, the principal resistive force being the nonhydrostatic pressure.

4. A primary balance between the negative buoyancy and the nonhydrostatic pressure leads to stresses within the downgoing slab that agree with the stresses inferred from earthquake mechanisms if there is a barrier to vertical motions at or near the 650 km seismic discontinuity.

5. The balance of forces on nonsubducting plates are more uncertain because there are less constraints, but Mendiguren & Richter (1977), using intraplate stresses inferred from South American earthquakes, find that "ridge pushing," the stresses on the thrust faults, and the local shear stress on the base of the plate due to the flow induced by the downgoing slab (see Figure 4) are all important in the overall force balance of such a plate.

With the exception of the arguments involving the nonhydrostatic pressure, the conclusions listed above are not particularly new or even dependent on an explicit plate model. The role of the simple plate models is to provide an idealized but self-consistent formulation for discussing the overall balance of forces maintaining plate tectonics.

THE CONCEPT OF TWO SCALES OF MOTION

Convection as seen in laboratory experiments is characterized by almost equidimensional cells. This is true not only of Rayleigh-Benard convection, but also of convection with internal heat sources (de la Cruz 1970) or variable viscosity (Richter, in preparation). Mantle convection involving the plates, on the other hand, is characterized by a multiplicity of scales, all of which are large compared to the depth extent if the flow cannot penetrate below the 650-km seismic discontinuity. The dichotomy between what is seen in the laboratory and what is inferred about convection in the natural system has led to the suggestion that convection in the mantle may be occurring on two distinct horizontal scales simultaneously (Richter 1973b). A large-scale flow consisting of the plates and a return flow below them is implied by plate tectonics. A second, smaller scale consisting of equidimensional cells similar to those seen in the laboratory is proposed to exist under the moving plates. Figure 5 is a schematic illustration of the two-scale concept in the special case when the small-scale flow is in the form of two-dimensional rolls aligned in the direction of plate motion and the depth extent of both the large- and small-scale flow is much smaller than the size of the plates.

The concept of two scales of motion has been tested using laboratory experiments on Rayleigh-Benard convection under a moving boundary that simulates a spreading plate (Richter & Parsons 1975). Figure 6 is an example of one such experiment. The experiments are idealized in that the large-scale flow is externally imposed and Rayleigh-Benard convection plays the role of the small-scale flow, which in the earth will have the added complications of internal heating and variable viscosity. Despite this idealization, several fairly robust conclusions can be drawn from the experiments. Most directly relevant is that we have found no reason why the existence of moving plates would suppress the small-scale flow below them. At most, the effect of fast-moving plates will be to change the planform of convection from the spoke pattern to two-dimensional longitudinal rolls with axes aligned in the spreading direction (Figure 6). Particularly when the small-scale flow takes the form of longitudinal rolls, it will make no contribution to maintaining the motion of the overlaying plate, which is one reason why in the earlier discussion of the driving mechanisms the only driving terms included were associated with plate boundaries.

The small-scale flow is still speculative in that there is no geophysical observation that unequivocably requires its existence. There is indirect evidence, such as the

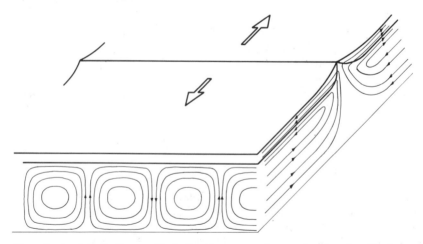

Figure 5 A schematic diagram illustrating the concept of two scales of motion. The large-scale flow directly involving the plates is taken from a calculation such as that given in Figure 4. The small-scale flow under the moving plates consists of equidimensional cells such as those seen in convection experiments and in this particular illustration are in the form of longitudinal rolls.

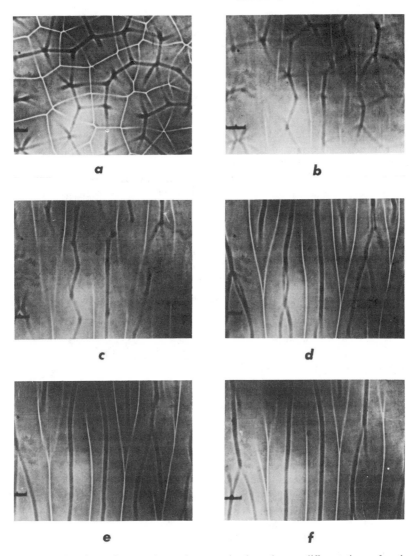

Figure 6 Plan views of convection under a moving boundary at different times after the boundary begins to move. A shadowgraph technique is used to visualize the planform. The light colored areas represent cold, sinking fluid and the darker color marks the rising of warm fluid. The top-left photograph is the typical "spoke" pattern of convection between isothermal, stationary boundaries. The distance between points on the marker seen near the lower left corner of each photograph is half the depth of the layer. Photographs *b–f* show the evolution of the planform as the upper boundary is made to move (from top to bottom in the photograph). The final form of convection under the moving boundary (meant to simulate a spreading plate) is longitudinal rolls. For details see Richter & Parsons (1975).

relatively shallow bathymetry of the old ocean basins, which implies a significant heat flux into the base of the oceanic plates (Parsons & Sclater 1977). The small-scale flow could supply this heat by enhancing the heat transfer in the upper mantle. A more definitive test should result from the improved satellite gravity measurements that are expected in the next few years.

THE STATE OF THE ART

Put briefly, we have learned a great deal about the effect of material properties thought to be important in the earth, and we also now have some viable ideas about how present day plate motions are maintained. On the other hand, many first-order questions remain. We do not understand how plate tectonics is initiated either locally or globally, or what accounts for the multiplicity of horizontal scales of the plates and their one-sided subduction. A good measure of the present state of affairs is to note that the concept of two scales of motion arises, at least in part, from the inability of all the model systems to reproduce many of the obviously important features of plate tectonics. The two-scale model in fact uses the differences between the models and the observations to suggest a convection system more complicated than the geophysical observations alone specifically require. The numerical models of convection with material properties appropriate to the upper mantle are most directly applicable to the small-scale flow. Ironically, we may understand more about the hypothesized small-scale flow than we do about the observed large-scale flow.

ACKNOWLEDGMENT

This work was supported by the National Science Foundation under Grant Number NSF EAR75-17170. The calculations shown in Figure 3a–d were kindly provided by Dan McKenzie.

Literature Cited

Cathles, L. M. 1975. *The Viscosity of the Earth's Mantle.* Princeton, N.J.: Princeton Univ. 386 pp.
de la Cruz, S. 1970. Asymmetric convection in the upper mantle. *Rev. Union Geofis. Mex.* 10:49–56
Forsyth, D., Uyeda, S. 1975. On the relative importance of the driving forces of plate motion. *Geophys. J. R. Astron. Soc.* 43:163–200
Isacks, B. L., Molnar, P. 1971. Distribution of stresses in the descending lithosphere from a global survey of focal mechanism solutions of mantle earthquakes. *Rev. Geophys. Space Phys.* 9:103–74
McKenzie, D. P. 1969. Speculations on the causes and consequences of plate motions. *Geophys. J. R. Astron. Soc.* 18:1–32
McKenzie, D. P. 1977. Surface deformation,

gravity anomalies and convection. *Geophys. J. R. Astron. Soc.* 48:211–38
McKenzie, D. P., Roberts, J. M., Weiss, N. O. 1974. Convection in the earth's mantle: towards a numerical simulation. *J. Fluid Mech.* 62:465–38
McKenzie, D. P., Weiss, N. O. 1975. Speculations on the thermal and tectonic history of the earth. *Geophys. J. R. Astron. Soc.* 42:131–74
Mendiguren, J. A., Richter, F. M. 1977. On the origin of compressional intraplate stresses in South America. *Earth Planet. Sci. Lett.* In press
Minear, J. W., Toksoz, M. N. 1970. Thermal regime of a downgoing slab. *Tectonophysics* 10:367–90
Minster, J. B., Jordan, T. H., Molnar, P., Haines, E. 1974. Numerical modeling of

instantaneous plate tectonics. *Geophys. J. R. Astron. Soc.* 36: 541–76

Parmentier, E. M., Turcotte, D. L., Torrance, K. E. 1976. Studies of finite amplitude non-Newtonian thermal convection with application to convection in the Earth's mantle. *J. Geophys. Res.* 81: 1839–46

Parsons, B., Sclater, J. G. 1977. An analysis of the variation of ocean floor bathymetry and heat flow with age. *J. Geophys. Res.* 82: 803–27

Richter, F. M. 1973a. Dynamical models for sea-floor spreading. *Rev. Geophys. Space Phys.* 11: 223–87

Richter, F. M. 1973b. Convection and the large-scale circulation of the mantle. *J. Geophys. Res.* 78: 8735–45

Richter, F. M. 1977. On the driving mechanism of plate tectonics. *Tectonophysics* 38: 61–88

Richter, F. M., Daly, S. F. 1977. Convection models having a multiplicity of horizontal scales *J. Geophys. Res.* In press

Richter, F. M., McKenzie, D. P. 1977. Simple plate models of mantle convection. *J. Geophys.* In press

Richter, F. M., Parsons, B. 1975. On the interaction of two scales of convection in the mantle. *J. Geophys. Res.* 80: 2529–41

Schubert, G., Yuen, D. A., Turcotte, D. L. 1975. Role of phase transitions in a dynamic mantle. *Geophys. J. R. Astron. Soc.* 42: 705–35

Torrance, K. E., Turcotte, D. L. 1971. Thermal convection with large viscosity variations. *J. Fluid Mech.* 47: 113–25

Tozer, D. C. 1972. The present thermal state of the terrestrial planets. *Phys. Earth Planet. Int.* 6: 182

Ann. Rev. Earth Planet. Sci. 1978. 6:21–42

THE PROBABLE METAZOAN ✕10085 BIOTA OF THE PRECAMBRIAN AS INDICATED BY THE SUBSEQUENT RECORD

J. Wyatt Durham

Department of Paleontology, University of California, Berkeley, California 94720

INTRODUCTION

In recent years Precambrian sedimentary rocks have been receiving much attention in our efforts to trace the beginnings and understand the development of the diverse biota known from the Phanerozoic. Because shelly fossils suddenly appeared at the beginning of the Cambrian, most of them without recognizable ancestors in the immediately preceding Precambrian, many investigators, such as Cloud (1968, 1976) and Stanley (1973, 1976), have suggested that eucaryotes and subsequently metazoans did not appear until relatively late in the earth's history. Cloud (1976) concluded that the first metazoans appeared about 700 m.y. ago, and the eucaryotic cell probably about 1300 m.y. ago. In contrast to these late dates he further suggests that the first autotrophs may have appeared as early as about 3800 m.y. ago.

Only a few workers have suggested that the Metazoa appeared earlier than concluded by Cloud. Glaessner (1972, p. 46), after noting that the Ediacaran metazoan fauna has been shown to have a time range of about 600 to 700 m.y. BP, commented, ". . . their differentiation indicates a long history of coelomates preceding that date" and "The well defined biostratigraphic, chronostratigraphic and evolutionary identity of the Ediacara fauna confirms the view that a sequence of three stages in the early history of the Metazoa has to be explained rather than the supposed sudden appearance of most of the known phyla at the beginning of Cambrian or of Phanerozoic time." I have argued (Durham 1971a) on theoretical grounds that the deuterostomate metazoans may have appeared "between 800 million years as a minimum and about 1700 million years as a maximum before the present . . . ." Thus two strongly contrasting viewpoints are held. The scarcity of conventional evidence of metazoans in pre-Ediacaran sediments has overawed most investigators into concluding that no metazoans were present. Their conclusions became widely accepted when this evidence was added to the prevalent ideas about

21

0084-6597/78/0515-0021$01.00

low levels of atmospheric oxygen in the Precambrian, as expressed by Berkner & Marshall (1965) and supported by Cloud's conclusion that the extensive Precambrian "banded iron formation" required low levels of atmospheric oxygen for its formation (see discussion in Cloud 1976, pp. 362–63, and references therein). These data are suggestive but not compelling. Recently the concept of low levels of Precambrian atmospheric oxygen has been seriously challenged by Dimroth & Kimberly (1976), and on different grounds by Margulis, Walker & Rambler (1976). If the arguments presented in these two papers are upheld by further study, the role played by oxygen in the Precambrian needs extensive re-evaluation—low atmospheric oxygen levels could not have been a factor limiting Precambrian organic diversity.

Since World War II there has been much discussion about the placement of the Cambrian/Precambrian boundary (see Cowie & Glaessner 1975). I have previously, in several places, stated that I was utilizing the first appearance of trilobites (implying body fossils) as marking the base of the Cambrian. However, the arguments advanced by Rozanov et al (1969), Daily (1972), and Cowie & Glaessner (1975) are persuasive, and I am tentatively utilizing the base of the Russian Tommotian Stage as the boundary. Trilobites (body fossils) apparently do not appear until about the top of the Tommotian or the base of the overlying Atdabanian Stage, so this position of the boundary is slightly older than that I have used previously. This position is radiometrically placed at about 570 m.y. BP (Cowie & Glaessner 1975). The Tommotian and Atdabanian Stages are based on the extensively studied Early Cambrian sequences of the Siberian Platform area, with supporting data from the Baltic region. The stages in the Siberian sequence are, from oldest to youngest, Tommotian, Atdabanian, Botomian, and Lenian. Comparable sequences in other parts of the world have not been as intensively studied and in part do not seem to be as fossiliferous. As a result, the Siberian sequence is being widely accepted as a standard of reference (see Rozanov & Debrenne 1974, Figure 3, for a correlation of differing usages).

Throughout this discussion I am emphasizing the "known" record. By "known" I mean that it has been published in such a manner that the information is freely available to others. Material or data in collections or investigators' unpublished notes may remain unknown to others and thus are not available.

THE STATE OF KNOWLEDGE OF PAST BIOTAS

Our poor knowledge of past biotas has often been attributed to inadequacy of the fossil record, with emphasis being laid upon the statistically low chances for any individual organism to be fossilized. The ultimate result of this inadequacy, of course, is that in most instances a taxon is represented in the described fossil record by far fewer individuals than were living at any one time and that only a limited representation of the taxonomic diversity of any moment in time is known. There are further causes for our deficient knowledge (see Durham 1967, pp. 461–63), the most significant of which is the lack of sufficiently intensive sampling. Another important cause is biased collecting, i.e., selecting samples for trilobites only, or only a cherty lithology. Only in a few very specialized instances such as the Paris

Basin Eocene do we have good comprehension of all the components of past biotas that are easily fossilized (let alone the soft-bodied members).

The recent tabulation of described fossil invertebrate species by Raup (1976a,b) emphasizes the inadequacy of our knowledge. By 1970 a total of only about 190,000 fossil invertebrates had been described. Valentine (1970) conservatively estimated that there are about 100,000 species of "easily fossilizeable" living marine invertebrates and also concluded that the average species duration of marine invertebrates is between 5 and 10 m.y. Inasmuch as Zhuravleva (1970) noted that about 2800 species had been recorded from the Early Cambrian, it is obvious that the recorded number of fossil invertebrates for the last 570 m.y. is far below the diversity that actually existed. The occurrence of fossils such as the bizarre *Tullymonstrum* (Johnson & Richardson 1969) and the enigmatic *Etacystis* (Nitecki & Schram 1976) in the Pennsylvanian of Illinois, each with no recognized ancestors, underscores the incompleteness of our knowledge of past biotas, and in particular the Metazoa.

Schopf (1975, pp. 214–19) has stressed the incompleteness of contemporary knowledge of the Precambrian premetazoan biosphere, noting that it is based on only 20 significant microbiotas and that at the time he wrote, only two of these had been studied monographically. He further emphasized that all but three of these were preserved in a cherty lithology, a rock facies that is characteristic of a very specialized environment. There were many other contemporary facies in the Precambrian, and thus we have only an inkling of the contemporary microbiotas inhabiting them. It is clear that our knowledge of the possible microbiota of the Precambrian is very limited.

The famed Australian Ediacara metazoan fauna was likewise entombed in a very specialized situation. Wade (1968) discussed in considerable detail the preservation of the soft-bodied fauna. She notes that the fossiliferous portion of the Pound Quartzite has been interpreted as having been deposited in an environment of lesser hydrodynamic intensity (i.e. lesser wave and current action) than that of the main sedimentary facies of the formation. A further and most significant conclusion (p. 266) is "The constancy of constituents of the bodily preserved fauna throughout the hydrodynamic changes . . . suggests strongly the introduction of this fauna to the environment of deposition. . . . In general . . . most of the animals were transported to the spots where they were buried. . . ." In other words, the Ediacara body fossils are a biased sample of the contemporary biota, being composed of those animals that were easily susceptible to transportation by the lessened hydrodynamic forces of the moment. This interpretation is underscored by the abundance of medusoids (most modern medusoids are planktonic) in the fauna. Animals (with nonmineralized parts) firmly attached to, or imbedded in, the substrate, or with heavy dense bodies, would not usually be so transported. Thus those who have accepted the Ediacara body fossil fauna as fully representative of the metazoan biota of that time will have a distorted concept. Utilizing the Ediacara body fossil biota in such a noncritical manner is comparable to characterizing the Cambrian biota by the soft-bodied metazoans of the Middle Cambrian Burgess Shale.

However, in addition to the well-publicized allochthonous Ediacara body fossils discussed above, it is well to note that there was also an indigenous biota at the site of deposition, now apparently represented (with one possible exception) only by numerous trace fossils (Wade 1968, p. 266, Glaessner 1969). Seemingly only one of these has been formally named (*Pseudorhizostomites howchini* Sprigg), although Glaessner (1969, pp. 379–82, Figure 5), having illustrated five of them, differentiated six common types, and stated, "None of them can be related, with any probability, to any of the known bodily preserved animals." He concludes, "These animals were all worm-like sediment feeders or detritus feeders, either living in the sand, or on the mud-sand interface, or grazing on the mud surface." The presence of these trace fossils, none of which can be related to the body fossils without any accompanying body fossils of the animals that made them, emphasizes the specialized nature of the Ediacara body fossil assemblage. On somewhat different grounds Cowie (1967) reached a similar conclusion. Why aren't the bodies of some of the Ediacara trace fossil makers represented in the body fossil assemblage? It seems probable that there was an even more complex depositional history than has been suggested.

In summary, the data available demonstrate that our knowledge of the Precambrian biota, both microfossil and metazoan, is inadequate and that sweeping generalizations about its diversity and relationships are unwarranted. Our knowledge of the Precambrian biosphere is much poorer than our knowledge of the Phanerozoic biosphere, which in turn is demonstrably very incomplete.

IMPLICATIONS OF THE DOCTRINE OF EVOLUTION

Application of the doctrine of evolution to the organic world about us implies descent from a common ancestor for a diversity of organisms, and in general, a progression from the simple to the more complex. A commonly overlooked corollary to the doctrine is that any multicellular organism must have ancestors, although it is not necessary to have descendents. Any metazoan fossil we find had a progression of ancestors from the ancestral protozoan to the descendent, a fact that is largely ignored in considering the composition of past biotas. As well as subscribing to the doctrine of evolution, most of us utilize (or attempt to) the biological species concept when we study and classify sexually reproducing metazoans. This concept carries with it the corollary that parent and offspring cannot be so different from one another that they cannot interbreed.

The above corollaries to evolution and the biological species concept along with the known fossil record place major constraints on our inferences about the diversity and time of origin of the Metazoa. There must have been a continuously ramifying stream of populations through time from the first metazoan to all the diverse known taxa that have been found in the Phanerozoic. Further, the first primitive metazoan must have come into existence at a time sufficiently prior to the inception of the Cambrian that all the genetic changes necessary to produce a complex echinoderm or protochordate could have occurred. By inference at least, Cloud and others who date the first appearance of the Metazoa at about 130 m.y. prior to the Cambrian believe that the necessary genetic changes could have occurred very rapidly. As noted previously, Valentine (1970) concluded that the average species duration for marine invertebrates in the Phanerozoic is between 5 and 10 m.y. Others usually have suggested an average interval somewhat greater. Using Valentine's lesser estimate (5), an average of only about 22 species durations would have intervened between the complex biotas of the Early Cambrian and their unicellular ancestor. This number seems highly improbable in view of all the characteristics that had to be acquired. These characteristics would include, among others, diploblastic and triploblastic development, a coelom, a throughgoing digestive tract, a nervous system, a haemal system, a water vascular system, a reproductive system, and mineralized skeletons of both ectodermal and mesodermal origin. Even though there may have been many vacant "niches" in the Precambrian biosphere that would provide the opportunity for "explosive evolution" to occur, it does not seem probable that the rate of evolution at that time could have been radically faster than during the Phanerozoic. The Phanerozoic interval for which Valentine's estimates were developed includes several episodes of so-called explosive evolution, which are therefore involved in the calculations that produced his results. It is obvious that something is wrong with these estimates of the time of origin of the Metazoa and the rates of evolution. The errors may lie in either estimate, but it seems more probable that both are incorrect. That is, the Metazoa must have originated considerably earlier and the rates of evolution must have been faster.

THE PRECAMBRIAN ENVIRONMENT

Watson (1976, p. 148 and Figure 1) has reviewed the character of the earth's crust in the Precambrian and has observed that the hydrological cycle possibly was in existence prior to 4000 m.y. ago and certainly by 3800 m.y. BP. She notes that in Laborador (Hebron Gneiss, possibly over 4000 m.y. old) the altered sedimentary rocks associated with granites "... preserve well defined bedding lamination or contain rounded pebbles, features which point to the existence of bodies of standing water and to deposition from moving water..." Thus the aquatic environment necessary for the origin of life was in existence by that time. Further, Garrels & Mackenzie (1974, p. 193), summarizing the results of their study of post-depositional changes in sedimentary rocks, state, "The materials entering the ocean today are similar in their heterogeneity, chemical and mineralogic composition and rates of addition to those of the past two billion years of Earth history" and "... it is likely that seawater composition of today is much like that of at least one-half of geologic time. Owing to their long residence times and significant oceanic masses, chloride, sulfate, sodium, and magnesium may have varied in oceanic concentration in the past by 10% or more from their present values. No data, however, are available to prove such changes have occurred." It is to be emphasized that these conclusions are stated to apply to only the last two billion years (no data bearing on the earlier history was presented). Thus the general oceanic chemistry was not unfavorable to the development of types of life similar to those that have inhabited this environment since the Precambrian.

There has been much consideration of the time at which significant free oxygen appeared in the earth's atmosphere. Many investigators, the foremost of whom is

Preston Cloud, have concluded that the early atmosphere lacked free O_2 and that significant levels of O_2 (which allows an O_3 shield against ultraviolet radiation and permits fully oxidative metabolism) were not present until relatively late in the Precambrian. Cloud (1972 and elsewhere) has argued that the extensive deposits of banded iron formation and absence of substantial red beds previous to about 1900 m.y. "... imply that the atmosphere prior to that time contained only trivial and transient amounts of free oxygen." Cloud suggests that the buildup of free oxygen was slow and that it was not until about 700 m.y. ago that the free O_2 reached a level sufficiently high for the emergence of the Metazoa (Cloud 1976, p. 379). Recently, however, Dimroth & Kimberly (1976) have presented persuasive arguments to indicate that free oxygen was present in the atmosphere long before the time suggested by Cloud, while Margulis, Walker & Rambler (1976) have re-evaluated the roles of oxygen and ultraviolet light in the evolution of eucaryotic organisms in the Precambrian and independently arrived at a similar conclusion.

Dimroth & Kimberly (1976) were led into their re-evaluation of the Precambrian atmosphere because exploration for uranium ores in the Middle Precambrian of northern Quebec had been unsuccessful, apparently because the putative uraniferous gravels had experienced oxidative diagenesis. At the same time prospecting for Archean copper-zinc ores was very successful, despite the fact that they were interpreted as analogues of Phanerozoic deposits formed in the presence of an oxydizing atmosphere. As a result Dimroth & Kimberly reconsidered the occurrence of sedimentary carbon, sulfur (in the form of sulfides), uranium, and iron and concluded, "... uniformitarian models of chemical sedimentation may be applied to the Precambrian, and ... there is no evidence of orders-of-magnitude changes in average atmospheric or hydrospheric abundance of chemically reactive organic species. However the proportions of distinct sedimentary environments and the forms and compositions of life have profoundly changed." Cloud has previously (1973, 1976, p. 362) noted some dilemmas in his favored hypothesis for the deposition of the banded iron formation in an oxygen-free atmosphere and suggested two possible mechanisms to explain them, both of which involve the presence of limited amounts of free oxygen. Dimroth & Kimberly interpret the banded iron formations, on the "... basis of textures and stratigraphic relationships to be early diagenetic replacements of dominantly aragonitic sediment ..." and note that differences in iron formations of different ages are "... more readily attributable to changes in tectonic setting and organic activity than to compositional changes in the atmosphere or hydrosphere." They also observe that their conclusions are applicable as far back in geological-time as the oldest well-preserved sedimentary rocks. They suggest (1976, p. 1169–70) that an oxydizing atmosphere had appeared at least as early as 3750 m.y. BP (Isua Complex iron formation).

Margulis, Walker & Rambler (1976) observe that from the viewpoint of the release of oxygen, photosynthesis in the Cyanophytes (or Cyanobacteria) is the same as green plant photosynthesis. Inasmuch as Cyanophytes may appear about 3800 m.y. BP (Cloud 1976, Table 1, p. 353), this implies that free oxygen began to appear at that time. This is in general agreement with the date suggested by Dimroth & Kimberly on mineralogic grounds. Margulis, Walker & Rambler conclude that the amount of oxygen released from this source would increase with time and would eventually reach a point where the release of oxygen to the atmosphere would exceed the rate of release of hydrogen (from a volcanic source) and permit aerobic metabolism. They observe (p. 621) that eucaryotes are aerobes that are very uniform in their response to oxygen: "... there are no metazoan animals or embryophyte green plants that complete their entire life cycles in the total absence of oxygen ..." and "... all eukaryotic anaerobes seem to be derived secondarily from aerobic ancestors." They also observe that the process of mitosis as well as mitochondrial metabolism requires oxygen, which makes it likely that full aerobiosis was distinctive of even primitive eucaryotes, and they suggest that "... close to modern quantities of oxygen are now and always have been required for eukaryotes." Summing up their discussion one might say that the eucaryotes could have appeared once the free oxygen supply was sufficient and not necessarily as late as about 1300 m.y. BP, the date preferred by Cloud (1976, p. 373). However, it should be noted that Cloud (1976, Figures 1 & 2) recognizes the possibility that they might have been able to appear as early as 2000 m.y. BP.

In summary, the evidence and new analyses of data presented in the last few years suggest that an atmosphere containing sufficient free oxygen for aerobic metabolism was present much earlier than previously hypothesized. This suggests that eucaryotes and eventually the metazoa could have appeared correspondingly earlier.

THE KNOWN BIOTAS

Knowledge of the diversity of biotas of the past, even of those organisms of Cambrian and younger times that could be fossilized easily is relatively poor, while information on those of the Precambrian is very poor and spotty. Some very significant but often overlooked reasons for these deficiencies are discussed above in the section on the state of knowledge of past biotas. Mineralized "skeletons" characterize many components of the younger biotas and consequently our knowledge for this interval is largely but not exclusively based on them. The biotas of the Precambrian seemingly lacked components with mineralized "skeletons" and thus are in large part as yet unknown.

For the most part there is no difficulty in recognizing whether or not a particular Cambrian or younger structure is biogenic, but in the Precambrian this distinction is not always clear. Among others, Cloud (1976), Hofmann (1971), and Schopf (1975) have reviewed the evidence that can be used to evaluate the biogenic nature and affinities of putative Precambrian fossils. Rigid application of the criteria discussed by these and other authors casts doubt upon the biogenic authenticity and/or affinities of many Precambrian fossils. On the other hand an object should not be assigned to the inorganic category simply because there is some doubt about its biogenic affinities. It should not be considered as unqualifiedly inorganic until equally rigid criteria for "inorganicness" have been applied. A similar methodology should likewise normally be used in determining the systematic placement of undoubted organisms. However sometimes the volume of suggestive evidence may be so great as to make a conclusion contrary to that indicated by one of the "rigid" criteria almost inescapable. Use of the "rigid" methodology should not be one-sided, for we are far from attaining omniscience.

The Cambrian Biota

The known Cambrian biota has two distinct aspects. The organisms of the Late Cambrian are easily related to those of subsequent epochs but the Early Cambrian complex of organisms includes a considerable number of types that cannot be easily compared to those of the post-Cambrian. The most notable apparent absentees from the currently known biota are the vertebrates and vascular plants. However, with respect to the vertebrates it is well to note that Conway Morris (1977) has concluded that *Pikaia gracilens* Walcott from the Middle Cambrian Burgess Shale is a chordate (he includes the protochordates within the Chordata) inasmuch as the segments are very suggestive of myotomes and there is a notochord-like structure. Likewise a protochordate (?) has been reported (Durham 1971b, Firby 1972) from the Lower Cambrian Poleta Formation of California and compared with the possible protochordate *Emmonsaspis* (Resser & Howell 1938) from the Lower Cambrian of Vermont. *Pikaia* and *Emmonsaspis* are both *Amphioxus*-like animals.

Three conodont-like fossils (*Hertzina, Protohertzina,* and *Fomitchella*) have been found in the Lower Cambrian (Tommotian Stage) of the Siberian Platform (see Matthews & Missarzhevsky 1975, p. 299; Repina et al 1974, pp. 82–83). *Protohertzina* has also been found (Matthews & Missarzhevsky 1975, p. 293) in the Manikay horizon of latest Precambrian age. Some authors have suggested that conodonts are parts of some vertebrate or that they represent some unidentified chordate (sensu lato), and the discovery of a conodont-bearing animal (with some resemblance to *Amphioxus*) in the Mississippian led Melton & Scott (1973) to propose the subphylum (of the Chordata) Conodontochordata for them. They did not compare *Emmonsaspis* and *Pikaia* with the Conodontochordata but there is a suggestive similarity that supports the assignment of the Cambrian fossils to the Chordata (sensu lato). If this conclusion is valid and the Russian fossils are valid conodonts, then there are late Precambrian chordates (sensu lato). Further, Müller (1977) has described a probable ascidian or urochordate (*Palaeobotryllus*) from the Upper Cambrian of Nevada, thus adding to the diversity of probable chordates in the Cambrian. These findings suggest that it is not improbable that vertebrates may eventually be discovered in the Cambrian.

In recent years the graptolites have usually been assigned to the phylum Hemichordata. To date none have been recorded from the Early Cambrian but several dendrograptid genera occur in the late Middle Cambrian of Siberia, Tasmania, Spain, and possibly Norway (Berry & Norford 1976), suggesting that their common ancestors lived earlier. The systematic position of the pogonophorans has been uncertain. They have been assigned to either the Hemichordata or Protochordata or given separate phylum rank, but they are always placed in a high systematic position. *Sabellidites* and similar chitinoid fossils as well as the Hyolithelminthids from the latest Precambrian and Early Cambrian have sometimes been assigned to the Pogonophora (Zhuravleva 1970, Matthews & Missarzhevsky 1975). If this advanced systematic position is confirmed for these fossils, it adds to the totality of such types already present at the inception of the Cambrian.

A major diversity of invertebrates is present in the Early Cambrian and earliest Middle Cambrian. The paper by Zhuravleva (1970) provides a good summary of the major higher categories of about 2800 species then known from the Early Cambrian. However, it should be noted that the systematic placement of some of them has since been changed. Additional data are presented in papers by Daily (1972), Durham (1971a), Handfield (1969), Henderson & Shergold (1971), Jell & Jell (1976), Korkutis (1971), Matthews & Missarzhevsky (1975), Pojeta & Runnegar (1976), Repina et al (1974), Rowell (1977), Rozanov & Debrenne (1974), Rozanov et al (1969), Runnegar & Jell (1976), and Zhuravleva & Rozanov (1974). Among the echinoderms, the Eocrinoidea and Helicoplacoidea occur with the earliest trilobites in California; elsewhere the Edrioasteroidea, Camptostromoidea and Lepidocystoidea are found in the middle part of the Early Cambrian; and the Crinoidea, Ctenocystoidea, Cyamoidea, Cycloidea, Cyclocystoidea, Holothuroidea and Stylophora occur in the early Middle Cambrian. There are thus many kinds of echinoderms that are known or can be expected in the Early Cambrian. Diversified trilobites represented by body fossils are world-wide in the Atdabanian Stage and its correlatives while their trace fossils (*Rusophycus, Cruziana,* etc) are present in the Tommotian Stage and equivalents (Alpert 1976, Glaessner 1969). Phyllocarid crustaceans occur in the Early Cambrian in several places. Zhuravleva (1970, p. 419, Plate 2, Figure 7) noted that ostracods have been reported from the Early Cambrian, but the genus *Fordilla* which she illustrated is now considered (Pojeta et al 1973) to be an early pelecypod.

Recent discoveries have greatly expanded our knowledge of early Mollusca. They are discussed in Pojeta & Runnegar (1976) and Runnegar & Jell (1976). In the classification used by these authors, Pelecypoda, Rostroconchia, Monoplacophora, and Gastropoda are all present in the Tommotian Stage and thus indicate that there were ancestral molluscs in the Precambrian. Runnegar & Jell (1976, p. 125–26) have suggested that they would be of "very small" size and thus might escape casual observation. The hyolithids have often been included within the Mollusca but Runnegar et al (1975) have concluded that they are best considered a separate phylum, Hyolitha. Over 30 genera are known from the Early Cambrian (Rozanov et al 1969).

Inarticulate brachiopods are common in the Early Cambrian. The data in Rowell (1977) and elsewhere suggest that more than nine genera have been recognized, but seemingly none have been reported from the Tommotian. Interestingly, the oldest (*Fallotaspis* Zone) brachiopod recognized in the California sequence is an articulate, possibly referrable to *Nisusia.* Two other Early Cambrian nisusids are known (Siberia, Virginia), as well as an unidentified articulate from Australia. The Bryozoa have not been recognized with certainty in Early Cambrian deposits but some species of *Chancelloria* (usually considered a poriferan) have been suggested to be bryozoans (Zhuravleva 1970), and Gangloff (1976, Plate 3, Figures 4–5) has illustrated a possible trepostomate bryozoan.

The Coelenterata are represented in the Early Cambrian by several major types. The earliest conulariid *Palaeoconularia* Tchudinova has been described from Siberia (Zhuravleva 1970). The genus *Conulariella* Boucek occurs in the Middle Cambrian. Handfield (1969) has described the genus *Tabulaconus* from the Early Cambrian of northwest Canada, referred it to the family Gastroconidae Korde, and considered it as a coral-like coelenterate. Korde proposed a new class, Hydroconozoa, including three families and four genera of Early Cambrian age (Zhuravleva 1970). Recently Jell & Jell (1976) have described three species (assigned to two genera) of stony corals from the earliest Australian Middle Cambrian and tentatively referred one genus to the Tabulata and the other to the Rugosa. The genus *Bija* Vologdin, now known to be of Early Cambrian age, has also been suggested to be an early tabulate (Zhuravleva 1970). The medusoid *Velumbrella* Stasinka from the Early Cambrian of Poland (Zhuravleva 1970) represents a link between the Precambrian Ediacaran Scyphozoans and those of later times. Zhuravleva also notes that several Early Cambrian problematic forms have been suggested to be stromatoporoids but this interpretation has not been widely accepted.

Spicules of several types of sponges (Zhuravleva 1970) occur in the Early Cambrian. Some of those described as *Chancelloria* are found in many parts of the world and may prove to be useful stratigraphically. In western North America they are widespread in Lower Cambrian rocks.

The Archaeocyatha are common elements in the more calcareous facies of the Lower Cambrian around the world, appearing before the earliest body fossil trilobites and becoming extinct at about the Lower-Middle Cambrian boundary. Rozanov & Debrenne (1974) have suggested a fourfold subdivision of the Lower Cambrian based on archaeocyathids. Over 140 genera and perhaps 700 species (Zhuravleva 1970) were known by 1970. Additional genera and species have been established since then. In Siberia about nine genera occur in the basal zone of the Tommotian Stage (Rozanov et al 1969, 1973, Zhuravleva & Rozanov 1974), thus indicating that there were Precambrian members of the phylum. The Cribrocyathea (3 orders, 30 genera) are a distinctive group of small (rarely up to several millimeters) fossils of distinctive aspect included within the Archaeocyatha. No well-authenticated foraminifera or radiolaria have been recorded from the Early Cambrian (Zhuravleva 1970), but inasmuch as typical arenaceous foraminifera are known from Middle Cambrian rocks it is very probable that they were present earlier.

The distinctive small conical Early Cambrian fossils *Salterella* and *Volborthella* have often been assigned to the Cephalopoda, incerta sedis, but studies during the past few years have shown that they are not referrable to that class. Yochelson (1977) has recently proposed the phylum Agmata for them, maintaining that they are not closely related to other organisms, while Glaessner (1976) has proposed that they are annelid worms. Whatever their affinities they are distinctive and useful stratigraphically. *Platysolenites* is another characteristic slender tubular fossil of uncertain affinities [although stated by Glaessner (1976) to be a foraminiferan] present in the Lower Cambrian. The last two genera are most widely known in the Baltic region (Korkutis 1971), but *Platysolenites* has recently been found in California (Firby & Durham 1974, Figure 2) and both were long ago recognized in

New Brunswick by Matthew (Patel 1976). *Platysolenites* is characteristic of part of the Tommotian Stage in the Baltic area (Rozanov 1973).

No well-documented predaceous organisms, unless some trilobites occupied this niche, are known from the Early Cambrian, although Durham (1971b) reported a large, elongate coprolite (50 by 150 mm) from California. The specimen contains echinoderm plates, fragments of trilobites and probable brachiopods. The animal that made it is unknown; it could have been a predator, but certainly it was not a small or sessile organism.

The interpretation and affinities of the assemblage of small conical denticle-like fossils (*Campitius titanius*) described and interpreted by Firby & Durham (1974) as a molluscan radula, probably referrable to the Cephalopoda, has been challenged. Yochelson (1977) believes the assemblage to be an artifact of sedimentation and considers the generic name (incorrectly cited as *Campites*) a synonym of *Volborthella* Schmidt, which he refers to his new phylum Agmata. Glaessner (1976) likewise tentatively considers *Campitius* to be a synonym of *Volborthella* but unlike Yochelson refers both *Salterella* and *Volborthella* to the Annelida, comparing them with the embryonic growth stages of modern Sabellarid worms. Morphological comparison (other studies are still in progress) of *Campitius titanius* with a suite of specimens of *Volborthella tenuis* from the Baltic Cambrian, furnished by A. Yu. Rozanov, demonstrates that whatever the generic level relationships, at the specific level *C. titanius* is distinct from *V. tenuis*. If *Campitius* is ultimately shown to be closely related to *Volborthella*, it is then removed from the status of a probable carnivore as suggested by Firby & Durham. The Siberian *Anabarites* Missarzhevsky and the Cribricyathida (considered by some to be archeocyathids and by others to belong to a separate phylum) are likewise suggested by Glaessner to be annelids comparable to the modern serpulids.

In summary, the fauna of the Early Cambrian includes representatives of most major groups of metazoans, including some of the lower chordates. No vertebrates are known and predaceous cephalopods are seemingly absent (unless *Campitius* is ultimately shown to be unrelated to *Volborthella*). Stony corals are mostly absent—probably the highly diversified archaeocyathids occupied most of their putative habitats. Predaceous animals are unknown unless they are represented by a large coprolite from California. The metazoan fauna seems to be composed mostly of filter feeders, grazers, detritus feeders, and burrowers.

The Earliest Cambrian Faunas and their Significance

The most extensively studied Early Cambrian faunas are those of the Siberian and Baltic regions. Biotas of comparable sequences in other parts of the world have not been as intensively examined. Thus the fauna of the Tommotian Stage and especially of the basal zone, the *Aldanocyathus sunnaginicus—Tiksitheca licis* Zone (to clarify possible confusion, the type species of *Aldanocyathus* is *Ajacicyathus sunnaginicus* and thus the zonal name was changed) is of special significance in evaluating the probable composition of the immediately preceding faunas. Despite suggestions and assumptions to the contrary, both the bottom and the top of the Tommotian Stage in the Siberian platform area appear to be marked by a change in facies. Savitsky (in Cowie & Glaessner 1975, pp. 233–35) and Daily (1972, p. 20) have commented on this, and the restriction to this stage of numerous taxa, often at the generic level (e.g. Rozanov et al 1969, Tables 7, 8, 9) emphasizes the limited character of the known Tommotian fauna. The fauna of the basal zone is even more limited in character than that of the succeeding zones, but nevertheless indicates that there were ancestors in the preceding Precambrian. Forty-one genera, probably representing ten or more high-level taxa, occur in the *A. sunnaginicus—T. licis* Zone (Table 1). Inasmuch as the underlying "Manikay horizon" exposed on the Kotui River of the Anabar shield area has only three "skeleton" bearing

Table 1 Genera present in fauna of *Aldanocyathus sunnaginicis—Tiksitheca licis* Zone of Tommotian. Data from Rozanov 1973, Rozanov et al 1969, Zhuravleva & Rozanov 1974

ARCHAEOCYATHA

Aldanocyathus Voronin 1971	*Monocyathus* Bedford & Bedford 1934
Archaeolynthus Taylor 1910	*Nochoroicyathus* Zhuravleva 1951
Cryptoporocyathus Zhuravleva 1960	*Okulitchicyathus* Zhuravleva 1960
Dokidocyathus Taylor 1910	*Robustacyathus* Zhuravleva 1960

PORIFERA

Chancelloria Walcott 1920

HYOLITHA

Allatheca Missarzhevsky 1969	*Kugdatheca* Missarzhevsky 1969
Circotheca Syssoiev 1958	*Laratheca* Missarzhevsky 1969
Conotheca Missarzhevsky 1969	*Majatheca* Missarzhevsky 1969
Crossbitheca Missarzhevsky 1974	*Spinulitheca* Syssoiev 1968
Egdetheca Missarzhevsky 1969	*Tiksitheca* Missarzhevsky 1969
Korilithes Missarzhevsky 1969	*Turcutheca* Missarzhevsky 1969
Kotuyitheca Missarzhevsky 1974	

GASTROPODS (including MONOPLACOPHORA)

Aldanella Vostokova 1962	*Igorella* Missarzhevsky 1969
Anabarella Vostokova 1962	*Latouchella* Cobbold 1921
Bemella Missarzhevsky 1969	*Purella* Missarzhevsky 1974

TOMMOTIIDA

Lapworthella Cobbold 1921	*Tommotia* Missarzhevsky 1966

HYOLITHELMINTHES

Hyolithellus Billings 1871	*Torolella* Holm 1893

POGONOPHORA

Sabellidites Yanishevsky 1926

CONODONTOCHORDATA

Fomitchella Missarzhevsky 1969

INCERTAE SEDIS

Anabaritellus Missarzhevsky 1974	*Coleolus* Hall 1879
Anabarites Missarzhevsky 1969	*Platysolenites* Eichwald 1860
Cambrotubulus Missarzhevsky 1969	*Sunnagina* Missarzhevsky 1969
	Tumulduria Missarzhevsky 1969

metazoans (*Sabellidites, Anabarites,* and *Protohertzina*) according to Matthews & Missarzhevsky (1975, p. 293), and inasmuch as its apparent correlative the "Moty horizon" of the Irkutsk amphitheater has only the sabellitid *Paleolina* (Sokolov 1973), it is obvious that something has happened. Either there is a considerable hiatus at the contact, or there is a marked change in facies, or the change from nonmineralized "skeletons" to mineralized "skeletons" began at this time. Systematically the "skeletal" material in the "Manikay and Moty horizons" represents the Conodonto-chordata, the Pogonophora, and an incertae sedis *Anabarites,* although Missarzhevsky (in Zhuravleva & Rozanov 1974, p. 186) has questionably assigned it to the Coelenterata, and Glaessner (1976) considers it a probable annelid. The first two taxa are usually considered to be high on the evolutionary scale.

One aspect of this earliest Cambrian fauna, and especially the taxa from Siberia, needs special consideration. Many of the fossils, except for some of the hyolithids and archaeocyathids, are small to very small, ranging up to a few mm in major dimensions (Matthews & Misserzyhevsky 1975). It appears probable that this phenomenon may be a function of the extensive use of acid etching techniques for preparation, which tends to make the smaller components of the biota easily available, whereas in mechanical preparation they are often overlooked. In the California sequence specimens of *Platysolenites* range to at least 40 mm in length and micaceous worm tubes (cf. *Proterebella,* Firby & Durham 1974) in beds of about the same age may be a meter or more long. In view of the larger sizes of many of the late Precambrian trace and body-imprint fossils, it appears that the small size of the reported Siberian platform fossils is not typical of all components of the earliest Cambrian metazoan biota from Siberia and elsewhere. In addition to the reported biota, there must have been a considerable number of larger-sized organisms (possibly without mineralized "skeletons"), the ancestors of the metazoan biotas of the Tommotian and Atdabanian. At this point it is well to note that if the Lower Cambrian lasted 30 m.y., then by interpolation (four stages, with four zones and subzones for the Tommotian) the *A. sunnaginicus*—*T. licis* Zone had a time span of about 2 m.y., and was a short interval in the geological sense. It appears possible that this short interval spans the time during which mineralization of the "skeleton" of many of the metazoans occurred and that the Tommotian as a whole (perhaps 7.5 m.y.) is the interval during which mineralization of the "skeleton" of all major groups of metazoans took place.

Whatever the explanation of the above situation, when the total known mineralized "skeleton" biota of the Tommotian and Atdabanian Stages is considered, and when it is recognized that even the later members of the biota had immediate ancestors approximately contemporary with the earliest mineralized ones, it is obvious that the total metazoan biota of the earliest Cambrian had a much greater diversity than that now known and that there was a nearly equal diversity in the late Precambrian.

Known Precambrian Metazoan Biota

Data on Precambrian metazoans is widely scattered in the literature. Much emphasis has been placed on the Ediacara soft-bodied fauna, but a considerable number of taxa have been described from European Russia (including the Baltic

Table 2   Precambrian metazoan body fossils (mostly imprints) and their occurrence. Al. = Alaska; Aus. = Australia; Eu. R. = European Russia, including Baltic area; G. Brit. = Great Britain; N. F. = Newfoundland; S.E. U.S. = southeastern United States; Sib. = Siberia; Swed. = Sweden; S.W. Afr. = southwest Africa; S.W. U.S. = southwestern United States.

MEDUSOIDS

Albumares Fedonkin 1976	Eu. R.
Asterosoma von Otto 1854 (for *Brooksella canyonensis* Bassler)	S. W. U. S.
Beltanella Sprigg 1947	Aus.
Beltanelliformis Menner 1974 (1968?)	Eu. R.
Bronicella Sokolov 1973	Eu. R.
Charniodiscus Ford 1958	Eu. R., G. Brit., N. F.
Chondroplon Wade 1971	Aus.
Conomedusites Glaessner & Wade 1966	Aus.
Cyclomedusa Sprigg 1947	Aus., N.F., Sib., Swed. (?)
Ediacaria Sprigg 1947	Aus.
Hallidaya Wade 1969	Aus.
Kimberia Glaessner & Wade 1966	Aus.
Lorenzites Glaessner & Wade 1966	Aus.
Mawsonites Glaessner & Wade 1966	Aus.
Medusinites Glaessner & Wade 1966	Aus.
Ovatoscutum Glaessner & Wade 1966	Aus., Sib.
Planomedusites Sokolov 1972	Eu. R.
Rugoconites Glaessner & Wade 1966	Aus.
Skinnera Wade 1969	Aus.
Suvorovella Vologdin & Maslov 1960	Sib.
Tirasiana Palij 1969 (?)	Eu. R.

PENNATULACEAN

Arborea Glaessner & Wade 1966	Aus.
Charnia Ford 1958	G. Brit., N.F.
Pteridinium Gurich 1930	Aus., Eu. R., Sib., S.W. Afr.
Rangea Gurich 1930	Aus., Eu. R., Sib., S.W. Afr.

PLATYHELMINTHES

Brabbinthes Allison 1975	Al.

PORIFERA

Spicular impressions (Wade 1970)	Aus.

ANNELIDA

Dickinsonia Sprigg 1947	Aus., Eu. R.
Spriggina Glaessner 1948	Aus.
Vermiforma Cloud 1976	S.E. U.S.

MOLLUSCA (?)

Wyattia Taylor 1966	S.W. U.S.
Cloudina Germs 1972	S.W. Afri.

Table 2 (*continued*)

ARTHROPODA

Onega Fedonkin 1976	Eu. R.
Praecambridium Glaessner & Wade 1966	Aus.
Vendia Keller 1969	Eu. R.
Vendomia Keller 1976	Eu. R.

POGONOPHORA

Paleolina Sokolov 1967	Sib.
Sabellidites Yanishevsky 1926	Eu. R.

CONODONTOCHORDATA

Protohertzina Missarzhevsky 1973	Sib.

ECHINODERMATA (?)

Tribrachidium Glaessner 1959	Aus.

INCERTAE SEDIS

Anabarites Missarzhevsky 1969	Sib.
Baikalina Sokolov 1972	Sib.
Nemiana Palij 1969 (?)	Eu. R.
Parvancorina Glaessner 1958	Aus.
"17 genera" Anderson (1976)	N. F.

Table 3   Precambrian metazoan trace fossils and their occurrence. Aus. = Australia; Can. = Canada; Eu. R. = European Russia, including Baltic area; Sib. = Siberia; S.W. Afr. = southwest Africa; S.W. U.S. = southwestern U.S.; W. Can. = western Canada.

Archaeichnium Glaessner 1963	S.W. Afr.
Bunyerichnus Glaessner 1969	Aus.
Chondrites von Sternberg 1833	Can.
?Cochlichnus Hitchcock 1858	Aus.
?Curvolithus Fritsch 1908	Aus.
Didymaulichnus Young 1972	W. Can., Eu. R.
Gordia Emmons 1844	Aus.
Harlaniella Sokolov 1972	Eu. R.
Helminthoidichnites Fitch 1850	Aus.
Margaritichnus Bandel 1973 (pro *Cylindrichnus* Bandel 1967)	Aus., Eu. R., Sib.
Nenoxites Fedonkin 1976	Eu. R.
?Phycodes Richter 1850	Aus.
Plagiogmus Roedel 1929	Aus., Eu. R., S.W. U.S.
Planolites Nicholson 1873	Aus., W. Can., Eu. R., S.W. U.S.
Pseudorhizostomites Sprigg 1949	Aus.
Rugoinfractus Palij 1974	Eu. R.
Skolithus Haldeman 1840	Aus., Can.
Squamodictyon Vialov & Golev 1960	Aus., Sib.
Suzmites Fedonkin 1976	Eu. R.
Torrowangea Webby 1970	Aus.

area) and the Siberian Platform (Sokolov 1973, Keller et al 1974). Elsewhere metazoans have been recorded from England, Newfoundland, North Carolina and the Cordilleran trough of western North America. The Nama Group of southwest Africa has received considerable attention, but doubt has been expressed as to whether it is of Precambrian or Cambrian age. Recently Kaever & Richter (1976) have described an archaeocyathid (n. gen., n. sp.) from the Buschman Klippe Formation of the lower part of the Nama Group and have concluded that it indicates an Early Cambrian age. In accordance with this data, most of the metazoans from southwest Africa are not included in the present lists (Tables 2 and 3), although some elements of the Ediacara fauna occur in the Nama Group. However, the occurrence of several archaeocyathids in the basal zone of the Tommotian Stage in Siberia indicates that there were late Precambrian ancestors and casts some doubt on the validity of this criterion. It is probable that the age of the Nama Group needs further investigation before the question finally can be resolved.

No single paper presents a complete summary of Precambrian metazoans but papers by Glaessner (1971), Keller (1976), Keller et al (1974), Sokolov (1973), and Stanley (1976) include many data and generalizations. The papers by Allison (1975), Anderson (1976), Cloud et al (1976), Germs (1972), Glaessner & Wade (1966), Keller & Fedonkin (1976), and Wade (1970, 1971) include more data on body fossils (mostly imprints). Trace fossils are noted in the papers by Alpert (1976), Fedonkin (1976), Glaessner (1969, 1976), Keller & Fedonkin (1976), Palij (1974a, 1974b), Sokolov (1973), Wade (1970), Webby (1970), and Young (1972).

In contrast to the earliest Cambrian and younger faunas with mineralized skeletons, the known Precambrian metazoans, except for *Anabarites, Paleolina, Protohertzina,* and *Sabellidites* from the very late Manikay horizon of Siberia, *Brabbinthes* from Alaska, *Cloudina* from Africa, and *Wyattia* from California, were soft-bodied. The majority of body fossil impressions are coelenterate medusoids, although there are three annelids, four arthropods, a possible echinoderm, and a number of forms of uncertain affinity. In addition a diversity of types is represented by trace fossils. The relatively common occurrence of imprints of soft-bodied medusoids and the scarcity of imprints of higher metazoans (although they may be represented by trace fossils) that presumably had firmer bodies suggests that there was something unusual about the manner of fossilization, or that the other types were less abundant. The pecularities of preservation of the Ediacara fauna (see section on the state of knowledge of past biotas) suggests the first possibility, but before that explanation is accepted the faunal composition of the trace fossil record needs more analysis.

All the genera listed in Tables 2 and 3, except *Rugoinfractus* Palij, are from beds that have Ediacaran or Vendian (sensu Sokolov 1973) age, a time range of around 100 m.y. The base of this interval has an age of between 650 and 680 m.y., probably nearer the latter in the Russian areas. However, Glaessner (1971) has suggested that the lower limit of the Ediacara fauna is between 680 and 700 m.y. Thus the duration of the Ediacaran-Vendian is about equivalent to that of the combined Ordovician-Silurian, or slightly greater than that of the combined Permian-Triassic. If the time of origin of the Metazoa is around 700 m.y. BP as suggested by Cloud, the rate of evolution in this interval is far faster than at any subsequent time. Several authors (see review in Stanley 1976) have observed that there is an apparent increase in trace fossil diversification in this interval as the base of the Cambrian is approached. In view of the special conditions necessary for the preservation of trace fossils and the relatively limited study accorded sediments of this interval by qualified investigators, it seems doubtful that this generalization is warranted as yet.

One striking aspect of the spectrum of metazoans represented by body fossil imprints (Table 2) in the Ediacaran-Vendian is the lack of ancestors for most of the Early Cambrian biota. If the soft-bodied medusoids, pennatulaceans and annelids, groups which are poorly represented in the known Phanerozoic fossil record although they are still extant, and the few very latest Precambrian taxa with mineralized "skeletons" are eliminated from consideration, the representation of the ancestors of the common Phanerozoic types is exceedingly poor. *Tribrachidium* probably is an ancestral echinoderm while *Onega, Praecambridium, Vendia,* and *Vendomia* seem to be ancestral arthropods. Both the Arthropoda and the Echinodermata are high on the evolutionary scale and the recorded taxa must have had ancestors. The morphological-organizational distance between these taxa and the Early Cambrian arthropods and echinoderms is considerable, but the distance between them and their protozoan ancestors is even greater. The absence of body fossil imprints of ancestral archaeocyathids, hyolithids, hyolithelminthids, "gastropods", pelecypods, rostroconchs, brachiopods, coelenterate polyps, and pogonophorans is strange. The trace fossils *Didymaulichnus* and *Bunyerichnus* occur in rocks that are well below the top of the Ediacaran-Vendian interval, and in the case of the last-named genus only a little above the Australian late Precambrian tillite. They probably are the trackways of primitive molluscs and neither was very small (*Didymaulichnus* trackways are over 25 mm wide and that of *Bunyerichnus* over 30 mm). Why haven't body fossil imprints of these two animals been recognized when those of medusae have been?

The "skeletonized" genera *Anabarites, Brabbinthes, Cloudina, Paleolina, Protohertzina, Sabellidites,* and *Wyattia* occur in beds that are very near the Precambrian-Cambrian boundary and apparently represent platyhelminthids, molluscs (sensu latu), pogonophorans, conodonts, and an incertae sedis. Why haven't body fossil imprints of their ancestors been recognized?

The trace fossil *Rugoinfractus* Palij (1974a) merits special attention. Morphologically it is comparable with *Didymaulichnus*. It was described from the Tovkach Formation of the Ovruch Series of Riphean age in the Ukraine. Salop (1968) places the Ovruch Series in his Neoprotozoic Group and states that the late Precambrian tillites occur in his younger Epiprotozoic Group. As correlated by Salop the Ovruch Series would fall within the Middle Riphean of Keller. There is a radiometric date (Palij 1974a) of 1100–1400 m.y. BP from the effusive rocks of the immediately underlying Zbrankov Formation (also part of the Ovruch Series). It is unfortunate that there is no radiometric date available from immediately overlying beds but the fact that the Series is older than the late Precambrian tillites lends credence to the suggestion that *Rugoinfractus* has an age near that cited above. *Rugoinfractus* is similar to the younger *Didymaulichnus* with a trackway width of up to 12 mm, and presumably was made by a somewhat similar animal.

The Riphean age of *Rugoinfractus* makes it the oldest undoubted trace of metazoan activity. However, recently Kauffman & Steidtmann (1976) have called attention (unfortunately as yet only in an abstract) to a diverse suite of structures in the upper Medicine Peak Quartzite of Wyoming that have the characteristics of structures of undoubted biogenic origin. Subsequent metamorphism in underlying and overlying rocks is dated at around 1600 m.y. BP, while the suggested age of the quartzite is around 2000 m.y. BP. Unfortunately more details on this occurrence have not yet been published, but the photographic illustrations accompanying the presentation of the paper in Denver (1976) were very suggestive of an organic origin for these structures. If they are of organic origin their size and diversity indicate the presence of a diverse metazoan biota at around 2000 m.y. BP. Further analysis of these structures is urgently needed. Another structure that needs more consideration was reported by Faul (1950) from the Ajibik Quartzite of Michigan. This quartzite has a well-established Middle Huronian age and there is a helium radiometric age of 1200 m.y. BP for magnetite from one of the overlying iron formations. Thus the structure is as old or older than *Rugoinfractus*. As interpreted by Faul, it is a burrow, but the illustration suggests a surface trackway rather than a burrow. Faul notes that two specimens were about 50 cm long. The illustrations suggest that the "burrows" are of organic origin rather than being mudcracks or similar inorganic structures, but they need more study. If their organic origin were verified, their size would indicate that they were made by metazoans and would tend to lend more credence to the organic origin of the Medicine Peak Quartzite structures.

The 20 different trace fossil genera listed in Table 3 represent diverse Precambrian metazoans, and some have several "species" assigned to them. In view of the fact that none of them can be related with certainty to any of the imprints of the Ediacaran-Vendian body fossils, it is obvious that the metazoan diversity of this interval was greater than the combined total of trace fossils and body fossils. A few of the trace fossils, notably *Cochlichnus*, *Plagiogmus*, some *Planolites*, *Phycodes*, and *Skolithus* represent infaunal activities. The others are seemingly epifaunal in origin. Thus there was a diversity of modes of life represented in the late Precambrian.

The Precambrian Non-Metazoan Biota

The presence of autotrophic organisms as early as 3800 m.y. BP has been suggested by Cloud (1976), and the presence of microorganisms is well documented by the stromatolites of the Bulawayan Group (about 3100 m.y. BP) of southwestern Rhodesia (Schopf 1975). However, the time of appearance of eucaryotic cells is the subject of much debate. Cloud believes that the oldest demonstrable eucaryotes are about 1300 m.y. old, but he notes that there are some possible forms from the Transvaal sequence of South Africa that may be more than 2250 m.y. old. Tappan (1976) believes that *Huroniospora* and *Palaeocryptidium* are probably unicellular red algae (eucaryotic) and suggests that the eucaryotic cell appeared at least by 1900 m.y. BP. Walter et al (1976) restudied the megascopic "trails" from the Greyson Shale of the Belt Supergroup of Montana referred by Walcott to *Helminthoidichnites*. They recognize five different taxa of megascopic algae in the material, "probably related" to the brown, red, or green algae, all eucaryotic groups. The specimens range in length from about 9 to 125 mm. The age of the Greyson Shale is well established at about 1300 m.y. BP. Despite the uncertainties surrounding the identification of eucaryotic cells noted by Cloud (1976) and Schopf (1975), it appears possible that they may have appeared at least as early as 2250 m.y. BP, and probably prior to 1900 m.y. BP. Once the eucaryotic cell appeared and as long as the atmosphere contained O_2 as suggested by Dimroth & Kimberly (1976), then metazoans could have appeared rapidly.

SUMMARY

From the viewpoint of the doctrine of evolution and biological species it is clear that every metazoan organism had ancestors. Inasmuch as our knowledge of the record of past life is very incomplete, the known fossil metazoans have very significant implications about the time of appearance of multicellular animals. From this perspective, examination of the known fossil record yields the following significant points.

1. Knowledge of the total diversity of past biotas is poor, even for those with mineralized "skeletons."

2. Knowledge of Precambrian premetazoans is likewise poor, being largely limited to those organisms that inhabited the specialized environment that produced cherty sediments.

3. The Australian Precambrian Ediacaran body fossil imprints represent a very unusual situation. All the body fossil organisms were transported to the site of burial while none of the trace fossils (autochthonous) occurring in the same beds can be confidently related to the body fossils. Why aren't there imprints of the organisms that produced the trace fossils?

4. Contrary to most past interpretations, Dimroth & Kimberly have suggested, on the basis of chemical-mineralogic studies, that the Precambrian atmosphere contained O_2 as early as 3750 m.y. BP. Margulis et al have reviewed photosynthesis in the Cyanophytes and concluded that they release oxygen during photosynthesis. Inasmuch as Cyanophytes may appear about 3800 m.y. BP, this supports Dimroth & Kimberly's conclusion.

5. The Early Cambrian biota includes organisms as highly evolved as the Conodontochordata and Protochordata. Most major groups of organisms except predaceous cephalopods and the vertebrates are represented.

6. The basal zone of the basal stage (Tommotian) of the Cambrian of the Baltic and Siberian areas has a recorded metazoan fauna with mineralized "skeletons" of at least 40 genera that must have had ancestors.

7. The latest Precambrian immediately underlying the Tommotian Stage has four "skeletonized" fossils (*Anabarites, Paleolina, Protohertzina,* and *Sabellidites*).

8. The basal zone of the Tommotian had a duration of perhaps 2 m.y. and seemingly spanned the interval during which most metazoans acquired mineralized "skeletons."

9. Over 40 metazoan genera, mostly soft-bodied, are known as body fossils from the late Precambrian Ediacaran-Vendian (duration about 110 m.y.). In addi-

tion about 19 trace fossil genera are known from the same interval. Most of the body fossils are soft-bodied coelenterates, types that are rarely represented in the subsequent record. Strangely, few ancestors of the diverse Early Cambrian types with mineralized "skeletons" can be recognized, but putative ancestral types include four probable arthropods and an apparent echinoderm. Nevertheless, probable molluscan trackways (*Bunyerichnus* and *Didymaulichnus*) occur well down in the Ediacaran-Vendian interval.

10. The trace fossil *Rugoinfractus*, an unequivocal metazoan trackway, occurs in Middle Riphean rocks of the Ukraine. It is older than the tillites at the base of the Vendian and younger than a radiometric date of 1100–1400 m.y. BP.

11. An unnamed probable metazoan trackway occurs in the Ajibik Quartzite (age more than 1200 m.y. BP) of Michigan, while possible organic structures are found in the Medicine Peak Quartzite (age about 2000 m.y. BP) of Wyoming.

12. The Metazoa appeared earlier than the about 700 m.y. BP date preferred by Cloud and others. They certainly appeared prior to the time of the trace fossil *Rugoinfractus* (Riphean, younger than a 1100–1400 m.y. radiometric date) and possibly prior to 2000 m.y. BP if the biogenic origin of the Medicine Peak Quartzite structures is confirmed.

13. The diversity of the Early Cambrian metazoan fauna implies that there was a much greater metazoan diversity in the late Precambrian than that currently recorded. The body fossil imprints of the Ediacara fauna of Australia are not representative of the metazoan biota of that time.

Literature Cited

Allison, C. W. 1975. Primitive fossil flatworm from Alaska: New evidence bearing on ancestry of the Metazoa. *Geology* 3: 649–52

Alpert, S. T. 1976. Trace fossils of the White-Inyo Mountains. In *Pacific Coast Paleogeography Field Guide 1*, ed. J. N. Moore, A. E. Fritsche, pp. 42–48. Los Angeles, Pacific Sect., Soc. Econ. Paleont. & Mineral. 69 pp.

Anderson, M. M. 1976. Fossil Metazoa of the Late Precambrian Avalon fauna, Southeastern Newfoundland. *Geol. Soc. Am. Abstr. with Programs* 8: 754

Berkner, L. V., Marshall, L. C. 1965. History of major atmospheric components. *Proc. Natl. Acad. Sci. USA* 53: 1215–25

Berry, W. B. N., Norford, B. S. 1976. Early Late Cambrian Dendroid Graptolites from the Northern Yukon. *Geol. Surv. Can. Bull.* 256: 1–12

Cloud, P. 1968. Pre-metazoan evolution and the origins of the Metazoa. *Evolution and Environment*, ed. E. T. Drake, pp. 1–72. New Haven: Yale Univ. Press

Cloud, P. 1972. A working model of the primitive earth. *Am. J. Sci.* 272: 537–48

Cloud, P. 1973. Paleoecological significance of the banded iron-formation. *Econ. Geol.* 68: 1135–43

Cloud, P. 1976. Beginnings of biospheric evolution and their biogeochemical consequences. *Paleobiology* 2: 351–87

Cloud, P., Wright, J., Glover, L. III. 1976. Traces of animal life from 620-million-year-old rocks in North Carolina. *Am. Sci.* 64: 396–406

Conway Morris, S. 1977. Aspects of the Burgess Shale fauna, with particular reference to the non-arthropod component. *J. Paleontol.* 51, Suppl. to No. 2, Pt. 3, pp. 7–8

Cowie, J. W. 1967. Life in Pre-Cambrian and Early Cambrian times. In *The Fossil Record*, ed. W. B. Harland et al, pp. 17–36. London Geol. Soc.

Cowie, J. W., Glaessner, M. F. 1975. The Precambrian-Cambrian Boundary: A Symposium. *Earth Sci. Rev.* 11: 209–51

Daily, B. 1972. The base of the Cambrian and the first Cambrian faunas. *Univ. Adelaide, Centre for Precambrian Res., Spec. Pap. 1*: 13–42

Dimroth, E., Kimberly, M. M. 1976. Precambrian atmospheric oxygen: evidence in the sedimentary distributions of carbon, sulfur, uranium, and iron. *Can. J. Earth Sci.* 13:1161–85

Durham, J. W. 1967. The incompleteness of our knowledge of the fossil record. *J. Paleontol.* 41:559–65

Durham, J. W. 1971a. The fossil record and the origin of the Deuterostomata. *Proc. North Am. Paleontol. Conv.* Pt. 4:1104–32

Durham, J. W. 1971b. Some Late Precambrian and Early Cambrian fossils from the White-Inyo Mountains of California. *Geol. Soc. Am. Abstr. with Programs* 2:114–15

Faul, H. 1950. Fossil burrows from the Precambrian Ajibik Quartzite of Michigan. *J. Paleontol.* 24:102–06

Fedonkin, M. A. 1976. The multicellular animal traces in Valdai series rocks. *Proc. USSR Acad. Sci., Geol. Ser. 1976* (4):129–32 (In Russian)

Firby, J. B. 1972. Possible cephalochordates from the Lower Cambrian. *Geol. Soc. Am.* Abstr. with Programs 4:504

Firby, J. B., Durham, J. W. 1974. Molluscan radula from earliest Cambrian. *J. Paleontol.* 48:1109–19

Gangloff, R. A. 1976. Archaeocyatha of eastern California and western Nevada. See Alpert 1976, pp. 19–30

Garrels, R. M., Mackenzie, F. T. 1974. Chemical history of the Oceans deduced from post-depositional changes in sedimentary rocks. *Soc. Econ. Paleont. Mineral Spec. Publ.* 20:193–204

Germs, G. J. B. 1972. New shelly fossils from Nama Group, Southwest Africa. *Am. J. Sci.* 272:752–61

Glaessner, M. F. 1969. Trace fossils from the Precambrian and basal Cambrian. *Lethaia* 2:369–93

Glaessner, M. F. 1971. Geographic distribution and time range of the Ediacara Precambrian Fauna. *Geol. Soc. Amer. Bull.* 82:509–14

Glaessner, M. F. 1972. Precambrian palaeozoology. *Univ. Adelaide, Centre for Precambrian Res., Spec. Pap.* 1:43–52

Glaessner, M. F. 1976. Early Phanerozoic annelid worms and their geological and biological significance. *J. Geol. Soc. London.* 132:259–75

Glaessner, M. F., Wade, M. 1966. The Late Precambrian fossils from Ediacara, South Australia. *Palaeontology* 9:599–628

Handfield, R. C. 1969. Early Cambrian coral-like fossils from the northern Cordillera of western Canada. *Can. J. Earth Sci.* 6:782–85

Henderson, R. A., Shergold, J. H. 1971. *Cyclocystoides* from early Middle Cambrian rocks of Northwestern Queensland, Australia. *Palaeontology* 14:704–10

Hofmann, H. T. 1971. Precambrian fossils, pseudofossils, and problematica in Canada. *Geol. Surv. Can. Bull.* 189:1–146

Jell, P. A., Jell, J. S. 1976. Early Middle Cambrian corals from western New South Wales. *Alcheringa* 1:181–95

Johnson, R. G., Richardson, E. S. Jr. 1969. Pennsylvanian invertebrates of the Mazon Creek area, Illinois: The morphology and affinities of *Tullimonstrum*. *Fieldiana Geol.* 12:119–49

Kaever, M., Richter, P. 1976. *Buschmannia roeringi* n. gen., n. sp. (Archaeocyatha) aus der Nama-Gruppe Sudwestafrikas. *Palaeontol. Z.* 50:27–33

Kauffman, E. G., Steidtmann, R. 1976. Are these the oldest metazoan trace fossils? *Geol. Soc. Am. Abstr. with Programs* 8:947–48

Keller, B. M. 1976. The animal remnants in the Riphean stratigraphy. *Proc. USSR Acad. Sci., Geol. Ser. 1976* 8:68–77 (In Russian)

Keller, B. M., Fedonkin, M. A. 1976. New findings of fossils in Precambrian Valday series rocks (Sjuz'ma river valley). *Proc. USSR Acad. Sci., Geol. Ser. 1976* 3:38–44 (In Russian)

Keller, B. M., Menner, V. V., Stepanov, V. A., Chumakov, N. M. 1974. Recent Metazoa finds in the Russian platform Vendomian strata. *Proc. USSR Acad. Sci., Geol. Ser 1974* 12:130–134 (In Russian)

Korkutis, V. 1971. The Cambrian Deposits of the Baltic Basin's Area. *Lithuanian Sci. Res. Geol. Surv. Inst., Trans.* 12:1–173 (In Russian)

Margulis, L., Walker, J. C. G., Rambler, M. 1976. Reassessment of roles of oxygen and ultraviolet light in Precambrian evolution. *Nature* 264:620–24

Matthews, S. C., Missarzhevsky, V. V. 1975. Small shelly fossils of late Precambrian and early Cambrian age: A review of recent work. *J. Geol. Soc. London* 131:289–304

Melton, W., Scott, H. W. 1973. Conodont-bearing animals from the Bear Gulch Limestone, Montana. *Geol. Soc. Am. Spec. Pap.* 141:31–66

Müller, J. K. 1977. *Palaeobotryllus* from the Upper Cambrian of Nevada—a probable ascidian. *Lethaia* 10:107–18

Nitecki, M. H., Schram, F. R. 1976. *Etacystis communis*, a fossil of uncertain affinities form the Mazon Creek Fauna (Pennsylvanian of Illinois). *J. Paleontol.* 50:1157–61

Palij, V. M. 1974a. On finding of the trace fossil in the Riphean deposits of the Ovruch Ridge. *Rep. Acad. Sci. Ukr. SSR,*

Ser. B. Geol., Geophys., Chem., Biol., Jahrg. 36(1): 34–37 (In Ukrainian)

Palij, V. M. 1974b. Double traces (Bilobites) in the deposits of the Baltic Series in the Dniester area. Rep. Acad. Sci. Ukr. SSR, Ser. B. Geol., Geophys., Chem., Biol., Jahrg. 36(6): 499–503 (In Ukranian)

Patel, I. M. 1976. Lower Cambrian of southern New Brunswick and its correlation with successions in northeastern Appalachians and parts of Europe. Geol. Soc. Amer., Abstr. with Programs 8 : 243

Pojeta, J. Jr., Runnegar, B. 1976. The paleontology of Rostroconch Mollusks and the early history of the phylum Mollusca. US Geol. Surv. Prof. Pap. 968 : 1–88

Pojeta, J. Jr., Runnegar, B., Kriz, J. 1973. Fordilla troyensis Barrande: The oldest known pelecypod. Science 180 : 866–68

Raup, D. M. 1976a. Species diversity in the Phanerozoic: A tabulation. Paleobiology 2 : 279–88

Raup, D. M. 1976b. Species diversity in the Phanerozoic: An interpretation. Paleobiology 2 : 289–97

Repina, N. L., Lazarenko, N. P., Meshkova, N. P., Korshunov, V. I., Nikiphorov, N. I., Aksarina, N. A. 1974. Biostratigraphy and fauna of the Lower Cambrian of Charaulach (ridge Tuora-Sis). Acad. Nauk. USSR, Siberian Branch, Trans. Inst. Geol. Geophys. 235 : 1–299 (In Russian)

Resser, C. E., Howell, B. F. 1938. Lower Cambrian Olenellus Zone of the Appalachians. Geol. Soc. Am. Bull. 49 : 195–248

Rowell, A. J. 1977. Early Cambrian brachiopods from the Southwestern Great Basin of California and Nevada. J. Paleontol. 51 : 68–85

Rozanov, A. Yu. 1973. Regularities in the morphological evolution of regular Archaeocyathean and the problems of the Lower Cambrian Stage Division. Akad. Nauk USSR, Trans. Geol. Inst. 241 : 1–164 (In Russian)

Rozanov, A. Yu., Debrenne, F. 1974. Age of Archaeocyathid assemblages. Amer. Journ. Sci. 274 : 833–48

Rozanov, A. Yu., Missarzhevsky, V. V., Volkova, N. A., Voronova, L. G., Krylov, I. N., Keller, B. M., Korokyuk, I. K., Lendzion, K., Michniak, R., Pychova, N. G., Sidorov, A. D. 1969. Tommotian Stage and the Cambrian Lower Boundary problem. Akad. Nauk. USSR, Trans. Geol. Inst. 206 : 1–380 (In Russian)

Runnegar, B., Jell, P. A. 1976. Australian Middle Cambrian molluscs and their bearing on early molluscan evolution. Alcheringa 1 : 109–38

Runnegar, B., Pojeta, J. Jr., Morris, N. J., Taylor, J. D., Taylor, M. E., McClung, G.

1975. Biology of the Hyolitha. Lethaia 8 : 181–91

Salop, L. L. 1968. Pre-Cambrian of the U.S.S.R. Int. Geol. Cong., XXIII Sess., Czech. 4 : 61–73

Schopf, J. W. 1975. Precambrian paleobiology: Problems and perspectives. Ann. Rev. Earth Planet. Sci. 3 : 213–49

Sokolov, B. S. 1973. Vendian of Northern Eurasia. In Arctic Geology, Am. Assoc. Petrol. Geol. Mem. 19 : 204–18

Stanley, S. M. 1973. An ecological theory for the sudden origin of multicellular life in the Late Precambrian. Proc. Natl. Acad. Sci. USA 70 : 1486–89

Stanley, S. M. 1976. Fossil data and the Precambrian-Cambrian evolutionary transition. Am. J. Sci. 276 : 56–76

Tappan, H. 1976. Possible eucaryotic algae (Bangiophycidae) among early Proterozoic microfossils. Geol. Soc. Am. Bull. 87 : 633–39

Valentine, J. W. 1970. How many marine invertebrate fossil species? A new approximation. J. Paleontol. 44 : 410–15

Wade, M. 1968. Preservation of soft-bodied animals in Precambrian sandstone at Ediacara, South Australia. Lethaia 1 : 238–67

Wade, M. 1970. The stratigraphic distribution of the Ediacara fauna in Australia. Trans. R. Soc. S. Australia 94 : 87–104

Wade, M. 1971. Bilaterial Precambrian Chondrophores from the Ediacara fauna, South Australia. Proc. R. Soc. Victoria 84 : 183–88

Walter, M. R., Oehler, J. H., Oehler, D. Z. 1976. Megascopic algae 1300 million years old from the Belt Supergroup, Montana: A reinterpretation of Walcott's Helminthoidichnites. J. Paleontol. 50 : 872–81

Watson, J. 1976. The earth's crust in Precambrian times. Proc. Yorkshire Geol. Soc. 41(2): 145–62

Webby, B. D. 1970. Late Precambrian trace fossils from New South Wales. Lethaia 3 : 79–109

Yochelson, E. L. 1977. AGMATA, a proposed extinct phylum of Early Cambrian age. J. Paleontol. 51 : 437–54

Young, F. G. 1972. Early Cambrian and older trace fossils from the Southern Cordillera of Canada. Can. J. Earth Sci. 9 : 1–17

Zhuravleva, I. T. 1970. Marine faunas and Lower Cambrian stratigraphy. Am. J. Sci. 269 : 417–45

Zhuravleva, I. T., Rozanov, A. Yu., eds. 1974. Biostratigraphy and Paleontology of the Lower Cambrian of Europe and Northern Asia. Moscow: Nauka. 311 pp. 41 plates (In Russian)

Ann. Rev. Earth Planet. Sci. 1978. 6 : 43–74

PHOTOCHEMISTRY AND DYNAMICS OF THE OZONE LAYER

×10086

Ronald G. Prinn, Fred N. Alyea, and Derek M. Cunnold

Department of Meteorology, Massachusetts Institute of Technology, Cambridge, Massachusetts 02139

1 INTRODUCTION

Ozone is found in trace amounts throughout the atmosphere, but the largest concentrations are located in a well-defined layer in the lower stratosphere between the altitudes of 15 and 30 km. The first theory advanced to explain the occurrence of this layer was due to Chapman (1930), who proposed a static pure-oxygen photochemical steady-state model. Subsequent observations and research on the photochemistry and circulation of the upper atmosphere have led us to a picture of considerably greater complexity; the chemistry has proven to be highly sensitive to the presence of certain atoms and free radicals in extremely low concentrations, while the circulation appears to be strongly influenced by mechanical forcing resulting from upwardly propagating tropospheric disturbances.

The ozone layer has recently received national and international attention because of the possibility of pollution of the layer by exhaust gases from supersonic aircraft, by chlorofluoromethanes used as refrigerants and aerosol-can propellants, and by increased production of nitrous oxides in fertilized soil by denitrifying bacteria. These pollutants result directly or indirectly in the production of the reactive species NO, NO_2, ClO, and Cl, which can catalytically destroy ozone.

Although horizontal winds in the lower stratosphere can be substantial, a number of special characteristics including the presence of a temperature inversion, the absence of significant local heating, and the presence of summer easterly winds and strong winter westerly winds combine to make the lower stratosphere much more stable to vertical mixing than the troposphere. Consequently, once destructive species are introduced into the 20–30-km altitude region it takes roughly 1–3 years before they can be removed again by mixing down to the ground. This long stratospheric residence time combined with the catalytic nature of the ozone destruction caused by NO, NO_2, ClO, and Cl makes the ozone layer particularly sensitive to increases in their concentrations.

Depletion of stratospheric ozone looms as an environmental hazard of significant

43

0084-6597/78/0515-0043$01.00

proportions, for this ozone plays a critical role in the biosphere by absorbing ultraviolet radiation with wavelengths between 2400 and 3200 Å which would otherwise be transmitted to the Earth's surface. This radiation is lethal to simple unicellular organisms (algae, bacteria, protozoa) and to the surface cells of higher plants and animals. It also damages the genetic material of cells (DNA) and is responsible for sunburn (erythema) in human skin. Even more important, the incidence of skin cancers (particularly squamous and basal-cell carcinomas) has been statistically correlated with the observed surface intensities of the ultraviolet wavelengths between 2900 and 3200 Å, which are not totally absorbed by the ozone layer (National Academy of Sciences 1975).

Ozone also plays an important role in the meteorology of the upper atmosphere through absorption of solar ultraviolet and visible wavelengths below 7100 Å and thermal infrared wavelengths around 9.6 μm. The consequent heating increases the static stability of stratospheric air by causing the temperature to rise steadily from about 220 K at the tropopause (8–16-km altitude) to about 280 K at the stratopause (50-km altitude). This ozone absorption furthermore provides a significant energy source for driving the circulation of the mesosphere and for forcing atmospheric tides, which have considerable amplitudes in the upper mesophere and thermosphere. The sensitivity of local heating to ozone concentrations and in turn of ozone concentrations to local temperature also results in a dramatic increase in the radiative damping rate of upper atmospheric waves. There is even a small contribution by O_3 absorption to the total energy budget of the earth and thus conceivably to the tropospheric and surface climate.

A complete theory of the ozone layer requires an analysis of both the chemistry and dynamics of this region of the atmosphere. We therefore begin in the next section with a brief review of the relevant observations of the composition and circulation of the stratosphere. This is followed by sections outlining the important photochemical and dynamical theories. Our understanding of the ozone-layer phenomenon has been advanced through the construction of appropriate models. Because of the coupling of chemistry and dynamics through ozone heating and through the temperature dependence of ozone chemistry we focus our discussion of models on a three-dimensional model that simultaneously predicts both ozone concentrations and atmospheric circulation. We complete the review with a brief statement of unsolved problems and the future direction of research and regulation concerning the ozone layer.

2   OBSERVATIONS

The ozone layer is a highly variable phenomenon. If we look first at the ozone column abundance that governs the ultraviolet dosages at the earth's surface we see that there are surprisingly large seasonal and latitudinal variations (Figure 1a; Dutsch 1971). The largest amounts in the Northern Hemisphere are found at polar latitudes in the spring. This spring maximum in the Southern Hemisphere occurs at mid-latitudes rather than the poles. The buildup of polar ozone during early spring is particularly dramatic (Newell et al 1973). Column ozone increases of up to

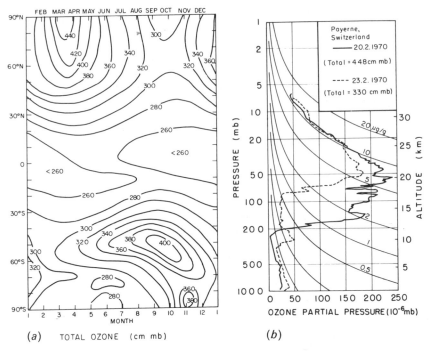

Figure 1 (*a*) Ozone column abundance as a function of latitude and time of year, redrawn from Dutsch (1971). (*b*) Vertical profiles of ozone partial pressure taken 3 days apart over Switzerland, redrawn from Dutsch (1974). Lines of constant ozone mixing ratio are also indicated.

50% can occur over a time span as short as 10 days. At the same time a strong correlation between northward and downward air motions appears to accompany the ozone buildup; a point we return to later. Daily variations in the ozone column are also evident, and these appear to be due in part to synoptic-scale travelling waves. These waves produce variations up to 25% in the O_3 column and have dispersion characteristics strongly suggestive of Rossby waves (Reiter & Lovill 1974).

Rapid temporal variations also occur in ozone concentrations at particular altitudes. For example in Figure 1b (Dutsch 1974) we see factor of 8 changes in the O_3 density occurring over a 3-day period in the 13–19-km altitude region. There is at the same time a substantial trend in the variation of the seasonally averaged ozone concentrations with latitude (Figure 2; Johnston 1975). Maximum ozone concentrations occur at about 25-km altitude in the tropics and 16-km altitude in the polar regions.

In addition to ozone variability over short periods and at various localities the long term trend in the total mass of ozone in the atmosphere is of considerable importance because of recent predictions of ozone depletion by various stratospheric pollutants. Derivation of this trend is difficult because the ground stations are irregularly located, intermittent in operation, and report different long-term trends. There is also a quasi-biennial oscillation in column ozone, particularly above tropical stations, which must be filtered from the data. Unfortunately absolute calibration of satellite ozone observations is not yet sufficiently accurate for application to the ozone trend problem. Angell & Korshover (1976) present data suggesting that total ozone decreased by $\sim 1\%$ in 1960–1962, increased by $\sim 3\%$ in 1962–1970, and decreased by $\sim 1\%$ in 1970–1972 and $\sim 2\%$ in 1970–1974. These workers also note the apparent correlation between total ozone and the solar cycle as measured by the sunspot number; maximum solar activity appears to correspond roughly to maximum total ozone. Similar results are reported by Hill & Sheldon (1975). There is in particular no unambiguous evidence for pollution-induced ozone depletion in recent years.

The chemistry of ozone is governed by a complex array of gas-phase minor constituents. In recent years the detection of these constituents has become possible from balloon, aircraft, and rocket platforms utilizing spectroscopic, chemical, and returned sample techniques. The temporal and spatial distributions of these compounds in the stratosphere are in general poorly characterized in comparison to ozone. Vertical profiles at several latitudes are available for the longer-lived species HNO_3 (Lazrus & Gandrud 1974), $CFCl_3$ and CCl_4 (Krey et al 1977), N_2O (Schmeltekopf et al 1977), and CF_2Cl_2. Less extensive measurements are available for HCl, CH_4, H_2O, H_2, NO_2, NO, and CH_3Cl (see National Academy of Sciences

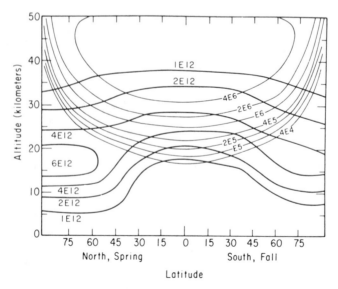

Figure 2 Ozone concentrations ($1E12 \equiv 10^{12}$ molecules cm^{-3}) during equinox as a function of latitude and altitude. Also shown are ozone production rates ($1E6 \equiv 10^6$ molecules cm^{-3} sec^{-1}). After Johnston (1975).

1976 for a review). In Figure 3 we present specific observations of $O(^3P)$ (Anderson 1975), Cl and ClO (Anderson 1976a), OH (Anderson 1976b), $CFCl_3$ and CF_2Cl_2 (Heidt et al 1975), NO, NO_2, and HNO_3 (Evans et al 1976), HCl (Williams et al 1976), and N_2O (Schmeltekopf et al 1977). We caution that these measurements were in general taken at different times and locations and therefore should not be superimposed (Prinn et al 1976). They are presented here only to provide examples of typical measurements rather than illustrate variability that for species such as HNO_3, N_2O, $CFCl_3$, NO, and NO_2 appears to be substantial. The measurements of the fluorocarbons $CFCl_3$ and CF_2Cl_2 also indicate that these species are presently building up in the atmosphere [at the rate of about 13% per year for $CFCl_3$ (Pack et al 1977)].

Meteorological observations of the stratosphere from radiosondes, rockets, and

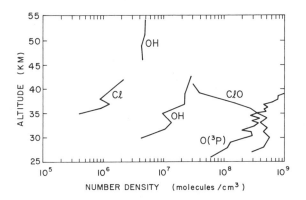

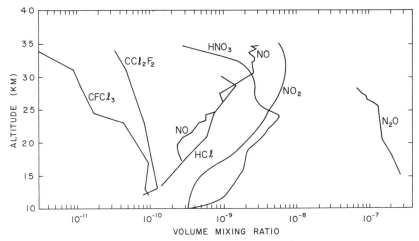

Figure 3 Vertical profiles of the concentration (top) or mixing ratio (bottom) of several free radicals and molecules important in ozone chemistry (see text for references).

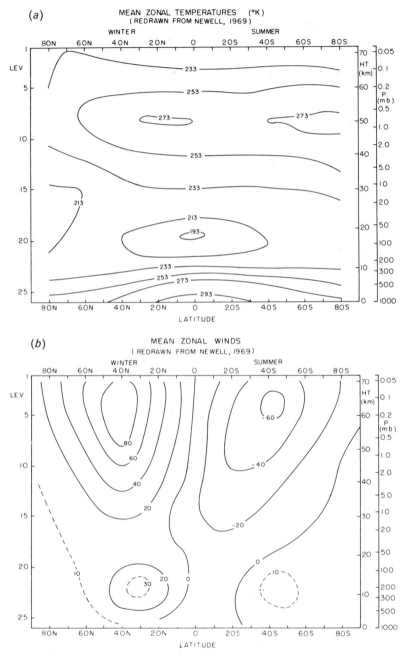

Figure 4 (a) Mean zonal temperatures for Northern Hemisphere winter and summer as a function of latitude and altitude. (b) Same as (a) but for mean zonal winds. Redrawn from Newell (1969).

remote-sensing satellites have provided knowledge of the main characteristics of the circulation of this region of the atmosphere. Observations of the zonal-mean temperature, $\overline{T}$, and zonal wind, $\bar{u}$, are summarized in Figure 4 (Newell 1969). The gross vertical temperature variations are explained by the differential deposition of solar energy resulting from (a) ultraviolet absorption by O_2 above the mesopause (80 km), (b) ultraviolet absorption by O_3 peaking at the stratopause (50 km), and (c) absorption at or near the ground of the bulk of the solar visible and infrared radiation. Horizontal temperature variations are less easily explained. In particular, the coolest temperatures are found at the equator in the lower stratosphere and at the summer pole in the lower mesosphere. As we discuss later these anomalous meridional temperature gradients must be maintained by dynamical heat flows.

The principal features of the zonal winds are the presence of the westerly upper tropospheric jet stream in both summer and winter and the strong westerly *polar night jet* that peaks in the winter mesosphere and is replaced by strong easterlies in the summer. During the equinoxes stratospheric zonal winds are changing direction in both hemispheres. Note that the mean zonal wind and temperature fields in Figure 4 are approximately related by the thermal wind relation, that is

$$\frac{\partial \bar{u}}{\partial lnP} = \frac{R}{f}\left(\frac{\partial \overline{T}}{\partial y}\right)_P, \tag{1}$$

where P is pressure, R is the gas constant, f is the Coriolis parameter, and y is the meridional coordinate. In addition to the annual cycle in $\bar{u}$ depicted in Figure 4, there are two other stratospheric oscillations superimposed. The semiannual oscillation consists of easterlies in summer and winter and westerlies in fall and spring and has maximum amplitudes ~ 30 m sec^{-1} near the equatorial stratopause (Belmont et al 1974). The quasi-biennial oscillation consists of interchanging easterlies and westerlies with a 26–30 month period and has maximum amplitudes ~ 20 m sec^{-1} in the 20–40-km region of the tropical stratosphere (Wallace 1973). This latter oscillation has been explained by interaction of the zonal flow with eastward-propagating equatorial Kelvin waves and westward-propagating mixed Rossby-gravity waves (Holton & Lindzen 1972).

The mean meridional circulation is more difficult to assess from observations because the zonal mean meridional wind $\bar{v}$ and vertical wind $\bar{w}$ have small amplitudes and must therefore generally be deduced indirectly. Studies (e.g. Vincent 1968) indicate that for each hemisphere the typical tropospheric 3-cell structure appears to become a predominantly 2-cell structure in the stratosphere. Rising motions occur above the area of the tropospheric intertropical convergence zone and at high latitudes while sinking motions occur at mid-latitudes. Referring to Figure 4a, we see that in the winter this stratospheric circulation involves relatively cold air being forced to rise in tropical and polar regions and warm air being forced to sink in mid-latitudes; that is, the cells are indirectly operating to convert kinetic energy to potential energy.

Differences between the local and zonal-mean values of dynamical quantities are referred to as wave or eddy motions. Such motions, which are not apparent in the observations discussed above, are of considerable importance in the stratosphere.

Here we define a "steady" (as opposed to "transient") wave as one whose amplitude is roughly constant over the period of a season and a "geostationary" (as opposed to "travelling") wave as one whose zonal phase speed is roughly zero throughout the same time period. Figures 5a and 5b are Northern Hemisphere maps that show seasonal mean temperatures and geopotential heights (height of a P surface) at the 30-mb (25-km) level and illustrate the main steady geostationary features. The summer is dominated by a highly symmetric, anticyclonic, easterly circumpolar vortex, and the tendency for isotherms and isobars to be closely parallel implies very weak poleward advection of heat by geostrophic motions. The winter on the other hand is dominated by a deep polar depression distorted by a high-pressure

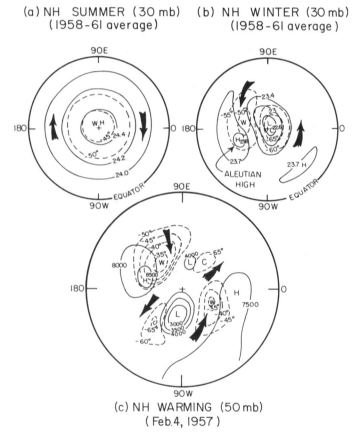

(a) NH SUMMER (30 mb)   (b) NH WINTER (30 mb)
(1958-61 average)   (1958-61 average)

(c) NH WARMING (50 mb)
(Feb.4, 1957)

Figure 5 (a) Northern Hemisphere mean summer temperatures (dashed lines, °C) and geopotential heights (solid lines, km) at 30-mb level. Redrawn from Hare (1968). (b) Same as (a) but for winter. (c) Northern Hemisphere winter warming at 50-mb level on Feb. 4, 1957, redrawn from Reed et al (1963). Dashed lines are temperatures (°C) and solid lines are geopotential height (ft). Large solid arrows indicate rough directions of geostrophic winds.

region centered over the North Pacific (the *Aleutian high*). The warm temperatures to the west of this high imply that poleward advection of heat by the cyclonic winter circumpolar westerly jet is occurring. Steady geostationary waves are much less obvious in Southern Hemisphere data.

There are also some well-defined travelling waves observed in the stratosphere. The mixed Rossby-gravity waves have periods ~ 4–5 days, zonal wave number ~ 4, and westward phase speed ~ 23 m sec^{-1}. The Kelvin waves have periods ~ 15 days, zonal wave number ~ 1–2, and eastward phase speeds ~ 25 m sec^{-1} (Wallace 1973). These two waves play roles in forcing the equatorial regions, but are not of great importance in mid-latitude transport of ozone. Thermal tidal motions with periods ~ 1 day and phase speeds ~ 450 m sec^{-1} have measureable amplitudes in the stratosphere but appear unimportamt in stratospheric dynamics. Of possible great importance for ozone transport are observed stratospheric travelling waves with periods of 1–4 weeks, zonal wave numbers ~ 1–3, and eastward or westward phase speed ~ 10 m sec^{-1} (Deland 1973, Leovy & Webster 1976).

A dynamical phenomenon of considerable interest in the upper atmosphere is the *Sudden Stratospheric Warming* that occurs about once every 1–3 years in the Northern Hemisphere winter. It is characterized by a large growth in amplitude of the geostatjonary wave numbers 1 or 2 over a period of 2 weeks and a concomitant temperature increase at high latitudes of up to 80 K at the 40-km level. This heating is sufficient to reverse the usual high-latitude meridional temperature gradient (see Figure 4a) and as suggested by Equation (1) the usual westerlies at high latitudes temporarily reverse to easterlies. The manner in which these planetary scale eddies transport heat polewards during these warmings is seen dramatically in Figure 5c (Reed et al 1963). If we imagine winds flowing parallel to the isobars it is apparent that warm air is moving northwards to the west of the highs and cool air southwards to the east. We would need a similar arrangement of phases for ozone as seen for temperature in Figure 5c if such large-scale eddies were also transporting ozone polewards. Measurements of ozone in the lower stratosphere are unfortunately not extensive enough to determine if this actually occurs.

Transport processes across the tropopause are of particular concern for quantitative studies of the ozone layer. From observations of radioactivity, potential vorticity, and ozone, Danielson (1968) illustrated the role played by tropopause folding in the jet-stream region in tropospheric-stratospheric exchange. Reiter (1975) estimates that during 1 year about 20% of the total stratospheric air mass is exchanged with the troposphere by such processes. In addition, about 40% is exchanged through the tropical Hadley cell and 10% through seasonal changes in tropopause heights. These exchange processes are all rather intermittent. Jet-stream exchanges can be related to tropospheric cyclonic activity, the upward flow in the Hadley cell occurs at least partly through localized cumulus convection, and tropopause positions change more widely in winter than summer.

3 PHOTOCHEMICAL THEORY

The production of ozone in the upper atmosphere is initiated almost exclusively by photodissociation of molecular oxygen to produce, either directly or indirectly, ground-state oxygen atoms. The pertinent reactions occur mainly above 25-km altitude and are

$$O_2 + hv \quad \begin{aligned} &\to O(^1D) + O \ (\lambda \lesssim 1750 \text{ Å}; z \gtrsim 80 \text{ km}), \\ &\to O + O \quad (\lambda \lesssim 2420 \text{ Å}; z \gtrsim 25 \text{ km}), \end{aligned} \right\} \tag{2}$$

$$O(^1D) + M \to O + M, \tag{3}$$

where M is any molecule. The O atoms may then produce O_3 by combination with O_2

$$O + O_2 + M \to O_3 + M. \tag{4}$$

Ozone has a very short lifetime during the day because the photodissociation reactions

$$O_3 + hv \to O_2 + O(^1D) \ (\lambda \lesssim 3100 \text{ Å}; z \gtrsim 20 \text{ km}), \\ \to O_2 + O \quad (\lambda \lesssim 7100 \text{ Å}; z \geqq 0 \text{ km}), \right\} \tag{5}$$

are very rapid. If we define the *odd oxygen* concentration as the sum of the O and O_3 concentrations, then odd oxygen is produced by reaction (2) and in the early Chapman (1930) theory is removed by

$$O_3 + O \quad \to O_2 + O_2 \ (z \lesssim 90 \text{ km}), \tag{6}$$

$$O + O + M \to O_2 + M \ (z \gtrsim 90 \text{ km}). \tag{7}$$

Clearly reactions (4) and (5) do not effect the odd oxygen concentration but merely define the ratio of O to O_3. Because the rate of reaction (4) decreases with altitude while that for reaction (5) increases with altitude, most of the odd oxygen below 60 km is in the form of O_3, while above 60 km it is in the form of O. The rate of production of O_3 below 45 km may therefore be equated with twice the rate of reaction (2). Computations of the zonally averaged O_3 production during equinox are shown in Figure 2 (Johnston 1975).

The essential result from research studies subsequent to Chapman (1930) is that reactions (6) and (7) can be catalyzed by trace constituents. Thus Johnston (1975) estimates that (6) is presently responsible for removal of only about 18% of the O_3 produced below 45 km. A good deal of the removal appears to be caused instead by catalysis of (6) by nitric oxide (NO) and nitrogen dioxide (NO_2) as first proposed independently by Crutzen (1970, 1971) and Johnston (1971). The catalytic cycle is

$$NO + O_3 \to NO_2 + O_2, \tag{8}$$

$$NO_2 + O \to NO + O_2, \tag{9}$$

and the cycle is partially short-circuited in the daytime because (8) is usually followed by

$$NO_2 + hv \to NO + O \ (\lambda < 3950 \text{ Å}), \tag{10}$$

which regenerates odd oxygen.

Nitrogen oxides are produced naturally in the stratosphere from nitrous oxide

(N_2O) by the reaction (McElroy & McConnell 1971)

$$O(^1D) + N_2O \rightarrow NO + NO. \tag{11}$$

Nitrous oxide is itself produced at least partly by denitrifying bacteria through reduction of nitrates in oxygen-poor soil and water. Increased use of nitrate fertilizers could therefore conceivably increase atmospheric N_2O concentrations and thus ozone destruction rates (Crutzen 1974a, McElroy et al 1976). The principal require-ment in assessing this possible future ozone depletion mechanism is an accurate definition of the nitrogen cycle and in particular of the sources and sinks of the water-soluble N_2O over the land and in the oceans. At the moment the principal recognized irreversible sink is photodissociation in the stratosphere (Bates & Hays 1967),

$$N_2O + h\nu \rightarrow N_2 + O \ (\lambda \lesssim 3150 \ \text{Å}), \tag{12}$$

which results in an atmospheric lifetime for nitrous oxide of about 118 years (McElroy et al 1976). Any irreversible surface sink for N_2O which would result in a lifetime less than this latter value would therefore significantly decrease the adverse effects of possible future growth in the N_2O production by denitrification. In this respect Junge (1974) estimates the atmospheric lifetime of N_2O from actual observations to be only about 8 years; a sink much more potent than (12) is implied.

Nitrogen oxides may also be injected directly into the stratosphere by supersonic aircraft such as the Anglo-French Concorde and Russian Tupolev-144 (both with cruise altitudes ~ 17 km) and the proposed but now cancelled American Boeing 2707 (cruise altitude ~ 20 km). These aircraft produce NO and NO_2 in their engines by thermal decomposition of the N_2 and O_2 in air. After upward transport to altitudes $\gtrsim 25$ km these additional nitrogen oxides will then contribute to ozone destruction (Crutzen 1971, Johnston 1971). Computations in a one-dimensional chemical-diffusive model (McElroy et al 1974) indicate that 320 Concordes operating 7 hours per day would lead to a 1% decrease in the vertical column amount of ozone while a fleet of 3000 Concordes would lead to a 10% decrease. However, one-dimensional models do not permit a complete picture, particularly in view of the fact that the flight paths are expected to lie predominantly at mid-latitudes in the Northern Hemisphere. Significant hemispherical, latitudinal, and seasonal variations in ozone depletion are therefore expected and have been quantitatively predicted (Alyea et al 1975, Cunnold et al 1977). We return to this point when we discuss multidimensional modelling in a subsequent section.

There are several other sources of nitrogen oxides that are generally considered to be of lesser importance than the N_2O source. Firstly, cosmic rays contribute to NO production over the altitude range from 10 to 30 km and particularly in regions poleward of 45° latitude (Warneck 1972, Brasseur & Nicolet 1973). These rays cause ionization reactions that lead to production of N atoms from N_2. Nitric oxide is then produced by reactions of N with O_2 and O_3. Although this source is not important on a global scale it probably constitutes the major chemical source of stratospheric NO during the polar night. Cosmic ray intensities and thus NO pro-duction rates vary by a factor of 2 over the 11-year cycle, with maximum values occurring during solar minimum. Secondly, ammonia produced at the surface by ammonifying bacteria during decay of protein has also been proposed as a source of stratospheric nitrogen oxides (McConnell 1973). In the stratosphere NH_3 is decomposed by photodissociation ($\lambda \lesssim 2300$ Å) and by reaction with OH. The problem here is that NH_3 is very soluble in water and there is no evidence that sufficient amounts survive rainout in the troposphere for NH_3 to become a significant source of odd nitrogen in the stratosphere. There is much less doubt that atmospheric tests of thermonuclear weapons inject NO and NO_2 produced by thermal decomposition of air directly into the lower stratosphere. Temporary ozone reductions of several percent as a result of the multimegaton bomb tests during 1961 and 1962 have been predicted (Johnston et al 1973). Although there is some indication from ground-based ozone measurements that a reduction did in fact occur at that time a definite relation between cause and effect is difficult to establish (Bauer & Gilmore 1975). Finally, NO is also produced in the ionosphere by ionic reactions leading to excited $N(^2D)$ atoms that react rapidly with O_2 to form NO. This NO could then conceivably diffuse down to the stratosphere. However, rapid predissociation ($\lambda \lesssim 1900$ Å) of NO at levels above the stratopause appears to prevent this high-altitude NO from contributing to the stratospheric odd nitrogen budget (Strobel 1971, Brasseur & Nicolet 1973).

The destructive effect of NO and NO_2 is to some extent diminished because of the reversible formation of nondestructive hydrogen nitrate in the stratosphere by the reactions

$$OH + NO_2 + M \rightarrow HNO_3 + M, \tag{13}$$

$$HNO_3 + h\nu \quad \rightarrow OH + NO_2 \ (\lambda \lesssim 3450 \text{ Å}), \tag{14}$$

$$OH + HNO_3 \quad \rightarrow H_2O + NO_3. \tag{15}$$

The amount of HNO_3 formed depends on the OH concentration, which depends in part on the fact that the reaction of NO with HO_2 yields OH and NO_2. This latter reaction also competes with (8) and accurate values of its rate at stratospheric temperatures are therefore required. Of lesser importance as temporary storages for NO and NO_2 are nitrogen pentoxide (N_2O_5) and nitrogen trioxide (NO_3). Although these storage mechanisms become much more effective at night and during the polar winters, ozone destruction during these latter periods is of little consequence to the total ozone budget.

If we define the *odd nitrogen* family to consist of the species NO, NO_2, HNO_3, N_2O_5, and NO_3, then most of the odd nitrogen above 25 km is in the form of NO_2 and NO, while below 25 km it is in the form of HNO_3. Downward transport of HNO_3, NO_2, and NO into the lower troposphere followed by rainout of these water-soluble gases appears to be the principal mechanism for removing odd nitrogen from the stratosphere.

In addition to the pure oxygen reactions (6) and (7) and the nitrogen oxide reactions (8) and (9), a significant fraction of the ozone produced in the stratosphere is currently believed to be removed by catalytic reactions involving the *odd hydrogen* species H, OH, and HO_2 (Johnston 1975). The importance of the latter species at least above the stratopause was first emphasized by Bates & Nicolet (1950). Hunt (1966) later proposed that the discrepancy between observed stratospheric ozone concentrations and those computed from the Chapman (1930) reactions was principally due to reactions involving odd hydrogen. Current theories still have odd hydrogen species playing an important chemical role in the stratosphere. Their influence on ozone is due in large part to the reactions

$$OH + O_3 \rightarrow HO_2 + O_2, \tag{16}$$

$$HO_2 + O \rightarrow OH + O_2, \tag{17}$$

$$H + O_3 \rightarrow OH + O_2, \tag{18}$$

$$OH + O \rightarrow H + O_2, \tag{19}$$

$$HO_2 + O_3 \rightarrow OH + 2O_2. \tag{20}$$

Odd hydrogen is produced in the stratosphere principally from H_2O and CH_4, which move up from the troposphere and are then destroyed mainly by

$$O(^1D) + H_2O \rightarrow OH + OH, \tag{21}$$

$$O(^1D) + CH_4 \rightarrow OH + CH_3. \tag{22}$$

The CH_3 radical in (22) ultimately yields two or three odd hydrogen species together with carbon monoxide (McConnell et al 1971). The reactions

$$HO_2 + HO_2 \rightarrow H_2O_2 + O_2, \tag{23}$$

$$H_2O_2 + h\nu \rightarrow OH + OH \ (\lambda \lesssim 3000 \ \text{Å}) \tag{24}$$

enable temporary storage of H, OH, and HO_2 as H_2O_2. If the odd hydrogen family is defined as consisting of H, OH, HO_2, HNO_3, and H_2O_2, most of the odd hydrogen below 40 km is in the form of H_2O_2 and HNO_3, while at higher altitudes OH and ultimately H become the dominant forms.

Odd hydrogen is finally destroyed in the stratosphere by formation of H_2O in the reactions

$$OH + HO_2 \rightarrow H_2O + O_2, \tag{25}$$

$$OH + H_2O_2 \rightarrow H_2O + HO_2, \tag{26}$$

$$OH + HNO_3 \rightarrow H_2O + NO_3. \tag{27}$$

The rate of reaction (25) is not known to great accuracy and in view of the importance of odd hydrogen in storing NO and NO_2 as HNO_3 and in the direct destruction of ozone this uncertainty is of considerable current concern.

With about 1% of the ozone produced in the stratosphere being removed by downward transport and ultimate destruction at or near the ground, we have a synopsis of current understanding of the earth's ozone budget at least until a few years ago. Considerable attention has recently been given to the possible impending role of chlorine compounds in stratospheric chemistry. Stolarski & Cicerone (1974) suggested that if Cl atoms were injected into the stratosphere they could destroy odd oxygen in a rapid catalytic cycle

$$Cl + O_3 \rightarrow ClO + O_2, \tag{28}$$

$$ClO + O \rightarrow Cl + O_2. \tag{29}$$

This cycle is partially short-circuited by the reactions

$$ClO + NO \rightarrow Cl + NO_2, \tag{30}$$

$$NO_2 + hv \rightarrow NO + O, \tag{31}$$

which reform odd oxygen. At that time, a sufficiently large, stratospheric, gaseous chlorine mixing ratio (about 10^{-9} would be significant) from natural sources such as HCl from volcanoes and acidification of marine aerosols was not evident. Interestingly, the chemistry of HCl, Cl, ClO and ClOO had already been considered on another planet, namely Venus, where HCl had been observed with a much larger mixing ratio of 10^{-6} but where O_2 and O_3 concentrations were extremely small (Prinn 1971).

The importance of the chlorine chemistry became apparent only when Molina & Rowland (1974a; see also Rowland & Molina 1975) proposed that the man-made chlorofluoromethanes $CFCl_3$ (Freon-11) and CF_2Cl_2 (Freon-12) would ultimately provide a very significant source of Cl atoms in the stratosphere. If release of the chlorofluoromethanes into the atmosphere continued at contemporary rates they emphasized that the only known significant sink for these remarkably inert compounds involved photodissociation in the stratosphere

$$CFCl_3 + hv \rightarrow CFCl_2 + Cl \ (\lambda \lesssim 2500 \text{ Å}; z \gtrsim 20 \text{ km}), \tag{32}$$

$$CF_2Cl_2 + hv \rightarrow CF_2Cl + Cl \ (\lambda \gtrsim 2300 \text{ Å}; z \gtrsim 20 \text{ km}), \tag{33}$$

with subsequent reactions appearing to result in all the chlorine in these two compounds eventually being released into the upper atmosphere. Of the total $CFCl_3$ and CF_2Cl_2 produced in 1973, 55% was used as a propellant in aerosol cans, 29% in refrigerators and air conditioners, and the remaining 16% in plastics, resins, solvents, blowing agents in plastic foams, and other related applications (National Academy of Sciences 1976).

Once injected into the atmosphere the potential effects of Cl and ClO are diminished by temporary storage mechanisms involving reversible formation of hydrogen chloride,

$$Cl + CH_4 \rightarrow HCl + CH_3, \tag{34}$$

$$Cl + HO_2 \rightarrow HCl + O_2, \tag{35}$$

$$OH + HCl \rightarrow Cl + H_2O, \tag{36}$$

and chlorine nitrate (Rowland et al 1976)

$$ClO + NO_2 + M \rightarrow ClONO_2 + M, \tag{37}$$

$$ClONO_2 + hv \rightarrow ClO + NO_2 \ (\lambda \lesssim 4600 \text{ Å}). \tag{38}$$

Other chlorine species such as OClO, HOCl, and Cl_2 appear to play less critical roles as storage mechanisms. If we consider inorganic chlorine as being composed of Cl, ClO, HCl, $ClONO_2$, OClO, HOCl, and Cl_2, then HCl is predicted to be the dominant inorganic chlorine species at all altitudes except in the vicinity of 30 km where $ClONO_2$ could be of comparable importance. Downward transport of the water-soluble HCl followed ultimately by rainout in the lower troposphere is the principal recognized sink for stratospheric inorganic chlorine.

The atmospheric lifetimes for the chlorofluoromethanes may be very long. If stratospheric photodissociation [reactions (32) and (33)] is the only sink, latest estimates indicate lifetimes of 50 and 100 years respectively for $CFCl_3$ and CF_2Cl_2 (National Academy of Sciences 1976). Tropospheric or surface sinks resulting in time constants less than these values would therefore be very significant and have been actively sought. Dissolution in the oceans followed by chemical or biological destruction, formation of clathrates in polar ice, electron capture followed by disproportionation, and reactions in the troposphere with stable cosmic-ray-produced ions such as hydrated H_3O^+ and O_2^- have all been considered and deduced to be ineffective sinks (National Academy of Sciences 1976). Photostimulated destruction of chlorofluoromethanes adsorbed on silicates and metal oxides in desert regions is another possibility that should be investigated.

A number of one-dimensional chemical-diffusive models have been used to assess the possible future reductions in ozone resulting from continued use of chlorofluoromethanes. Crutzen (1974b) estimated that indefinite use at 1972 production rates would ultimately result in a 10% steady-rate reduction in the ozone column. Cicerone et al (1974), Rowland & Molina (1975), and Wofsy et al (1975a) studied a number of time-dependent cases that illustrated three further points. Firstly, because it takes several years before fluorocarbons released in the troposphere are mixed up into the stratosphere, they will continue to build up in the stratosphere for this period of time even after injection into the troposphere has ceased. Secondly, if the atmospheric residence time for the fluorocarbons is as long as hypothesized, fluorocarbon concentrations would remain significant for a further 100 years after their usage ceased. In this respect, an assessment of the actual residence time based on fluorocarbon release data and a long-term measurement program would prove invaluable. Thirdly, if production of fluorocarbons were allowed to steadily grow at modest rates (e.g. 10% per year) the theoretical ozone reduction reaches a value as large as 20% before the year 2000.

A number of other halogen compounds are also entering the stratosphere and decomposing to yield halogen atoms and halogen oxides. However, their apparent role in ozone depletion appears to be of less immediate interest than that of $CFCl_3$ and CF_2Cl_2. For example, carbon tetrachloride (CCl_4) is used in the manufacture of the chlorofluoromethanes and may contribute to about 1% of the present ozone destruction rate (Molina & Rowland 1974b). This is similar to estimates of the contemporary contribution to ozone depletion by $CFCl_3$ and CF_2Cl_2. Two other chlorofluoromethanes, namely $CFHCl_2$ (Freon-21) and CF_2HCl (Freon-22), which are both manufactured mainly for use as refrigerants, and also CH_3Cl, which is derived from both industrial and marine biological sources, are all readily decomposed by OH radicals in the troposphere and stratosphere. Consequently, they appear to be presently only minor sources of stratospheric Cl and ClO (Cicerone et al 1975, Robbins & Stolarski 1976). Solid-fueled rockets using ammonium perchlorate as an oxidizer (Minuteman Missiles, NASA Space Shuttle Booster) inject small amounts of HCl directly into the stratosphere. Computations suggest however that even with the projected 50 Space Shuttle flights per year, the ozone depletion will only be a few tenths of a percent (Whitten et al 1975, National Academy of Sciences 1976). A contribution to ozone depletion by bromine compounds has been conjectured (Wofsy et al 1975b). Catalytic odd oxygen destruction by Br and BrO appears to be similar to but faster than Cl and ClO. Increased use of CH_3Br (an agricultural fumigant) and $C_2H_4Br_2$ (a gasoline additive) could lead to an increased role by bromine in the future ozone budget.

4   DYNAMICAL THEORY

The net radiative heating of the stratosphere and mesosphere is dominated by the opposing effects of local heating due to absorption of solar radiation at wavelengths less than 7100 Å by O_3 and local cooling due to emission of thermal radiation by CO_2 in its 15-μm and 4-μm fundamentals. Of additional importance in the lower stratosphere is heating due to absorption of outgoing planetary thermal radiation by O_3 and CO_2 in their 9.6-μm and 15-μm bands respectively. Emission in the O_3-9.6-μm band also plays a role in cooling, particularly near the stratopause, while H_2O emission in its extensive rotational bands and 6-μm vibrational band contributes to cooling throughout the stratosphere. Computations of the net radiative heating of the atmosphere below 30 km (Dopplick 1972) are shown in Figure 6.

Comparison of Figure 6 with Figure 4a leads to an interesting conclusion in the lower stratosphere (10–200 mb). The coolest regions near the equator are being weakly heated by radiation while the warmest regions in the winter mid-latitude and summer polar regions are being weakly cooled by radiation. It is therefore clear that radiation does not play the dominant role in determining the anomalous meridional temperature gradient in this portion of the atmosphere. The dynamics must instead lead to a substantial poleward flux of heat.

Some insight into the dynamical mechanisms involved may be obtained by considering conversions between the kinetic energy of the zonal mean flow $\bar{K}$, the kinetic energy in eddy motions K', the available potential energy in the zonal mean temperature field $\bar{P}$, and the available potential energy in the eddy temperature field P' (Lorenz 1955). The annual mean Northern Hemisphere energy cycles have been computed for the 30–100-mb region (Oort 1964) and the 10–100-mb region poleward of 20°N (Dopplick 1971). In Oort's (1964) energy cycle for the 30–100-mb region the stratosphere is apparently *forced* by an upward flux of K' (i.e. by wave motions) mainly from the troposphere. Some of this K' is then converted to $\bar{K}$ (i.e. wave motions force the stratospheric jets), while the rest is converted to P' by cold air being forced to rise (and cool adiabatically) and warm air being forced to sink (and heat adiabatically). A northward eddy transport of heat from the cool equator to the warm mid-latitudes then causes conversion of P' to $\bar{P}$ giving rise to the observed anomalous meridional temperature gradient. Finally, radiation

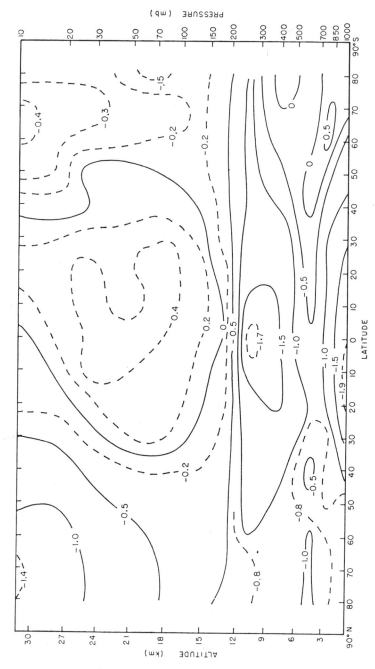

Figure 6 Mean total radiative heating rates (°C day^{-1}) for December–February as a function of latitude and altitude, after Dopplick (1972).

leads to a cooling of the high latitudes and warming of the equator, thus destroying $\bar{P}$.

The stratospheric cycle of Dopplick (1971) differs from that of Oort (1964) in some respects. Forcing is due to fluxes of both K' and $\bar{K}$, again mainly from the troposphere. However, conversion of both K' and $\bar{K}$ to $\bar{P}$ is principally accomplished through cool air rising at low and high latitudes and warm air sinking at mid-latitudes in the zonal mean circulation (i.e. $K' \rightarrow \bar{K} \rightarrow \bar{P}$ rather than $K' \rightarrow P' \rightarrow \bar{P}$). In addition there is a significant conversion of $\bar{P}$ to P' rather than P' to $\bar{P}$ and $\bar{K}$ is much greater and $\bar{P}$ smaller than in Oort (1964). Presumably these differences are partly due to the fact that the Dopplick study omits the 0–20°N region where a good deal of the $\bar{P}$ exists and where northward eddy heat transport up the temperature gradient (see Figure 4a) is occurring (i.e. $P' \rightarrow \bar{P}$). At the same time the Oort study omits the 10–30-mb region where a good deal of the $\bar{K}$ exists (in the jets) and where northward eddy heat transport down a temperature gradient (see Figure 4a) is occurring during the winter at high latitudes (i.e. $\bar{P} \rightarrow P'$).

Comparison of the stratospheric energy cycle with that in the troposphere shows several important differences. First there is the obvious fact that the troposphere is forced by differential solar heating producing $\bar{P}$ while the stratosphere is mechanically forced, mainly by the troposphere producing $\bar{K}$ and K'. In some sense the lower stratosphere acts like a refrigerator using kinetic energy to create temperature contrasts and thus potential energy. Secondly, the principal energy conversion in the troposphere involves $\bar{P} \rightarrow P' \rightarrow K'$, which occurs in growing baroclinic waves for which we have little evidence in the lower stratosphere, at least on the average. Thirdly, total K' is about 15–20 times greater in the troposphere than the lower stratosphere. Since the volumes involved are roughly equal it appears that the eddy or wave kinetic energy *density* in the stratosphere is much less than that in the troposphere. If tropospheric waves were able to propagate energy and phase vertically without dissipation we would expect these densities to be comparable. Clearly most of the kinetic energy associated with the travelling cyclones and anticyclones in the troposphere must be trapped in the lower atmosphere.

The observations discussed in Section 2 showed considerable daily and seasonal variations in the stratosphere. The same variability is seen in $\bar{K}$, K', $\bar{P}$, and P' as shown in Figure 7 (Dopplick 1971). Particularly noteworthy is the fact that zonal wave motions (K') and zonal temperature waves (P') have negligible amplitudes

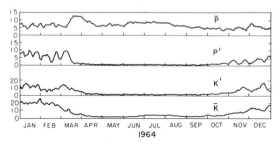

Figure 7   Daily variations of energy per unit area (10^7 erg cm^{-2}) in the form of $\bar{P}$, P', K', and $\bar{K}$ between the 10 and 100-mb levels, after Dopplick (1971).

throughout the late spring, summer, and early fall. Maximum wave activity occurs during the late winter months, which is also the time in which the largest poleward transports of heat and ozone leading to the observed general spring warming and spring maximum in the ozone column must occur. The marked seasonal changes and also the large scales of the wave motions are evident in Figure 5. The distortion of the low pressure field in winter is due to the presence of planetary wave numbers 1 and 2, which constructively interfere to produce the Aleutian high. The predominance of these very long quasi-stationary waves in the stratosphere is in marked contrast to the troposphere, where the largest amplitudes are found in significantly shorter travelling waves (wave number $\gtrsim 7$).

Many aspects of the dynamics of the stratosphere can be studied using the simplified quasi-geostrophic equations in place of the more complete primitive equation set. In this approximation the prognostic zonal and meridional momentum equations are replaced by a single prognostic potential vorticity equation and a diagnostic "balance" equation. This approximation can be rationalized for extra-tropical motions using scaling arguments, including the fact that the Rossby number is small and the motions are quasi-nondivergent (see Holton 1975, for a review). Internal gravity waves including thermal tides are not represented in this approximation, but these latter motions usually have significant amplitudes (e.g. meridional winds $> 5 \, \mathrm{m \, sec^{-1}}$) only above the stratopause (see Lindzen 1971, for a review). The equatorial Kelvin waves (which have essentially no meridional velocities) are also not represented, although the equatorial mixed Rossby-gravity waves are.

The principal feature of steady quasi-geostrophic motions which appears to lead to the significant trapping of cyclone-scale waves in the troposphere mentioned earlier was first elucidated by Charney & Drazin (1961). To study wave motions the quasi-geostrophic potential vorticity equation can be conveniently separated into two equations; one describing the zonal mean vorticity and the other the perturbation or eddy vorticity. In the latter equation we then neglect all nonlinear terms. This neglect can often be justified *a priori* if eddy velocities are much less than zonal mean velocities and/or zonal phase velocities, but in other cases the neglect must be rationalized *a posteriori*. The basic question to be asked is which quasi-geostrophic waves (including the familiar Rosssby waves) can propagate out of their forcing region in the troposphere. Substitution of a steady wavelike solution for the stream function ψ' of the form

$$\psi' = A(z) \exp[i(kx + ly - kct)] \exp\left[\frac{z}{2H}\right], \tag{39}$$

into the linearized eddy potential vorticity equation with all diabatic forcing terms omitted leads us to conclude that the amplitude function $A(z)$ must satisfy a "vertical structure" equation

$$\frac{\partial^2 A}{\partial z^2} + \left[\frac{N^2}{f^2}\left(\frac{\beta}{\bar{u} - c} - k^2 - l^2\right) - \frac{1}{4H^2}\right]A = 0. \tag{40}$$

Here k and l are respectively the zonal and meridional wave numbers. For purposes of illustration it has been assumed here that the mean zonal wind $\bar{u}$, the atmospheric density scale height H, and the Brunt-Väisälä frequency N are constant in the region under consideration and that the variation of the Coriolis parameter f with latitude can be approximated by the mid-latitude "β-plane" approximation in which $\beta = \partial f / \partial y$ (Holton 1975, p. 71). Oscillatory or internal wave solutions to (40) clearly require that

$$O < \bar{u} - c < \frac{\beta}{\dfrac{f^2}{4N^2H^2} + k^2 + l^2} = U_c, \tag{41}$$

where U_c is referred to as the "Rossby critical velocity." Thus if the Doppler-shifted wind velocity $\bar{u} - c$ is either easterly or is westerly but exceeds U_c then the waves cannot propagate phase (and therefore energy) in the vertical. Note that U_c decreases rapidly as k and l increase; that is, as the zonal and meridional wavelengths decrease. Charney & Drazin (1961) concluded that $U_c \lesssim 38$ m sec^{-1} for the long wavelengths typical of mid-latitude mountain-forced waves; later calculations by Dickinson (1968) for a sphere rather than a mid-latitude β-plane imply $U_c \lesssim 62$ m sec^{-1} for the same wavelengths. For cyclone-scale mid-latitude waves (41) implies $U_c \lesssim 1$ m sec^{-1}. For mountain-forced waves $c \simeq 0$, while for typical mid-latitudes Rossby waves $c \lesssim 10$ m sec^{-1}. Referring now to Figure 4b for the observed $\bar{u}$ values in the upper atmosphere, the Charney-Drazin theorem summarized in (41) implies that only the long planetary-scale waves can propagate vertically during winter where $0 < \bar{u} < 80$ m sec^{-1} while no waves propagate during summer where $-60 < \bar{u} < 0$ m sec^{-1}. This picture is modified somewhat when there are lateral and vertical shears in $\bar{u}$ (Matsuno 1970). In particular the long waves that do propagate are guided into the region of maximum westerlies. There is also an important role played by radiative transfer in damping these long waves, particularly during the equinoxes when $\bar{u} \sim 0$ (Dickinson 1969). Quasi-geostrophic theory certainly appears to provide at least a qualitative explanation for the observations discussed earlier.

Interactions between the wave and zonal-mean fields are also important. When waves are being radiatively or viscously damped momentum conservation obviously implies that concurrent accelerations of the mean flow become possible. However, when $\bar{u} \neq c$ and the diabatic forcing and damping terms are assumed zero, examination of the zonal-mean and eddy potential-vorticity equations leads immediately to the conclusion that steady quasi-geostrophic waves cannot transport potential vorticity and cannot grow or decay by exchange of energy and momentum with the zonal-mean flow. This theorem was first derived for internal gravity waves by Eliassen & Palm (1960). However, at a so-called "critical level" where $\bar{u} = c$ the wave is apparently totally absorbed whenever dissipative processes dominate non-linear effects (Booker & Bretherton 1967) and reflected whenever the nonlinear effects dominate (Benney & Bergeron 1969). The former appears to be what occurs in the upper atmosphere. Thus whenever a wave encounters a critical level there is a very large convergence of the wave momentum flux that results in acceleration of the mean zonal flow. Such a wave–mean-flow interaction is expected to occur, for example when mountain-forced geostationary waves reach the boundary between westerlies and easterlies. Such interactions also form the basis of Matsuno's (1971) explanation for the Sudden Stratospheric Warmings in which growing quasi-stationary tropospheric wave numbers 1 and 2 propagate into the winter stratosphere. Poleward eddy heat fluxes associated with these growing waves induce an equatorward mean flow and thus an easterly Coriolis torque that decelerates the westerly flow. Once $\bar{u} \to 0$ then critical level absorption of the waves occurs, resulting in total absorption of the wave energy. The meridional temperature gradient reverses and a temporary easterly circumpolar vortex develops as observed.

The forcing mechanisms for the vertically propagating quasi-geostrophic waves are also of considerable interest. Particularly in the Northern Hemisphere, flow over large-scale topographic features (e.g. Himalayas, Alps, Rockies) and zonal differential heating due to ocean-land temperature contrasts appear to be the major forcing mechanisms for planetary-scale quasi-stationary waves. For the travelling long waves Geisler & Garcia (1977) have recently shown from a study of baroclinic instability that the slowly growing planetary-scale "Green" modes (Green 1960) exhibit wind and temperature amplitudes typically an order of magnitude greater in the middle and upper stratosphere than in the troposphere. These may be contrasted with the faster-growing cyclone-scale "Charney" modes (Charney 1947), which have large amplitudes but only in the troposphere. For a given value of the zonal wind shear the Green modes appear at longer zonal wavelengths than the Charney modes, and in view of the many observations of travelling long waves in the stratosphere the possible role of Green modes in forcing the upper atmosphere needs to be fully investigated.

One further point of dynamical theory originally made by Fjortoft (1951) is particularly relevant here. Considerations of continuity lead to the conclusion that the mean meridional circulation tends to act in a direction to diminish the transport of heat and momentum by the eddies. This partial cancellation effect is seen rather dramatically in the primitive-equation general circulation model of the troposphere and lower stratosphere developed by Manabe & Mahlman (1976). At the 10-mb (30-km) level they show at each latitude that the convergences of angular momentum by eddies and mean flows have almost equal magnitude but opposite sign. A similar partial cancellation is illustrated for convergence of heat at the 110-mb (16-km) level, but the results are modified somewhat by other important heat transfer mechanisms, particularly radiation. In the next section we show that this same type of cancellation even appears in ozone transport.

5   A THREE-DIMENSIONAL DYNAMICAL-CHEMICAL MODEL

Most of the purely chemical theory of the ozone layer has been developed using one-dimensional models that use empirically derived eddy-diffusion coefficients in order to simulate globally-averaged vertical transport. This particular use of the diffusion concept is difficult to justify on dynamical grounds but has been rationally defended by its users as a reasonable approximation in studies focussing on the chemistry. However, in this review we wish to emphasize the interplay between the dynamics and the chemistry and are therefore primarily interested in models that simultaneously predict both these processes. In order to incorporate dynamical predictions such models must of necessity be three-dimensional.

As alluded to in the Introduction the link between ozone chemistry and dynamics results from two properties of ozone. Firstly, absorption by ozone of ultraviolet, visible, and infrared radiation leads to a temperature inversion in the upper atmosphere and to thermal forcing of motions (Leovy 1964). Secondly, the temperature-dependence of the ozone chemistry causes the O_3 density to decrease when the temperature increases, which leads to an increase in the damping rate of temperature waves (Lindzen & Goody 1965). Thus interactive models of the upper atmosphere need to incorporate both these properties.

Early models of this type for the stratosphere include that of Hunt (1969), who utilized a primitive-equation model with 18 levels in the vertical covering the 0–38-km altitude range, and that of Clark (1970), who utilized a quasi-geostrophic model with 6 vertical levels in pressure coordinates. The chemistry in these models is now outdated, but they served to illustrate the important photochemical-dynamical interactions and also the roles played by large-scale eddies and zonal-mean motions in the poleward transport of ozone.

More recently, we have developed at M.I.T. a detailed quasi-geostrophic model with 26 levels in the vertical equally spaced in logarithmic pressure coordinates between the ground and 72-km altitude. The chemical scheme incorporates the important odd nitrogen, odd hydrogen, and odd oxygen chemistry but is simplified in the sense that it requires specification of the distributions of NO_2, OH, and HO_2. The prognostic equations are the vorticity equation, the perturbation thermodynamic equation, and the global mean and perturbation continuity equations for ozone. Diagnostic equations include the hydrostatic equation, the balance condition, and the mass continuity equation. Predicted quantities on each horizontal level are expressed in terms of a rhomboidally truncated series of spherical harmonics with maximum zonal wave number 6 and 6 degrees of freedom in latitude. Surface topography and tropospheric diabatic heating rates are also expressed in series form. Small-scale vertical diffusion is incorporated into the model but is usually assumed small in the lower stratosphere so that vertical transport there is dominated by large-scale motions, as suggested by observation (Panofsky & Heck 1974). Heating due to ultraviolet absorption by ozone is computed exactly while radiative heating and cooling due to infrared radiation in the upper atmosphere are computed using a Newtonian cooling approximation. Chemical and heating computations are carried out at a series of grid-points chosen such as to minimize errors during transformation from spectral to grid-point representations. This quasi-geostrophic spectral model is computationally extremely efficient in comparison with existing primitive-equation grid-point models; it requires only 40 sec on an IBM 360/95 to integrate over 1 day.

Several runs of this model have been performed. Run 12 provided the first realistic simulation of the observed latitudinal and seasonal variations in total ozone and is fully described by Cunnold et al (1975). Run 17 provided an even better total ozone simulation (Alyea et al 1975). Total ozone predictions from this latter run are shown in Figure 8, $\overline{T}$ and $\bar{u}$ predictions in Figure 9, and the predicted mean meridional circulation pattern in Figure 10. For total ozone, comparison between Figures 1 and 8 shows that the observed spring maximum is correctly predicted but with slightly less amplitude than observed. For $\overline{T}$ and $\bar{u}$, comparison of Figures 4 and 9 shows that agreement is surprisingly good in the stratosphere. However, the predicted polar night and summer jets fail to decrease in intensity in the mesosphere and in accordance with (1) the meridional temperature gradient has the wrong sign at the top of the model. Also in the troposphere the predicted jet-streams are too close to the equator. Thus geostationary waves propagating upward from the model winter jet-stream encounter weak easterlies aloft rather than weak westerlies, and (41) would imply a resultant underprediction of stratospheric geostationary wave

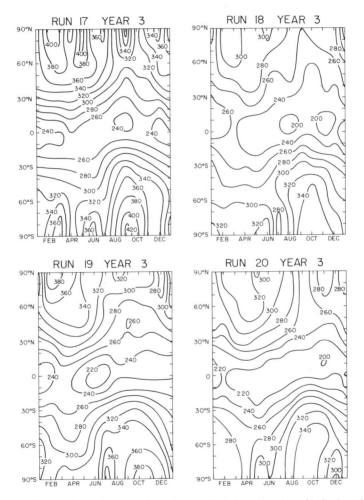

Figure 8 Ozone column abundance (cm mb) as predicted as a function of latitude and time in the M.I.T. model. Run 17 is the natural stratosphere while Runs 18, 19, and 20 depict the stratosphere in the presence of a fleet of supersonic aircraft flying at 20 km and 45°N, 17 km and 45°N, and 20 km and 10°N respectively. After Cunnold et al (1977).

66

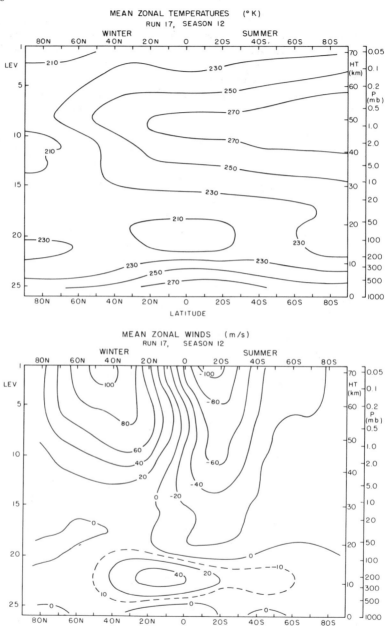

Figure 9 Northern Hemisphere winter and summer mean zonal temperatures (top) and winds (bottom) as predicted in Run 17 of the M.I.T. model.

amplitudes. These inadequacies do not appear to have adversely affected stratospheric ozone transports and could possibly be removed by altering the rigid-lid boundary condition at the top, by including neglected eddy-flux terms in the thermodynamic equation, and by increasing horizontal resolution in the model.

As illustrated in Figure 2, maximum ozone production occurs in tropical regions at altitudes of 30–50 km. However, maximum concentrations appear at about 18 km in polar regions. The way in which ozone is carried polewards and downwards in the model described above is therefore of particular interest. The mean meridional circulation shown in Figure 10 would result in a poleward flux of ozone in equatorial regions but an equatorward flux at mid-latitudes. The mid-latitude ozone flux is in fact poleward at mid-latitudes because as illustrated in Figure 11 the eddy flux of ozone is poleward and of significantly greater amplitude than the opposing meridional flux. In this same diagram the significant difference between the summer where meridional and eddy ozone transports are both weak and in winter where they are both strong is very evident.

In view of the importance of eddy fluxes of ozone it is useful to have a physical picture of how such fluxes might develop. Two points are particularly important. Firstly, the dynamical terms in the continuity equation for ozone dominate the

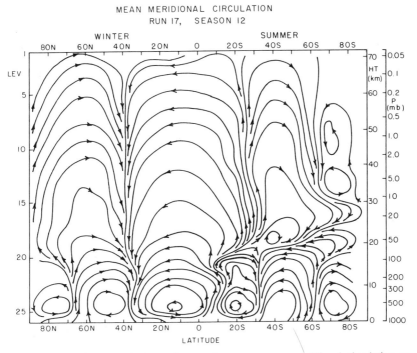

Figure 10 Northern Hemisphere winter and summer mean meridional circulation as predicted in Run 17 of the M.I.T. model (after Alyea et al 1975).

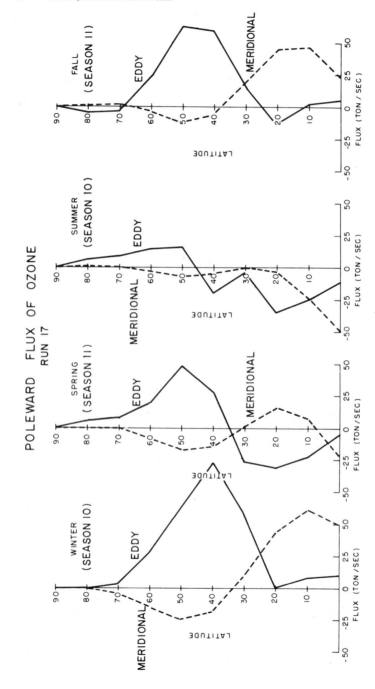

Figure 11 Seasonally-averaged poleward fluxes of ozone due to the mean meridional circulation (dashed lines) and to the eddies (solid lines) as predicted in Run 17 of the M.I.T. model.

chemical terms below about 25 km in both the summer and in the winter equatorward of 50° latitude. In the polar night the dynamical terms dominate at all stratospheric altitudes. Secondly, as illustrated in Figure 1(b), the ozone mixing ratio increases rapidly up to about 25-km altitude. Thus we expect that below 25-km air parcels that have suffered downward displacements will have positive perturbations in ozone mixing ratio, while those that have suffered upward displacements will have negative perturbations. Since potential temperature also increases rapidly upwards in the stratosphere and motions are approximately adiabatic we expect a similar behavior for temperature perturbations. Thus a poleward transport of heat and ozone is expected when poleward moving air is displaced downwards relative to equatorward moving air. The situation for ozone eddy fluxes would therefore be similar to that for eddy heat fluxes illustrated in Figures 5(b) and 5(c). Fluctuations or waves in total ozone would also accompany these poleward fluxes. Regions in which air below the ozone peak has been displaced downwards would be expected to have a large total ozone column, while those with upward displacements would have a small column.

Analyses of the M.I.T. model for the mid-latitude lower stratosphere in winter tend to support this simple picture. Strong correlations exist between perturbations in ozone mixing ratio and temperature, between poleward and downward ozone fluxes, and between perturbations in ozone mixing ratio at say 18 km and total ozone. The dynamical forcing that leads to these particular phase relations has not however been investigated in any great detail. Green (1960) studied the growth of linear baroclinic waves in a simple model with a stable stratosphere overlying a less stable troposphere. In the stable stratosphere it was found that the remnants of the tropospheric waves resulted in warm air moving poleward and downward just to the east of troughs, while cold air moved equatorward and upward just to the east of ridges. This type of study however needs to be further refined before a complete understanding is found.

Detailed three-dimensional models of the type we have described here are particularly useful for studies of perturbations to the ozone layer resulting from the various pollutants discussed in Section 3. Unlike simple one-dimensional models they enable realistic simulation of trace constituent transports, and accurate estimates of the effects of localized sources of stratospheric pollution such as supersonic aircraft and the Space Shuttle. Such models also illustrate ozone depletion as a function of latitude, hemisphere, and season so that accurate estimates of ultraviolet dosage for highly populated latitudes can be made. Finally, they enable incorporation of the dynamical feedback resulting from the fact that ozone contributes substantially to heating in the upper atmosphere.

In order to apply the M.I.T. model to a study of the impact of supersonic aircraft on the ozone layer it was necessary to develop an auxiliary two-dimensional model (Prinn et al 1975) in which the detailed computations of the chemistry were carried out. In this model atmospheric motions were simulated using mean vertical and meridional winds and eddy diffusion coefficients deduced from upper atmospheric observations. A set of 27 chemical reactions involving odd oxygen, odd nitrogen, and odd hydrogen was included. Continuity equations for all long-lived species were integrated using grid-points spaced 5° in latitude and 2 km in altitude until annual cycles roughly repeated themselves. The lower boundary condition for water-soluble species such as HNO_3 was formulated in terms of tropospheric transport and rainout times. Despite the simplifications inherent in a model of this type the agreement between computed and observed concentrations of odd nitrogen compounds was remarkably good (Figure 12; Evans et al 1976).

The three hypothetical supersonic aircraft fleets considered in the M.I.T. model were assumed to continuously inject 1.8×10^6 tons yr^{-1} of NO_2 into the stratosphere. Injections at 45°N and 20-km altitude (Run 18) were designed to simulate mid-latitude flights of several hundred of the Boeing 2707 aircraft and resulted in ozone depletions of ~16% in the northern hemisphere and ~8% in the southern. Injections at 45°N and 17-km altitude (Run 19) were designed to simulate a few thousand of the Concorde or Tupolev 144 aircraft and resulted in ozone depletions of about one-half those in Run 18. Finally, injections at 10°N and 20 km (Run 20), which were designed to show the advantage or disadvantage of equatorial flight corridors, resulted in ozone depletions similar to those in Run 18. Results for total ozone from

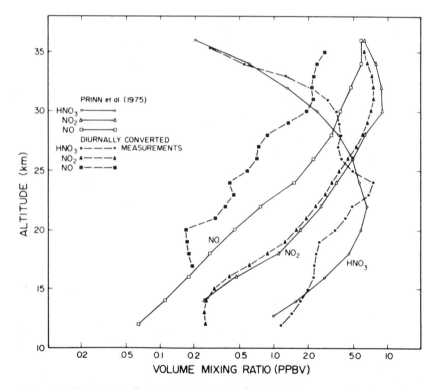

Figure 12   Comparison of results of M.I.T. 2D model for 60°N, summer with observations of NO, NO_2, and HNO_3 at 59°N, July 22, 1974. The observed values have been adjusted to allow for diurnal variations. After Evans et al (1976).

these 3 runs are shown in Figure 8 (Cunnold et al 1977). As a "rule of thumb" the percentage increase in the erythemal or sunburn-producing dosage of ultraviolet radiation resulting from these depletions can be obtained approximately by multiplying the predicted percentage reductions in column ozone by 2. As is evident from Figure 8 these increases in erythemal doses show considerable variations with latitude and time of year; a factor only brought out by accurate multidimensional modelling.

6 CONCLUDING REMARKS

Our knowledge of the ozone layer has been growing at a very rapid pace in the last decade. It would be foolhardy however to claim that either the chemical or dynamical processes that lead to the formation of this layer are now entirely understood. The new measurements of OH, Cl, HCl, and ClO presented earlier have not yet been reconciled with the theoretical models. There are still a number of crucial rate constants (e.g. for $OH + HO_2 \rightarrow H_2O + O_2$ and $NO + HO_2 \rightarrow NO_2 + OH$) over which there is still a great deal of controversy as to their correct values. We presently do not have enough measurements of many species to ensure that natural spatial and temporal variability is taken into account in their interpretation. A quantitative understanding of the sources and sinks of N_2O is still lacking. We do not know if fluorocarbons can be effectively destroyed on mineral surfaces. Observations of stratospheric winds and temperatures are insufficient to validate dynamical models, particularly above the 10-mb level. The dynamics of travelling long waves in the stratosphere and their possible role in atmospheric mixing are poorly understood. The question of whether perturbations to the upper atmosphere can effect tropospheric climate has not been satisfactorily answered. We could continue this list of inadequacies much longer. However, the point has hopefully been made that a rational long-term program of research into the chemistry and dynamics of the upper atmosphere is required and indeed fortunately appears to be emerging.

Against this background of research effort we have the concurrent need to respond to questions concerning regulation of various anthropogenic gases. It is clear that the response of scientists should be based on a detailed theory of the upper atmosphere that has been fully validated by observations and laboratory measurements. The better grounded the theory used for predictions the less chance of the type of errors which ultimately reflect on credibility.

ACKNOWLEDGMENTS This work was supported by the National Aeronautics and Space Administration under NASA Grant NSG-2010 to the Massachusetts Institute of Technology.

Literature Cited

Alyea, F., Cunnold, D., Prinn, R. 1975. Stratospheric ozone destruction by aircraft-induced nitrogen oxides. *Science* 188:117–21

Anderson, J. G. 1975. The absolute concentration of $O(^3P)$ in the earth's atmo-

sphere. *Geophys. Res. Lett.* 2:231–34

Anderson, J. G. 1976a. A simultaneous measurement of Cl and ClO in the earth's stratosphere. Presented at *Int. Conf. Stratos. Relat. Probl., Logan, Utah*

Anderson, J. G. 1976b. The absolute concentration of OH $(X^2\Pi)$ in the earth's stratosphere. *Geophys. Res. Lett.* 3: 165–68

Angell, J., Korshover, J. 1976. Global analysis of recent total ozone fluctuations. *Mon. Weather Rev.* 104: 63–75

Bates, D., Hays, P. 1967. Atmospheric nitrous oxide. *Planet. Space Sci.* 15: 189–97

Bates, D., Nicolet, M. 1950. The photochemistry of atmospheric water vapor. *J. Geophys. Res.* 55: 301–27

Bauer, E., Gilmore, F. 1975. Effect of atmospheric nuclear explosions on total ozone. *Rev. Geophys. Space Phys.* 13: 451–58

Belmont, A., Dartt, G., Nastrom, G. 1974. Periodic variations in stratospheric zonal wind from 20 to 65 km at 80°N to 70°S. *Q. J. R. Meteorol. Soc.* 100: 203–11

Benney, D., Bergeron, R. 1969. A new class of nonlinear waves in parallel flows. *Stud. Appl. Math.* 48: 181–204

Booker, J., Bretherton, F. 1967. The critical layer for internal gravity waves in a shear flow. *J. Fluid Mech.* 27: 513–39

Brasseur, G., Nicolet, M. 1973. Chemospheric processes of nitric oxide in the mesosphere and stratosphere. *Planet. Space Sci.* 21: 939–61

Chapman, S. 1930. A theory of upper atmospheric ozone. *Mem. R. Meteorol. Soc.* 3: 103–25

Charney, J. G. 1947. The dynamics of long waves in a baroclinic westerly current. *J. Meteorol.* 4: 135–62

Charney, J., Drazin, P. 1961. Propagation of planetary scale disturbances from the lower into the upper atmosphere. *J. Geophys. Res.* 66: 83–109

Cicerone, R., Stedman, D., Stolarski, R. 1975. Estimate of late 1974 stratospheric concentration of gaseous chlorine compounds. *Geophys. Res. Lett.* 2: 219–22

Cicerone, R., Stolarski, R., Walters, S. 1974. Stratospheric ozone destruction by manmade chlorofluoromethanes. *Science* 185: 1165–67

Clark, J. H. E. 1970. A quasi-geostrophic model of the winter stratospheric circulation *Mon. Weather Rev.* 98: 443–61

Crutzen, P. J. 1970. The influence of nitrogen oxides on the atmospheric ozone content. *Q. J. R. Meteorol. Soc.* 96: 320–25

Crutzen, P. J. 1971. Ozone production rates in an oxygen-hydrogen-nitrogen oxide atmosphere. *J. Geophys. Res.* 76: 7311–27

Crutzen, P. J. 1974a. Estimates of possible variations of total ozone due to natural causes and human activities. *Ambio* 3: 201–10

Crutzen, P. J. 1974b. Estimates of possible future ozone reductions from continued use of fluorochloromethanes. *Geophys. Res. Lett.* 1: 205–08

Cunnold, D., Alyea, F., Phillips, N., Prinn, R. 1975. A three-dimensional dynamical-chemical model of atmospheric ozone. *J. Atmos. Sci.* 32: 170–94

Cunnold, D., Alyea, F., Prinn, R. 1977. Relative effects on atmospheric ozone of latitude and altitude of supersonic flight. *Am. Inst. Aeronaut. Astronaut. J.* 15: 337–45

Danielson, E. F. 1968. Stratospheric-tropospheric exchange based on radioactivity, ozone, and potential vorticity. *J. Atmos. Sci.* 25: 502–18

Deland, R. J. 1973. Analysis of Nimbus 3 SIRS radiance data: Travelling planetary waves in the stratosphere temperature field. *Mon. Weather Rev.* 101: 132–40

Dickinson, R. E. 1968. On the exact and approximate linear theory of vertically propagating planetary Rossby waves forced at a spherical lower boundary. *Mon. Weather Rev.* 96: 405–15

Dickinson, R. E. 1969. Vertical propagation of planetary Rossby waves through an atmosphere with Newtonian cooling. *J. Geophys. Res.* 74: 929–38

Dopplick, T. G. 1971. The energetics of the lower stratosphere including radiative effects. *Q. J. R. Meteorol. Soc.* 97: 209–37

Dopplick, T. G. 1972. Radiative heating of the Global Atmosphere. *J. Atmos. Sci.* 29: 1278–94

Dutsch, H. 1971. Photochemistry of atmospheric ozone. *Adv. Geophys.* 15: 219–322

Dutsch, H. 1974. *Regular* ozone soundings at the aerological station of the Swiss Meteorological Office at Payern, Switzerland, 1968–1972. *Lapeth 10*, Lab. Atmos. Phys., ETH, Zurich. 337 pp.

Eliassen, A., Palm, E. 1960. On the transfer of energy in stationary mountain waves. *Geophys Publ.* 22: No. 3. 23 pp.

Evans, W., Kerr, J., Wardle, D., McConnell, J., Ridley, B., Schiff, H. 1976. Intercomparison of NO, NO_2, and HNO_3 measurements with photochemical theory. *Atmosphere* 14: 189–98

Fjortoft, R. 1951. Stability properties of large-scale atmospheric disturbances. In *Compendium of Meteorology*, ed. T. Malone, p. 454. Boston: Am Meteorol. Soc. 1334 pp.

Geisler, J., Garcia, R. 1977. Baroclinic instability at long wavelengths on a β-plane. *J. Atmos. Sci.* 34: 311–21

Green, J. S. A. 1960. A problem in baroclinic stability. *Q. J. R. Meteorol. Soc.* 86: 237–51

Hare, F. K. 1968. The arctic. *Q. J. R. Meteorol. Soc.* 94: 439–59

Heidt, L., Lueb, R., Pollock, W., Ehhalt, D. 1975. Stratospheric profiles of $CFCl_3$ and CF_2Cl_2. Geophys. Res. Lett. 2: 445–47

Hill, W., Sheldon, P. 1975. Statistical modelling of total ozone measurements with an example using data from Arosa, Switzerland. Geophys. Res. Lett. 2: 541–44

Holton, J. R. 1975. The Dynamic Meteorology of the Stratosphere and Mesosphere. Boston: Am. Meteorol. Soc. 218 pp.

Holton, J., Lindzen, R. 1972. An updated theory for the quasi-biennial cycle of the tropical stratosphere. J. Atmos. Sci. 29: 1076–80

Hunt, B. G. 1966. Photochemistry of ozone in a moist atmosphere. J. Geophys. Res. 71: 1385–98

Hunt, B. G. 1969. Experiments with a stratospheric general circulation model. III. Large-scale diffusion of ozone including photochemistry. Mon. Weather Rev. 97: 287–306

Johnston, H. 1971. Reduction of stratospheric ozone by nitrogen oxide catalysts from supersonic transport exhaust. Science 173: 517–22

Johnston, H. 1975. Global ozone balance in the natural stratosphere. Rev. Geophys. Space Phys. 13: 637–49

Johnston, H., Whitten, G., Birks, J. 1973. Effect of nuclear explosions on stratospheric nitric oxide and ozone. J. Geophys. Res. 78: 6107–35

Junge, C. E. 1974. Residence time and variability of tropospheric trace gases. Tellus 26: 477–88

Krey, P., Lagomarsino, R., Toonkel, L. 1977. Gaseous halogens in the atmosphere in 1975. J. Geophys. Res. 82: 1753–66

Lazrus, A., Gandrud, B. 1974. Distribution of stratospheric nitric acid vapor. J. Atmos. Sci. 31: 1102–08

Leovy, C. 1964. Simple models of thermally driven mesospheric circulation. J. Atmos. Sci. 21: 327–41

Leovy, C., Webster, P. 1976. Stratospheric long waves: comparison of thermal structure in the Northern and Southern hemispheres. J. Atmos. Sci. 33: 1624–38

Lindzen, R. 1971. Atmospheric tides. In Mathematical Problems in the Geophysical Sciences, ed. W. Reid, 2: 293. Providence: Am. Math. Soc. 370 pp.

Lindzen, R., Goody, R. 1965. Radiative and photochemical processes in mesospheric dynamics: Part I, models for radiative and photochemical processes. J. Atmos. Sci. 22: 341–48

Lorenz, E. N. 1955. Available potential energy and the maintenance of the general circulation. Tellus 7: 157–67

Manabe, S., Mahlman, J. 1976. Simulation of seasonal and interhemispheric variations in the stratospheric circulation. J. Atmos Sci. 33: 2185–2217

Matsuno, T. 1970. Vertical propagation of stationary planetary waves in the winter northern hemisphere. J. Atmos. Sci. 27: 871–83

Matsuno, T. 1971. A dynamical model of the stratospheric sudden warming. J. Atmos. Sci. 28: 1479–94

McConnell, J. 1973. Atmospheric ammonia. J. Geophys. Res. 78: 7812–21

McConnell, J., McElroy, M., Wofsy, S. 1971. Natural sources of atmospheric CO. Nature 233: 187–88

McElroy, M., Elkins, J., Wofsy, S., Yung, Y. 1976. Sources and sinks for atmospheric N_2O. Rev. Geophys. Space Phys. 14: 143–50

McElroy, M., McConnell, J. 1971. Nitrous oxide: a natural source of stratospheric NO. J. Atmos. Sci. 28: 1095–98

McElroy, M., Wofsy, S., Penner, J., McConnell, J. 1974. Atmospheric ozone: possible impact of stratospheric aviation. J. Atmos. Sci. 31: 287–303

Molina, M., Rowland, F. 1974a. Stratospheric sink for chlorofluoromethanes: chlorine catalyzed destruction of ozone. Nature 249: 810–12

Molina, M., Rowland, F. 1974b. Predicted present stratospheric abundances of chlorine species from photodissociation of carbon tetrachloride. Geophys. Res. Lett. 1: 309–12

National Academy of Sciences. 1975. Environmental impact of stratospheric flight. Washington, D.C.: Nat. Acad. Sci. 348 pp.

National Academy of Sciences. 1976. Halocarbons: effects on stratospheric ozone. Washington, D.C.: Nat. Acad. Sci. 352 pp.

Newell, R. 1969. Radioactive contamination of the upper atmosphere. Prog. Nucl. Energy Ser. 12. 2: 535–50

Newell, R., Boer, G., Dopplick, T. 1973. Influence of the vertical motion field on ozone concentration in the stratosphere. Pure Appl. Geophys. 106–108: 1531–43

Oort, A. H. 1964. On the energetics of the mean and eddy circulations in the lower stratosphere. Tellus 16: 309–27

Pack, D., Lovelock, J., Cotton, G., Curthoys, C. 1977. Halocarbon behaviour from a long time series. Atmos. Environ. 11: 329–44

Panofsky, H., Heck, W. 1974. Vertical dispersion near 20 km. In Proc. 3rd Conf. Clim Impact Assess. Prog., ed. A. Broderick, T. Hard, p. 102. Springfield, Va: Natl.

Tech. Inf. Serv. 660 pp.

Prinn, R. G. 1971. Photochemistry of HCl and other minor constituents in the atmosphere of Venus. *J. Atmos. Sci.* 28: 1058–68

Prinn, R., Alyea, F., Cunnold, D. 1975. Stratospheric distributions of odd nitrogen and odd hydrogen in a two-dimensional model. *J. Geophys. Res.* 80: 4997–5004

Prinn, R., Alyea, F., Cunnold, D. 1976. The impact of stratospheric variability on measurement programs for minor constituents. *Bull. Am. Meteorol. Soc.* 57: 686–94

Reed, R., Wolfe, J., Nishimoto, H. 1963. A spectral analysis of the energetics of the Stratospheric Sudden Warming of early 1957. *J. Atmos. Sci.* 20: 256–75

Reiter, E. R. 1975. Stratospheric–tropospheric exchange processes. *Rev. Geophys. Space Phys.* 13: 459–74

Reiter, E., Lovill, J. 1974. The longitudinal movement of stratospheric ozone waves as determined by satellite. *Arch. Meteorol. Geophys. Bioklim.* A23: 13–27

Robbins, D., Stolarski, R. 1976. Comparison of stratospheric ozone destruction by fluorocarbons 11, 12, 21, and 22. *Geophys. Res. Lett.* 3: 603–06

Rowland, F. S., Molina, M. J. 1975. Chlorofluoromethanes in the environment. *Rev. Geophys. Space Phys.* 13: 1–36

Rowland, F., Spencer, J., Molina, M. 1976. Stratospheric formation and photolysis of chlorine nitrate. *J. Phys. Chem.* 80: 2711–13

Schmeltekopf, A., Albritton, D., Crutzen, P., Goldan, P., Harrop, W., Henderson, W., McAfee, J., McFarland, M., Schiff, H., Thompson, T., Hofmann, D., Kjome, N. 1977. Stratospheric nitrous oxide altitude profiles at various latitudes. *J. Atmos. Sci.* 34: 729–36

Stolarski, R., Cicerone, R. 1974. Stratospheric chlorine: a possible sink for ozone. *Can. J. Chem.* 52: 1610–15

Strobel, D. F. 1971. Odd nitrogen in the mesosphere. *J. Geophys. Res.* 76: 8384–93

Vincent, D. G. 1968. Mean meridional circulations in the northern hemisphere lower stratosphere during 1964 and 1965. *Q. J. R. Meteorol. Soc.* 94: 333–49

Wallace, J. M. 1973. General circulation of the tropical lower stratosphere. *Rev. Geophys. Space Phys.* 11: 191–222

Warneck, P. 1972. Cosmic radiation as a source of odd nitrogen in the stratosphere. *J. Geophys. Res.* 77: 6598–91

Whitten, R., Borucki, W., Poppoff, I., Turco, R. 1975. Preliminary assessment of the potential impact of solid-fueled rocket engines in the stratosphere. *J. Atmos. Sci.* 32: 613–19

Williams, W., Kosters, J., Goldman, A., Murcray, D. 1976. Measurement of the stratospheric mixing ratio of HCl using infrared absorption techniques. *Geophys. Res. Lett.* 3: 383–85

Wofsy, S., McElroy, M., Sze, N. D. 1975a. Freon Consumption: implications for atmospheric ozone. *Science* 187: 535–37

Wofsy, S., McElroy, M., Yung, Y. 1975b. The chemistry of atmospheric bromine. *Geophys. Res. Lett.* 2: 215–18

Ann. Rev. Earth Planet. Sci. 1978. 6 : 75–91

PALEOMAGNETISM OF THE   ⋇10087
PERI-ATLANTIC PRECAMBRIAN

A. E. M. Nairn and R. Ressetar

Department of Geology, University of South Carolina, Columbia, South Carolina 29208

INTRODUCTION

Paleomagnetism, by providing independent evidence for large-scale relative motion of cratonic blocks, revitalized the concept that "Wegnerian drift" has operated since the early Mesozoic. Additionally, it presented data indicating hitherto unsuspected horizontal movements in still earlier ages. While many of the questions concerning younger crustal movements are more readily answered from studies of oceanic magnetic anomalies, recognition of pre-Mesozoic displacements remains the domain of paleomagnestim. Plate tectonic theory, by linking orogenic processes to plate movements (Dewey & Bird 1970), not only supports tentative suggestions of earlier periods of drift (e.g. Irving 1956, Nairn 1963, Wilson 1966, Briden 1967) but also provides a means of identifying margins of ancient plates (Burke, Dewey & Kidd 1976), although care must be exercised in recognizing them (e.g. Shackleton 1973, Nairn 1975). That attention has concentrated more upon the Precambrian than the Paleozoic is largely the result of data availability.

Provided there is adequate age control, paleomagnetism can fulfill a vital role in considerations of the Precambrian history of the continents now bordering the Atlantic Ocean. From the results emerges a pattern of movement, and the possibility of intercontinental comparison.

There have been several discussions of Precambrian continental drift (Spall 1971, 1972a,b, Burke, Dewey & Kidd 1976, Nairn 1975, Piper 1973a, 1976a,b, Piper, Briden & Lomax 1973, Irving, Emslie & Ueno 1974, Hurley & Rand 1969, McElhinny & McWilliams 1977, Brock & Piper 1972). If the conclusions are somewhat diverse, this is merely a reflection of the fact that data have been, and to a lesser extent continue to be, sparse, and so cause major difficulty in reaching clear interpretations. Before reviewing the data, it is necessary to discuss data selection and to comment upon the Cambrian-Precambrian boundary.

This paper is based upon a report presented to UNESCO/IGCP Working Group 2 during the 1976 Iberian Field Meeting.

DATA SELECTION

The very antiquity of Precambrian rocks accentuates the problems encountered in examination of material for paleomagnetic research. In general, the older the

75

0084-6597/78/0515-0075$01.00

rock is, the greater is the likelihood of a polyphase history and, in consequence, the greater the danger that secondary effects were impressed upon the original remanent magnetization. Detailed analyses such as those of Roy & Park (1974) indicate the complexity of the magnetic history of some rocks and the care needed in their interpretation (see comments on Cavanaugh & Seyfert 1977, p. 87). The high proportion in the Precambrian record of metamorphic rocks, which have on the whole received only superficial examination, becomes a matter of significance.

The question of age is important, for while relative age may be established on a local scale, the absence of diagnostic fossil evidence prohibits long-distance correlations. In consequence, we come to rely entirely upon radiometric dating. If by this means the paleomagnetic data from a single craton can be resolved into a time sequence and a geomagnetic polar wander path can be established, a valuable correlation tool is generated. Coeval data from different cratons may then be compared and contrasted for evidence of relative motion, as is done with data from the Phanerozoic.

As in other fields of endeavor, the advance of paleomagnetism has been marked by refinement in equipment and techniques, with the result that there exists a considerable range of data treatments, and many of the earlier works are more of historic interest. In the absence of internationally adopted standards on what constitutes "reliable" data, various minimal criteria have been specified from time to time. Comparison of the apparent geomagnetic polar wander paths of Spall (1971) and Irving & Park (1972) for North America is illustrative of the differences that may arise when different standards of data selection are used. Perhaps one of the more conservative systems, that of Hicken et al (1972), is the most widely accepted and is the one used in selecting class A data from the peri-Atlantic Precambrian. This system specifies a minimum sample size (10 separately oriented cores) and requires that standard magnetic cleaning techniques be applied. It further requires that the radius of the circle of 95% confidence about the mean magnetic direction (alpha 95) be less than 20°. A result qualifying as class A data may then be accepted as a parameter of a stable, single-component magnetization without prejudice as to whether it is a reliable indicator of the ancient geomagnetic field.

Collection of data listings (Tables 1 and 2) has been made immeasurably easier by the publication of catalogues of paleomagnetic directions and poles (e.g. Irving & Hastie 1975). The results reported are predominately from igneous bodies, which allow for radiometric dating, but include data from unmetamorphosed sedimentary sequences. Data from metamorphic rocks are still sparse, and the special problems posed by metamorphism require separate discussion.

METAMORPHIC ROCKS

Early in the process of trying to determine the long-term behavior of the geomagnetic field, metamorphic rocks, because of their obviously complex geological history, were excluded. Now that certain broad trends have been established, and as research expands into Precambrian rocks and mobile zones, metamorphic

Table 1 North American Precambrian poles listed by age

Pole #	Rock unit	Province[a]	Lat.	Pole Long.[b]	Age (m.y.)	Reference
1	Skead Group	S	22	44	2755	Ridler & Foster (1975)
2	Matachewan Dikes	S	−50	−112	2690	1–107[c]
3	Chibougamau Greenstone belt	S	61	93	2600	1–581
4	Dogrib Dikes	Sl	35	130	2692	1–491
5	Stillwater	W	62	112	2645	1–244
6	Kaminak Dikes, unmetamorphosed	C	24	58	2300	Christie, Davidson & Fahrig (1975)
7	Otto Stock, primary	S	69	−133	2500	1–543
8	Otto Stock, primary	S	60	−179	2470	Cavanaugh & Seyfert (1977)
9	Firstbrook Member, (Huronian redbeds "B")	S	62	158	2300–2600	Roy & Lapointe (1976)
10	X-Dikes	Sl	20	119	2300	1–493
11	Big Spruce "D"	Sl	67	−113	2218	1–555
12	Big Spruce "Z"	Sl	−10	−44	1800[d]	Irving & McGlynn (1976b)
13	Big Spruce "Y"	Sl	7	−78	1600[d] or 1900	Irving & McGlynn (1976b)
14	Nipissing Diabase "C"	S	42	−102	2150	Roy & Lapointe (1976)
15	Nipissing Diabase "D"	S	−15	−96	1800–1900[d]	Roy & Lapointe (1976)
16	Nipissing Diabase "E"	S	−9	−180	younger than 15	Roy & Lapointe (1976)
17	Nipissing Diabase	S	−14	−94	compare with 15	Symons & Lowndry (1975)
18	Thessalon Volcanics	So	65	−153	2288	Symons & O'Leary (1975)
19	Gowganda Formation	So	65	−123	2290	Symons (1975)
20	Abitibi Dikes major magnetization	S	27	−134	2147	1–109
21	Mugford Basalt	N	49	−143	2100	1–314
22	Indin Dikes	Sl	19	−76	2093	1–493
23	Indian Harbour Dikes	N	−6	−117	2080	1–316

Table 1 (*continued*)

Pole #	Rock unit	Province[a]	Lat.	Pole Long.[b]	Age (m.y.)	Reference
24	Owl Creek Dikes	W	−28	−112	1910–2090	1–389
25	Otish Gabbro	S	35	−107	1465–2300	1–467
26	Wind River Dikes	W	43	−121	1880–2060	1–326
27	Gunflint Formation	So	28	−94	1700–2300	1–191
28	Spanish River Complex	S	37	−96	1890	1–413
29	Kahochella Formation, Primary Magnetization	C	6	−72	1873	McMurry, Reid & Evans (1973)
30	Kaminak Dikes— Metamorphosed	C	20	−92	1800–1900	Christie, Davidson & Fahrig (1975)
31	Martin Formation	C	−9	−72	1830	1–468
32	Dubawnt Group	C	7	−83	1825	1–481
33	Cape Smith Volcanics	C	16	−107	1800	Fujiwara & Schwarz (1975)
34	Et-Then Red Beds	C	−1	−48	1630–1835	1–341
35	Tochatwi Formation	C	−18	−144	1630–1845	Evans & Bingham (1976)
36	Sparrow Dikes	C	12	−69	1700	1–410
37	Kahochella Formation, Secondary Magnetization	C	18	−94	1660?	McMurry, Reid & Evans (1973)
38	Nonacho Group	C	13	−86	1650?	1–411
39	Melville-Daly Bay Metamorphics	C	11	−101	1622	1–456
40	Western Channel Diabase	C	9	−115	1500	1–504
41	Michael Gabbro	G	10	163	1488	1–352
42	Crocker Island Complex	So	6	−143	1475	1–129
43	Sherman Granite	So	−8	−151	1410	1–172
44	Michikamau Anorthosite	N	−1	−145	1400	1–205
45	Sibley Red Beds	S	−21	−144	1370	1–473
46	Seal & Croteau Volcanics	G	5	152	1400–1520	1–466
47	Belt Series	C	−11	−152	1350	1–300
48	Seal Red Beds	N	6	−155	1300	1–464

Table 1 (*continued*)

Pole #	Rock unit	Province[a]	Pole Lat.	Pole Long.[b]	Age (m.y.)	Reference
49	Coppermine Lava	C	1	−177	1280	1–472
50	Muskox Intrusion	C	4	−175	1250	1–115
51	MacKenzie Diabase	C & S	1	−171	1200	1–453
52	South Trap Range Normal	So	10	−160	1200	1–374
53	South Trap Range Reversed	So	28	−128	1200	1–150
54	Tudor Gabbro	G	17	138	1125	1–458
55	Morin Complex. Hematite	G	−42	141	1124	1–482
56	Logan Diabase Reversed	S	48	−143	1100	1–309
57	Whitestone Anorthosite WW	W	−16	156	1100	1–503
58	Whitestone Anorthosite WZ	G	−22	146	1100	1–500
59	Logan Diabase Normal	S	33	−172	1100	1–82
60	Keweenawan Intrusives Combined	S	35	−172	1100	1–308
61	North Shore Volcanics Reversed	S	46	−161	1115	1–306
62	North Shore Volcanics Normal	S	33	−175	1115	1–393
63	Portage Lake Lavas	C	25	−176	1100	1–392
64	Copper Harbour Lavas	So	21	176	1046–1200	1–304
65	Mamainse Point Lavas	S	34	−176	1076	1–141
66	Umfraville Intrusive	G	−1	158	1000–1200	1–43
67	Freda & Nonesuch Shales	So	8	173	1046	1–303
68	Nemogosenda complex	S	43	−178	1036	1–412
69	Lac Allard Anorthosite	G	−39	140	1000	1–130
70	Mealy Mountain Anorthosite	G	8	179	?	1–415
71	Wilberforce Pyroxenite	G	−14	148	?	1–457
72	Morin Complex Magnetite	G	0	164	1000	1–483

Table 1 (*continued*)

Pole #	Rock unit	Province[a]	Pole Lat.	Long.[b]	Age (m.y.)	Reference
73	Whitestone Anorthosite, WY	G	−5	168	1000	1–502
74	Aillik Dikes	N	27	−136	995	1–353
75	El Paso Rocks	E	28	−160	953	1–167
76	Rama Diabase	E	2	174	935	1–391
77	Frontenac Dikes	G	−12	163	750–820	1–363
78	Franklin Diabase	C	8	167	675	1–162
79	Coronation Sills	C	−1	163	647	1–312

[a] Province abbreviations: S, Superior, Sl, Slave, C, Churchill, So, Southern, N, Nain, W, Wyoming, G, Grenville, E, Elsonian.
[b] North latitudes and east longitudes are positive.
[c] References such as 1–107 refer to the Ottawa catalogue of Irving & Hastie (1975) or Roy & Lapointe (1976).
[d] Age assignment made on the basis of proximity to other poles.

Table 2   African Precambrian poles listed by age

Pole #	Rock Unit	Province[a]	Pole Lat.	Long.	Age (m.y.)	Reference
1	Modipe Gabbro	S	33	−149	2630	1–166
2	Great Dike 1	S	−21	−118	2530	1–63
3	Great Dike 2	S	−11	−111	2530	1–64
4	Lower Ventersdorp Lavas	S	4	−120	2500	Henthorn (1973), Piper (1976a)
5	Gaberones Granite	S	−35	−76	2340	1–154
6	Upper Ventersdorp Lavas	S	−71	−7	2300	Henthorn (1973), Piper (1976a)
7	Obuasi Greenstone	W	−50	102	2000–2200	1–424
8	Tarkwa Dolerite	W	−53	36	2000–2200	1–425
9	Obuasi Dolerite	W	−56	69	2000–2200	1–426
10	Cunene Anorthosite	C	3	75	2070	Piper (1974)
11	Orange River Lavas	S	19	74	>1850	Piper (1975a)
12	Vredefort Rings	S	22	27	1970	1–477
13	Bushveld Gabbro	S	23	36	1920	1–7
14	Premier Mine Kimberlite	S	51	38	1750	1–128
15	Mashonaland Dolerite	S	7	−20	1430–1640	1–102
16	Waterberg Red Beds 1	S	36	48	<1950	1–138[d]
17	Waterberg Red Beds 2	S	8	10		1–137[b]

Table 2 (*continued*)

Pole #	Rock Unit	Province[a]	Pole Lat.	Pole Long.	Age (m.y.)	Reference
18	Waterberg Red Beds 3	S	3	−27		1–136
19	Waterberg Red Beds 4	S	41	33		1–135[b]
20	Waterberg Red Beds 5	S	67	44	>1400	1–134
21	Pilansberg Dikes	S	8	43	1290	1–6
22	Post-Waterberg Dikes	S	65	51	<1250	1–242
23	Barby Formation	S	63	30	1265	Piper (1975a)
24	Guperas Formation	S	63	−43	1250–1265	Piper (1975a)
25	Auborus Formation	S	43	−6	>990 <1250	Piper (1975a)
26	O'okeip Intrusions	S	15	−25	1070	McElhinny & McWilliams (1977)
27	Bukoban Sandstone	E	40	−43	1000	1–359
28	Kisii Series	E	6	−12	909–964; 1200	1–362 Piper (1975a)
29	Ikorongo Group	E	35	−96	900–1000[c]	Piper (1975b)
30	Klein Karas Dikes	S	15	−68	878	Piper (1975a)
31	Bukoban Instrusives	E	−11	−79	806	1–355
32	Mbala Dolerites	E	−9	−80	806[c]	Piper (1975b)
33	Gagwe Lavas	E	−29	−77	813	1–356
34	Umkando Dolerite	S	−64	−153	650–1150	1–124
35	Mbozi Complex	E	−72	−112	743	Piper (1975b)
36	Lower Buanji Series	E	−87	−97	Late pC[c]	Piper (1975b)
37	Pre-Nama Dikes	S	−85	53	653	Piper (1975b)
38	Ntonya Ring Complex	E	−28	165	600	1–127
39	Sijarira Group	S	−2	172	pre C to C	1–344

[a] Province abbreviations: S, South African, E, East African, W, West African, C, Congo.
[b] "B" quality data.
[c] Age assignment made on the basis of proximity to other poles

rocks are beginning to receive more attention. Following common geological practice, diagenetic processes are not normally considered as metamorphism. Thus, the important low-temperature hydrothermal alteration recognized in petrologic studies of igneous rocks is not considered, nor are such complex questions associated with sedimentary diagenesis such as the origin of red beds.

The effect of uniaxial (overburden) pressure on the remanent vector of so-called unmetamorphosed rocks is largely ignored. For, although it has been demonstrated experimentally that changes in intensity and susceptability with rotation of the remanent vector are associated with the application of uniaxial stress (Stacey 1960, Kern 1961, Kume 1962, 1965, Nagata & Kinoshita 1965, Ohnaka & Kinoshita 1968a,b), the effect appears to be reversible (Stott & Stacey 1960), at least on the scale of laboratory experiments (Domen 1975). The most convenient means of considering metamorphic rocks is in terms of a high-pressure, low-temperature suite where deformation fabrics may be expected to dominate and a high-temperature, low-pressure suite where thermoremanent effects can be anticipated. (However, it is important to realize that the two suites grade into one another.) The investigation of rocks with well-defined fabrics has in general shown that although the magnetic anisotropy of susceptibility is not great (Khan 1962), it does appear in some cases to result from tectonic deformation (Fuller 1964) and can result in the rotation of the remanent vector (Khan 1962, Hargraves 1959, Fuller 1964), although the degree of motion is sometimes small (Abouzakhm & Tarling 1975, Irving & Park 1973).

Correlations have also been found between anisotropy of susceptibility and flow fabrics in igneous rocks (King 1966, Stone 1963, Van der Voo & Klootwijk 1972) and sediments (Hamilton 1963), as well as with tectonic fabrics (Girdler 1961, Stone 1963, Yaskawa 1959, Buddington & Balsley 1958, Nagata & Shimizu 1959). It is premature to generalize their significance (see Henry 1973). Studies upon gneisses, which may have been subjected to higher temperatures, are less numerous and rather of a trial nature (Yaskawa 1959, Nagata & Shimizu 1959, Irving & Park 1973). They suggest that useful data may be retrieved, although considerable care may be needed in deciphering polyphase magnetic histories (Morris 1977, Beckmann 1976). The most important series of results from regionally metamorphosed rocks is that from the Grenville province, and its significance is discussed in some detail by Irving, Emslie & Ueno (1974) and Irving & McGlynn (1976a).

THE CAMBRIAN-PRECAMBRIAN BOUNDARY

No detailed paleomagnetic studies have yet been carried out in the peri-Atlantic region on rocks that appear to span this important boundary. Kirschvink (1977) has demonstrated that a long normal polarity interval followed by a reversed interval is characteristic of the Upper Precambrian in the Amadeus Basin of Australia, while the basal Cambrian occurs within a zone of mixed polarity. It seems desirable that more such studies be carried out, particularly in Morocco where well-exposed rocks are being considered for the type section of this boundary. It is known from relatively small collections that some lithologies within this time interval are suitable for paleomagnetic studies (Hailwood 1972, D. L. Martin et al, in preparation).

Some years ago, Harland (1964) proposed that widely spread glacial deposits of late Precambrian age might provide a useful time horizon in helping to identify the Cambrian-Precambrian boundary. While the glacial nature of some of the deposits has been questioned (Schermerhorn 1974, Schermerhorn & Stanton 1963,

Winterer 1964), the identification of others has been confirmed (Spencer 1971). Bidgood & Harland (1961) and Harland & Bidgood (1959) indicated that paleomagnetic measurements on glacial horizons from Norway and Greenland suggested deposition in low latitudes. This finding is consistent with the most detailed studies to date, those of Tarling (1974) in Scotland and work on Spitsbergen rocks by Piper (1973b).

The preceeding data have usually been interpreted as evidence of a near global ice age in late Precambrian time (Harland 1964) on the assumption that their occurrence was synchronous, an interpretation that Rudwick (1964) tried to expand to explain the evolutionary burst of the Cambrian. Williams (1975) offered an alternative explanation of low-latitude glaciation involving increased obliquity of the earth's axis rather than an ice age of global proportions. General reviews of the problems raised by Eocambrian (or Infracambrian) glaciations are to be found in Harland & Herod (1975) and Spencer (1975). A more cautious note is sounded by Morris (1977), who indicated that there may be several components of magnetization involved. In the case of the Precambrian Rapitan Group of glaciogenic rocks he found both steeply dipping and shallowly dipping components of magnetization and thus generalized that the data currently available are inconclusive.

APPARENT GEOMAGNETIC POLAR WANDER PATHS

The accepted interpretive device in paleomagnetism when dealing with a series of apparent poles is to link them in chronological succession by an apparent geo-. magnetic polar wander path. In the discussion of Phanerozoic data it is customary to do so in terms of current continental units. In the case of the Precambrian, with its several tectonic events, it must first be established that this device is legitimate. It is unfortunate that, relative to the duration of the Precambrian, the number of Precambrian data is small, and that this number may be further reduced according to the standard of data selection adopted.

Where data are few, and where the apparent geomagnetic poles lie in low latitudes, it may be impossible to discriminate between a North and a South Pole. The choice of a pole or its antipole may thus ultimately depend on an author's opinion (or prejudice), since widely spaced poles separated by long time intervals may make the usual conservative choice of minimum movement far from obvious. Thus, as Briden (1977) pointed out, the same data pool may lead to a spectrum of interpretations, depending on what data are selected, whether a pole or antipole is used, and whether each province is considered as an individual unit or as part of a larger cratonic area.

At the risk of burdening the literature with yet more "state of the art" apparent polar wander paths, new curves are produced here for Africa and North America as a means of illustrating the above points. Europe and South America, the other continents of concern in the peri-Atlantic region, are excluded because of the relative paucity of data. The data, upon which the discussions of the North American and African apparent polar wander paths are based, are listed in Tables 1 and 2. These data include all class A data and the Ottawa listings of Roy &

Lapointe (1976) for which "reliable" radiometric or stratigraphic ages are available. More recent results that qualify as class A data are identified by reference to the original work. The structural province from which the rocks were derived is given based upon Irving & Lapointe (1975) and Cavanaugh & Seyfert (1977) for North America and upon Piper (1975a) for African results. There are difficulties in the assignment of the best radiometric age, for the reported ages from a single unit may vary widely depending on the method used and the degree of metamorphism suffered. Further, as the tabulated error may be as much as 10% of the reported age, there may be errors in the sequence presented. There are also errors of measurement, and a dispersion due to secular variation, which result in a circle of confidence around the calculated geomagnetic pole position. In consequence, the apparent polar wander path should appear as a broad band rather than a line connecting individual points. To simplify the figures, a line has been drawn to pass through groups of poles spanning intervals of about 100 m.y.

North America

If we accept the Phanerozoic curve as it is usually drawn, we are then constrained to use the chronological sequence of geomagnetic poles or antipoles for the late Precambrian, which form two elongate loops in the region of the present northwestern Pacific back to about − 1475 m.y. The younger of these two is the "Logan Loop" of Robertson & Fahrig (1971).

Further back in time the path forms a third loop in the area of the present Caribbean (− 1800 m.y.) before passing back into the equatorial Pacific (− 1900 m.y.). The path sweeps northward and then westward through Siberia to the earliest geomagnetic poles in western Asia. The apparent simplicity of this section of the path may well be an artifact of the relatively few available poles.

This presentation, as a generalized curve, ignores small excursions such as that of Irving & Lapointe (1975) to include pole 29 (Kahochella Formation, − 1873 m.y.). Excursions of this type are probably real, but require better definition than is given by a single point. The path, since it was drawn from a compilation of all North American provinces except the Grenville (see Figure 1), carries the implicit assumption that these provinces have retained their present spatial distribution. The alternate hypothesis is presented in the interpretation of Cavanaugh & Seyfert (1977). Their polar wander path includes a loop in the western Pacific for the period − 2300 m.y. to − 1600 m.y. They claim that the configuration of this loop shows that suturing between the Slave and Superior provinces occurred at around − 1800 m.y. To reach this conclusion, however, they had to re-interpret four earlier reports. They had to assign as representing the primary magnetization of the Nipissing diabase dikes a direction regarded by the original authors (Roy & Lapointe 1976) as secondary; they had to disregard the significance of the age of the Abitibi dikes given as − 2147 m.y. by Gates & Hurley (1973); they had to use the antipole of the unmetamorphosed Kaminak dikes (Figure 1, pole 6) and assign them an age of about − 2100 m.y. rather than − 2300 m.y. or older (Christie, Davidson & Fahrig 1975); and finally they had to ignore a group of poles dating between − 1800 and 2220 m.y. (Figure 1, poles 21, 25–28) which are distributed through an area centered over the western United States.

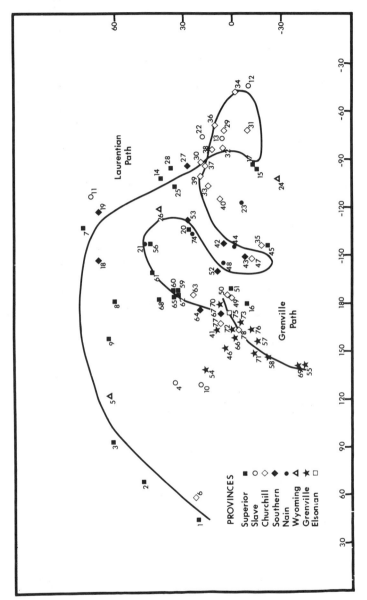

Figure 1 Mercator projection of the apparent polar wander path of North America during the Precambrian.

The question of the structural integrity of North America throughout the Proterozoic is again brought up in consideration of geomagnetic poles from the Grenville province. To date, they seem to provide the best case that can be made for independent movement of one Precambrian province, an interpretation first offered by Irving, Emslie & Ueno (1974). We have therefore chosen to present them as forming a separate grouping from the polar wander path of the rest of the North American poles. However, other interpretations of the Grenville data have been made. Piper (1976a) links the paths of the Grenville and the other North American provinces together as the movements of a single Proterozoic super continent. McElhinny & McWilliams (1977) attach the Grenville poles to the continued southern motion of the Logan loop on the grounds that the Grenville rocks were magnetized during post orogenic cooling. The poles, therefore, could not document movement of a separate province prior to collision. The problem is still unresolved because of the complex metamorphic history of the Grenville province and the lack of coeval poles from the remainder of North America.

Africa

Our interpretation of the apparent polar wander path for Africa is illustrated in Figure 2. The data upon which it is based are listed in Table 2 and follow the same conventions used in Table 1. Not only is the path more difficult to define because of the sparsity of data, but there is unfortunately a long unrepresented interval between the latest Precambrian and the oldest of the Phanerozoic results. Although the six poles available for the time prior to −2000 m.y. line up in a simple swath, the distances involved, particularly between poles 5 and 6, lead to some question as to how realistic the general path may be.

A concentration of results in the age range −2000 to −1900 m.y. suggests northward movement of the path to the present vicinity of the Arabian peninsula (poles 7–13). For the next 600 m.y., however, the path is not easy to define. The poles from progressively younger Waterberg red beds (poles 16–20) indicate an arc from near the Bushveld pole (13) toward that of the post-Waterberg dikes (22), but three of these poles are not class A data. With this interpretation, however, some of the presumed younger poles appear within the older segment of the arc. While some of the age assignments may be disputed (e.g. pole 28, Kisii series: −909 to −964 m.y. in the Ottawa listing, −1200 m.y. in Piper 1975a), others seem well documented. Some of the younger poles may be explained by crossing of the polar wander path as shown in Figure 2 to accomodate poles 26 and 28, though Piper (1976a) suggests that the arc closes and repeats. Possibly the period from −1900 to −1300 m.y. was characterized by relatively little apparent polar wander while the paucity of reported poles reflects a lack of orogenic activity. Clearly, at this stage the data are inconclusive.

For the poles subsequent to −1300 m.y. we have drawn a path that closely resembles Piper's (1976a) in possessing a closed loop, although our configuration differs from his in that the closure occurs in the vicinity of poles 15, 18, 26, and 28 rather than 19 and 25. This difference is based on new age data reported by McElhinny & McWilliams (1977). Their interpretation of this data is that the

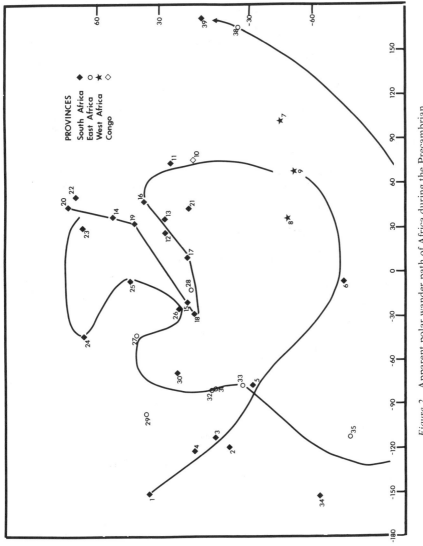

Figure 2   Apparent polar wander path of Africa during the Precambrian.

loop is not real, although to arrive at this conclusion they had to ignore poles 15 and 18. The similarities in age and shape between the African loop and the Logan loop of the North American polar wander path comprise a fundamental part of the evidence of Piper (1976a) for a Proterozoic supercontinent. Again, we feel more data are required to resolve this issue.

The poles younger than 900 m.y. suggest fitting a track across the southern geographic pole and terminating in the equatorial Pacific, although Piper (1976a) suggests that the age uncertainties of these poles are large.

DISCUSSION

From this brief review of the Precambrian paleomagnetic data, presented in terms of two apparent geomagnetic polar wander paths, several qualifications and some interesting speculations are possible.

1. It is clear that even with the most conservative data selection, the apparent polar wander paths are ambiguous. They should all, therefore, be regarded as provisional "state of the art" figures and should not be regarded as definitive. Indeed, it is clear in many cases (Piper 1976a, Irving & McGlynn 1976a, Irving, Emslie & Ueno 1974, McElhinny & McWilliams 1977) that this is how they are regarded by their authors. The great need for more data requires no further comment.

2. Taken at their face value, the paths do appear to indicate horizontal "plate" motions on a scale at least as great as during recent times.

3. The representative paths for North America and Africa can be considered in terms of superintervals (periods of uniform directional motion), interspersed with hairpins (times at which radical changes in direction occur). It has been noted previously that the hairpins appear to coincide with periods of maximum frequency of radiometric age dates (Irving & Park 1972, Nairn 1975). Irving interpreted this coincidence as showing an association between the hairpins and times of orogenic activity and plate reorganization.

4. With but a single exception (the poles from the Grenville province), we believe that the data presently available do not suggest differential movements of various tectonic provinces within a craton. The alternate view has been championed by Cavanaugh & Seyfert (1977) and by Burke, Dewey & Kidd (1976). The data from the Grenville province are not easy to fit to a reconstruction based upon the remaining zones, and the variety of possible interpretations has been discussed by Irving, Emslie & Ueno (1974). Time and more data will indicate whether one or other or some combination will provide the best tectonic interpretation.

5. The complex problems of obtaining meaningful results from rocks that have undergone metamorphism or more than one episode of magnetization are just beginning to receive adequate attention. If suitable data can be derived from such rocks, the opportunities for application of paleomagnetic techniques to Precambrian geology will increase greatly.

It is however clear that paleomagnetism already provides a powerful tool for use in the synthesis of Precambrian events.

Literature Cited

Abouzakhm, A. G., Tarling, D. H. 1975. Magnetic anisotropy and susceptibility of late Precambrian tillites from northwestern Scotland. *J. Geol. Soc. London.* 131:647–52

Beckmann, G. E. J. 1976. A palaeomagnetic study of part of the Lewisian complex, north-west Scotland. *J. Geol. Soc. London.* 132:45–59

Bidgood, D. E. T., Harland, W. B. 1961. Palaeomagnetism in some East Greenland sedimentary rocks. *Nature* 189:633–34

Briden, J. C. 1967. Recurrent continental drift of Gondwanaland. *Nature* 215:1334–39

Briden, J. C. 1977. Paleomagnetism and Proterozoic tectonics. *Tectonophysics* 38:167–68

Brock, A., Piper, J. D. A. 1972. Interpretation of Late Precambrian palaeomagnetic results from Africa. *Geophys. J.* 28:136–46

Buddington, A. F., Balsley, J. R. 1958. Iron-Titanium oxide minerals, rocks, and aeromagnetic anomalies of the Adirondack area, New York. *Econ. Geol.* 53:777–805

Burke, K., Dewey, J. F., Kidd, W. S. F. 1976. Precambrian palaeomagnetic results compatible with contemporary operation of the Wilson cycle. *Tectonophysics* 33:287–99

Cavanaugh, M. D., Seyfert, C. K. 1977. Apparent polar wander paths and the joining of the Superior and Slave provinces during early Proterozoic time. *Geology* 5:207–11

Christie, K. W., Davidson, A., Fahrig, W. F. 1975. The paleomagnetism of Kaminak dikes — no evidence of significant Hudsonian plate motion. *Can. J. Earth Sci.* 12:2048–64

Dewey, J. F., Bird, J. H. 1970. Mountain belts and the new global tectonics. *J. Geophys. Res.* 75:2625–47

Domen, H. 1975. Longterm effect of one directional moderate compression to the remanent magnetization of rocks. *Bull. Fac. Sci. Yamaguchi Univ.* 25:53–62

Evans, M. E., Bingham, D. K. 1976. Paleomagnetism of the Great Slave Supergroup, Northwest Territories, Canada: the Tochatwi Formation. *Can. J. Earth Sci.* 13:555–62

Fujiwara, Y., Schwarz, E. J. 1975. Paleomagnetism of the Circum-Ungava Proterozoic fold belt, I: Cape Smith komatiitic basalts. *Can. J. Earth Sci.* 12:1785–93

Fuller, M. D. 1964. On the magnetic fabrics of certain rocks. *J. Geol.* 72:368–76

Gates, T. M., Hurley, P. M. 1973. Evaluation of Rb–Sr dating methods applied to the Metachewan, Abitibi, Mackenzie and Sudbury dike swarms in Canada. *Can. J. Earth Sci.* 10:900–19

Girdler, R. W. 1961. Some preliminary measurements of anisotropy of magnetic susceptibility of rocks. *Geophys. J.* 5:197–206

Hailwood, E. A. 1972 *Paleomagnetic studies on rock formations in the High Atlas and Anti-Atlas regions of Morocco.* PhD. thesis. Univ. of Newcastle-upon-Tyne, England. 184 pp

Hamilton, N. 1963. Susceptibility anisotropy measurements on some Silurian siltstones. *Nature* 197:170–71

Hargraves, R. B. 1959. Magnetic anisotropy and remanent magnetism in hemo-ilmenite from ore deposits at Allard lake, Quebec. *J. Geophys. Res.* 64:1565–78

Harland, W. B. 1964. Evidence of late Precambrian glaciation and its significance, In *Problems in Paleoclimatology*, ed. A. E. M. Nairn. pp. 119–49. New York: Interscience, 705 pp.

Harland, W. B., Bidgood, D. E. T. 1959. Palaeomagnetism in some Norwegian Sporagmites and the late Precambrian. *Nature* 184:1860–62

Harland, W. B., Herod, K. N. 1975. Glaciations through time. In *Ice Ages: Ancient and Modern*, ed. A. E. Wright, F. Mosely, pp. 189–217. Liverpool: Seel House Press. 320 pp.

Henry, B. 1973. Studies of microtectonics, anisotropy of magnetic susceptibility and paleomagnetism of the Permian Dome de Barrat (France): Paleotectonic and paleosedimentological implications. *Tectonophysics* 17:61–72

Henthorn, D. L. 1973. Paleomagnetism of the Witwatersrand Trian, Republic of South Africa, and related topics. (Thesis Abstr.) *17th Ann. Rep. Res. Inst. Afr. Geol.* Univ. Leeds, England

Hicken, A., Irving, E., Law, L. K., Hastie, J. 1972. Catalogue of paleomagnetic directions and poles: first issue. *Dep. Energy, Mines Resour. Can. Earth Phys. Branch Pub.* 45, no. 1, 135 pp.

Hurley, P. M., Rand, J. R. 1969. Pre-drift continental nuclei. *Science* 164:1229–42

Irving, E. 1956. Palaeomagnetic and palaeoclimatological aspects of polar wandering. *Geofis. Pura. Appl.* 33:23–41

Irving, E., Emslie, R. F., Ueno, H. 1974 Upper Proterozoic paleomagnetic poles from Laurentia and the history of the Grenville structural province. *J. Geophys.*

Res. 79:5491–5502

Irving, E., Hastie, J. 1975. Catalogue of paleomagnetic directions and poles: second issue Precambrian results 1957–1974. *Dep. Energy Mines Resour. Can. Geomagn. Ser.* no. 3, 42 pp.

Irving, E., Lapointe, P. L. 1975. Paleomagnetism of Precambrian rocks of Laurentia. *Geosci. Can.* 2:90–98

Irving, E., McGlynn, J. C. 1976a. Proterozoic magnetostratigraphy and the tectonic evolution of Laurentia. *Philos. Trans. R. Soc. London Ser. A.* 280:433–68

Irving, E., McGlynn, J. C. 1976b. Polyphase magnetization of the Big Spruce Complex, Northwest Territories. *Can. J. Earth Sci.* 13:476–89

Irving, E., Park, J. K. 1972. Hairpins and superintervals. *Can. J. Earth Sci.* 9:1318–24

Irving, E., Park, J. K. 1973. Palaeomagnetism of metamorphic rocks: Errors owing to intrinsic anisotropy. *Geophys. J.* 34:489–93

Kern, J. W. 1961. Effects of moderate stresses on directions of thermoremanent magnetization. *J. Geophys. Res.* 66:3801–05

Khan, M. A. 1962. The anisotropy of magnetic susceptibility of some igneous and metamorphic rocks. *J. Geophys. Res.* 67:2873–85

King, R. F. 1966. The magnetic fabric of some Irish granites. *Geol. J.* 5:43–66

Kirschvink, J. L. 1977. The Precambrian-Cambrian boundary problem: Magnetostratigraphy of the Amadeus Basin, Central Australia. *Geol. Mag.* In Press

Kume, S. 1962. Sur des changements d'aimantation remanente de corps ferrimagnetiques saurcis a des pressions hydrostatiques. *Ann. Geophys.* 18:18–22.

Kume, S. 1965. Effect of unidirectional pressure on chemical remanent magnetization of α-haematite. *Geophys. J.* 10:51–57

McElhinny, M. W., McWilliams, M. O. 1977. Precambrian geodynamics—a paleomagnetic view. *Tectonophysics.* 38:137–59

McMurry, E. W., Reid, A. B., Evans, M. E. 1973. A paleomagnetic study of the Kahochella Group, N.W.T. Canada. *EOS (Am. Geophys. Union Trans).* 54:248 (Abstr.)

Morris, W. A. 1977. Paleolatitude of the upper Precambrian Rapitan Group and the use of tillites as chronostratigraphic marker horizons. *Geology* 5:85–88

Nagata, T., Kinoshita, H. 1965. Studies on piezo-magnetization (1) Magnetism of titaniferous magnetite under uniaxial compression. *J. Geomagn. Geoelectr.* 17:121–25

Nagata, T., Shimizu, Y. 1959. Natural remanent magnetization of Precambrian gneiss of Angul Islands in the Antarctic. *Nature* 184:1472–73

Nairn, A. E. M. 1963. Review of the variation in position of landmasses during geological times. *Uzita geofis.* 1:97–108

Nairn, A. E. M. 1975. Germanotype Tektonik und die Platten-tektonik-Hypothese. *Geol. Rundsch.* 64:716–27

Ohnaka, M., Kinoshita, H. 1968a. Effects of uniaxial compression on remanent magnetization. *J. Geomag. Geoelectr.* 20:93–99

Ohnaka, M., Kinoshita, H. 1968b. Effect of axial stress upon initial susceptibility of an assemblage of fine grains of Fe_2TiO_4–Fe_3O_4 solid solution series. *J. Geomag. Geoelectr.* 20:107–10

Piper, J. D. A. 1973a. Geological interpretation of palaeomagnetic results from the African Precambrian. In *Implications of Continental Drift to the Earth Sciences.* ed. D. H. Tarling, S. K. Runcorn, 1:19–31. London: Academic.

Piper, J. D. A. 1973b. Latitudinal extent of late Precambrian glaciations. *Nature* 244:342–44

Piper, J. D. A. 1974. Magnetic properties of the Cunene anorthosite complex, Angola. *Phys. Earth Planet. Int.* 9:353–63

Piper, J. D. A. 1975a. The palaeomagnetism of Precambrian igneous and sedimentary rocks of the Orange River Belt in South Africa and Southwest Africa. *Geophys. J.* 40:313–44

Piper, J. D. A. 1975b. Paleomagnetic correlations of Precambrian formations of east-central Africa and their tectonic implications. *Tectonophysics* 26:135–51

Piper, J. D. A. 1976a. Paleomagnetic evidence for a Proterozoic super continent. *Philos. Trans. R. Soc. London. A* 280:469–90

Piper, J. D. A. 1976b. Definition of pre-2000 m.y. apparent polar movements. *Earth Planet. Sci. Lett.* 28:470–78

Piper, J. D. A., Briden, J. C., Lomax, K. 1973. Precambrian Africa and South America as a single continent. *Nature* 245:244–48

Ridler, R. H., Foster, J. H. 1975. Archean paleomagnetic stratigraphy of the Skead Group, Kirkland Lake area. *Geol. Soc. Am. Abstr. with Programs* 7:844

Robertson, W. A., Fahrig, W. F. 1971. The great Logan paleomagnetic loop—the polar wandering path from Canadian Shield rocks during the Neohelikian Era. *Can. J. Earth Sci.* 8:1355–72

Roy, J. L., Lapointe, P. L. 1976. The paleomagnetism of Huronian red beds and Nipissing diabase; post-Huronian igneous events and apparent polar path for the interval -2300 to -1500 Ma for Laurentia. Can. J. Earth Sci. 13: 749–73

Roy, J. L., Park, J. K. 1974. The magnetization process of certain red beds: vector analysis of chemical and thermal results. Can. J. Earth Sci. 11: 437–71

Rudwick, M. J. S. 1964. The Infra-Cambrian glaciation and the origin of Cambrian fauna. See Harland 1964, pp. 150–54

Schermerhorn, L. J. G. 1974. Late Precambrian mixtites: glacial and/or non-glacial? Am. J. Sci. 274: 709–13

Schermerhorn, L. J. G., Stanton, W. I. 1963. Tilloids in the West Congo geosyncline. J. Geol. Soc. London 119: 201–41

Shackleton, R. M. 1973. Correlations of structures across Precambrian orogenic belts in Africa. See Piper 1973a, 2: 1091–95

Spall, H. 1971. Precambrian apparent polar wandering: Evidence from North America. Earth Planet. Sci. Lett. 10: 273–80

Spall, H. 1972a. Paleomagnetism and Precambrian continental Drift, Int. Geol. Congr. 24th 3: 172–79

Spall, H. 1972b. Did Southern Africa and North America drift independently during the Precambrian? Nature 236: 209–11

Spencer, A. M. 1971. Late Pre-Cambrian glaciation in Scotland. Mem. Geol. Soc. London, No. 6. 100 pp.

Spencer, A. M. 1975. Late Precambrian glaciation in the North Atlantic region. See Harland & Herod 1975, pp. 217–40

Stacey, F. D. 1960. Stress induced magnetic anisotropy in rocks. Nature 188: 134–35

Stone, D. B. 1963. Anisotropic magnetic susceptibility measurements on a phonolite and on a folded metamorphic rock. Geophys. J. 7: 375–90

Stott, P. M., Stacey, F. D. 1960. Magnetostriction and palaeomagnetism of igneous rocks, J. Geophys. Res. 65: 2419–24

Symons, D. T. A. 1975. Huronian glaciation and polar wander from the Gowganda Formation, Ontario. Geology 3: 303–6

Symons, D. T. A., Lowndry, J. W. 1975. Tectonic results from paleomagnetism of the Aphebian Nipissing Diabase at Gowganda, Ontario. Can. J. Earth Sci. 12: 940–48

Symons, D. T. A., O'Leary, R. J. 1975. Paleomagnetism of the Thessalon Volcanics and Huronian polar wander. Geol. Soc. Am. Abstr. with Programs 7: 867

Tarling, D. H. 1974. A palaeomagnetic study of Eocambrian tillites in Scotland. J. Geol. Soc. London 130: 163–77

Van der Voo, R., Klootwijk, C. T. 1972. Paleomagnetic reconnaisance study of the Glamanville granite with special reference to the anisotropy of its susceptibility. Geol. en Mijn. 51: 609–17

Williams, G. E. 1975. Late Precambrian glacial climate and the Earth's obliquity. Geol. Mag. 112: 441–65

Wilson, J. T. 1966. Did the Atlantic close and then reopen? Nature 211: 676–81

Winterer, E. L. 1964. Late Precambrian pebbly mudstone in Normandy, France: tillite or tilloid? See Harland 1964, pp. 159–78

Yaskawa, K. 1959. Remanent magnetism of dynamo-metamorphic rocks. Mem. Coll. Sci. Univ. Kyoto ser. B. 26: 225–27

Ann. Rev. Earth Planet. Sci. 1978. 6: 93–125

THE GALILEAN SATELLITES ✻10088 OF JUPITER: FOUR WORLDS

Torrence V. Johnson

Jet Propulsion Laboratory, California Institute of Technology, 4800 Oak Grove Drive, Pasadena California 91103

INTRODUCTION

In 1610 Galileo published his discovery of the satellites of Jupiter, (Galilei 1610), saying, in part: "But that which will excite the greatest astonishment by far, and which indeed moved me to call the attention of all astronomers and philosophers, is this, namely, that I have discovered four planets, neither known nor observed by any one of the astronomers before my time." These "four planets" now collectively bear their discoverer's name, the Galilean satellites (Galileo called them the "Medician stars"). They are, in order of their distance from Jupiter, Io (J1), Europa (J2), Ganymede (J3) and Callisto (J4). The discovery of these new planetary objects, obviously moving about another celestial body, was an important event in establishing the Copernican view of the solar system. Since Galileo's day, studies of the Galilean satellites have run as a continuous thread through the history of astronomy. The satellites were used to obtain the first accurate measurement of the speed of light in 1656. Michelson used his newly developed interferometric techniques to measure their diameters in 1891. At the beginning of this century mathematical studies of the satellites' motions developed a classical field of celestial mechanics and emphasized the importance of resonant phenomena in the solar system. In the last half of the twentieth century, as we begin the initial exploration of the outer solar system with spacecraft, new interest is centering on the satellites, first as individual planetary bodies, each a separate, intriguing world in its own right, and second as members of a system that in many ways resembles a miniature solar system and that may hold clues to the conditions and processes that led to the formation of the solar system. The purpose of this article is to review the very considerable amount of knowledge we have gained about the satellites, primarily through ground-based observations and the initial reconnaissance performed by the Pioneer 10 and 11 spacecraft, and to describe some of the exciting areas of current research. From this review will emerge three principal points: (a) the Galilean satellites are profoundly different in many important respects from the terrestrial planets, the Moon, Mercury, and Mars; (b) the satellites differ among themselves greatly; (c) the location of the satellites, orbiting the giant planet,

93

Jupiter, deep within an extensive and active magnetosphere, has a significant effect on their current state and the processes now acting on them and may well have been of major importance in their origin and evolution.

The four Galilean satellites (and Amalthea, J5, lying inside Io's orbit) have nearly circular, prograde orbits lying almost exactly in the plane of Jupiter's equator. Figure 1 shows a perspective view of the system. The other satellites of Jupiter have eccentric, highly inclined orbits, some retrograde. The Galilean satellites with Amalthea form one of three known "regular satellite systems," the other two being the Saturn system (including the Rings) and the Uranus system. These systems have been frequently referred to as miniature solar systems, and most models of solar-system origin consider, at least in passing, the formation of such systems as byproducts of planet formation, scaled-down versions of the larger system. We shall see that current research on the properties of the satellites themselves has led many workers to believe that the similarities of the satellite systems to the solar system may well go deeper than the obvious morphological resemblances.

PHYSICAL PROPERTIES

Size

The great distance of the Jovian system from Earth has made precise measurements of satellite radii difficult. At times of exceptional atmospheric "seeing" the satellites can be seen to have visible disks. This shows them to be lunar sized or somewhat larger. However, the angle subtended by a satellite's diameter as seen from Earth is on the order of only one second of arc. Since the blurring of astronomical images caused by the Earth's atmosphere is of the same order, accurate direct measurements of the angular size of the satellites' disks is rendered very difficult. Despite these intrinsic difficulties, a number of methods have been used during the last hundred years to measure satellite diameters [for a brief review of historical measurements see Morrison, Cruikshank & Burns (1977)]. These measurements show considerable scatter and are all affected to some degree

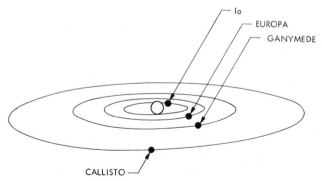

Figure 1   Galilean satellite system seen from above the satellites' orbital plane.

by systematic errors. They have established that Io is similar in size to Moon, Europa somewhat smaller and Ganymede and Callisto distinctly larger, similar to Mercury in size.

In the last decade, our knowledge of satellite radii has improved significantly. The most important advances occurred when Io (Taylor et al 1971) and Ganymede (Carlson et al 1973) occulted reasonably bright stars. Accurate timings of these occultation events allowed the radii of these satellites to be calculated with un-precedented precision. These observations also placed important limits on the satellites' atmospheres. The radii of Europa and Callisto remain less well determined but analyses of mutual occultation data improved these values as well [see Table 1 from Morrison, Cruikshank & Burns (1977)].

Masses

Deriving the masses of the satellites from their motions has been a classical problem in celestial mechanics. The problem is tractable only because of the resonant commensurabilities of the satellites' orbits, which enhance certain mutual perturbations (see Brouwer & Clemence 1961). Sampson (1921) and de Sitter (1931) independently solved this problem and the values thus derived remained the most accurate satellite mass determinations until the fly-by of Pioneers 10 and 11 in 1973. Accurate tracking of the Pioneer spacecraft allowed a precise solution for satellites' masses (to about 1%). These new values (Anderson, Null & Wong 1974) are given in Table 1 and differ by only about 20% in the worst case from de Sitter's values.

Densities

With the new values for satellite masses our knowledge of satellite densities is now limited by the accuracy of the radii determinations discussed above. Values for the densities are given in Table 1, and even with the remaining uncertainties in the radii these allow us to draw some interesting conclusions concerning the satellites' bulk properties. It is clear that the two inner, lunar-sized satellites, Io and Europa, have lunar-like densities (greater than 3 gm cm^{-3}), while the two outer, larger satellites, Ganymede and Callisto, have extraordinarily low densities for solid planetary bodies (less than 2 gm cm^{-3}). This difference is even more striking when compared to other terrestrial planets. Figure 2 shows a plot of radius versus density for the inner planets, Moon, the Galilean satellites and Titan (the fifth satellite of Saturn). The error bars shown arise primarily from the errors in the radii. The close similarity in bulk properties of Io, Europa, and the Moon and their difference from the other bodies is clearly shown. As we shall see, however, Io and Europa differ profoundly from the Moon in surface properties.

The three large, low-density satellites appear to form a class of bodies distinctly different from either the smaller terrestrial planets (such as Mercury or Mars) or the other satellites. These satellites' densities, below 2 gm cm^{-3}, argue strongly for significant amounts of low-density, nonrock material in their bulk composition. This material is most probably water in some form on Ganymede and Callisto (see section on evolution and interior structure).

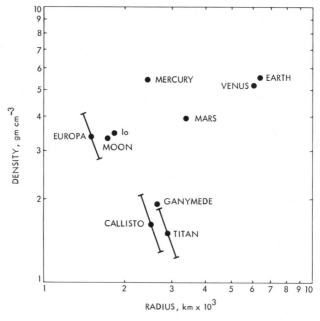

Figure 2   Log radius vs. log density for the inner planets, Galilean satellites and Titan.

Rotation

The satellites all display systematic variations in brightness as they orbit Jupiter (see section on surface markings) with periods identical to their individual orbit periods. These variations were first measured accurately by Stebbins & Jacobsen (1928) and have been observed to remain virtually unchanged in amplitude and phase since that time. This is strong evidence for synchronous rotation (i.e. each satellite forever keeping one hemisphere toward Jupiter). In addition, theoretical studies indicate that the time scales for "despinning" the satellites by tidal torques into the synchronous state are very short (see Peale 1977). Positions on the satellites' surfaces are frequently referred to with respect to orbital motion (e.g. "leading" or "trailing" sides).

SURFACE PROPERTIES

Albedo

Albedo is a measure of the reflectivity of a surface, or the fraction of incident light reflected. There are several different types of albedo that can be defined. One common albedo is the "Bond albedo," which is the ratio of total energy reflected in all directions to total incident energy. The Bond albedo is very useful, but also very difficult to measure in practice, particularly if the surface in question is very

Table 1   Physical properties

	Orbital Radius		Period (days)	Radius (km)	Mass (10^{23} gm)	Density (g cm^{-3})
	(10^3 km)	(Planetary radii)				
Io	421.6	5.95	1.769	1820 ± 10	891	3.52 ± 0.10
Europa	670.9	9.47	3.551	1500 ± 100	487	3.45 ± 0.75
Ganymede	1070.	15.1	7.155	2635 ± 25	1490	1.95 ± 0.08
Callisto	1880.	26.6	16.689	2500 ± 150	1065	1.62 ± 0.34

far away. The geometric albedo is defined as the brightness of a planet seen at zero phase (sun-planet-Earth angle equal to zero) compared to a uniformly diffusing white disk of the same size. Note that this definition means that a bright backscattering surface can have a geometric albedo greater than unity (see Allen 1973).

Since the brightness of a satellite is reasonably easy to measure, the primary uncertainty in satellite geometric albedos, as with their densities, has been in the radii determinations. Even taking the full range of pre-occultation values led to the strong suggestion that several of the satellites, at least, were much brighter than the Moon, Mercury or Mars (see Johnson 1971). The recent radii values given in Table 1 confirm that, at visual wavelengths ($\lambda \sim 0.55\,\mu$m), Io and Europa are five to ten times brighter than the Moon, Ganymede only slightly darker and Callisto, the darkest, still almost twice as bright as the Moon (see Table 2). Such bright surfaces combined with the satellites' cold environments naturally suggest ice-covered surfaces. This idea is amply confirmed in some cases by other data but in the case of Io some other explanation of the high visual albedo appears to be required (see section on composition).

From the Earth we can only observe the satellites at phase angles between $0°$ and about $12°$. Over this range all the satellites except Callisto are considerably less backscattering than the Moon, which is generally consistent with their higher albedos. We can get indirect information about the phase integral, (the value relating Bond to geometric albedo, usually designated q) from comparing infrared

Table 2   Photometric properties

	Geometric albedo, p	Phase integral, q	Bond albedo $A = pq$	T_{max} °K
Io	0.63	0.9 ± 0.2	0.56 ± 0.12	141 ± 11
Europa	0.64	1.1 ± 0.2	0.58 ± 0.14	139 ± 12
Ganymede	0.43	1.0 ± 0.3	0.38 ± 0.11	154 ± 6
Callisto	0.17	0.8 ± 0.4	0.13 ± 0.06	167 ± 3
Moon	0.12	0.58	0.07	

and reflected light data. Current estimates of q from this method are given in Table 2 along with the resulting expected maximum surface temperatures (see Morrison 1977).

Surface Markings

We have three types of information concerning surface markings on the satellites, photometric observations of rotational variations in brightness and color, visual observations, and images. The most extensive and quantitative of these data types is the set of photoelectric observations dating back to 1928. These data show a remarkable degree of hemispheric variation in surface brightness and color. Figure 3

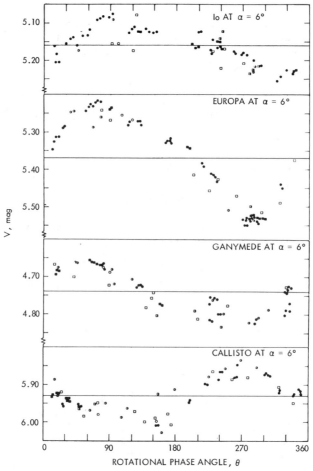

Figure 3 Variation of brightness (V magnitude at 6° phase) with rotational phase (from Morrison & Morrison 1977).

(from Morrison & Morrison 1977) shows the variation in visual brightness as a function of orbital angle (equivalent to rotational angle, measured around the satellites' orbit from the point of superior geocentric conjunction, where the satellite is directly behind Jupiter as seen from the Earth). The satellites' brightnesses at about 0.55 μm are given in stellar magnitudes (2.5 magnitudes equals a factor of 10 in brightness, 0.1 mag being equivalent to about a 10% change in brightness). The variations evident in Figure 3 are substantial, amounting to 15% or greater for Io, Ganymede, and Callisto (although Callisto's light curve is dependent on solar phase), and nearly 30% for Europa. These variations are associated with large changes in color as well (see Johnson 1971 and section on composition).

For purposes of comparison, we can look to the Moon. About 40% of the near side of the Moon is covered with maria having an albedo approximately half that of the uplands, which predominate on the far side. If we could stand back from the Earth-Moon system and measure the Moon's light curve, we would see about a 20% variation in total brightness from one side to the other. Lunar color variations would be much smaller, however. The satellites are thus in some sense at least as "varied" as the Moon in surface markings.

Observations of brightness variations give information only about global properties, primarily variations in contrast as function of satellite longitude. We must refer to visual observations and images to obtain an idea of how the contrast documented by the global brightness variations is distributed on the satellites' surfaces. As mentioned in the above section on size, the satellites can be seen as distinct disks on nights of exceptional "seeing"; and observers have reported and mapped markings on their surfaces by various techniques. These maps [see for example Lyot in Dollfus (1961) and Murray (1975)] are of necessity very poorly defined, but they do suggest some broad characteristics of the satellites' surface markings. Io appears to have dark polar regions, somewhat redder than the rest of the satellite. Europa has bright poles, or perhaps more accurately, a dusky equatorial zone marked by some contrast. Ganymede presents a mottled visual appearance, and Callisto shows a generally low-contrast disk. It should be noted in passing that very difficult visual observations of this sort have an undeservedly low reputation, due in large part to the infamous "canals" of Mars and the legacy of Percival Lowell. Critics frequently overlook the fact that the large-scale dark and light features mapped by the visual observers of the nineteenth and early twentieth centuries were remarkably accurate (see Cutts 1971). As long as the limitations of visual observations relative to fine detail are kept in mind such data can provide a reliable guide to the general longitudinal and latitudinal variations on satellite surfaces.

Ground-based images of the satellites are limited, as are the visual observations, by the optical properties of the Earth's atmosphere. Photographs taken under very good conditions show markings consistent with the visual observations (Minton 1973). A very fine image of Io was obtained by the Stratoscope balloon-born telescope by Danielson and Tomasko [see reproduction in Morrison & Cruikshank (1974)]. This image has a resolution of about 0.1 arc second and clearly shows Io's dark poles. Images of all the satellites with roughly similar quality were obtained through analysis of Pioneer Imaging Photo Polarimeter data (see Gehrels 1977). Photometric analysis of these data confirm that very high contrasts exist on the satellite surfaces, at least a factor of two in brightness for some of the dark areas on Ganymede, for example.

Composition

The very high albedos of the Galilean satellites relative to other solar-system objects indicate that they are covered by material very different from the rocks characteristic of the Moon, Mars, and Mercury. The low densities of Ganymede and Callisto and the stability of water ice on satellite surfaces strongly suggest that condensed water may exist on at least some of the satellites.

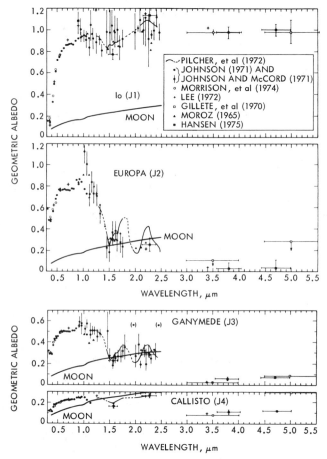

Figure 4   Geometric albedos as a function of wavelength for the Galilean satellites and the Moon (from Johnson & Pilcher 1977).

The most direct evidence concerning the satellites' surface compositions comes from analysis of the spectral characteristics of sunlight reflected from them. The variation of albedo as a function of wavelength, or the "spectral reflectance," can be affected by composition in several ways, of varying importance and diagnostic power: (*a*) Molecular absorptions. These are primarily important for condensates of gaseous species (H_2O, CO_2, CH_4, etc.) and minerals containing water, carbonate, sulfate, or nitrate groups. (*b*) Electronic transitions. These are important for minerals containing transition metal ions. The exact location and character of these absorptions is frequently diagnostic of crystal structure. The most important of this class of feature are the absorptions due to iron ions in the structure of the major mafic minerals olivene and pyroxene, occurring in the 1.0 and 2.0 μm spectral regions. (*c*) Charge transfer absorptions. These occur primarily at short wavelengths and are generally very broad features affecting the continuum "color" of the material. For recent discussions of the uses of spectral reflectance in remote sensing and the strengths and weaknesses of the technique see Johnson & Pilcher (1977) and Adams (1974).

Figure 4 shows a compilation of spectral albedo data for the satellites from the ultraviolet (0.3 μm) to the infrared (5.0 μm), which is essentially the entire available range for measurement of reflected sunlight (beyond 5.0 μm most radiation from the satellites is thermally emitted, below 0.3 μm only spacecraft data are available). Also shown is the Moon's average spectral reflectance. Several important features of these curves are immediately apparent, (*a*) the large absorptions at wavelengths longer than 1.0 μm in the spectra of Europa and Ganymede, (*b*) the decreasing reflectances in the ultraviolet characteristic of all the satellites, (*c*) the remarkably high infrared albedo of Io, in strong contrast to the spectra of Europa and Ganymede, and (*d*) the great differences between any of the satellite curves and the Moon's spectral reflectance. Additionally, the spectra of the satellites vary with rotation, most of the variation occurring in the ultraviolet. This variation suggests that the surface markings are highly colored compared with the Moon (see Johnson & Pilcher 1977).

The large infrared absorptions in Europa's and Ganymede's spectra are characteristic of molecular absorptions. High-resolution spectra in the 1.0 to 3.0 μm region have been obtained using interferometric spectrometers. The absorptions seen in those data are identified as being due to water molecules (Pilcher, Ridgeway & McCord 1972, Fink, Dekkers & Larson 1973). Figure 5 shows a comparison of such spectra with laboratory water-frost spectra taken at typical satellite temperatures. The close correspondence of the absorption features is obvious, including the small absorption near 1.66 μm that is seen only in low-temperature frost spectra (see Fink & Larson 1975).

The infrared spectra provide strong evidence for abundant water in the surface materials of Europa and Ganymede. Pure water frosts, however, are white at visible wavelengths, while the satellites have decreasing reflectances in the blue and ultraviolet. This visual color and the evidence for surface markings suggest that the water is mixed with some other material. This mixture may be on a large scale, that is patches of water frost and patches of some other material, or it may be extremely intimate, such as might be produced by cooling very hydrated minerals to the satellites' surface temperatures (about 120 K).

In addition to a lower albedo, Callisto's spectrum lacks the well-defined water absorptions near 1.4 and 1.8 μm evident in Europa's and Ganymede's spectra. However, the broadband data shown in Figure 4 suggest that Callisto does have a low reflectance in the 3.0 μm region, characteristic of one of the fundamental stretching modes of H_2O. Higher resolution data show very similar absorptions for both Ganymede and Callisto in this region, suggesting substantial water on Callisto's surface as well (Lebofsky 1977). The nonwater darker component of the surface material apparently masks the shorter wavelength absorptions.

Io's spectrum is unique in its combination of high albedo and lack of prominent infrared water absorptions. It appears to be clearly different in surface composition from any of the other satellites, at least as regards the hydrated phase or phases. The spectral reflectance for Io also drops much more sharply in the ultraviolet than does that of the other satellites. With few sharp, diagnostic absorptions to work from, it has been difficult to determine the composition of Io's surface with any certainty, although some categories of material are clearly ruled out, such as lunar basalts and water ice. Among reasonable materials studied in the laboratory, sulfur appears the most attractive candidate for producing the very sharp drop in reflectance at short wavelength. Spectra of mixtures of sulfur with bright, non-absorbing materials are in reasonable agreement with Io's spectrum (see Wamsteker, Kroes & Fountain 1974, Fanale, Johnson & Matson 1974, Nash & Fanale 1977).

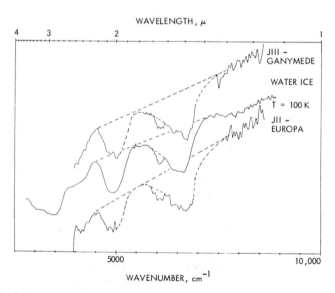

Figure 5   Ratio spectra of Europa and Ganymede with the Moon. Low temperature water ice spectrum is show for comparison (from Fink & Larson 1975).

Constraints on the nonsulfur phase do not uniquely determine the total composition. Both very bright feldspathic-type material or even metal oxides and evaporite minerals have been investigated [see Fanale, Johnson & Matson (1977b) and the section on models for origin]. New infrared data in the 4.0 to 5.0 μm region may ultimately provide the necessary clues. Absorptions in this range have been reported by Cruikshank, Jones & Pilcher (1977) and J. Pollack (personal communication). Although detailed assignments have not been made, such absorptions are suggestive of evaporite minerals such as carbonates and nitrates and not of oxides or refractory silicates.

In addition to the above compositional considerations, the possible effects of radiation from Jupiter's magnetosphere on the optical properties of the satellite surfaces must be kept in mind. A variety of possible optically active substances have been suggested. Also, experiments with proton irradiation have shown that color-center production may affect the visible and ultraviolet spectra of many of the above substances (see Fanale, Johnson & Matson 1974, Nash & Fanale 1977).

In summary then, we can characterize the satellites' surfaces as follows: None of the satellites appears to have on its surface material resembling lunar mare basalts or anorthositic uplands rocks, meteorites, or terrestrial igneous rocks. Three of the satellites, Europa, Ganymede, and Callisto, show spectral evidence for solid water on their surface along with varying amounts of an unknown, nonwater substance (e.g. either "dirty ice" or "icy dirt"). Io, despite its high albedo, does not appear to have water or frost on its surface. Its surface composition is uncertain, but good candidates include sulfur and evaporite minerals.

Physical State

We have little direct evidence concerning the physical state of the satellites' surfaces. By analogy to the inner planets, meteoroid bombardment is expected to play a role in surface modification, but there may be major differences in some processes. In addition to possible differences in the flux of impacting bodies with time, the abundance of relatively volatile material on the surfaces may significantly modify the ultimate condition of the satellite regoliths (for example, recent images of craters and other surface morphology on Mars suggest that the presence of water has greatly modified the evolution of the surface).

The degree of polarization of light reflected from a planet as a function of phase angle is quite sensitive to both the albedo and surface texture of the surface. Polarization data for the satellites show that they have particulate, scattering surfaces. The details of the polarization versus phase curves are generally consistent with the satellites' albedos and what is known of polarization from solid surfaces from laboratory data and the study of other planets (for a recent review of satellite photometry, phase functions, and polarization behavior see Veverka 1977a,b).

The most striking feature of satellite polarization data is the pronounced difference in polarization characteristics displayed by the leading and trailing hemispheres of Callisto. The leading hemisphere has a polarization-phase curve characteristic of relatively bare rock with some dust or microstructure on its surface, while the trailing hemisphere shows the chacteristics of a finely divided particulate regolith (see Dolfus 1975, Morrison et al 1975).

Thermal properties of the upper surface have been probed by infrared observations of the satellites' cooling during eclipses. These data suggest the presence of extremely insulating material in the upper few millimeters. Although a completely consistent thermal model has not been worked out, values for the thermal inertia derived from simple models are even larger than those for the Moon, formally one of the most insulating surfaces known [see Morrison (1977) for a recent review.]

At larger scales, radar observations have added yet another set of unusual properties to the satellites' list. Radar returns indicate that satellites have higher radar cross sections and are "rougher" than the terrestrial planets. A considerable degree of multiple scattering at centimeter wavelengths is implied by these results (see Goldstein & Morris 1975, Campbell et al 1977).

IO'S ATMOSPHERE

Background

Investigation of the atmosphere of Io is a very recent and active area of satellite research. Although the surface gravity for all the satellites is similar to the Moon's, the low temperatures at Jupiter's distance from the Sun makes them at least potential candidates for possession of atmospheres (see Kuiper 1952). However, numerous searches for spectral absorptions associated with atmospheric gases proved unsuccessful. The stellar occultation of 1971 provided an upper limit to Io's atmosphere of 10^{-6} bars. It was therefore somewhat of a surprise when R. Brown discovered in 1972 atomic emission lines emanating from the vicinity of Io (Brown 1974). This was followed quickly by the discovery that the sodium emission comes from a large region of space around Io, extending to at least ten Io diameters (Trafton, Parkinson & Macy 1974). The neutral sodium atoms responsible for this extended emission are not gravitationally bound to Io, although definitely associated with it. In December 1973 Pioneer 10's Ultraviolet Photometer Experiment discovered yet another extended "cloud" of neutral atoms, this time of hydrogen atoms (Carlson & Judge 1974, 1975). Pioneer 10 also performed a radio occultation of Io and discovered a reasonably dense ionosphere, with peak electron densities close to 10^5 cm^{-3} (Kliore et al 1974). Since that time neutral potassium emission has been detected (Trafton 1975b), as well as emissions from ionized sulfur in the inner magnetosphere that may be associated with Io (Kupo, Mekler & Eviatar 1976).

In discussing Io's "atmosphere" there are thus at least two types of phenomena to deal with: first, the material within about one Io radius of the surface, which may be at least loosely bound to Io, and second, the large, unbound clouds of atoms surrounding Io and extending many tens of thousands of kilometers around Io's orbit. The following sections describe some of the properties of the atmosphere and review current models and thoughts concerning the implications of these for Io's composition and its interaction with the magnetosphere.

Sodium Cloud

EXCITATION MECHANISM   The first step in characterizing the processes responsible for the sodium cloud is to understand the mechanism that excites the observed emission. The emissions that Brown discovered are the familiar D lines at 5890 Å and 5896 Å, well known to generations of physics students from the flame spectra of matches (indeed, Brown received considerable ribbing from his colleagues about "smoking in the dome" before his observations were confirmed). These lines are excited very easily; in the terminology of classical physics they have a large oscillator strength. Figure 6 shows a typical spectrum of the sodium emission taken with the spectrograph's slit including the disk of Io so that the emission is superimposed on the reflected solar continuum. It can be seen that the emission is very prominent at this resolution. The apparent strength of this emission combined with indications of temporal variability led to the initial suggestion that the emission originates in Io's atmosphere through an auroral mechanism. As the extent of the emitting region became better defined, however, resonant scattering

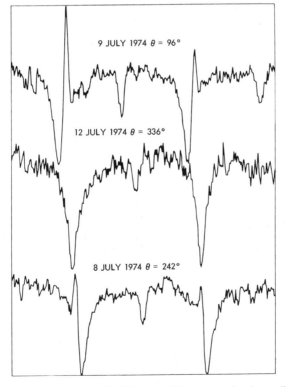

Figure 6   Spectra of Io taken at Table Mountain Observatory showing sodium emission features (from Bergstralh, Matson & Johnson 1975).

of incident sunlight emerged as a reasonable candidate mechanism (see Trafton, Parkinson & Macy 1974, Matson, Johnson & Fanale 1974).

Resonant scattering was shown to be the dominant mechanism through a detailed study of the temporal variation of the emission strength. As Io orbits Jupiter its line-of-sight velocity relative to the sun varies by ± 17 km sec^{-1}, sufficient to cause a variation in the Doppler shift of the incident sunlight of $\sim \pm 0.350$ Å. Thus, as Io orbits Jupiter the rest frequency of the sodium D lines shifts from one wing of the Fraunhofer absorptions, through the core of the lines and onto the other wing, causing the energy available for scattering to vary by nearly an order of magnitude (see Figure 6). Systematic measurements of the sodium emission strength were carried out in 1974. These showed that the emission varies with orbital position in precisely the way predicted for resonant scattering (Bergstralh, Matson & Johnson 1975). Figure 7 shows these results, the dashed lines being the expected variation due to the Doppler effect described above.

SUPPLY AND LOSS MECHANISMS   Two key questions are: How is sodium supplied to the cloud? and How is it lost? The two questions are closely coupled since together the answers determine the mass balance of the system. It seems virtually certain that Io itself must be the immediate source of the sodium atoms. The fact that

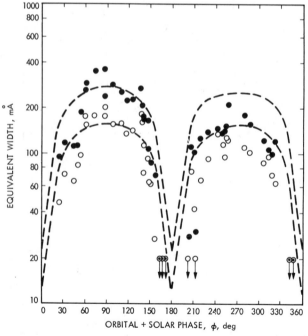

Figure 7   Sodium emission strength as a function of orbital position. Open circles are D_1 data, filled are D_2. Dashed line is expected variation based on resonant scattering theory (from Bergstralh, Matson & Johnson 1975).

most of the observed sodium has escaped from Io's gravity places an important constraint on candidate supply mechanisms, namely that they must be quite energetic. The escape velocity from Io is about 2.5 km sec^{-1} and sodium is by no means a volatile gas, particularly at temperatures of 150 K or less. Thus, some considerable source of energy is required to impart escape velocity to the sodium atoms. The most obvious place to search for such an energy source is Jupiter's magnetosphere and its energetic trapped particles. To date, all proposed mechanisms for sodium escape have appealed to this source in one way or another.

Since only neutral atoms scatter the observed resonant lines, ionization is the major mechanism responsible for removing sodium from the observed cloud. Ionized sodium ceases to emit D-line radiation and is immediately swept away from the cloud by the Jovian magnetic field (moving with a velocity of 56 km sec^{-1} relative to Io). While some ionized sodium may conceivably be reacquired by Io and neutralized, most will be lost by diffusion in the magnetosphere; we assume here that, once ionized, sodium is essentially "lost" to the system. Photoionization by solar ultraviolet radiation yields a lifetime of $\sim 10^6$ sec for sodium. This is too slow a process to account for the high concentration of sodium near Io. Carlson, Matson & Johnson (1975) have shown that electron impact ionization by thermal plasma electrons is capable of satisfying the observed sodium distribution if the plasma has characteristics compatible with both Pioneer plasma and hydrogen cloud measurements. Typical lifetimes against ionization for this process are about 10^5 sec, or the same order as Io's orbital period. Observations of rapid changes in sodium intensity associated with magnetic-field position have been interpreted by Trafton & Macy (1975) as requiring even more rapid ionization in portions of the magnetosphere near the magnetic equator.

Although the current levels of sodium emission could in principle have been detected anytime since about 1910, high-resolution spectra of small objects such as satellites were not taken frequently and there is no conclusive evidence as to how long sodium emission has been going on. Continuation of the systematic observations that established the resonant scattering mechanism has shown that, at least since 1974, the cloud has remained in essentially steady state. The approximate supply rate necessary to maintain this steady-state density can be calculated, and there is no reason to believe that it has not existed for most of geologic time. Approximately 10^{24} to 10^{25} atoms sec^{-1} are required by this calculation (see Brown & Yung 1976). This is equivalent to 10^7 or 10^8 atoms cm^{-2} sec^{-1} from Io's surface. If the ultimate source of sodium atoms is Io's surface this rate implies the removal of tens to hundreds of meters of material over geologic time, depending primarily on sodium concentration (one to ten percent was taken here). This scale of erosion, while important locally, would be virtually undetectable even with the highest resolution images from currently planned reconnaissance missions.

Although the above discussion indicates that there is at this time no compelling need to look beyond Io's surface for a source of sodium, other possible sources for the sodium have been suggested. These involve supply of sodium to Io's surface and its subsequent removal to the cloud. One possible source is infalling meteorites. Although little is known in detail about the meteoritic environment around

Jupiter, current estimates based on Pioneer micrometeoroid data suggest that even fluxes considerably higher than that for the Moon cannot supply sufficient sodium (see Brown & Yung 1976).

Another potential source is magnetospheric trapped particles. In a sense, this hypothesis merely removes the enigma of the source of sodium one step, since very large concentrations of sodium ions in the magnetosphere would be required and a source found for them. Cloutier et al (1977) have suggested that Jupiter's ionosphere may supply heavy ions to Io where they are neutralized and escape into the cloud. In this model large currents ($\sim 10^7$ amps) are driven by the fields set up by Io's motion through Jupiter's magnetic field; one part of this current system consists of ions derived from Jupiter's ionosphere. In the limiting case this mechanism could supply the necessary sodium, but only if 10% or more of the current were composed of sodium ions, a stringent requirement on Jupiter's ionosphere and the details of the current mechanisms.

The problem of the energy required to remove sodium atoms from the surface and give them escape velocity is common to most theories, since virtually all possibilities involve some residence time for sodium on Io's surface. The only mechanism that has been explored in any detail to date is sputtering by magnetospheric ions. This process, in which atoms are "blasted" off surfaces by ion impacts, was proposed by Matson, Johnson & Fanale (1974) as a method of using the energy available in the magnetosphere to eject atoms from Io into the cloud. Velocities of sputtered atoms are generally above Io's escape velocity. Depending on yields, ion mass, surface sodium concentration, etc ion fluxes of 10^8 to 10^{10} cm^{-2} sec^{-1} would be required to supply the observed sodium. While these are fairly large fluxes, they are within an order of magnitude of measured high-energy fluxes (> 0.4 MeV) and within the range of proposed thermal plasma (~ 100 eV) proton fluxes.

The fate of atoms liberated from Io's surface by sputtering or other energetic processes depends critically on the neutral atmosphere. If the atmosphere is very tenuous (less than about 10^{-10} bar surface pressure), atoms ejected from the surface will escape essentially unimpeded. If the atmosphere is thick enough for an atom to suffer multiple collisions, material removed from the surface will be thermalized in the atmosphere. A further process, such as direct ion-atom collisions, is then needed to supply the escape velocity. Another constraint on this possibility is that nonvolatile atoms such as sodium must have very short residence times in the atmospheric "reservoir" since they would remain on the surface if allowed to return to it. Current interpretations of ionosphere and atmosphere models range from "thin" to "thick" (see section on ionosphere).

CLOUD DYNAMICS AND DISTRIBUTION   The shape of the cloud and the motions of atoms in different parts of it are important clues to the processes that cause it. The fundamental fact underlying the dynamics of the neutral atoms in the cloud is that Jupiter's huge gravitational field dominates the situation. Atoms escaping Io's modest gravity at 2.5 km sec^{-1} or faster go into orbit about Jupiter unless they have velocities greater than about 60 km sec^{-1}. If their velocities are not far above Io's escape velocity, the escaping atoms will go into orbits not very different from Io's and will "drift" away from proximity to Io after several orbits. The theory governing such a situation was worked out by McDonough & Brice (1973) for the hypothetical case of hydrogen escape from Saturn's satellite, Titan. They showed that, if no process (or a very slow one) removes atoms from the system, the resulting steady-state distribution is a flattened tòroid centered on the satellite's orbital track, a huge "doughnut" of co-orbiting debris. If a rapid loss process such as ionization operates, the orbits of escaped atoms will not have time to diverge far from the satellite's orbit and the "doughnut" will appear to be partially or mostly "eaten." This appears to be what is happening around Io.

Direct information about the velocities of the sodium atoms comes from high spectral resolution studies of the individual emission lines. These lines are broadened by the Doppler effect when atoms moving at different velocities are included in the field of view. Such high-resolution spectral observations have shown that the sodium lines are relatively broad (~ 0.100 Å), indicating a broad velocity distribution. The maximum of the emission is near Io's mean velocity. However, Trafton (1975a) discovered that the lines are also highly asymmetric, with a "skirt" of emission shifted by velocities of 10 km sec^{-1} or more from the mean velocity of Io at the time of observation. This skirt was observed at long wavelengths when Io was moving away from the Earth and at short wavelengths when moving toward Earth. Figure 8 shows line profile data from Trafton (1975a) illustrating these effects.

A simple interpretation of the line profile data is that there are more atoms moving in the direction of Io's orbital motion, but faster, while far fewer atoms are moving rapidly in the other direction with respect to Io. What is required to create such a situation? It is relatively simple to construct a computer model of the cloud; the position of each escaping atom can be calculated based on orbital dynamics and assumptions of initial velocity (speed and direction), ionization rate, and escape geometry (isotropic or directional). These models suggest strongly that isotropic escape of sodium from Io cannot produce the observed velocity distri-

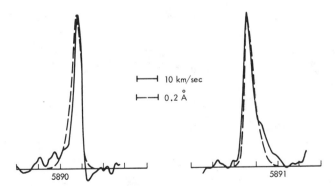

Figure 8   Sodium emission lines at high resolution, illustrating asymmetric line profiles. Data are from Trafton (1975a); dashed line is a fit from a model by Carlson et al 1977.

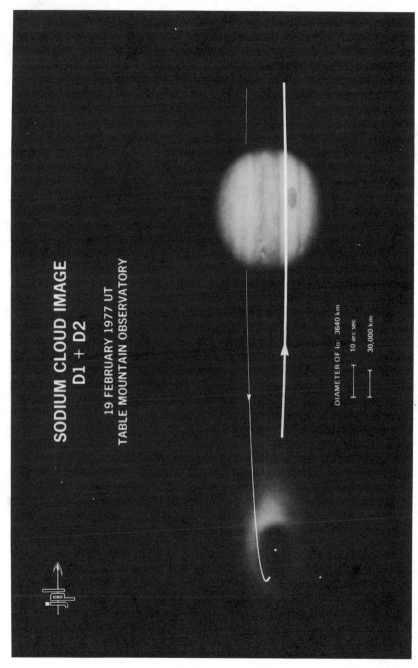

Figure 9 Image of sodium cloud ($D_1 + D_2$) taken at Table Mountain Observatory. Jupiter and projection of Io's orbit are drawn to scale. Dark area around Io's position is the shadow of an occulting disk (from Matson et al 1977).

butions. Preferential escape of atoms in the direction of Io's orbital motion, however, can satisfy the data reasonably well.

In addition, these model calculations can constrain the initial velocity distribution of the escaping sodium. A velocity distribution centered above $2.5 \, \text{km sec}^{-1}$ but having a significant fraction of the atoms with velocities of $10 \, \text{km sec}^{-1}$ or greater is required. The dashed lines in Figure 8 show calculated profiles from a model having preferential escape from Io's leading hemisphere with velocity distribution characteristic of sputtering (from Carlson et al 1977). Such fits are not unique for any given parameter (for instance it is possible that other processes than sputtering could produce a similar velocity distribution, such as collision with a heavy ion plasma), but these models illustrate the way in which important parameters can be constrained by theory and observation.

What is the observed shape of the cloud? It might be expected that asymmetrical processes such as those suggested by the line profiles would have important effects on the morphology of the cloud as well. Any successful model must surely satisfy the observations of emission intensity as well. The faintness of the emission and its time variability make it difficult to obtain a synoptic view of the cloud geometry. The patrol data in Figure 7 show an asymmetry, with greater intensity when Io is moving toward us than away. This was interpreted by Bergstralh, Matson & Johnson (1975) as indicating more sodium atoms preceding Io than trailing it in orbit.

Direct imaging of the cloud through narrow filters has been frustrated by the problems of scattered light from Io and Jupiter (a filter $0.1 \, \text{Å}$ wide would be required to reduce the reflected sunlight from Io to the same levels as the sodium emission). However, approaches using the imaging properties of spectrographs and observations with Fabry-Perot interferometers are now providing us with our first images of the cloud. Münch & Bergstralh (1976) used multiple slits as well as images of Fabry-Perot fringes to observe the cloud, showing that it appeared consistent with the flattened toroid suggested by theory. Goody and his co-workers at Harvard are currently using a very fine grid of slits to produce contour maps of the emitting region near Io.

If the light from Io is blocked by some occulting device (such as an aluminized dot on a glass plate), then slits can be dispensed with entirely and a "slitless spectrum" will show the two images of the cloud in the light of each D line. Slitless images of this sort have been obtained using a television tube detector at Table Mountain Observatory (Matson et al 1977). Figure 9 shows an image of the cloud taken this way, along with a picture of Jupiter and Io's orbit, projected and drawn to scale. This image clearly shows that there is indeed more material in front of Io than behind it and confirms that the cross section of the cloud is that of a partial torus.

We can use models to investigate what circumstances can produce such an asymmetrical cloud. Isotropic escape of sodium produces a spacial distribution which is essentially a symmetrical torus and cannot satisfy the cloud-image data any better than the line-profile data. Escape in a generally forward direction does produce a model sodium distribution that matches the data well. Figure 10

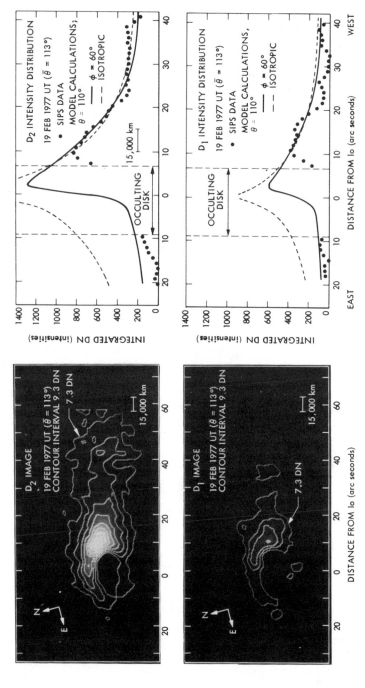

Figure 10 Integrated East-West intensity profiles through the cloud image in Figure 9, compared with model calculations (from Matson et al 1977).

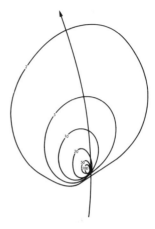

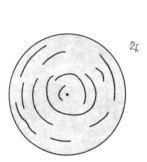

Figure 11   Cloud shape as seen from above Jupiter's North pole, calculated from model used in Figure 10 (Carlson, personal communication).

shows an intensity profile through the cloud compared with isotropic and preferential escape models (from Matson et al 1977). Figure 11 shows the shape of the cloud as seen from above, predicted using this model. ·

What processes could produce the asymmetries required by both the line profile and image data? This is one of the important problems under current investigation. Asymmetries in the interaction of Io with the magnetosphere are an obvious candidate. Plasma sheath models (see Gurnett 1972, Hubbard Shawhan & Joyce 1974) suggest very different conditions at different places on Io, possibly leading to differences in the ion impact fluxes. In addition, the Pioneer 10 radio occultation found very different ionospheric densities over the two parts of the satellite probed, suggesting possible differences in the atmosphere and/or its interaction with the magnetospheric plasma. At the moment no model is totally satisfactory.

Hydrogen Cloud

Figure 12 shows typical data from the Ultraviolet Photometer on Pioneer 10, illustrating the signal from the extended hydrogen cloud (from Carlson & Judge 1974). This cloud is excited by resonant scattering of sunlight. As with sodium, the primary loss mechanism is ionization and the cloud's extent requires a more rapid process than photoionization. In this case the most likely source of ionization is charge-exchange reactions with thermal plasma protons (see Carlson & Judge 1975). Plasma values within the range suggested by Pioneer experiments can satisfy the hydrogen geometry as well as the sodium ionization rate requirements.

From estimates of total cloud hydrogen population and the ionization rate deduced by Carlson & Judge, a supply rate of $\sim 10^{28}$ atoms $\sec^{-1}$ is calculated. This is equivalent to $\sim 10^{11}$ atoms cm^{-2} $\sec^{-1}$ from Io's surface. Several possible sources for this hydrogen have been suggested. Io's atmosphere is one obvious possibility. Hydrogen does not suffer from the difficulty of nonvolatility as sodium does. Once released in the atmosphere hydrogen will escape Io's gravity fairly easily despite the low temperatures. Among the sources of atmospheric hydrogen considered are photolysis of hydrogen-bearing molecules, particularly NH_3 (McElroy, Yung & Brown 1974), and dissociative recombination of hydrogen-bearing species (Johnson, Matson & Carlson 1976). In these models the main problem is the buildup of the heavy element(s) attached to the hydrogen-bearing molecule. If the heavy residue is also gaseous, an efficient escape mechanism must be postulated to avoid a rapid buildup of a thick atmosphere, in violation of the observations [see Brown & Yung (1976) for a discussion]. In some cases, such as H_2S, the residue might be nonvolatile and the model then predicts a buildup of material on the surface. Over geologic time a large quantity of material could be involved, representing the loss of one to ten kilometers of surface material or a similar deposition, depending on the fate of the heavy component.

Other suggested sources for the hydrogen are the plasma itself and Jupiter's ionosphere. The flux of protons at Io inferred from plasma densities compatible with Pioneer 10 data, the sodium cloud, and the required hydrogen charge-exchange rate is a factor of 50 to 100 too small to balance the hydrogen escape rate (see Brown & Yung 1976). Likewise, estimates of the maximum Birkland current supplying ions from Jupiter's ionosphere to Io also fall about two orders of magnitude short of the flux necessary to replace the hydrogen (see Cloutier et al 1977). Currently, then, Io's atmosphere appears to be the most likely source for the hydrogen cloud, but our ignorance of the exact plasma conditions and interactions in the inner Jovian magnetosphere do not allow us to rule out completely possible magnetospheric or Jovian sources.

We do not yet have a picture of the hydrogen cloud, but the Pioneer 10 data suggests that more hydrogen trails Io in its orbit than leads it, the opposite of the

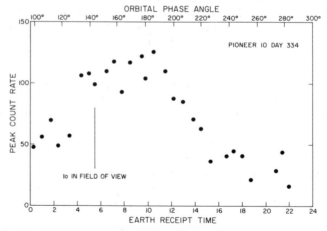

Figure 12 Ultraviolet Photometer data from Pioneer 10 encounter (Carlson & Judge 1975).

situation with sodium. Models similar to those described for sodium indicate that this geometry requires escape of hydrogen primarily from the side of Io facing away from Jupiter, the "outer hemisphere," (Carlson & Judge 1976). The difference in the inferred escape geometries for sodium and hydrogen suggests that different processes are responsible for the escape of each species. This may be due to the great volatility of hydrogen compared with sodium. In any case, this circumstance is evidence against some models, such as those involving "blow off" of a hydrogen-bearing atmosphere that drags heavy species with the light (see Brown & Yung 1976) and those involving direct collisional sweeping of all atmospheric constituents by plasma ions.

Other Species

After the discovery of the sodium cloud, the questions were immediately posed, "Why sodium? Why do we see nothing else?" In retrospect, given that a cloud (or clouds) of neutral atoms surrounds Io, it is obvious that sodium would be one of the most easily detected species. This expectation arises from the combination of the very strong oscillator strength of the sodium resonant D lines and sodium's relatively high abundance in likely cosmic source materials. If these factors are taken into account, only calcium, potassium, and possibly aluminum are likely candidates to produce similarly strong resonant emission features in the visible and near infrared parts of the spectrum. Potassium has been discovered (Trafton 1975b), although little is yet known of its distribution or characteristics. Sensitive searches have not detected calcium and limits to emissions from a number of other elements have been established (Trafton 1976). Fanale, Johnson & Matson (1974) interpreted the absence of calcium emission at strengths expected for solar-type compositions as evidence that the cloud, and probably Io's surface, is calcium poor.

Fanale, Johnson & Matson (1977b) have analysed existing information about cloud composition in terms of possible Io surface compositions. They conclude that surface/cloud fractionation is probably not a dominant factor and that the cloud composition allows us to place constraints on surface composition. According to this analysis, surface materials similar to chondritic meteorites and common terrestrial igneous rocks are ruled out (see also Trafton 1976). Evaporite assemblages and possibly some higher temperature compositions are compatible with the current data.

Finally, Pioneer 10 found as yet unidentified ultraviolet emissions from around Io (Judge et al 1976). This signal is smaller than the hydrogen channel signal and at shorter wavelengths. Possible sources for such a signal include ionized sodium, helium, and possibly oxygen.

Sulfur Cloud

In addition to the neutral clouds discussed above, an extensive region of emission due to singly ionized sulfur has been discovered recently (Kupo, Mekler & Eviatar 1976). The emission is not concentrated around Io but is generally associated with the region bounded by Io's orbit. There are two major differences between this cloud and the neutral clouds. First, the emission lines are due to an ionized species. This means that instead of following Keplerian orbits as the neutrals do, the motion of the emitting ions is controlled by the magnetic field of Jupiter. Second, the emission line observed is not a resonant line due to scattered sunlight but is excited by collisions with plasma electrons. The characteristics of the sulfur emission are an indirect probe of plasma density, temperature, and composition. Analysis of the line strengths and emission profiles indicates that the plasma is denser and cooler than that inferred near Io from Pioneer measurements (see Brown 1976, Kupo, Mekler & Eviatar 1976).

Io is a possible source for the sulfur ions; and, as with the neutral clouds, the possibility of other, as yet unobserved, heavy ions remains open. The neutral clouds must certainly provide a source for some heavy ion species. Evaluation of the possible ion sources and theoretical modeling of the diffusion of ions in the magnetosphere are currently underway (see Siscoe 1978).

Ionosphere

The radio occultation of Pioneer 10 by Io detected the presence of an appreciable ionosphere. The electron density profiles derived from analysis of the radio tracking data showed two highly interesting characteristics, first a high electron density of $\sim 10^5\,\mathrm{cm}^{-3}$ and second a large difference between the profiles obtained from entry and exit data (Kliore et al 1974, 1975). Figure 13 shows both profiles as a function of altitude.

To deduce anything about the neutral atmosphere from the electron density

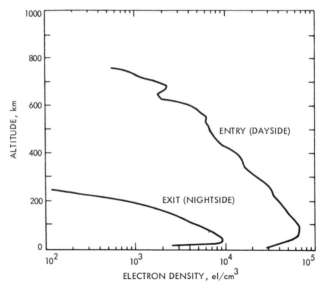

Figure 13 Electron density as a function of altitude in Io's ionosphere (from Kliore et al 1975).

data it is necessary to have a model relating neutral atmospheric density and composition to electron density through ionization processes, rates, atmospheric dynamics, etc. A simple scaling of ionospheric models used for Mars and Venus yielded estimates of a relatively "thick" neutral atmosphere with a surface pressure of 10^{-8} to 10^{-10} bars. Atomic atmospheric models required fewer neutrals (Whitten, Reynolds & Michelson 1975). A detailed analysis of many models by McElroy & Yung (1975) showed that molecular atmosphere models where photo-ionization is the major ionization mechanism require neutral densities of $\sim 10^{11} cm^{-3}$ (or surface pressures of $\sim 10^{-9}$ bars). The other possible source of ionization is charged particle irradiation. McElroy & Yung showed that for charged particle models a wide range of neutral densities were possible, depending on the flux and energy of the incoming particles. Johnson, Matson & Carlson (1976) investigated models dominated by electron impact ionization, using electron energies and fluxes compatible with the plasma characteristics suggested by Pioneer 10 data and the sodium and hydrogen clouds. These models provide good fits to the electron densities for "thin" atmospheres ($\sim 10^{-10}$ bar).

All of the above models are simple to the extent that they assume no interaction between the ions in the ionosphere and the magnetosphere. A model with a dynamical, comet-like ionosphere has been proposed by Cloutier et al (1977). They point out that an Earth-like ionosphere is unlikely to be stable against the dynamic pressure of the corotating plasma and the large corotational electric field. Their model proposes both critical velocity ionization [a mechanism proposed by Alfvén (1957) for comets] and electron impact ionization. In the Cloutier et al model the ionosphere is being continually swept "downstream" (i.e. ahead of Io) by the corotating plasma and must be continually replenished. They propose Birkland currents from Jupiter's ionosphere to supply all or part of the atmosphere, although other sources in the atmosphere or surface of Io are possible (see discussion in sections on sodium cloud and hydrogen cloud).

In summary, the neutral atmosphere and ionosphere of Io are still subjects of debate and much ongoing research. The estimates of neutral atmospheric pressure from current models range over several orders of magnitude, from about 10^{-9} bars to 10^{-11} bars. Theoretical studies are only just beginning to take account of even a small number of the forces and processes that must be at work in Io's extraordinary environment, and the best available models suggest fascinating mixtures of planetary and cometary atmospheric characteristics.

ATMOSPHERES OF THE OTHER SATELLITES

We have very little direct information about the atmospheres of Europa, Ganymede, and Callisto. Observation of a stellar occultation by Ganymede (Carlson et al 1973) suggested the possible presence of a very tenuous atmosphere, but the circumstances were very unfavorable compared with the Io occultation of 1971 and no detailed analysis was possible. Pioneer 10's Ultraviolet Photometer detected signals from the vicinity of Europa, similar to those discovered at Io but much weaker (Judge et al 1976; Carlson, personal communication). These data suggest a possible hydrogen cloud of limited extent and some other component, also seen in the short wavelength channel around Io (see section on other species).

Despite this lack of observational data, the presence of condensed water on some satellites' surfaces is an argument for expecting interesting, although tenuous, atmospheres for the other satellites. Although water has a very low vapor pressure at satellite surface temperatures, some water molecules will definitely sublime from the satellites; it is thus virtually certain that there will always be some number of water molecules, hydrogen, oxygen and hydroxyl groups present above the surfaces of these bodies. How significant these are for forming an atmosphere depends on the rate of supply of water (possibly enhanced by sputtering, e.g. Brown et al 1977), dissociation, and loss processes. These possibilities have been discussed by Carlson (1976) and Yung & McElroy (1976). Yung & McElroy suggest that a significant ($\sim 10^{-6}$ bar) atmosphere consisting in large part of oxygen could conceivably be maintained by Ganymede.

MODELS FOR ORIGIN

As pointed out in the introduction, the Jovian system has been regarded since its discovery as a "miniature" solar system. The increasingly detailed knowledge about the satellites reviewed in the previous sections is now being used to test and improve hypotheses concerning the origin and processes of formation of Jupiter and the other planets. In this and the next section I discuss a few of the main points of these hypotheses and the consequences of these ideas and the known properties of the satellites for the satellites' evolution and interior structures.

The regular satellite systems all occur in the outer solar system, beyond the asteroid belt. The Jovian system, at 5.2 au from the Sun, is the closest to Earth. The large outer planets, particularly Jupiter and Saturn, appear to have retained essentially a solar composition, based on their densities and what we know of their atmospheric compositions. The inner planets appear to have "lost" most of their volatile elements [see Lewis (1974) for a review]. We expect that the outer planets' satellites may also retain more volatiles than their counterparts in the inner solar system.

The asteroid belt is an important dividing line in terms of present conditions. Watson, Murray & Brown (1963) showed that beyond the heliocentric distance of the asteroid belt, water ice is stable on unprotected (atmosphereless) surfaces for periods comparable to the age of the solar system. Thus, in the outer solar system, we expect volatiles, especially water, to be important in the bulk composition of planets and to have played a major role in their evolution. The Galilean satellites are our first chance to study such planetary bodies in detail.

As we have seen, water is indeed important on several of the satellites; and the ice-plus-rock compositions suggested by Ganymede's and Callisto's densities are exactly what one would expect based on equilibrium condensation models of solar-system formation (see Lewis 1974). The higher density rocky satellites Europa and Io do not fit this simple picture, however. Even before the current accurate values of the satellites' densities were known, this discrepancy and its similarity to the case of the planets (where higher density objects are generally closer to the Sun) was noted (see Kuiper 1952).

Qualitatively, it is obviously tempting to suggest that Jupiter played a role similar to the Sun's in the formation of the system. Increased knowledge of Jupiter from observations and theoretical studies in the last decade have allowed researchers to explore this similarity of Jupiter and the Sun in new detail. Jupiter is composed primarily of hydrogen and helium and is not much less massive than some small stars; the notion of the Sun and Jupiter as an "almost" double star system is common. In 1969 it was discovered that Jupiter emits more energy (in the far infrared) than it receives from the Sun (Aumann et al 1969). This observation, combined with new evolutionary models of small stars in their "pre-nuclear burning" stages, sparked the development of the first quantitative, albeit still simple, models of Jupiter's early history. These models, which take the current infrared flux as a boundary condition (see Grossman et al 1972), suggest that early in its history Jupiter became very luminous due to the release of gravitational energy as the mass of the forming planet contracted toward Jupiter's center. The central portions of the planet became very hot but did not reach the critical values of temperature and pressure required to light the thermonuclear fires that would have transformed Jupiter into a star. Since that time, in this picture, Jupiter has been cooling through convection, conduction, and radiation, and the observed excess infrared flux is the present day expression of that early stage of formation (see Hubbard 1977).

The high luminosities (10^{-2} to 10^{-4} of the present solar luminosity) that Jupiter may have achieved, according to the current models, are sufficient to have significantly altered the conditions under which the satellites formed from the ambient solar nebular conditions. Pollack & Reynolds (1974) have investigated some of the consequences of these models for satellite formation. They find that at a time when temperatures at the distance of Ganymede's orbit were cold enough to condense water ice, the temperatures at Io's and Europa's positions could have remained high enough to prevent ice formation for a considerable period of time. Although the general picture of Jupiter's effect on local conditions follows from our current understanding of Jupiter's evolution, the precise temperature at any position in the circum-Jovian nebula and its variation with time are, of course, subject to many uncertainties. Cameron & Pollack (1976) have, for instance, investigated the effect of cloud opacity on the temperature distribution around Jupiter. Also, the effects of dynamical considerations on the time scale and energy of satellite formations are still being studied (e.g. Harris & Kaula, 1975).

The particular effects of these models on the possible formation conditions for Io have been investigated by Fanale, Johnson & Matson (1977a,b). They note that although Io must form at temperatures high enough to exclude the incorporation of large quantities of ice ($\geq \sim 170\,K$), most of the models also require that it form at temperatures low enough ($\leq \sim 500\,K$) that significant amounts of water would have been included in the form of chemically bound water in hydrated silicates. At the extreme range of models with high opacity in the circum

Jovian nebula, formation temperatures at Io may have been just beyond the limits for formation of hydrated compounds, leading to the possibility of "high-temperature" condensates on Io coexisting in the system with the volatile-rich outer satellites (see Fanale, Johnson & Matson 1977b).

EVOLUTION AND INTERIOR STRUCTURE

What are the implications of the various formation models and the known physical properties of the satellites for their evolution? John Lewis (1971) pointed out that, for the low-density satellites, the combination of relatively large size and the presence of large quantities of a low melting point condensate permits some surprising first-order conclusions to be drawn with very few assumptions. He showed that decay of radionuclides in the silicate portion of an icy satellite with the size and density of Ganymede or Callisto would rapidly heat up the interior

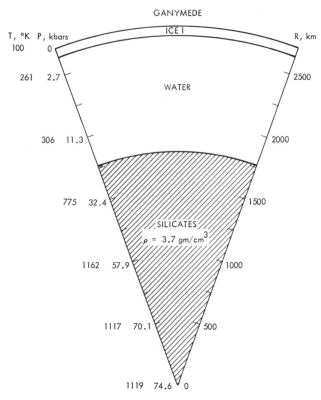

Figure 14 Model for Ganymede's present interior structure (from Consolmagno & Lewis 1976).

to the melting temperature of the ice. Thus, no matter what the initial conditions or structure, he suggests that large portions of the interiors of such satellites will be melted and that the higher density silicates will sink to form a "muddy" core. Although Lewis used chondritic radionuclide abundances for his calculation, the end result is insensitive to these details and, in fact, the only way to avoid melting is to postulate very low initial temperatures and silicates completely depleted in all radioactive elements. Any energy available from accretion or tidal dissipation only adds to the strength of these conclusions.

The consequences of Lewis' ideas for the current interior structure of the satellites have been investigated in more detail by Consolmagno & Lewis (1976) and Fanale, Johnson & Matson (1977a). These studies suggest that both Ganymede and Callisto should have extensive liquid water mantles and relatively thin ice crusts. Figure 14 shows a typical example of the structures predicted by Consolmagno & Lewis for icy satellites.

There are several interesting geophysical considerations suggested by these models. First, even a solid ice-rich body of satellite dimensions cannot sustain significant departures from a figure of hydrostatic equilibrium (Johnson & McGetchin 1973). If interior temperatures are at or near the melting temperature of ice it is thus virtually certain that the large icy satellites will have hydrostatic figures [which will be triaxial due to the combination of Jovian tide and rotational flattening—see O'Leary & van Flandern (1972)]. Second, if large liquid mantles exist, then any structures larger than the scale of the ice crusts (tens to hundreds of kilometers depending on the satellite and assumptions) must be isostatically supported, raising the possibility of continent-sized "icebergs." Third, surface temperatures are nearly half the melting temperature of ice, and temperatures will increase with depth to the point of transition to the liquid mantle. Under these conditions even small-scale structures, such as impact craters, may be modified by ice deformation or creep (see Johnson & McGetchin 1973). Finally, the presence of extensive liquid mantles and thin crusts in rotating bodies raises the interesting possibility that large-scale convection patterns will develop and that something analogous to plate tectonics may be important on Ganymede or Callisto.

The histories of the mostly rocky satellites, Io and Europa, are more difficult to determine. The absence of surface materials in any way similar to lunar, terrestrial, or meteoritic basalts or other igneous rocks indicates either that igneous materials never reached the surfaces of these satellites or that the surfaces were subsequently modified or covered by some other material. Fanale, Johnson & Matson (1974, 1977a) suggest that the possible presence of hydrated silicates in Io and Europa may be the key to understanding their evolution. They point out that, as with the icy satellites, modest heating can produce significant effects. Hydrated minerals will lose their water at temperatures of 600 K or greater. They suggest that radioactive heating or accretional energy could have defluidized much of Io's interior and resulted in the transport of salt-rich solutions to the surface where a layer of evaporite deposits would be produced following the escape of some water to space. On Europa, they suggest, this process may not have proceeded as far, leaving some significant layer of ice on the surface (it is not known whether the ice on Europa is millimeters or kilometers thick). Alternatively, the water on Europa may be the residue of the very final stages of accretion after temperatures had dropped to a point allowing condensation of ice at Europa's position.

CONCLUSION

I have tried in this article to convey some of the broad scope, interdisciplinary nature, and excitement of the rapidly developing field of satellite planetology. We have learned an impressive amount about Galileo's "four planets," which even under the best conditions are mere dots of light in our telescopes. The satellites are interesting individually as planets, ranging from an entirely new class of planet represented by ice-rich Ganymede and Callisto to Io and Europa, which are ostensibly similar to the Moon but have apparently evolved in significantly different directions. Taken together, as a system, the satellites exhibit systematic variations in density and composition that are being used as important tests of hypotheses concerning solar system formation.

As is generally the case, what we know about the satellites has raised many more, as yet unanswered, questions. We do not know what the satellite surfaces look like, whether there are craters, mountains, rifts, or new and unexpected tectonic features. We do not understand the composition of their atmospheres or their interactions with their magnetospheric environment in any detail. The answers to these, and many other questions, await detailed examination of the system by spacecraft following the truly pioneering flights of Pioneers 10 and 11, supported by continuing ground-based observations and theoretical studies. The two Voyager spacecraft, scheduled to arrive at Jupiter in 1979, will provide our first close-up study of the satellites. The cameras on board these spacecraft will return detailed, multicolor images of the satellite surfaces, with resolutions of a few kilometers. These pictures should establish the major morphological forms, provide evidence for tectonic activity, and map the variations in colour across the surfaces. Ultraviolet, visible, and infrared spectral measurements will be utilized to probe the composition of the atmospheres and study the composition of surfaces at high resolution. Finally, Voyager will build on the magnetospheric information obtained by Pioneer, providing a more complete picture of the low-energy-charged particles and plasma characteristics and directly investigating the region of Io-magnetosphere interaction (the "flux tube"). Hopefully, the next generation of orbiting spacecraft will allow even more detailed exploration of the satellites' surfaces and perform geophysical measurements of magnetic and gravity fields impossible from a flyby.

ACKNOWLEDGMENTS

In this article, a combination of review and instruction intended primarily for scientists in other fields, I have been unable to mention every piece of important satellite research or even everyone who has worked on a particular topic. I apologize to those who may have been inadvertently left out and to those whose work I have not heard of yet—it is a rapidly moving field, send me a preprint! This work represents one phase of research carried out at the Jet Propulsion Laboratory of the California Institute of Technology under NASA Grant NAS 7–100.

Literature Cited

Adams, J. B. 1974. Uniqueness of visible and near infrared diffuse reflectance spectra of pyroxenes and other rock forming minerals. In *Infrared and Raman Spectroscopy of Lunar and Terrestrial Minerals*, ed. C. Karr. New York: Academic. 375 pp.

Alfven, H. 1957. On the theory of comet tails. *Tellus* 9: 92–96

Allen, C. W. ed. 1973. *Astrophysical Quantities*, p. 142. London: Univ. London, Athlone Press. 310 pp. 3rd ed.

Anderson, J. D., Null, G. W., Wong, S. K. 1974. Gravitational parameters of the Jupiter system from the doppler tracking of Pioneer 10. *Science* 183: 322–23

Aumann, H. H., Gillespie, C. M., Jr., Low, F. J. 1969. The internal powers and effective temperatures of Jupiter and Saturn. *Ap. J.* 157: L69–72

Bergstralh, J. T., Matson, D. L., Johnson, T. V. 1975. Sodium D-line emission from Io: synoptic observations from Table Mountain Observatory. *Ap. J.* 195: L131–35

Brouwer, D., Clemence, G. M. 1961. Orbits and masses of planets and satellites. In *Planets and Satellites*, ed. G. P. Kuiper, B. M. Middlehurst, pp. 31–94. Chicago: Univ. Chicago Press

Brown, R. A. 1974. Optical line emission from Io. In *Exploration of the Planetary System*, ed. A. Woszczyk, C. Iwaniszewski, pp. 527–31. Dordrecht, Holland: Reidel

Brown, R. A. 1976. A model of Jupiter's sulfur nebula. *Ap. J.* 206: L179

Brown, R. A., Yung, Y. L. 1976. Io, its atmosphere and optical emissions. In *Jupiter*, ed. T. Gehrels, pp. 1102–45. Tucson: Univ. Arizona Press. 1254 pp.

Brown, W. L., Lanzerotti, L. J., Poate, J. M., Augustyniak, W. M. 1977. Sputtering of ice by mev light ions and relevance to planetary astronomy. *EOS* 58: 423

Cameron, A. G. W., Pollack, J. B. 1976. On the origin of the solar system and of Jupiter and its satellites. In *Jupiter*, ed. T. Gehrels, pp. 58–84. Tucson: Univ. Arizona Press. 1254 pp.

Campbell, D. B., Chandler, J. F., Pettengill, G. H., Shapiro, I. I. 1977. Galilean Satellites of Jupiter: 12.6 cm radar observations. *Science* 196: 650–53

Carlson, R. W. 1976. Atmospheres of the outer planets' satellites. In *Exploration of the Solar System—AIAA Progress in Astronautics and Aeronautics* ed. E. W. Greenstadt, M. Dryer, D. S. Intriligator, Vol. 50. Cambridge: MIT Press

Carlson, R. W., Bhattacharyya, J. C., Smith, B. A., Johnson, T. V., Hidayat, B, Smith, S. A., Taylor, G. E., O'Leary, B. T., Brinkmann, R. T. 1973. An atmosphere on Ganymede from its occultation of SAO-186800 on 7 June 1972. *Science* 182: 53–55

Carlson, R. W., Judge, D. L. 1974. Pioneer 10 ultraviolet photometer observation at Jupiter encounter. *J. Geophys. Res.* 79: 3623–33

Carlson, R. W., Judge, D. L. 1975. Pioneer 10 UV photometer observations of the Jovian hydrogen torus: the angular distribution. *Icarus* 24: 395–99

Carlson, R. W., Judge, D. L. 1976. *The problem of hydrogen at Io*. Presented at the Ann. Meet. Div. Planet. Sci., 7th, Austin, TX.

Carlson, R. W., Matson, D. L., Johnson, T. V. 1975. Electron impact ionization of Io's sodium emission cloud. *Geophys. Res. Lett.* 2: 469–72

Carlson, R. W., Matson, D. L., Johnson, T. V., Bergstralh, J. T. 1977. Sodium D-line emission from Io: comparison of observed and theoretical line profiles. *Ap. J.* In press.

Cloutier, P. A., Daniell, R. E., Dessler, A. J., Hill, T. W. 1977. A cometary ionosphere model for Io. *Astrophys. Space Sci*, Submitted for publication

Consolmagno, G. J., Lewis, J. S. 1976. Structural and thermal models of icy Galilean satellites. In *Jupiter*, ed. T. Gehrels. pp. 1035–51. Tucson: Univ. Arizona Press. 1254 pp.

Cruikshank, D. P., Jones, T. J., Pilcher, C. B. 1977. *Absorptions in the spectrum of Io, 3.0–4.2 microns*. Presented at Ann. Meet. Div. Planet. Sci., 8th, Honolulu.

Cutts, J. 1971. Martian spectral reflectivity properties from Mariner 7 observations. Ph.D. thesis. Calif Inst. Technol. Pasadena. 92 pp.

de Sitter, W. 1931. Jupiter's Galilean satellites. *MNRAS*. 91: 706–38

Dollfus, A. 1961. Visual and photographic

studies of the planets at the Pic du Midi. In *Planets and Satellites*, ed. G. P. Kuiper, B. M. Middlehurst, pp. 534–71. Chicago: Univ. Chicago Press.

Dollfus, A. 1975. Optical polarimetry of the Galilean satellites of Jupiter. *Icarus* 25: 416–31

Fanale, F. P., Johnson, T. V., Matson, D. L. 1974. Io: a surface evaporite deposit? *Science* 186: 922–24

Fanale, F. P., Johnson, T. V., Matson, D. L. 1977a. Io's surface and the histories of the Galilean satellites. In *Planetary Satellites*, ed. J. A. Burns, pp. 379–405. Tucson: Univ. Arizona Press, 598 pp.

Fanale, F. P., Johnson, T. V., Matson, D. L. 1977b. Io's surface composition: observational constraints and theoretical considerations. *Geophys. Res. Lett.* 4: 303–6

Fink, U., Dekkers, N. H., Larson, H. P. 1973. Infrared spectra of the Galilean satellites of Jupiter. *Ap. J.* 179: L155–5

Fink, U., Larson, H. P. 1975. Temperature dependence of the water ice spectrum between 1 and 4 microns: application to Europa, Ganymede and Saturn's rings. *Icarus* 24: 411–20

Galilei, G. 1610. *Sidereus Nuncius*. Venice, Italy

Gehrels, T. 1977. Picture of Ganymede. See Fanale, Johnson & Matson 1977a, pp. 406–11

Goldstein, R. M., Morris, G. A. 1975. Ganymede: Observations by radar. *Science* 188: 1211–12

Grossman, A. S., Graboske, H. C., Pollack, J. B., Reynolds, R. T., Summers, A. 1972. An evolutionary calculation of Jupiter. *Phys. Earth Planet. Inter.* 6: 91–98

Gurnett, D. A. 1972. Sheath effects and related charged-particle acceleration by Jupiter's satellite Io. *Ap. J.* 175: 525–33

Harris, A. W., Kaula, W. M. 1975. A co-accretional model of satellite formation. *Icarus* 24: 516–24

Hubbard, R. F., Shawhan, S. D., Joyce, G. 1974. Io as an emitter of 100 keV electrons. *J. Geophys. Res.* 79: 920–28

Hubbard, W. 1977. The Jovian surface condition and cooling rate. Preprint

Johnson, T. V. 1971. Galilean satellites: narrowband photometry 0.30 to 1.10 microns. *Icarus* 14: 94–111

Johnson, T. V., Matson, D. L., Carlson, R. W. 1976. Io's atmosphere and ionosphere: new limits on surface pressure from plasma models. *Geophys. Res. Lett.* 3: 293–96

Johnson, T. V., McGetchin, T. R. 1973. Topography on satellite surfaces and the shape of asteroids. *Icarus* 18: 612–20

Johnson, T. V., Pilcher, C. B. 1977. Satellite spectrophotometry and surface compositions. See Fanale, Johnson & Matson 1977a, pp. 232–68

Judge, D. L., Carlson, R. W., Wu, F. M., Hartmann, U. G. 1976. Pioneer 10 and 11 ultraviolet photometer observations of the Jovian satellites. See Brown & Yung 1976, pp. 1068–1101

Kliore, A., Cain, D. L., Fjeldbo, G., Seidel, B. L., Rasool, S. I. 1974. Preliminary results on the atmosphere of Io and Jupiter from Pioneer 10 S-band occultation experiment. *Science* 183: 324–29

Kliore, A. J., Fjeldbo, G., Seidel, B. L., Sweetnam, D. N., Sesplaukis, T. T., Woiceshyn, P. M., Rasool, S. I. 1975. Atmosphere of Io from Pioneer 10 radio occultation measurements. *Icarus* 24: 407–10

Kuiper, G. P., ed. 1952. *The atmospheres of the Earth and planets*. Chicago: Univ. Chicago Press

Kupo, I., Mekler, Y., Eviatar, A. 1976. Detection of ionized sulfur in the Jovian magnetosphere. *Ap. J.* 205: L51–3

Lebofsky, L. 1977. Callisto: evidence for water frost. *Nature* Submitted for publication

Lewis, J. S. 1971. Satellites of the outer planets: thermal models. *Science* 172: 1127–28

Lewis, J. S. 1974. The chemistry of the solar system. *Sci. Am.* 230 (3): 50–65

Matson, D. L., Goldberg, B. A., Johnson, T. V., Carlson, R. W. 1977. Images of Io's sodium cloud. *Science* In press

Matson, D. L., Johnson, T. V., Fanale, F. P. 1974. Sodium D-line emission from Io: sputtering and resonant scattering hypothesis. *Ap. J.* 192: L43–46

McDonough, T. R., Brice, N. M. 1973. New kind of ring around Saturn? *Nature* 242: 513

McElroy, M. B., Yung, Y. L. 1975. The atmosphere and ionosphere of Io. *Ap. J.* 196: 227–50

McElroy, M. B., Yung, Y. L., Brown, R. A. 1974. Sodium emission from Io: implications. *Ap. J.* 187: L127–30

Minton, R. B. 1973. The polar caps of Io. *Comm. Lunar Planet. Lab.* 10: 35–39

Morrison, D. 1977. Radiometry of satellites and of the Rings of Saturn. See Fanale, Johnson & Matson 1977a, pp. 269–301

Morrison, D., Cruikshank, D. P. 1974. Physical properties of the natural satellites. *Space Sci. Rev.* 15: 641–739

Morrison, D., Cruikshank, D. P., Burns, J. A. 1977. Introducing the satellites. See Fanale, Johnson & Matson 1977a, pp. 3–17

Morrison, D., Jones, T. J., Cruikshank, D.

P., Murphy, R. E. 1975. The two faces of Iapetus. *Icarus* 24:157–71

Morrison, D., Morrison, N. D. 1977. Photometry of the Galilean satellites. See Fanale, Johnson & Matson 1977a, pp. 363–78

Münch, G., Bergstralh, J. T. 1976. Sodium D-line emission from Io: spatial brightness distribution from multislit spectra. *PASP* 89:232–37

Murray, J. B. 1975. New observations of surface markings on Jupiter's satellites. *Icarus* 25:397–404

Nash, D. B., Fanale, F. P. 1977. Io: surface composition model based on reflectance spectra of sulfur/salt mixtures and proton irradiation experiments. *Icarus* 31:40–80

O'Leary, B. T., van Flandern, T. C. 1972. Io's triaxial figure. *Icarus* 17:209–15

Peale, S. J. 1977. Rotation histories of the natural satellites. See Fanale, Johnson & Matson 1977a, pp. 113–56

Pilcher, C. B. Ridgeway, S. T., McCord, T. B. 1972. Galilean satellites: identification of water frost. *Science* 178:1087–89

Pollack, J. B., Reynolds, R. T. 1974. Implications of Jupiter's early contraction history for the composition of the Galilean satellites. *Icarus* 21:248–53

Sampson, R. A. 1921. Theory of the four great satellites of Jupiter. *Mem. R. Astron. Soc.* 63:1–270

Siscoe, G. L. 1978. Toward a comparative theory of magnetospheres. *Solar System Plasma Physics—A 20th Anniversary Review*, ed. C. F. Kennel, L. J. Lanzerotti, E. N. Parker. Amsterdam: North Holland

Stebbins, J., Jacobsen, T. S. 1928. Further photometric measures of Jupiter's satellites and Uranus, with tests for the solar constant. *Lick Obs. Bull.* 13:180–95

Taylor, G. E., O'Leary, B. T., van Flandern, T. C., Bartholdi, P., Owen, T., Hubbard, W. B., Smith, B. A., Smith, S. A., Fallon, F. W., Devinney, E. J., Oliver, J. 1971. Occultation of Beta Scorpii C by Io on May 14, 1971. *Nature* 234:405–6

Trafton, L., 1975a. High resolution spectra of Io's sodium emission. *Ap. J.* 202:L107–12

Trafton, L., 1975b. Detection of a potassium cloud near Io. *Nature* 258:790–92

Trafton, L. 1976. A search for emission features in Io's extended cloud. *Icarus* 27:429–37

Trafton, L., Macy, W. 1975. An oscillating asymmetry to Io's sodium emission cloud. *Ap. J.* 190:L85–89

Trafton, L., Parkinson, T., Macy, W., Jr. 1974. The spatial extent of sodium emission around Io. *Ap. J.* 190:L85–89

Veverka, J. 1977a. Photometry of satellite surfaces. See Fanale, Johnson & Matson 1977a, pp. 171–209

Veverka, J. 1977b. Polarimetry of satellite surfaces. See Fanale, Johnson & Matson 1977a, pp. 201–231

Wamsteker, W., Kroes, R. L., Fountain, J. A. 1974. On the surface composition of Io. *Icarus* 23:417–24

Watson, K., Murray, B. C., Brown, H. 1963. The stability of volatiles in the solar system. *Icarus* 1:317–27

Whitten, R. C., Reynolds, R. T., Michelson, P. F. 1975. The ionosphere and atmosphere of Io. *Geophys. Res. Lett.* 2:49–51

Yung, Y. L., McElroy, M. B. 1976. Stability of an oxygen atmosphere of Ganymede. *Icarus* 30:97

Ann. Rev. Earth Planet. Sci. 1978. 6: 127–43

EFFECTS OF WASTE DISPOSAL OPERATIONS IN ESTUARIES AND THE COASTAL OCEAN

×10089

M. Grant Gross

Chesapeake Bay Institute, The Johns Hopkins University, Baltimore, Maryland 21218

INTRODUCTION

Since about 1800, human activities have significantly altered shorelines, bottom topography, sediment characteristics, and marine life in estuarine and coastal waters. The first man-induced changes of the ocean were restricted to nearshore waters and most were small, scattered, and primarily associated with food production or port development (Klimm 1956). Sediment deposition caused by erosion of agricultural lands resulted in extensive delta building and shoreline changes in the Persian Gulf, the Adriatic Sea, and the Mississippi Delta, to name a few examples (Davis 1956). Diking and draining of wetlands, shallow ocean areas, and lakes to form agricultural land has greatly altered shorelines in the Netherlands and England (Davis 1956).

Not all wastes are placed directly in the ocean. Some are brought there by normal sediment transport processes. For example, mining is a prolific sediment producer and has also caused extensive changes in wetlands and shorelines due to downstream sediment deposition. Perhaps the best studied case is the hydraulic mining of gold in California's Sierra Nevada (Gilbert 1917). Between 1850 and 1914, 1.8×10^9 m^3 of debris was mobilized by mining and erosion in the San Francisco Bay drainage system. About 1.1×10^9 m^3 was deposited in the bay system or on wetlands (Gilbert 1917). Movement and deposition of this material apparently persisted for approximately 50 years after cessation of mining (Smith 1965).

More recently, rapid growth of coastal cities and associated industry has led to greatly increased construction, demolition, and dredging, and the disposal of wastes produced from these activities has emerged as a geologic process causing significant changes in coastal areas.

Because waste solids have caused the most obvious geologic changes, this review deals primarily with these materials. Effects of dissolved wastes such as nutrients have been extensively discussed elsewhere (NAS 1969, Likens 1972).

127

0084-6597/78/0515-0127$01.00

Increasing use of the coastal ocean for disposal of waste solids is causing changes in the seafloor topography and sediment deposits tens of kilometers at sea on the continental shelf. In fact, waste disposal operations have outstripped rivers and littoral drift as sediment transport agents in many urbanized coasts (Gross 1972). Despite the widespread occurrence of waste disposal in the ocean, scientific studies of waste solids have only recently been undertaken. Most of the existing data come from a few heavily used regions, such as the New York Bight and the continental shelf between New Jersey and Long Island, New York on the U.S. Atlantic coast (Gross 1976).

ESTUARIES AND THE COASTAL OCEAN

To understand the effect of waste disposal on the ocean waters and the sea floor, it is necessary to understand the basic circulation process and other processes that transport wastes in the ocean.

Estuaries are semi-enclosed basins where fresh water from the land mixes with salt water from the ocean (Pritchard 1955). This mixing process creates a circulation pattern that tends to retain wastes, especially waste solids, in the estuary. This pattern can be established in a typical estuary by averaging water movements over many tidal cycles. There is a net seaward flow of less saline waters near the surface and a deeper net flow of more saline waters from the ocean toward the head of the estuary. There is also a net flow from the deeper layers to the surface layers which mixes salt with the fresh waters (Pritchard 1955).

The coastal ocean is the relatively shallow area that lies over the continental shelf. Freshwater runoff from land and local winds dominate the movements and mixing of nearshore waters, generally causing the currents to parallel the shoreline. Where the shelf is narrow, coastal ocean currents may extend beyond the edge of the shelf. Where the shelf is wide, coastal currents may occur only over the part of the shelf nearest the shore. Coastal water movements, being driven by winds and river discharges, vary markedly in space and time and increase the difficulty of predicting the behavior of wastes released in nearshore waters. Tidal currents are usually the strongest currents near the shore; they typically parallel the coast and may be either oscillatory or rotary in character. In general, waste materials introduced into the nearshore zone are transported away from the point of discharge by currents parallel to the shoreline. Mixing across the shelf is slower and the processes involved are poorly known.

The estuarine circulation pattern also prevails in most of the coastal ocean where the amount of fresh water from rain, river runoff, and sewers exceeds local evaporative loss. Thus, over the continental shelf, surface waters generally have a net motion seaward, while deep waters move toward the shore as well as along the coast. The net effect of estuarine circulation is recirculation of particulate and biologically active constituents (such as nutrients) in the coastal ocean to the extent that mixing with open ocean waters beyond the edge of the continental shelf is inhibited. Sediments are also trapped and may be moved toward the shoreline under certain circumstances (Meade 1969).

Processes controlling movements of bottom deposits are poorly known. There is strong evidence that most sediment movement on the continental shelf occurs during relatively infrequent storms (Swift et al 1976). Presumably waste solids would behave in much the same way as ordinary bottom sediments. The less dense constituents of wastes are likely to be more readily put into motion and to move farther than the denser constituents.

TYPES AND SOURCES OF WASTE SOLIDS

It will be useful to describe briefly some types of waste involved in disposal operations, though many are poorly characterized and highly variable in their properties (Table 1).

Dredged wastes are solids removed from waterways, generally to improve navigation. Typically they consist of sand, silt, or clay mixed with wastes discharged by industrial plants or municipal sewage treatment facilities and with locally produced organic matter, in the case of maintenance dredging. Waste from new construction normally involves uncontaminated rock, soil, and sediment.

Sewage sludges are slurries of solids removed from sewage during waste-water treatment and usually containing mixtures of solids from human wastes, street runoff, eroded soils, and industrial wastes.

Table 1 Some common urban wastes, their sources, composition, and usual disposal areas (after Gross 1972, 1976)

Wastes	Sources	Major constitutents	Minor constitutents	Disposal area
Municipal	Domestic, industrial	Paper and wood (50%) Food wastes (12%)	Glass, stones (10%) Metals (8%)	Landfill Incineration
Dredged wastes	Harbor and channel construction and maintenance	Sand, shell, gravel River sediment	Sewage solids Industrial wastes	Ocean
Rubble	Construction and demolition	Stone, concrete, steel		Landfill Ocean
Sewage solids	Municipal sewage systems and treatment plants	Organic matter (50%) Alumino-silicates (50%)	Industrial wastes	Ocean Harbor
Coal ash	Coal combustion, primarily power generation	Quartz, mullite	—	Ocean
Fermentation wastes	Pharmaceutical industry	Organic matter	—	Ocean
Waste acids	Metal processing	—	—	Ocean
Waste alkalis	Petrochemical industry	—	—	Ocean
Steel-making slag	Steel production	FeO (25%) CaO (40%) SiO_2 (17%)	MgO (9%) MnO (4%) Al_2O_3 (1%)	Landfill Harbor fill

Industrial wastes are mill and industrial process wastes, such as the extracted ores and waste acids from titanium dioxide production or the alkaline wastes from aluminum production. Other wastes are mostly liquid process wastes; wastes from chemical and pharmaceutical manufacturing and residues from petroleum refining and petrochemical processing.

Construction and demolition wastes consist of soil, masonry, tile, stone, plumbing, glass, tar, plaster, and other construction and demolition debris.

Chemical wastes are mostly liquid process wastes; many are known or suspected to be toxic to organisms or humans. These include wastes from chemical manufacturing and residues from petroleum refining and petrochemical processing.

Dredging and Disposal Operations

Aside from large construction projects or land reclamation activities, dredging and disposal of dredged materials are among the most important activities affecting estuaries and the coastal ocean in the United States (see, for example, Table 2 for data on waste disposal in the Middle Atlantic region). Both dredging and disposal involve movements of large volumes of materials (Boyd et al 1972). In the United States, about 2×10^8 m^3 are dredged each year from existing channels (maintenance dredging) and about 6×10^7 m^3 are dredged during construction of new facilities (new work) at an annual cost of $150,000,000 (1970 dollars).

Most of the materials dredged (about 1.90×10^8 $m^3 yr^{-1}$) are discharged to open water sites in bays, estuaries, rivers, and lakes. Nearly equal amounts are discharged in freshwater (primarily rivers) and marine sites. Maintenance dredging (dredging to remove deposits from navigation facilities) accounts for 80% of the dredging. Half of the annual maintenance dredging is done in the Gulf of Mexico with the Southern Atlantic region of the U.S. and Delaware Bay also contributing significant quantities (Boyd et al 1972).

Sediment moved during maintenance dredging operations is poorly characterized. About half (1.2×10^8 m^3) is mixed silt and sand. Another 6×10^7 $m^3 yr^{-1}$ is finer grained mud, clay, and silt while 4×10^7 $m^3 yr^{-1}$ is coarse grained sand, gravel, and shell (Boyd et al 1972). While it is even more difficult to determine the concentration of pollutants in the dredged materials, about one third of the materials moved during maintenance dredging were considered polluted (Boyd et al 1972). Discharge of

Table 2 Wastes discharged to the U.S. Middle Atlantic Continental Shelf, 1974 [Data from EPA (1975) for Regions II and III]

Type of Waste	Volume (10^6 m^3)	(%)	Solids (10^6 t)	(%)
Dredged wastes	9.9	47.0	5.0	73.6
Sewage sludges	6.1	29.0	0.3	4.4
Rubble	0.6	3.0	1.2	17.6
TiO$_2$ wastes	3.1	14.8	0.3	4.4
Other chemical	1.3	6.2	?	—

Table 3   Sediment yield from various areas in Maryland (Wolman & Schick 1967)

Areas or activities (number of observations)	Sediment Yield ($t\,km^2\,yr^{-1}$)	
	Range	Median value
Wooded areas (20)	5.3–73	50
Rural areas (8)	82–320	180
Strip mine (1)	1,040	1,000
Urban and development (4)	37.5–820	600
Housing and industrial construction (6)	2,000–28,000	10,000

materials dredged from heavily polluted areas like New York Harbor is a major source of sediment, metals, and nutrients to the coastal ocean (Mueller, Anderson & Jeris 1976).

Major sources of the materials removed during dredging in urban areas are the natural sediment load of rivers and littoral drift. Part of this sediment load is polluted because of waste discharges from upriver cities or runoff from agriculture or mine drainage. Some sediment, such as beach sand, is initially clean but mixes with other wastes when deposited in industrial harbors or in dredged navigation channels. Such deposits are eventually incorporated into the urban waste stream and require disposal (Gross 1972).

Agriculture and mining are major sediment contributors to rivers everywhere (Kenahan 1971). But in urban areas the principal sediment source is erosion of construction sites left bare during construction (Wolman 1967, Wolman & Schick 1967). As Table 3 indicates, erosion of construction sites yields 10 to 100 times more sediment per unit area than either mining or agriculture.

Sediment yield to U.S. rivers from agriculture has decreased markedly from its peak in 1900–20 owing to a reduction in land areas farmed and to better conservation practices. For example, sediment yield from an area in the Southern Piedmont (South Carolina, Alabama, Georgia) decreased from about 200 t km^{-2} yr^{-1} in 1910–34 to about 30 t km^{-2} yr^{-1} in 1967–72 (Meade & Trimble 1974).

Despite the present decreased sediment yield to rivers, soils eroded from farm lands in past decades no doubt reach urban estuaries today, especially during floods. Only about 5% of the soil eroded from upland slopes since European settlement has reached the ocean (Trimble 1975). The remainder is deposited temporarily in stream channel and bank sediments (Wolman 1967, Wolman & Schick 1967) or impounded by dams (Meade & Trimble 1974). These deposits are scoured during high river flow and carried downstream, and some of this sediment load is deposited in urbanized estuaries. For example, the Susquehanna River transported an estimated 30×10^6 t of sediment during the floods associated with tropical storm Agnes in 1972 (Schubel 1974), whereas the river normally transports 0.5 to 1.0×10^6 t yr^{-1} (Schubel 1968, 1972, Gross et al 1977).

Erosion accompanying conversion of croplands or woodlands to housing developments is another major sediment source in urban areas (Wolman 1967). The sediment eroded from a small construction site can equal 28,000 t km^{-2} yr^{-1} (Wolman & Schick 1967). For the Baltimore-Washington metropolitan area, Wolman & Schick (1967) estimated that 630–1600 t of sediment are mobilized per 1000 increase in population. Assuming comparable erosion in the New York-New Jersey Metropolitan Region we can estimate that as much as 2×10^5 t of sediment were mobilized by the increase in 200,000 persons between 1971 and 1972 as approximately 75 km^2 of agricultural land were converted to suburbs (Tri-State Regional Planning Commission 1973). Such large sediment yields cause extensive alteration of stream channels because of sediment deposition.

Another natural source of solids is littoral drift along ocean beaches. Littoral drift along beaches may have been reduced slightly by human intervention through seawalls and jetties (Caldwell 1966, Yasso & Hartman 1975) or because of dredging and removal of sand deposited in inlets. In areas where beaches are restored by adding sand (beach replenishment), there may be a slight increase in littoral drift but the amounts of sand involved usually are small.

Municipal Sewage

Discharge of municipal sewage contributes large volumes of waste solids to coastal and estuarine waters. Discharge of untreated sewage has been the common practice for many areas, and sewage treatment plants themselves discharge large amounts of waste—sludges and liquid effluents—into harbors and coastal areas. The plants treat sewage (see Table 4), so that the effluents contain suspended solids, typically 50 ppm. In addition, the nutrients (nitrogen compounds, phosphates) in the sewage discharges stimulate the growth of phytoplankton. When these minute plants die, they contribute an unknown amount of organic matter to the sediment accumulating in harbors and adjacent waterways.

Sewage treatment plants also produce, thick semiliquid slurries (sludges) removed from sewage before the plant effluents are discharged. Sewage sludges constitute only a small part of the total waste load from most coastal metropolitan regions but they have been investigated intensively because of their significance for both recreational activities and marine resources (Gross 1970a).

Table 4 Municipal wastewater treatment (American Chemical Society 1969)

Primary treatment level
  Particulate matter settles out from raw sewage through sedimentation; removes half to two thirds of suspended solids.
Secondary treatment level
  Aerobic biological process takes organic matter from primary effluents; removes about 90% of suspended solids and oxygen-demanding substances.
Tertiary treatment level
  Physical or chemical treatment of secondary effluents removes residual organics, nutrients (nitrogen, phosphorus), chlorine, colors. Typical processes include: lime addition to remove phosphorous, filtering to remove solids; activated charcoal treatment to remove organic matter; disinfection.

Urban Storm-water Runoff

In urbanized areas precipitation (rainfall, snow) commonly flows into sewage systems together with domestic sewage, since a large portion of urban land is covered by streets and buildings, reducing infiltration of rain and snow into the soil. During a storm, runoff combined with domestic sewage generally exceeds the capacity of the sewage treatment plants. The mixture of runoff and sewage then flows untreated into the nearest waterway. From 40% to 80% of the total precipitation in the New York Metropolitan Region flows through sewers to waterways (Federal Water Pollution Control Administration 1969).

Although rainwater is initially relatively pure (Table 5), it picks up pollutants from the urban atmosphere and streets. Among the many sources of contaminants are: vehicular wastes, including oil and grease; atmospheric fallout; combustion wastes (incinerator fly ash); animal wastes; sewage deposits from collection systems; and plant debris. Thus street runoff is far from clean (Weibel 1969, Southern California Water Research Project 1973).

Using data on estimated sewage overflows in the Hudson River area (Federal Water Pollution Control Administration 1969) and on typical urban runoff from low-density areas, we can estimate, for example, that the sewered portion (3,000 km^2 or 1,200 mi^2) of the New York urban region discharges street runoff into the Bight at an average rate of 390 m^3 sec^{-1} or 9,000 million gallons per day. While this is only a crude estimate (Gross 1976), it suggests that urban storm-water runoff in the region exceeds the sewage system's dry weather flow of 114 m^3 sec^{-1} or 2,500 million gallons per day (Tri-State Regional Planning Commission, unpublished data). Using the composition of the urban storm-water runoff (Weibel 1969), we can calculate that the solids discharged with storm-water from the New York Urban Region could be as much as 18 million metric tons per year. In general runoff waters are

Table 5 Runoff constituents and concentrations (after Weibel 1969, Biggar & Corey 1969)

Constituent	Average Concentrations (mg/l)			
	Rainfall[a]	Untreated sewage[b]	Urban stormwater runoff[c]	Rural land runoff[d]
Suspended solids	13	200	227	310
Chemical oxygen demand	16	350	111	—
Total nitrogen, as N[e]	1.3	40	3.1	9
Inorganic nitrogen, as N[f]	0.7	30	1.0	5
Total phosphate[g]	0.08	10	0.4	0.6

[a]Measured in Cincinnati, Ohio, August and December 1963 (Weibel 1969).
[b]Generalized composition of domestic sewage (Weibel 1969).
[c]Residential-light commercial section of Cincinnati, 27 acres (11 hectares) (Weibel 1969).
[d]After Biggar & Corey (1969).
[e]Total of four forms of nitrogen.
[f]Total of ammonia, nitrite, and nitrate.
[g]Total acid-hydrolyzable phosphate.

discharged into local waterways, and whether any significant fraction of the entrained solids reach the Bight in an average year—except through dredging of navigation channels—is questionable. During major storms and floods, however, deposits of wastes in estuaries, rivers, and in navigation channels can be eroded and carried out of the estuary to the ocean.

Industrial Wastes

Industries handling bulky materials are commonly located in coastal areas because of the relative ease and low cost of transporting large volumes of raw materials and handling finished products. The large volume of waste generated by these industries is traditionally discharged in nearby coastal sites or coastal ocean areas, again to minimize transportation costs. Often alternative disposal sites are expensive and difficult to acquire.

The steel industry is one that produces large amounts of wastes. For instance, a large steel mill manufacturing about 6×10^6 t of steel per year produces about 2×10^6 t of steel-making slag. Typically 40 to 80% of the slag produced is recycled at the plant or sold for other uses, leaving the remainder (1.5×10^5 to 6×10^5 m^3 yr^{-1}) to go into landfill or coastal or estuarine waters. In waters 5 m deep, an area of about 9 hectares (22 acres) per year is required to accommodate the waste produced.

Coal was a major fuel in coastal cities between 1850 and 1960 (Borchert 1967), and coal ash and cinder are widespread in sediment deposits near cities. The ash content of typical U.S. coals is between 5 and 15% (Abernathy, Peterson & Gibson 1969). Coal ash from the New York area (Table 6) had median grain diameter of about 20 μm and a grain density of 2.2 g cm^{-3}; typical settling velocities were around 0.015 cm sec^{-1} (Gross 1972). Chemical composition of coal ash resembled shales and sandstone (Bowen 1966), except for slight enrichment in titanium, potassium,

Table 6   Some physical and chemical properties of waste solids, New York Metropolitan Region (after Gross 1972)

	LOI[a] (%)	Carbon concentrations (%)			Grain density (g cm^{-3})	Solid content (g liter^{-1})
		Total	Carbonate	Oxidizable		
Sewage Sludges (6)	50.6	29.5	1.64	22.7	1.81	56.2
Duck sludge,						
Long Island, N.Y.	11.8	6.1	0.31	2.9	2.64	n.d.[b]
Fermentation wastes	50.7	21.9	2.51	16.7	1.49	31.9
Coal ash	n.d.	6.0	0	3.6	2.2	700.0
Sewage sludge ash	n.d.	1.03	n.d.	n.d.	2.71	n.a.[c]
Waste alkali	n.d.	2.6[d]	n.d.	1.0[d]	n.d.	120.0
Harbor sediments						
Sands	1	0.3	0.1	0.2	2.65	750.0
Silts	11	5.5	0.3	5.2	2.53	500.0

[a]LOI: Loss on Ignition.
[b]n.d.: not determined.
[c]n.a.: not applicable.
[d]Measured on liquid samples. (all other values on a dry weight basis)

cobalt, and lead. The ash consisted of a mixture of quartz and hollow spheres of mullite, a mineral formed during combustion and not found in normal marine sediments.

Of the estimated 3.0×10^8 tons of "fly ash" produced in the United States between 1945 and 1970, only about 3% was utilized for any purpose (Capp & Spencer 1970). Coal ash was dumped in the New York Bight for many years (Gross 1972). The amount averaged about 0.1 million tonnes a year between 1960 and 1968, ranging from 0.46 in 1965 to 0.18 in 1963. Because of substantial reduction in coal use in the New York metropolitan region, and increased use of coal ash for various purposes (Environmental Science and Technology 1970), little coal ash was dumped in the New York Bight in the 1970s. Because of high sulfur contents, and attendant air pollution problems, coal is now used primarily for steam electric power generation where SO_2 in stack gases can be removed by scrubbers. However, the scrubbing processes also produce large volumes of wastes, some of which will possibly go to marine disposal sites in the future. And as coal replaces oil as a major fuel for steam electric power generation and other uses, coal ash disposal will become an increasingly important problem in coastal cities. Some ash will likely go into estuaries and the coastal ocean.

Acid-iron waste from titanium-pigment production is another large-volume industrial waste (Peschiera & Freiherr 1968) consisting of suspended titanium ore residues in dilute iron-rich sulfuric acid. This material is dumped at sea in the New York Bight (Redfield & Walford 1951) and at other locations. These wastes have a solid content of 10% by weight of the waste liquid, based on the analysis of a single sample (Table 6). Except for unusually high concentrations of Fe and Ti, titanium pigment wastes are chemically quite similar to average shales. The Fe and Ti in the wastes may locally alter the chemical composition of suspended particles enough so that they can serve as tracers for particle movements (Biscaye & Olsen 1976).

Reduction of bauxite to alumina also produces a large volume of alkaline wastes, called "red mud" (Wilson & Blackman 1974). These extracted ores comprise approximately half the amount of ore produced for aluminum production. Such wastes are commonly discharged in coastal areas producing bauxite.

Construction and Demolition Debris

Waste from construction of new buildings and demolition of old ones has been dumped in oceans, harbors, or wetlands when no other disposal sites were available. When large landfill projects are underway, these wastes are often used as fill materials.

The diverse origins and heterogeneous composition of the wastes make it impossible to identify them with confidence, but the available data (Pararas-Carayannis 1973) indicate that they consist principally of excavated earth and rock, broken concrete, rubble, and other nonfloatable debris.

Waste Deposits

Identification and mapping of waste deposits usually involve two approaches. The simplest is to identify artifacts or physical properties. For instance, tomato seeds or watermelon seed, which pass through sewage treatment plants little altered, are used to identify sewage sludge deposits. Human artifacts, such as cigarette filters, bandages, sanitary napkins, and prophylactics are removed at the treatment plant but remain in the sludge dumped at waste disposal sites and accumulate in those deposits (Pearce 1972).

The second approach is to compare the chemical composition of wastes (e.g. dredged sediment and sewage sludges) with that of natural substances likely to accumulate in estuaries or on continental shelves. Such analyses indicate that physical properties, such as grain size, and the abundance of certain chemical constituents from industrial or domestic wastes, such as carbon and metals like lead and silver, are useful tags for some waste deposits (Gross 1970a, b, Carmody, Pearce & Yasso 1973).

Changed chemical characteristics aid in mapping waste deposits in areas of sandy sediment (Gross 1976). They have less application in areas where the sediment deposits naturally covering the bottom are similar in carbon content and physical properties to many wastes. Presence of sewage solids is still a useful indicator of some waste deposits, regardless of when they are deposited.

Using these criteria to map waste accumulation, one can easily show that deposits of carbon-rich, metal-rich wastes are widespread in some urbanized estuaries, such as New York (Gross 1972) and Baltimore Harbors (Villa & Johnson 1974).

In shallow waters, waste solids accumulate in the disposal sites unless their physical characteristics—low density, fine particle size—permit them to be transported by currents beyond the disposal site before settling to the bottom. For example, in New York Bight, waste deposits are most common in the disposal site near the head of Hudson Channel (Gross 1976, Carmody et al 1973)—over 15 m (50 ft) deep in the axis (Williams & Duane 1974)— where waste disposal has been carried out since the late nineteenth century.

It appears that some wastes are moved by currents south-southeast, down Hudson Channel (Carmody et al 1973). Deposits from the deep parts of the channel south of the disposal sites include carbon-rich, metal-rich sediments. Total carbon concentrations in shelf deposits indicate that carbon-rich waste deposits move southward down the channel. Because of this previously undetected southerly extension of the waste disposal area, the waste-affected area in the Bight was estimated to be about 150 km^2 (60 mi^2).

Available data on sediment deposits and water quality at bathing beaches provide no convincing evidence that large masses of sludge solids move long distances across the continental shelf onto either the Long Island or New Jersey beaches. Some materials may be moved with the water outside the disposal sites, but the multiplicity of sources makes it difficult to be certain of a specific origin. Also the coastal ocean deposits may change seasonally making it difficult to characterize both natural and waste deposits on the continental shelf (Harris 1976).

Ocean areas now used for waste disposal may be valuable as future sources of sand and gravel, petroleum, or as navigational channels for deep-draft vessels or for large off-shore facilities such as power plants or airports. Thus presently forming waste deposits may have to be moved and again disposed of at some future time.

Some solid wastes deposited in the coastal ocean have been used beneficially. For example, construction rubble and other wastes have been used for construction of artificial fishing reefs in many coastal areas (Jensen 1975). Some carefully selected types of wastes may be useful for rehabilitating parts of the coastal ocean as well as for covering over badly polluted bottom areas, although this has not yet been tested.

EFFECTS OF SOLID WASTE DISPOSAL

The impact of solid waste disposal on continental shelves can be appreciated only when we realize that—except for a few major rivers—nearly all sediment transported by rivers is trapped and normally deposited in estuaries, bays, and harbors (Emery 1965, Meade 1969). Consequently, metropolitan regions are not only supplying a large quantity of waste solids but are dumping them on the continental shelf where little other sediment is being deposited to dilute or bury the wastes (Gross 1970c, 1972).

Waste dumping operations are localized and usually involve releasing many tons of solids whenever a barge or dredge is emptied. Thus, the ocean bottom in the disposal area may receive a relatively thick layer of waste solids almost instantaneously. Designated disposal grounds are usually small (a few square kilometers), although it is probable that dumping actually occurs over much larger areas owing to navigational errors, adverse weather conditions, and illegal dumping.

Studies of the behavior of sewage sludges discharged have pinpointed some of the probable effects of sludge disposal in coastal ocean waters (NOAA 1975, 1976). Sludge particles are fine grained and low density, and thus they remain suspended in water. When sewage sludges are discharged in near-surface waters, visible plumes of discolored water move with the local surface currents. Slicks of surface-active materials and accumulations of floatable substances also form in disposal areas, causing not only aesthetic problems but also possible public health hazards associated with atmospheric transport of materials from the air-sea interface. Wave action and currents scouring the ocean bottom can resuspend and move sludge deposits in near-bottom waters (Harris 1976, Swift et al 1976). These processes are known to be important but their frequency and duration are not known.

Sludge deposits are often rich in organic carbon and metal content (Segar & Cantillo 1976, Thomas et al 1976), but the effect of metal enrichment on the marine ecosystem is poorly understood. There is evidence that metals accumulate in bottom-dwelling organisms. The public health implications of such metal transfers through seafood to man have not been widely studied (Verber 1976).

In areas where large volumes of sludge have been dumped for a long time, bottom-dwelling communities have been transformed substantially. The abundance and number of different types of marine organisms was substantially reduced in the disposal areas of the New York Bight, for example (Pearce 1972). Observed changes could have been caused by altered physical properties of the bottom, by toxic metals and hydrocarbons associated with sludge solids, and by reduction in dissolved oxygen concentrations of near-bottom waters. Sewage sludge deposits have also been strongly implicated in the occurrence of diseases in marine organisms such as finrot—the erosion of fishes' fin tissue (Murchelano & Ziskowski 1976)—and shell erosion in crabs, lobsters, and other crustacea (Rosenfield 1976).

The presence of human pathogens (disease-causing agents) in sewage sludges makes ocean bottoms used for waste disposal unsuitable for production of shellfish for human consumption (Verber 1976) and less attractive for recreational fishing. Some bacteria in sludge deposits are also known to resist antibiotics, thus possibly complicating treatment of diseases from these pathogens.

Most waste disposal programs assume that wastes remain in the area of initial discharge. As long as no identifiable wastes wash up on beaches or interfere with recreation or commercial fishing, there is likely to be little immediate complaint. Although waste solids dumped in the ocean remain out of sight for a time, this does not necessarily mean that they do not pose problems for future generations.

Geological Effects of Waste Solid Disposal

Disposal of waste solids in continental shelf areas can alter bottom topography and sediment composition and texture (see Table 7). Dredged wastes deposited at the head of the Hudson Shelf Valley have partially filled one of the tributary arms (Williams 1975, Freeland et al 1976). Between surveys made in 1936 and 1973, deposits of dredged material more than 10 m thick accumulated at the site, amounting to 87% of the volume reportedly discharged (Freeland et al 1976). Previous disposal operations have formed hills that stand about 8 m above the surrounding shelf.

Table 7   Some results of waste solid disposal in estuarine and coastal ocean waters

Physical effects
  Changed bottom topography
    Changed circulation: shoaling; elimination of small stagnant basins; restriction of lateral and vertical water circulation
  Changed bottom type
    Changed substrate for benthic organisms: large solid blocks—rock, rubble—attachments for benthic organisms; movable bottom materials (sand and silts, for example) undesirable for attached organisms; burying undesirable deposits
  Increased turbidity
    Reduction of photosynthesis due to decreased light penetration

Chemical effects
  Leaching from deposits
    Addition of nutrients (NH_3) or other substances (NH_4^+ Mn^{2+}) to water
  Reactions with suspended particles
    Removal of materials from water by sorption onto particles
    Possible depletion of dissolved oxygen

Biological effects
  New habitats created
    Aquaculture: oyster bottom rehabilitated; artificial fishing reefs; lobster reefs
    Previous benthic communities covered and possibly destroyed
  Disease
    Finrot on bottom-dwelling fishes: shell erosion in crustacea
    Closed shellfish grounds due to pathogenic organisms

Some of these disposal operations may have been associated with building the Brooklyn Bridge in 1869–1883 (Williams 1975).

Some of the changes in sediment texture are quite distinctive. Artifact gravels—mixtures of bricks, concrete fragments, and rock fragments—are conspicuous in the surficial sediments accumulating in the area used for disposal of construction and demolition debris in the Hudson Shelf Valley (Freeland et al 1976). The fate of the fine-grained carbon-rich particles is less obvious. The deepest portions of the Shelf Valley are floored by fine-grained sediment (Freeland et al 1976). These deposits are enriched in metals typical of industrial wastes (Gross 1972, 1976, Harris 1976) and in organic carbon-total carbohydrates (Hatcher & Keister 1976) suggesting some deposition of fine-grained wastes. None of the available tests is adequate to indicate the amount of waste material mixed with sediments derived from natural sources.

Changes in bottom sediment characteristics are reflected by changes in the abundance and distribution of benthic organisms in the New York Bight. Pearce (1972) showed that the benthic populations near the designated disposal sites were relatively impoverished over an area of about 50 km^2. Those deposits were carbon-rich, had high metal contents, and were relatively fine grained. This suggests that the impoverishment of the fauna was caused by some constituent or property of the wastes. Fin rot, in fish (Murchelano & Ziskowski 1976), and shell rot in crustacea (Pearce 1972) are noted in the New York Bight apex, which receives most of the barged wastes and all the river-borne load of liquid and solid wastes.

Discharges of municipal sewage and industrial wastes have caused significant changes in water quality, sediment properties, and marine life in harbors, bays, and other estuaries. The Thames Estuary is a well-studied example where the waste load caused parts of the river and estuary to be anoxic (devoid of dissolved oxygen) for several months each summer (Department of Scientific and Industrial Research 1964). A long-term decrease in dissolved oxygen in near-bottom waters has also been reported in parts of the Baltic Sea (Dybern, 1972) since the 1890s.

Waste disposal can also significantly affect dissolved oxygen levels in coastal ocean waters. Again the New York Bight provides a striking example (Sharp 1976). Nutrients (nitrogen compounds and phosphorus) discharged by sewage treatment plants and by street runoff (see Table 5) stimulate photosynthetic activity. The carbon produced by phytoplankton in the surface layers is largely decomposed in water (Thomas et al 1976), thereby consuming dissolved oxygen, especially in near-bottom waters. The amount of dissolved oxygen used in the decomposition of the photosynthetically produced carbon greatly exceeds the composition of dissolved oxygen, exceeding barged disposal of carbon-rich dredged materials and sewage sludges (Segar & Berberian 1976).

During late summer when near bottom waters are isolated from surface waters by density stratification, the dissolved oxygen can be totally consumed. This occurred in summer 1976 (July-September) causing mass mortality in fin fish and shellfish (Segar & Berberian 1976). New York waters may naturally be subject to low oxygen events but the waste discharges to local rivers and the coastal ocean may increase the frequency of occurrence, the area affected, and the duration of the event.

Discharge of liquid and fine-grained particulate wastes can also alter the chemical composition of suspended matter in coastal ocean waters. Titanium-iron-rich wastes dumped at sea cause Ti-Fe hydroxide coatings on particles and contribute distinctive Ti-oxide particles (Biscaye & Olsen 1976).

REGULATION AND MONITORING OF DISPOSAL ACTIVITIES

Coastal nations and various international agencies have undertaken to control— in some cases to eliminate—disposal of wastes in estuaries and the coastal ocean since about 1970. The result in the United States has been the elimination of many smaller volume discharges to the oceans, principally of industrial wastes. Dredged materials and sewage sludges from U.S. coastal cities still go to sea because of the large volumes involved and the cost or difficulty of developing alternative disposal methods or sites.

The result has been to eliminate or reduce some materials with very high concentrations of toxic metals and certain potentially harmful organic constituents. Also the data have been greatly improved, providing better estimates of volumes of materials involved and their physical characteristics and chemical composition, as well as the exact disposal sites.

Monitoring of the effects of disposal operations and long-term effects of the waste sites are planned for many ocean areas still receiving large volumes of wastes. These monitoring operations, if carefully done, will provide useful information about the effects of the waste disposal activities. They should also provide valuable information about variability of conditions and processes in estuaries and coastal ocean areas.

CONCLUSION

Waste solids, including dredged materials, rubble, sewage sludge, and industrial wastes, are discharged in estuaries and coastal ocean areas. The largest volumes and the largest discharges of solids come from dredging operations and from sludges produced by sewage treatment plants. Because of the large volumes of materials and the many disposal operations at sea, waste deposits are of geological significance in urbanized coastal areas.

Deposits of wastes can be detected by their black color, human artifacts, high carbon content, and content of metals such as silver, copper, chromium, and lead.

Waste deposits form hills and cover large areas on the continental shelf. Submarine canyons have been filled by waste deposits.

Physical alterations of the bottom have caused obvious changes in abundance and distribution of bottom-dwelling organisms. Accumulations of sewage sludges on the ocean bottom are associated with diseases in crustacea and fin erosion in certain bottom-dwelling fishes. Low dissolved oxygen concentrations have occurred in coastal ocean areas during late summer, apparently caused by river-borne nutrients rather than disposal of barged wastes.

Regulation of ocean disposal of wastes has eliminated small-volume industrial discharges. In the United States large volumes of sewage sludge and dredged wastes are still discharged at sea.

Literature Cited

Abernathy, R. F., Peterson, M. J., Gibson, F. H. 1969. Major ash constituents in U.S. coals. *US Bur. Mines Rep. Invest. 7240.* 9 pp.

American Chemical Society. 1969. Cleaning our environment: the chemical basis for action. Washington, D.C. 249 pp.

Biggar, J. W., Corey, R. B. 1969. Agricultural drainage and eutrophication. In *Eutrophication: Causes, Consequences, Correctives*, pp. 404–45. Washington, D.C.: National Academy of Sciences, 658 pp.

Biscaye, P. E., Olsen, C. R. 1976. Suspended particulate concentrations and compositions in the New York Bight. *Am. Soc. Limnol. Oceanogr. Spec. Symp.* 2:124–37

Borchert, J. R. 1967. American metropolitan evolution. *Geogr. Rev.* 57:301–32

Bowen, H. J. M. 1966. Trace elements in biochemistry. New York: Academic. 241 pp.

Boyd, M. G., Saucier, R. T., Keeley, J. W., Montgomery, R. L., Brown, R. D., Mathis, D. B., Guice, C. J. 1972. Disposal of dredge spoil. *US Army Waterw. Exp. Stn. Tech. Rep. H-72-8.* Vicksburg, Miss. 121 pp.

Caldwell, J. M. 1966. Coastal processes and beach erosion. *J. Soc. Civ. Eng.* 53(2):142–57

Capp, J. P., Spencer, J. D. 1970. Fly ash utilization. *US Bur. Mines Inf. Circ. 8483.* 72 pp.

Carmody, D. J., Pearce, J. B., Yasso, W. E. 1973. Trace metals in sediments of New York Bight. *Mar. Pollut. Bull.* 4(9):132–35

Davis, J. H. 1956. Influences of man upon coast lines. In *Man's Role in Changing the Face of the Earth*. ed. W. L. Thomas, Jr., pp. 504–21. Chicago: Univ. Chicago Press. 1193 pp.

Department of Scientific and Industrial Research. 1964. Effects of polluting discharges on the Thames Estuary. *Water Pollut. Res. Tech. Pap. 11.* London: H.M. Stationery Office. 609 pp.

Dybern, B. I. 1972. Pollution in the Baltic. *Marine Pollution and Sea Life*, ed. M. Ruivo, pp. 15–23. London: Fishing News (Books) Ltd. 624 pp.

Emery, K. O. 1965. Geology of the continental margin off eastern United States. In *Submarine Geology and Geophysics*, ed. W. F. Whittard, R. Bradshaw, pp. 1–20. London: Butterworth.

Environmental Science and Technology. 1970. Fly ash utilization climbing steadily. *Environ. Sci. Tech.* 4:187–90

Federal Water Pollution Control Administration. 1968. *The cost of clean water*, Vol. 1. Washington, D.C.: GPO

Federal Water Pollution Control Administration, U.S. Department of the Interior. 1969. Conference in the matter of pollution of the interstate waters of the Hudson River and its tributaries—New York and New Jersey. Proc. 3rd Sess. 18–19 June 1969. Washington, D.C.: GPO

Freeland, G. L., Swift, D. J. P., Stubblefield, W. L., Cok, A. E. 1976. Surficial sediments of the NOAA-MESA areas in the New York Bight. See Biscaye & Olsen 1972, pp. 90–101

Gilbert, G. K. 1917. Hydraulic-mining debris in the Sierra Nevada. *US Geol. Surv. Prof. Pap 105.* 154 pp.

Gross, M. G. 1970a. New York metropolitan region—A major sediment source. *Water Resour. Res.* 6:927–31

Gross, M. G. 1970b. Preliminary analysis of urban wastes, New York metropolitan region. *Mar. Sci. Res. Cent. Tech. Rep. 5.* State Univ. New York, Stony Brook. 35 pp. Also in Congr. Rec. 116(31):S2885–2890

Gross, M. G. 1970c. Analyses of dredged wastes, fly ash and waste chemicals, New York metropolitan region. *Mar. Sci. Res. Cent. Tech. Rep. 7.* State University of New York, Stony Brook. 33 pp.

Gross, M. G. 1972. Geologic aspects of waste solids and marine waste deposits, New York Metropolitan Region. *Geol. Soc. Am. Bull.* 83:3163–76

Gross, M. G. 1976. Waste disposal. *MESA New York Bight Atlas, Monogr. 26.* Albany, NY: New York Sea Grant Inst. 32 pp.

Gross, M. G., Karweit. M., Cronin, W. B., Schubel, J. R. 1977. Suspended sediment discharge of the Susquehanna River to Northern Chesapeake Bay, 1966–1976. *Chesapeake Science* In press

Harris, W. H. 1976. Spatial and temporal variation in sedimentary grain-size facies and sediment heavy metal rations in the New York Bight apex. See Biscaye & Olsen 1976, pp. 102–23

Hatcher, P. G., Keister, L. E. 1976. Carbohydrates and organic carbon in New York Bight sediments as possible indicators of sewage contamination. See Biscaye & Olsen 1976, pp. 240–48

Jensen, A. C. 1975. Artificial fishing reefs. *MESA New York Bight Atlas Monogr. 18.* 23 pp.

Kenahan, C. B. 1971. Solid wastes—Resources out of place. *Environ. Sci. Technol.* 5:594–600

Klimm, L. E. 1956. Influence of man upon coast lines. See Davis 1956, pp. 522–41

Likens, G. E., ed. 1972. Nutrients and eutrophication: The limiting nutrient controversy. *Am. Soc. Limnol. Oceanogr. Spec. Symp.*, Vol. 1. 328 pp.

Meade, R. H. 1969. Landward transport of bottom sediment in estuaries of the Atlantic Coastal Plain *J. Sediment. Petrol.* 39:222–34

Meade, R. H., Trimble, S. W. 1974. Changes in sediment loads in rivers of the Atlantic drainage of the United States since 1900. *Int. Assoc. Sci. Hydrol. Publ.* 113:99–104

Mueller, J. A., Anderson, A. R., Jeris, J. S. 1976. Contaminants entering the New York Bight: Sources, mass loads, significance. See Biscaye & Olsen 1976, pp. 162–70

Murchelano, R. A., Ziskowski, J. 1976. Finrot disease studies in the New York Bight. See Biscaye & Olsen 1976, pp. 329–36.

National Academy of Sciences. 1969. *Eutrophication: Causes, Consequences, Correctives.* Washington, D.C. 661 pp.

National Oceanic and Atmospheric Administration (NOAA). 1975. Ocean dumping in the New York Bight. *NOAA Tech. Rep. ERL-321-MESA 2* 78 pp.

National Oceanic and Atmospheric Administration (NOAA). 1976. Evaluation of proposed sewage sludge dumpsite areas in the New York Bight. *NOAA Tech. Rep. ERL-MESA-11.* 212 pp.

Pararas-Carayannis, G. 1973. Ocean dumping in the New York Bight: an assessment of environmental studies. *US Army Corps Eng. Coastal Res. Cent. Tech. Mem. 39*

Pearce, J. B. 1972. The effects of solid waste disposal on benthic communities in the New York Bight. See Dybern 1972, pp. 404–11

Peschiera, L., Freiherr, F. H. 1968. Disposal of titanium pigment process wastes. *Water Pollut. Control Fed. J.* 40:127–31

Pritchard, D. W. 1955. Estuarine circulation patterns. *ASCE Proc.* 81(717). 11 pp.

Redfield, A. C., Walford, L. A. 1951. A study of the disposal of chemical waste at sea. *NAS-NRC Publ. 201.* 49 pp.

Rosenfeld, A. 1976. Infectious diseases in commercial shellfish on the Middle Atlantic coast. See Biscaye & Olsen 1976, pp. 414–23

Schubel, J. R. 1968. Suspended sediment discharge of the Susquehanna River at Havre de Grace, Maryland during the period 1 April 1966 through 31 March 1967. *Chesapeake Sci.* 9:131–35

Schubel, J. R. 1972. Suspended sediment discharge of the Susquehanna River at Conowingo, Maryland during 1969. *Chesapeake Sci.* 13:53–58

Schubel, J. R. 1974. Effects of Tropical Storm Agnes on the suspended solids of northern Chesapeake Bay. In *Suspended Solids in Water*, ed. R. J. Gibbs, pp. 113–32 New York: Plenum. 320 pp.

Segar, D. A., Berberian, G. A. 1976. Oxygen depletion in the New York Bight apex: Causes and consequences. See Biscaye & Olsen 1976, pp. 220–39.

Segar, D. A., Cantillo, A. Y. 1976. Trace metals in the New York Bight. See Biscaye & Olsen, pp. 171–98

Sharp, J. H., ed. 1976. Anoxia on the Middle Atlantic Shelf during the Summer of 1976. *Rep. from the Nat. Sci. Found. Int. Decade of Ocean Explor., Washington, D.C.* 122 pp.

Smith, B. J. 1965. Sedimentation in the San Francisco Bay system. In *Proc. Fed. Inter-Agency Sediment. Conf. Agric. Res. Serv. Misc. Publ.* 970:675–708. Washington, D.C. 933 pp.

Southern California Coastal Water Research Project (SCCWRP). 1973. The ecology of the Southern California Bight: Implications for water quality management. *SCCWRP TR104.* El Segundo, Calif. 531 pp.

Swift, D. J. P., Freeland, G. L., Gadd, P. E., Han, G., Lavelle, J. W., Stubblefield, W. L. 1976. Morphologic evolution and coastal sand transport New York–New Jersey shelf. See Biscaye & Olsen 1976, pp. 69–89

Thomas, J. P., Phoel, W. C., Steimle, F. W., O'Reilly, J. E., Evans, C. A. 1976. Seabed oxygen consumption—New York Bight apex. See Biscaye & Olsen 1976, pp. 354–69

Trimble, S. W. 1975. Denudation Studies: Can we assume steady state. *Science* 188: 1207–08

Tri-State Regional Planning Commission. 1973. *Ann. Rep.* 1973. New York, N.Y. 24 pp.

Verber, J. L. 1976. Safe shellfish from the sea. See Biscaye & Olsen 1976, pp. 433–41

Villa, O. Jr., Johnson, P. G. 1974. Distribution of metals in Baltimore Harbor sediments. *Tech. Rep. 59, Annapolis Field Off. Reg. III, EPA 90319–74–012*

Weibel, S. R. 1969. Urban drainage as a factor in eutrophication. See *Natl. Acad. Sci.* 1969, pp. 383–403

Williams, S. J. 1975. Anthropogenic filling of the Hudson River (shelf) Channel. *Geology* 3:597–600

Williams, S. J., Duane, D. B. 1974. Geomophology and sediments of the inner New York Bight continental shelf. *US Army Corps of Eng. Tech. Mem. 45.* Fort Belvoir, Va: Coastal Eng. Res. Cent.

Wilson, K. K., Blackman, R. A. A. 1974.

Red mud. *Mar. Pollut. Bull.* 5(8): 127–28

Wolman, M. G. 1967. A cycle of sedimentation and erosion in urban river channels. *Geografiska Annaler 49A* : 385–95

Wolman, M. G., Schick, A. P. 1967. Effects of construction on fluvial sediment, urban and suburban areas of Maryland. *Water Resour. Res.* 3 : 451–64

Yasso, W. E., Hartman, E. M. Jr. 1975. Beach forms and coastal processes. *MESA New York Bight Atlas, Monogr. 11.* 51 pp.

Ann. Rev. Earth. Planet. Sci. 1978. 6 : 145–72

AFAR   ✻10090

Paul Mohr

University College Galway, Galway, Ireland

INTRODUCTION

In the less than 20 years since I reapplied the term Afar (d'Abbadie 1873, Dreyfuss 1931) to an awesome, inhospitable, and barely explored desert sink of northeastern Ethiopia, that region has become a mecca for detailed and integrated studies of the tectonic, volcanological, and geophysical properties of a very hot spot (see Pilger & Rösler 1975). This unprecedented transition has been motivated by recognition, following the prophecies of plate tectonics, that three zones of active crustal extension converge on Afar, namely the Red Sea, Gulf of Aden, and African Rift System. Afar is thus a triple rift (RRR) junction, the meeting point for the Arabian, Nubian (or African), and Somali plates.

Afar is a unique and extraordinary geological entity, embellished for the field geologist by a ferocious climate and the reputation of its native peoples. It exposes, free of cover by the sea, the processes occurring at the birth of a new oceanic basin (Figure 1). Although the same exposure is found on Iceland, there the two separating plates are essentially oceanic in character, whereas the three Afro-Arabian plates each have a continental character up to the very margins of Afar. In both Afar and Iceland, the uplift above sea level has been accompanied by a modification and diffusion of the typical mid-oceanic ridge-rift process, though in extreme eastern and northern Afar sub-sea level terrestrial exposures reveal structures very similar indeed to mid-ocean rift valleys (Needham et al 1976).

In this review, I take a broad look at the geological evolution and present geophysics of Afar. Despite the intense research efforts of the past decade, several important lacunae remain and topics for controversy abound. In attempting to be objective, I may not be doing full justice to some of these controversies. However, no one writing today on Afar can avoid a major bone of contention: is continental crust, albeit attenuated, present under the floor of Afar? Many geophysicists say "yes," many geologists say "no." Obviously, the answer to this question affects the plate tectonic interpretation not only of Afar, but also of the three spreading zones meeting there. Furthermore, it can give insight into the initial process of formation of Atlantic-type continental margins.

0084-6597/78/0515-0145$01.00

Figure 1   Astronauts' eye view of Afar, looking west from 450 km altitude above the Gulf of Aden (Skylab 4 photograph 3385, courtesy of NASA). The southern Red Sea occupies the right foreground, the broad Awash valley of southern Afar lies to the mid-left, and the Ethiopian plateau forms the backdrop. In the left foreground, the Gulf of Tajura is partly cloud-covered but curvilinear faults are visible, fanning out northeastwards from the western end of the Gulf via the Asal graben (note the salt-ringed, sub–sea-level lake in this graben), and out into the complex graben system of northeastern Afar.

The calderas and cones of the Dubbi volcanic chain are clearly visible, straddling the Danakil sialic block in the right middle distance. Beyond the Dubbi chain, the white Salt Plain contains three dark fissure basalt shields that mark incipient development of oceanic crust: clockwise from the right, they are Erta-ali, Tat-ali, and Alaita.

The 2500 + m escarpment of the Ethiopian plateau runs prominently across the entire middle distance of the photograph. NNE-trending faults intersecting this escarpment are visible, and also the characteristic, narrow "marginal graben" between the escarpment and the Afar floor. The culminating mountains of the plateau, the glaciated Simien massif (4543 m), protrude behind the escarpment to the right of center.

An equally magnificent view of the southern margin of Afar and the "funnel" of the Ethiopian rift valley is provided by the consecutive Skylab 4 photograph 3384.

PHYSIOGRAPHY

Afar is a triangular-shaped depression, some 300 km wide in the south and about 600 km long from south to north. The mean elevation declines rather uniformly from ca 1000 m in the south to below − 100 m in the Salt Plain in the north. Superimposed young volcanic relief only exceptionally exceeds 1000 m. Intense crustal slicing in eastern Afar has produced abrupt relief of as much as 500 m with, for example, the Asal (lat. 11.7° N, long. 42.4° E) and Kimbiri (lat. 12.8° N, long 41.2° E) graben floors lying below sea level.

High plateaux border Afar to west and south. The Ethiopian (Western) plateau soars from the western edge of the Afar plains, some nonvolcanic plateau summits exceeding 3000 m elevation. Severe denudation has caused badlands averaging 40 km in width to develop between the floor of Afar and the crest of the present escarpment. Along the southern side of Afar, the denudation zone between the Afar floor and the ca 2600-m crest of the Somalian (Eastern) plateau is only some 15–20 km wide. Beyond the western and southern bounding escarpments of Afar there is a gentle topographic decline, into the Nile drainage system on the Ethiopian plateau and into the Webi Shebeli (Indian Ocean) drainage system on the Somalian plateau. Afar itself is a region of internal drainage.

The eastern margin of Afar is less clearly defined. In the north, the low hills (< 1400 m) of the Danakil block lie between Afar and the Red Sea. In the south, the even more subdued hills of the Aisha block have an indefinite boundary with the floor of Afar to the west. Between these two blocks, the Gulf of Tajura penetrates west from the Gulf of Aden towards internal Afar.

STRATIGRAPHY

Because of the low relief, allied to the essentially flat-lying nature of the strata in Afar, the deeper levels of the crust are not exposed, a fact that permits the ongoing controversy over the presence or absence of pre-Cenozoic rocks beneath Afar. However, old sialic rocks are exposed along the deeply eroded plateau margins bordering Afar, and it is here that the early history of Afar may be elucidated.

Granites within the "Basement Complex" of Ethiopia have yielded K-Ar ages in the range 975–415 m.y., the oldest ages applying to rocks from northern areas of the Danakil block (Frazier 1970) and Ethiopian plateau (Miller et al 1967). The basement schists and granites of the northern Ethiopian plateau are unconformably overlain by Paleozoic glacial strata (Dow et al 1971), but elsewhere around Afar the first post-basement rocks are marine Mesozoic sediments (Mohr 1962). The original thickness of these sediments totalled between 500 and 1500 m on the present Ethiopian, Somalian, and Yemeni plateaux, but reached double this amount on parts of the present Danakil block, thus hinting at a percursor to Afar subsidence (Hutchinson & Engels 1970).

Unconformable on the Mesozoic sedimentary strata lie the Trap Series stratoid basalts of the Ethiopian and Somalian plateaux, and possible thin equivalents on the Danakil and Aisha blocks. On both plateaux, the total thickness of these basalts increases towards the center of the Ethiopian uplift, where it exceeds 3000 m at the present rift margins (Mohr & Rogers 1966, Juch 1975). Along the margins of Afar, the Trap Series is broadly downwarped and now exposes a major zone of conjugate dike swarms (Mohr 1971a). The age range and internal stratigraphy of the Series, crucial to an understanding of the early evolution of Afar, are still not clearly known. Jones (1976) obtained an age range of 26–19 m.y. (excluding the giant Simien shield (lat. 13.1° N, long. 38.3° E) where the basal lavas are older than 36 m.y.), whereas the more detailed study by Zanettin et al (1974) of the lavas of the western margin of Afar yielded an age range of 32–16 m.y., excluding the as yet undated, basal Ashanghi basalts. Jones disputes the validity of Zanettin et al's older ages, but K-Ar dating of the Afar margin dikes by Megrue et al (1972) confirms an age range of 32–19 m.y., with some distinctly younger silicic and basaltic rocks in the 15–11 m.y. bracket. Superficial Quaternary sediments and very localized young basalt flows are the only rocks to lie upon the Trap Series of the plateaus, attesting to differential uplift of the plateaus from Afar in the Miocene and a subsequent major such episode in the late Pliocene-Pleistocene (Merla & Minucci 1938, Mohr 1967, McDougall et al 1975).

Turning to the Afar floor, both sedimentary and volcanic stratigraphies reveal a tendency for progressively younger formations to occur exposed as the interior of Afar is approached. Whether the older formations at the fringes of Afar continue beneath internal Afar is not presently known, though plate tectonic theory (see later) would lead one to suppose they are absent. Bearing this probable offlap in mind, the generalized stratigraphic sequence in northern Afar is presented in Table 1.

Table 1   Sedimentary sequence in northern Afar, grossly generalized[a,b,c,d]

Holocene	Afrera Group (6), lacustrine limestones, 10,000–5000 yr	0–15 m
Pleistocene	⎧ Upper terrace gravels and fanglomerates ⎨ Zariga Group (6), Enkafala Beds or White Series (3) ⎪ marine limestones, 200,000–?20,000 yr ⎩ Lower terrace gravels and fanglomerates	0–50 m
Mio-Pliocene	Danakil Group (6), Red Series (3) or Desset Series (1): upper part marine, lower part fluvio-lacustrine	0–1100 m
?Eo-Oligocene	Dogali Series (1), restricted to Red Sea margin	

[Jurassic-
  Cretaceous]
	Marine sandstones and limestones exposed on the
	  continental margins of Afar, not proven present
	  under Afar itself (2)
		2200–3300 m

[a] In the Salt Plain, a salt-gypsum facies of the Zariga Group reaches a maximum thickness of 2200 m and is dated between ?300,000 and 70,000 yr (4).

[b] In southwestern Afar, the Plio-Pleistocene Hadar Series (8) bears important hominid finds in near-shore lacustrine silts and mudstones. There is progressive offlap of younger units towards the tectonic axis of Afar, but the whole area is patchily covered with late Quaternary lacustrine sediments (5).

[c] At the western end of the southern margin of Afar, a thick volcanic sequence contains the Chorora Formation, comprising diatomaceous sediments up to 30 m thick and dated at 10.5–9.0 m.y. (7).

[d] References: (1) Filjak et al 1959, (2) Hutchinson & Engels 1970, (3) United Nations 1973, (4) Behle et al 1975, (5) Gasse 1975, (6) Kürsten 1975, (7) Sickenberg & Schönfeld 1975, (8) Taieb et al 1976.

The sedimentary history of Afar must have begun with the initiation of the original depression, which is placed by Barberi et al (1972, 1975) as early Miocene on the evidence of lava ages, but by Girdler & Styles (1974, 1976) as late Eocene. The latter authors however, whose case rests on the difficult identification of magnetic anomalies generated by sea-floor spreading in the Red Sea and Gulf of Aden, have since revised their date to mid-Oligocene (Girdler & Styles 1978). The Tertiary sediments of Afar have not received any detailed attention. There is certainly no equivalent to the Paleocene Mej-zir marine sediments of Yemen (Geukens 1960), and the oldest rocks so far identified are unfossiliferous sandstones and conglomerates interbedded with basaltic lavas and tuffs, reaching a maximum thickness of 3000 m at the northern tip of Afar and the Danakil block. These sediments comprise the supposedly Eocene-Oligocene Dogali Series of Filjak et al (1959), but are attributed a possible early-middle Miocene age by Hutchinson & Engels (1970).

What is more or less agreed on is that by the late Miocene, a marine basin had become established over northern Afar (presumably less wide then than now, according to plate tectonic dictates), and these conditions persisted until the isolation and desiccation of the Salt Plain arm of the sea, some 70,000 years ago (CNR-CNRS Afar Team 1973). No Cenozoic marine deposits of any age are exposed in central or southern Afar, a fact to recall when the magnetic lineations of these areas are discussed. Surface geology gives no indication that a proto-Gulf of Aden ever extended into Afar. In southern Afar, the oldest known sediments are middle-late Miocene diatomites (Sickenberg & Schönfeld 1975) localized along the southern margin, and Pliocene lacustrine silts along the Awash valley (Taieb et al 1976). The Quaternary lacustrine sedimentation of eastern and southern Afar has been discussed in detail by Gasse & Rognon (1973), and Rognon & Gasse (1973).

The volcanic stratigraphy of Afar summarized in Table 2 is based largely on radiometric dating and to a lesser extent on observations of superpositioning, and on unconformities whose regional significances are not well understood. There has been a definite episodicity of volcanism. The first episode, which was essentially basaltic, is recognized from the Danakil and Aisha blocks and coincides with the 26–19-m.y. age range of Jones (1976) for the plateau Trap Series. After a break of some 5 m.y. duration, volcanism on the two blocks resumed during the 14–10 m.y. interval, but is overwhelmingly rhyolitic in character. After another interval of 2 m.y., basaltic volcanism spread across Afar to the western margin during 8–6.3 m.y. ago. After a further interval of 2 m.y., the profuse flood lavas of the Afar Stratoid Series covered most of central, eastern, and southern Afar during 4.3–0.6 m.y. (the upper age limit is 3 m.y. in southern Afar). These lavas have effectively buried any earlier formations and thus limit our understanding of the earlier evolution of internal Afar. During the middle-late Quaternary, initially submarine eruptions in northern Afar built up subaerial axial ranges along lines of ongoing crustal separation, and axial ranges have also developed across central into eastern Afar. Erta-ali volcano has maintained an active lava-lake during historical times (Barberi & Varet 1970).

Afar volcanism shows interesting examples of synchroneity with volcanism in the Gregory rift of Kenya: the volcanics are much better dated in that rift than in the Ethiopian rift (Baker et al 1971). Thus the Afar rhyolites of 14–10 m.y. ago were erupted contemporaneously with the Kenya plateau phonolites of 15–11 m.y. ago, and there were extensive eruptions of basalt within the Gregory rift during the Pliocene (post-5 m.y.).

Table 2 Summary volcanic and tectonic history of Afar

Age (m.y.)	Volcanic episodes	Faulting episodes	(m.y.)
0 0.6 ↑ 1.2	Axial range basalts (5,8)		0
2	Afar stratoid basalts over all Afar except north (5,8,10)	Afar fault belts and WFB | and marginal graben	2
4 4.3		South, central, east Afar and western margin	4
6 6.3	Ghibdo, Dalha and Galemi basalts of eastern Afar (1,2,5,6,7,9)		6
8		Eastern Afar and Afar margins (espec. south)	8
10	Mabla, Damerkada, and Chinili rhyolites of eastern Afar; basalts in SW Afar (2,5,6,7,8,12)		10
12			12
14	Younger dikes of the Afar margins (4)		14
16		Eastern and ? western Afar	16
18			18
20	Adolei and Galile basalts (some ignimbrites of eastern Afar): plateau Trap Series (2,5,6,7,9,13)		20
22			22
24	Peralkaline granites (2,3,5)		24
26		?early faults of proto-Afar margins	26
[?38–28]	Older Trap Series basalts (Ashanghi and Aiba) of plateau-Afar margins (4,10,11)		

References: (1) Voute 1959, (2) Brinckmann & Kürsten 1971, (3) Black et al 1972a, (4) Megrue et al 1972, (5) Barberi et al 1975, (6) Black et al 1975, (7) Chessex et al 1975, (8) Christiansen et al 1975, (9) Civetta et al 1975, (10) Kunz et al 1975, (11) Zanettin & Justin-Visentin 1975, (12) Boucarut & Clin 1976, (13) Jones 1976.

SETTING, FORM, AND CHEMICAL FEATURES
OF AFAR VOLCANOES

Five major styles of igneous activity have occurred in Afar:

Peralkaline Granites

Peralkaline granites were intruded during the early Miocene re-emphasis of relief between the depression and the bordering plateaux, and are peculiarly restricted to a latitudinal zone crossing central Afar (Figure 2). From west to east occur: the Limmo granite (lat. 12.2° N, long. 40.1° E) and associated minor bodies of the Afar-Ethiopian plateau margin; the Affara Dara granite (lat. 12.6° N, long. 40.9° E), within Afar; the Asa-ali granite (lat. 13.5° N, long. 41.5° E), near the Afar-Danakil block margin. Tertiary granites have previously been recognized from the Yemen highlands on the opposing side of the Red Sea (Comucci 1933; Karrenberg 1957, Figure 3; Geukens 1960), and notably include the Jabal Sabir massif which towers to 3000 m elevation on the Ta'iz skyline (lat. 13.7° N, long. 44.1° E). No matching granites occur upon the Ethiopian plateau.

The Afar granites are usually markedly leucocratic, and portions of them show syenitic tendencies. Their peralkaline character is attested to in their chemistry (Barberi & Varet 1974), and mineralogically in the presence of riebeckite, though less alkaline portions contain hastingsite and biotite (Black et al 1972a). An intimate association with granophyric bodies suggests a very shallow emplacement of the granites in the upper crust, and indeed contemporaneous alkali trachyte and per-alkali rhyolite lavas occur in close proximity to the Limmo and Asa-ali granites (Bannert et al 1970, 1971). However, the Afar granites appear to be of greater magnitude than the likely size of silicic cupolas underlying the preserved silicic volcanic cones of Afar, and their singular character is not yet understood. The Limmo and Affara Dara granites were intruded into altered stratoid basalts not exposed anywhere else on the floor of Afar, and there has evidently been post-emplacement uplift of these light granitic massifs and their fringing country rock. Whether the intruded basalts are part of the plateau Trap Series, or are part of an Icelandic-style crust lacking underlying continental rocks, is also not yet known.

Stratoid Basalts

Very extensive fields of stratoid basalts were erupted from fissures situated along the margins of Afar during the late Oligocene-early Miocene, and again during a late Miocene episode (Megrue et al 1972, Mohr 1971a). These mid-Tertiary basalts belong to the Trap Series (Kazmin 1975). The exposed dike swarms show that the majority of the feeders ran parallel to the structural trend of the particular margin, though obliquity developed at the fringe of the Trap Series province (Mohr 1971a): both at the northern end of the western margin and at the eastern end of the southern margin, the swing is anticlockwise. At most localities, subordinate sets of dikes form a conjugate relationship with the main trend. Within the main trend, virtually all the dikes dip plateauward, nearly perpendicular to the lavas dipping Afarward. The modal thickness of the dikes is 1.5 m, as in Iceland.

Renewed and voluminous outpourings of stratoid basalts have occurred since 4.3 m.y. ago on the floor of all except northern Afar. The buried fissure-feeders of these younger basalts are rarely exposed, but appear to be located well inside Afar and generally to run parallel to the local fault trend (Christiansen et al 1975). This

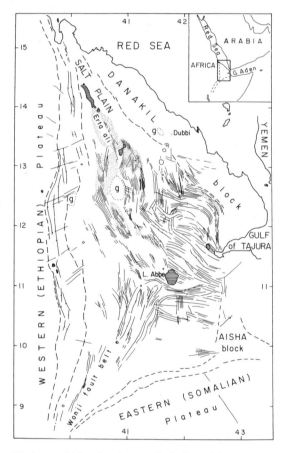

Figure 2 Simplified map of Afar, showing margin limits and floor fault belts. The margins with the Ethiopian and Somalian plateaux indicate both the present topographic plateau rim (outer dashed line) and the limit of the Afar plain (inner dashed line). The marginal grabens of the Ethiopian plateau-Afar margin are also shown. The floor fault belts are schematized (see Mohr 1972b for more detail). The lava aprons of the Quaternary axial ranges are stippled. Volcanic calderas are circled to scale, and important volcanic peaks are marked as a dot. Tertiary granites are marked "g." The NW–SE Marda fault zone of the Somalian plateau is shown in the bottom-right corner of the map.

younger pile of basalts contains a relatively high proportion of more silicic lavas, extending from hawaiite through mugearite to alkali trachyte (De Fino et al 1973). Christiansen et al have described a remarkable alternating sequence from Hert-ali (lat. 9.8° N, long. 40.3° E), southwestern Afar, where thick pantelleritic ignimbrites are interbedded with young flood basalts. Usually, in Afar, rhyolite volcanoes are restricted to end-stage cones built up on the main basaltic platform.

Both older and younger Afar stratoid "basalts" are in fact most often of hawaiite-basaltic icelandite composition, true basalts being subordinate (De Fino et al 1973, Mohr 1975a). This slightly evolved character is also shown in a typical iron enrichment, with total FeO between 14 and 20% (Barberi et al 1970, Mohr 1975a). In terms of alkalinity, these lavas tend to be transitional between mildly alkaline and tholeiitic, though more alkaline members occur in upper levels of both the older and younger piles.

Axial Ranges

Tazieff and co-workers (1969, 1972) first identified and named the axial volcanic ranges that overlie NW–SE trending zones of active crustal extension in northern Afar. These ranges are wholly of Quaternary age (Barberi et al 1975). The best-studied range is that of Erta-ali (lat. 13.6° N, long. 40.7° E) in the northern apex of Afar (Figure 2), 100 km long and with a lava apron 30–40 km wide that bisects the sub-sea level Salt Plain (Tazieff & Varet 1969). Although fissure eruptions have built up the bulk of this 500 km^3 range, seven later volcanic centers surmount its crest (Barberi & Varet 1970).

The early evolution of an axial range within a steady-state, mid-ocean-style rift is documented by Harrison et al (1975) and Delibrias et al (1975) for the Ghubat-Asal graben of eastern Afar. There, a 10-km-wide zone of faulting and subsidence is veneered with young anorthite-phyric basalts that were erupted subaqueously from fissures. Superimposed along the axis of this zone is a 2-km-wide graben containing very recent (7400–3000 years) tholeiitic basalts that are aphyric. The lava units have been transported laterally at a mean half-spreading rate of 0.75 cm yr^{-1}, and the graben margins have been uplifted relative to the floor axis at ca 2 mm yr^{-1}. The progressive age sequence from axis to margin in the Asal graben has not been swamped by excessive magmatism and range building as at Erta-ali.

Initially, the fluid-fissure basalts of axial ranges build up a low and extensive lava apron: these earliest, voluminous basalts are tholeiitic but have higher potash (ca 0.6% K_2O) compared with the Red Sea and Gulf of Aden floor tholeiites (ca 0.2% K_2O) (Barberi et al 1970, Mohr 1972a). Further volcanic activity results in the superposition of shields along the tectonic axis of the preceding apron, with short fissure sources now concentrated into discrete sectors along the range. The shield lavas include iron-enriched basaltic icelandite-hawaiite and fluid mafic trachyte. Finally, central volcanoes develop upon the shields, and silicic trachyte and pantellerite differentiates are erupted (Barberi et al 1970, 1974, CNR-CNRS Afar Team 1973). Episodic renewal of crustal extension can lead to a superimposition of a repeated fissure → shield → stratovolcano cycle on a time scale of possibly a million years.

Of all the Afar basalts, those of the axial ranges most closely resemble mid-ocean ridge basalts in their major element chemistry. Nevertheless, they occupy an intermediate position between Red Sea floor tholeiites (Chase 1969) and the transitional or mildly alkaline basalts of southern Afar and the Ethiopian rift valley (Mohr 1975a). This intermediate composition, in a region of clear crustal separation and generation, can be related to the "hot spot" regime of Afar as indicated by the LIL (light rare-earth element) enrichment in the basalts of this region (Schilling 1969, 1973).

Oblique Ranges

Oblique and transverse chains of volcanoes run, as their name implies, at a broad angle to the axial ranges and the zones of crustal extension in Afar. These chains are essentially restricted to the fringes of Afar, for example the Ma'alalta (lat. 13.0° N, long. 40.2° E) and Dabayra (lat. 12.3° N, long. 40.2° E) volcanic fields that have pierced continental crust at the western margin of central Afar, and the Assab (lat. 13.0° N, long. 42.7° E) and Gufa (lat. 12.5° N, long. 42.5° E) sets of volcanic cones situated upon the southern part of the Danakil block. They are marked by alignments of occasional fissures and dikes, basaltic cinder and lava cones, and small faults all typically trending between E–W and ENE–WSW (Barberi et al 1974, 1975, Mohr 1967).

The basalts of the oblique chains are mildly to strongly alkaline in composition, and are notably less voluminous than the weakly alkaline basalts of the axial ranges or the older stratoid basalts. Some of the most alkaline basalts bear ultramafic xenoliths (Hutchison & Gass 1971, Ottonello et al 1975). At Ma'alalta, a stratovolcano has built up in which the later products are peralkaline rhyolites; these rhyolites are more potassic than are the rhyolites of the axial ranges. Strontium isotope data (Tazieff & Varet 1969) show appreciably higher values for the trachytes of the oblique ranges (.708–.715) than for the axial range trachytes and rhyolites (.701–.705), though the basalts of the two settings are indistinguishable on this basis (.702–.705).

Singular amongst the Afar oblique chains is the Dubbi (or Bidu) volcanic line (lat. 13.6° N, long. 41.8° E). It comprises eleven major volcanic centers and extends for at least 135 km in a NNE direction from central Afar, across the mid-latitude of the Danakil block (Figure 2) to the Red Sea coast (Mohr 1967, Mohr & Wood 1976). Although fissure basalt sources lie on the Dubbi chain, notably in the north, the essential volcanic character is one of trachyte-pantellerite cones built 1700–1000 m above base level, the largest cones having calderas. The Dubbi volcanoes are rather similar to the southern Afar and Ethiopian rift valley volcanoes in their physical form, their chemistry, and their common structural alignment. Nevertheless, Tazieff et al (1972) and Barberi et al (1974) interpret this major volcanic feature as marking a pre-Afar tensional structure that makes a single unit with the Dabayra line of western Afar on a pre-drift reconstruction. More recently, Barberi & Varet (1975a) have proposed the Dubbi line to be a major, active transform fault linking the Red Sea with Afar. The Dubbi volcanoes appear to be successively younger, northwards [the most northerly, Dubbi itself, last erupted in 1861 (Gouin 1978)], and this may suggest a line of lithospheric weakness propagating in the same direction at a mean rate of ca 6 cm yr^{-1}.

Ethiopian Rift Volcanoes of Southwestern Afar

Ethiopian rift-style volcanism extends NNE from the Ethiopian rift valley for a further 350 km across southwestern Afar, in intimate association with the young Wonji fault belt (Mohr 1967, Mohr & Wood 1976). Trachyte-pantellerite lava and ashflow tuff stratovolcanoes have built 400–1000 m above base level, though again caldera collapse makes these minimum values (Gibson 1967, Cole 1969, Brotzu et al 1974). The volcanoes are spaced at remarkably regular intervals, close to 40 km and reflecting the thickness of the underlying lithosphere (Mohr & Wood 1976). Fissure basalts of mildly alkaline or transitional character have erupted from the flanks of most of the cones, attesting to episodes of renewed, regional crustal extension acting across the Wonji fault belt. A line of older, denuded silicic cones runs parallel and close to the southern margin of Afar. Afar lacks the thick late Miocene-Pliocene peralkaline silicic ignimbrites that fill the Ethiopian rift valley and mantle the rift margins.

STRUCTURE OF AFAR

The structural complexities of Afar are summarized here from the descriptions by Mohr (1967, 1972b) and the CNR-CNRS Afar Team (1973). Afar is a triple-rift junction, and so possesses a tectonic framework controlled by the three zones of crustal extension that converge there: NNE-trending structures of the Ethiopian rift valley continue into southwestern Afar; NW-trending structures with affinity to the Red Sea occur in northern and central Afar; and WSW-trending structures impinge on eastern Afar from the Gulf of Tajura, but swing immediately to WNW–NW as the floor of the active, extensional zone emerges above sea level (Figure 2).

However, the detailed picture of Afar tectonics is inevitably much more ravelled, owing to factors that include the influence of pre-existing basement structures on the Cenozoic fracture pattern of the Afar margins, variations of lithospheric rigidity and thickness across Afar due, for example, to the presence of mid-Tertiary intrusions, possible changes of the stress field with time especially in eastern Afar, and the interaction of differing contemporaneous stress fields near the "triple point" in central Afar. Additionally, on even the simplest microplate model of Afar, there occur "horsts" or blocks of attenuated continental crust within the triple junction region whose roles in the evolution of Afar remain poorly understood (Black et al 1972b, 1975, Barberi & Varet 1975a, Mohr 1970, 1972b).

Afar Margins

Evidence of the earlier structural history of Afar is preserved along the plateau margins, though the western margin in particular remains tectonically active. The western and southern Afar margins are direct continuations of the respective margins of the Red Sea and Gulf of Aden (Figure 2, inset), in conformity of course with the notion that the Yemen portion of the Arabian Peninsula once occupied the present Afar region (Laughton 1966). The early history of Afar can therefore be studied in the context of the early evolution of the Red Sea and Gulf of Aden basins (see below).

Taking the western margin first, we find that its structural elements trend NNW–SSE, but keep close to meridian 40°E along a 600-km length by means of offsets that appear related to transverse tectonics (Barberi et al 1974) more than to an enechelon pattern (CNR-CNRS Afar Team 1973). Connection at offsets is also made in some instances through NNE–SSW structural elements that follow the Precambrian "grain" (Mohr 1974). The dominant, NNW–SSE structures of the western margin are contained within a zone 25–50 km wide; this is a minimum value where there is overlap by young sediments of the Afar floor.

North of latitude 13°N, which is also the northerly limit of thick development of Trap Series basalts on the Ethiopian plateau, the margin has been formed by synthetic step faulting whose largest vertical displacements occur closest to the Afar floor (Mohr & Gouin 1967, Tazieff & Varet 1969). The total displacement, west of the Salt Plain, is 3000–4000 m. The step faulting is typically associated with block tilting directed away from Afar. Superimposed on this zone of synthetic faults is a discontinuous line of antithetic fault(s), nowhere yet mapped in detail, which has produced marginal graben (Merla & Minucci 1938, Gouin & Mohr 1964, Mohr 1967) within the now severely denuded Afar marginal zone. These marginal grabens, and the relatively uplifted block separating the graben from the Afar floor, are in each case some 7–10 km wide, and are younger features (or have been more recently reactivated) than the major synthetic faults. The marginal grabens tend to become peculiarly constricted at their southern terminations (Mohr 1975b).

South of latitude 13°N, within the central area of the Ethiopian uplift where the Trap Series flood basalts are thickly developed, the predominant structural style changes to that of a broadly antithetically faulted margin. The stratoid lavas here dip at angles of 20 to 50^g towards Afar. Studies of the feeder dikes suggest that this tilt of the lavas is largely, if not wholly, a post-depositional phenomenon (Mohr 1971a). It remains an open question whether the tilting merely reflects the result of block rotation caused by crustal attenuation (Morton & Black 1975, CNR-CNRS Afar Team 1973), or whether a monoclinal flexure is present (Abbate & Sagri 1969, Mohr 1971a, Justin-Visentin & Zanettin 1974, Christiansen et al 1975). The latter concept is endorsed by a hint of overprinting of mild flexuring on the synthetically faulted margin near latitude 14°N (Mohr & Gouin 1967). However, the need for a further explanation is provoked by the observation that the antithetic faulting at latitude 11°N is concentrated into discrete zones whose spacing increases eastwards, with each faulted block showing a renewed steepening of the east-dipping stratoid lavas immediately east of each fault zone (Mohr 1971a). This is a feature not accounted for by the otherwise attractive concept of Morton & Black (1975), which relates a progressive block tilting to a progressive attenuation of the crust.

The marginal grabens run close to the floor of Afar, north of latitude 13°N, but are adjacent to the main topographic escarpment of the plateau rim, south of latitude 13°N (Figure 2). No graben is developed south of latitude 10.5°N, where the western margin of Afar is controlled by a funneling out of the Ethiopian rift valley (Mohr 1975b). The discontinuous nature of the marginal grabens is expressed either through lateral offsets or by NNE-trending connecting faults. In either case there is usually a marginal volcanic center located close by: e.g. Ma'alalta volcano lies near the offset between the Sheket-Dergaha (lat. 13.2° N, long. 39.8° E) and Guf Guf (lat. 12.5° N, long. 39.7° E) grabens, and Werabo Abo volcano (lat. 11.2° N, long. 39.8° E) lies between the Menebay-Hayk (lat. 11.5° N, long. 39.7° E) and Borkenna grabens (lat. 10.7° N, long. 39.8° E).

The southern margin of Afar swings progressively clockwise eastwards, and in the same direction changes in character from a narrow zone of step faulting and plateau upwarping, at the rift valley funnel, to an antithetically faulted down-flexure that faces sectors of synthetic faulting along the plateau side of the marginal zone (Juch 1975, Figure 2). Further to the east, beyond longitude 41.5°E, synthetic faulting resumes as the only visible means of Afar-plateau displacement, but a flexure may occur to the north, now buried under the Afar plains (Juch 1975). The continuing plateauward dip of the dikes in the Dire Dawa (lat. 9.6° N, long. 41.9° E) region supports this concept. Topographic and structural expression of the southern Afar margin weakens abruptly at the junction with the Aisha block and the Marda fault zone of the plateau (Black et al 1974, Purcell 1976, Mohr 1967, 1975c), and there is no direct connection with the Gulf of Aden coastal structures (Mohr 1974).

The ages of the faults and associated structures of the Afar margins are still poorly known, because studies relating dated lavas and sediments to cutting or noncutting faults are lacking. A minimum age for the oldest faults is 25 m.y. (Barberi et al 1972), but the Ashanghi-Aiba unconformity in the lower part of the Trap Series indicates earlier regional tectonism (Mohr & Rogers 1966, Zanettin & Justin-Visentin 1975). On the basis of geomorphology alone, it is evident that the synthetic faults of the western Afar margin are older than the main antithetic faults, and it is possible that the two episodes are of mid-Tertiary and end-Tertiary age respectively (Table 2). The antithetic faulting itself may belong to at least two subepisodes: Juch (1975) considers that the southern margin of the Afar depression was already developed prior to a late-Miocene (ca 9 m.y. ago) episode of antithetic faulting and warping, during which this margin received essentially its present form, and that renewed faulting occurred during the late Pliocene–early Pleistocene. Along the western margin of Afar, the initiation of antithetic faulting may have been earlier than along the southern margin, perhaps in the ca 26 m.y. (and 15 m.y.?) episodes. Conversely, the big, Pliocene-Pleistocene antithetic faults of the western margin remain active along the marginal grabens: for example, there was a swarm of strong earthquakes [maximum M < 6.5] during 1961 in the northern sector of the Borkenna graben (latitudes 10°–11° N), with up to one meter renewal of the eastern boundary fault scarp of the graben (Gouin 1970, 1975, 1978).

Afar Floor

Intense fracturing of the Afar floor has sliced the lava pile of Pliocene-Pleistocene Afar stratoid basalts. The fracturing is concentrated into sinuous and curvilinear belts, 15–30 km wide and 50–150 km long. The belts show a variety of interrelationships, including bifurcation (the width of the two "branches" usually totals that of the "trunk"), tangential touching, abrupt abutment and termination of one belt obliquely against another, intersection of belts, and, within single belts, a change of style in the spacing and number of faults, or a lateral offset of the belt. The areas separating the fault belts take various forms: aprons of young basalts, resistant granitic intrusions, shallow sediment-filled depressions, and deep grabens. The resulting fault-belt complex occupies at least half the area of the Afar floor. Young sedimentary plains, 30–70 km wide, occupy the floor adjacent to the western and southern margins of Afar (Figures 1 and 2), especially where those margins are downwarped (Mohr 1967, 1972b, Bannert et al 1970, Brinckmann & Kürsten 1971, CNR-CNRS Afar Team 1973).

The fault belts of the Afar floor trend grossly parallel to the direction of the nearest of the three "rifts" converging on the Afar triple junction. Thus in northern Afar the belts run NNW–NW, parallel to the Red Sea trough; in southwestern Afar, the belts trend NNE–NE in association with the Wonji fault belt; and in eastern Afar, although the ENE trend of the Gulf of Aden margins is not imitated in the fault belt orientations, which run close to ESE, there is no disparity if account is taken of the northeasterly direction of crustal spreading in the Gulf of Aden.

The conjunction of the three fault-belt trends occurs at a postulated triple point in the Lake Abbe region of central Afar (Figures 1 and 2). A mutual adaptation of the "Red Sea" and "Gulf of Aden" fault belts is a response to the outward rotation of the Danakil block from Afar (Tazieff et al 1972, Mohr 1967, 1970). The fault belts of southwestern Afar terminate rather abruptly against the northern Afar fault belts, at latitudes 11–11.5°N: likewise, the southwestern fault belts themselves abruptly terminate the more southerly of the eastern Afar fault belts at longitude 41.5°N, though both adjusting and intersecting faults can be observed here. This singular relationship of the southwestern Afar fault belts to the other two sets of fault belts leads Tazieff et al (1972) to consider the Ethiopian rift as having played a separate and minor role in the opening of Afar.

Peculiar areas of annular faulting (Mohr 1968, CNR-CNRS Afar Team 1973) and intersecting fault belts (Tazieff 1971, Black et al 1972b, CNR-CNRS Afar Team 1973, Mohr 1967, 1975c) occur in eastern and central Afar. Their implications concerning the nature and extent of the causative stress fields have been addressed in the forementioned references. Very briefly, the annular faulting of northeastern Afar was originally identified and interpreted as due to a rotational shear response to minor but significant transcurrent components of motion along the extensional fault systems converging on this region. Subsidence of the 90-km-diameter basin enclosed by the annular faulting, and high gravity over this basin, were interpreted as due to relatively near-surface intrusion of high-density rock (Gouin & Mohr 1964, Mohr 1968). These interpretations have not been accepted by Tazieff et al (1972) and the CNR-CNRS Afar Team (1973), who envisage the annular patterns as a deflectional response of the stress field around massive intrusions. Differing interpretations have also been made of the intersecting fault belts of central Afar, but it may be that "instantaneous" conjugate fault systems and consecutive over-printing of fault systems are not mutually exclusive, particularly amidst the complexities of an RRR triple junction. Stress studies in Afar are in their infancy.

The individual faults of the Afar fault belts are more commonly disposed in ratchet than in horst-graben style. Thus in northwestern Afar, the direction of upthrow is usually to the SW side of the fault. The mean spacing of the faults is variable, but averages close to 1000 m. Likewise, the mean throw is of the order of 75, m, though varying over a wide range from less than 10 m to as much as 500 m or more where a belt is margined by a local graben, for example the Dobi (lat. 11.9° N, long. 41.7° E), Kimbiri (lat. 12.8° N, long. 41.2° E), and Gamarri grabens (lat. 11.4° N, long. 41.6° E) (Mohr 1971b). Fracture planes are generally steep at the surface, but block tilting suggests that the planes are gently curved and may flatten out near 10 km depth, the lower limit to strain accumulation in the crust of central Afar (Knetsch 1970, Morton & Black 1975). Crustal extension in the Afar floor has also taken the form of direct dilatation, with resulting gjar (open fissure) notably concentrated along the axial ranges (Barberi & Varet 1970, Harrison et al 1975) and the Wonji fault belt (Gibson 1967).

Attention can be briefly drawn to transcurrent components of movement on Afar faults. In the absence of suitable geological markers in a flat lava pile, transcurrent displacements are not readily evident and may be more common than observations presently suggest, despite the essentially extensional stress environment of Afar.

The best-established transcurrent displacements have occurred along NW–SE trending faults in central and northern Afar, and are sinistral. The evidence stems from earthquake ground fractures (Dakin et al 1971), seismic focal-plane solutions (McKenzie et al 1970, but Fairhead & Girdler 1970 select dextral slip on the ENE nodal plane, which is disputed by Gouin 1975), geological offsets (Tazieff 1970), and analysis of fault patterns in the Dobi graben region (Mohr 1968, 1971b). However, on the northern tangent of the annular fault zone in northeastern Afar, near latitude 12° 35′ N, dextral slip has been identified by the CNR-CNRS Afar Team (1973) on NNW-trending faults that cut and displace presumed earlier WNW faults. The significance of sinistral shear on Red Sea-trend faults in central and northern Afar, subordinate though it is to extension, is highlighted by application of Lensen's (1958) method of fault-slip analysis: for central Afar, this method yields a minimum compressive stress direction of N30^{g}E, which could well approximate the present vector of Arabia-Africa plate separation (Girdler & Styles 1974, 1976, Searle & Ross 1975).

GEOPHYSICS AND CRUSTAL STRUCTURE OF AFAR

Seismic Profiling

Seismic profiling reveals a structure to the Afar crust that differs not only from that of the Ethiopian plateau, the Gulf of Tajura, the Gulf of Aden, and the Red Sea, but also from the geologically comparable region of Iceland.

Five seismic refraction profiles have been made on the floor of Afar, each 120–250 km long (Berckhemer et al 1975). The upper surface of an anomalous mantle ($V_p = $ 7.3–7.6 km sec^{-1}) shoals from a depth of 28–25 km under southern Afar to 16 km under northern Afar, ie. crustal attenuation is most severe under the axial ranges of northern (and eastern) Afar. This anomalous mantle is estimated to be 15–40 km thick (Makris et al 1975), lying on normal mantle with a mean P-wave velocity of 7.95 km sec^{-1} (Searle & Gouin 1971).

The Afar crust comprises three layers, the lowest and thickest layer having a P-wave velocity of 6.6–6.8 km sec^{-1}. On the seismic evidence alone, several interpretations are possible. The layer could be the equivalent of the lower crustal layer (6.6 km sec^{-1}) under the Ethiopian plateau (see Figure 3): this is the interpretation made by Berckhemer et al (1975) themselves, but it requires the presence of only slightly attenuated continental crust under southern and central Afar. The layer could also be equated with layer 3 in the modified oceanic environment of Iceland (Christensen 1974), but there the the ca 5-km-thick, 6.5–6.7 km sec^{-1} layer is much thinner than under Afar. Pálmason & Saemundsson (1974) interpret the Icelandic layer 3 as a densely intruded flood basalt pile, affected by compaction and metamorphism in a region of elevated crustal isotherms. A third interpretation of the Afar layer identifies it as partially melted uppermost mantle (Barberi & Varet 1975b): this would reduce the total crustal thickness in Afar to only 6–7 km. Let us return to layer "3" after examining the other crustal layers of Afar.

Neglecting the superficial sediments comprising layer 1 in an oceanic scheme, the topmost 2–3 km of the Afar crust has an average P-wave velocity of 3.5 km sec^{-1}

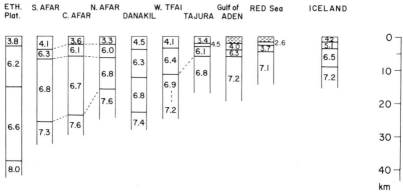

Figure 3 Crustal layers of Afar and related regions, established from seismic refraction profiling (numbers are average P-wave velocities in km sec^{-1}). The left five profiles are from Berckhemer et al (1975): "Eth. Plat." is from a profile across the Ethiopian plateau at latitude 9°N. The S(outhern) Afar profile follows the Wonji fault belt from Awash Station (lat. 8.9° N, long. 40.2 E) to Lake Hertali (lat. 9.9° N, long. 40.3° E). The C(entral) Afar profile extends on NNE from Lake Hertali to Mille (lat. 11.5° N, long. 40.9° E). The N(orthern) Afar profile traverses the Erta-ali axial range. "Danakil" refers to an ENE profile from central Afar to the Red Sea coast at Assab, thus crossing the southern part of the Danakil block. The Western Jibuti and Gulf of Tajura profiles are taken from Ruegg (1975), specifically from Lake Abbe (lat. 11.1° N, long. 42.0° E) to Jibuti (lat. 11.6° N, long. 43.1° E) (Djibouti), and along the northern coast of the Gulf of Tajura between Obock (lat. 12.0° N, long. 43.3° E) and Lake Asal (lat. 11.7° N, long. 42.4° E). The Gulf of Aden and Red Sea profiles are from Laughton et al (1970) and Drake & Girdler (1964) respectively, the first located along the central rift in the western reaches of the Gulf, the second along the axial trough of the Red Sea near latitude 16°N. The generalized Icelandic profile is taken from data in Pálmason & Saemundsson (1974).

under the Salt Plain of northern Afar, increasing to 4.1 km sec^{-1} under southern and eastern Afar. Between this layer and the lowermost layer "3" occurs a 2–5-km-thick layer with velocities of 6.0–6.3 km sec^{-1}. The recognition of this layer has tended to pit geophysicists against geologists! On the basis of seismic velocities alone, this middle layer in the Afar crustal sandwich has no equivalent in the Red Sea, Gulf of Aden, or Iceland (Figure 3). Berckhemer et al (1975) interpret it as attenuated upper continental crust, equivalent to the 12-km-thick, 6.2 km sec^{-1} layer under the central Ethiopian plateau, and to a less attenuated 8-km-thick, 6.2–6.5 km sec $^{-1}$ layer under the southern part of the Danakil block and in Jibuti. On this contentious issue we can note two points. First, the 6.0–6.3 km sec^{-1} Afar layer thickens from south to north, the converse of the underlying layer "3" and also of the geologically inferred degree of crustal attenuation. Second, Ruegg (1975) identifies the 6.1–6.4 km sec^{-1} layer in Jibuti with oceanic layer 3, the appreciably lower P-wave velocity relative to the Gulf of Aden and Red Sea being explained in terms of a presumed higher geothermal gradient. This explanation concurs with the interpretation of Barberi & Varet (1975b) for Afar, but it reduces the crustal thickness to less than 5 km for Jibuti, which is less even than in the Gulf of Aden (Laughton et al 1970). It also requires identification of the 6.8 km sec^{-1} layer under Jibuti with the more typical anomalous mantle velocities of 7.2 km sec^{-1} in the Gulf of Aden and Red Sea.

Magnetotelluric measurements in southwestern Afar (Berktold et al 1975) show resistivities indicative of temperatures of 800–1000°C at 15-km depth, very similar to Iceland (Hermance & Grillot 1970) and implying a high thermal gradient of ca 60°C km^{-1}. The very different geothermal gradients and resulting metamorphic and mineralogical sequences (Gehlen et al 1975) inferred for Afar and the Ethiopian plateau point to the riskiness of equating crustal layers from one tectonic domain with those of another on the basis of seismic velocities alone. The analysis of Berckhemer et al (1975) must be reconciled with geological constraints, and the latter make it improbable that the required thickness of continental crust remains under Afar. On the other hand, the interpretation of Ruegg (1975) and Barberi & Varet (1975b) as applied to Afar requires (a) an unusually thin crust, and (b) a thickening from north to south of the 6.8 km sec^{-1} layer, assumed to be partially melted upper mantle. Both recent axial volcanism and estimated rate of crustal extension (Schaefer 1975, Mohr 1972b) decrease in this direction.

Insight into the nature of the Afar crust comes from a realization of the importance of metamorphic processes in determining seismic layering in oceanic crust (DeWit & Stern 1976, Clague & Straley 1977). Very briefly, I suggest that the 6.8 km sec^{-1} Afar layer is equivalent to oceanic layer "3A," composed of serpentinized and amphibolitized gabbro, resting on fresh gabbros of the 7.3–7.6 km sec^{-1} layer "3B." The excessive thickness of the 6.8 km sec^{-1} Afar layer compared with oceanic or even Icelandic crust is consistent with the slow crustal extension in Afar, and thus more protracted and episodic magmatic processes (Sleep 1975) at the spreading axes above the Afar hot spot.

The middle crustal layer of Afar (V_p = 6.0–6.3 km sec^{-1}) can now be identified with layer "2B" of the oceans (5.8–6.0 km sec^{-1}), a sheeted dike complex that has been metamorphosed to greenschist facies (DeWit & Stern 1976) and does not have to be a sialic layer. The thickening of postulated layer "2B" at the expense of the overlying fresh or zeolitized basalts of layer "2A," from southern to northern Afar, is consistent with increase in the rate of crustal extension.

Afar is surely an ideal candidate for deep borings to elucidate crustal accretion processes above a mantle hot spot.

Gravity

Gravity data from Afar and the bordering plateaus (Gouin & Mohr 1964, Gouin 1970, Makris et al 1975) show that when Afar is compared with Iceland, for any given elevation there is a Bouguer anomaly difference of ca 40 m.gal., being more positive for Iceland (Makris 1975). As there are similar seismic travel-time delays through the mantle under Iceland and Ethiopia, the gravity difference must stem from the lithosphere. We have already seen that seismic profiling reveals an exceptionally thick lower crust under Afar, 10 km thick in the north increasing to 20 km in the south.

Crustal-upper mantle density models for the Ethiopian plateau, Afar, and the Danakil block, constrained by the seismic results, have been presented by Makris et al (1975). A good match of the modelled and observed gravity field ensues when the middle crustal layer is given a density of 2.78 g cm^{-3}, the lower ($V_p = 6.8$ km sec^{-1}) layer a density of 2.90 g cm^{-3}, the anomalous mantle 3.20 g cm^{-3}, and the normal mantle 3.35 g cm^{-3}. Makris et al model a thinned continental crustal layer under Afar that is continuous with thick continental crust under the Ethiopian plateau, but this is not critical to the gravity matching.

Gravity-elevation plots have been discussed by Gouin & Mohr (1964) and Gouin (1970). Afar and the bordering plateaus would seem not to be far from isostatic equilibrium.

Magnetics and Sea-Floor Spreading

The aeromagnetic map of Afar (Girdler & Hall 1972: these authors have not published their map but it is reproduced in United Nations 1973, Figure 66) has been constructed from profiles flown at ca 2000-m elevation and spaced 10 km apart. The contour interval is 100 nT. Lineations of short (< 100 km) length with maximum amplitude ca 1000 nT are immediately evident. The linear pattern is most strongly developed in southern and central Afar, with a grain oriented WNW–W, though elements of WSW, "Gulf of Aden" trend are present in parts of central Afar.

The one and only surficial tectonic feature paralleling the magnetic lineations in southern and central Afar is the belt of closely spaced, WNW-trending faults containing the Gobad (lat. 10.9° N, long. 41.8° E) and related grabens, immediately south of Lake Abbe. Otherwise, the Wonji fault belt, which dominates the tectonics of southwestern Afar, runs almost perpendicular to the lineations. This observation leads to the tantalizing corollary that the magnetic lineations are parallel to the admittedly short offsets in the Wonji fault belt (Mohr 1967, Mohr & Wood 1976), and also near-parallel (but see Wise 1974!) to W–WSW trending thermal structures under southwestern Afar (Berktold et al 1975). Indeed, the strongest, shortest wavelength lineation in all southern Afar matches a structural line identified by Christiansen et al (1975) and Mohr (1967) that runs WNW (330^g) from near Dire Dawa, directly through the Amoissa-Ayalu offset (lat. 10.1° N, long. 40.8° E) in the Wonji fault belt, to the margin of the Ethiopian plateau.

In the ocean basins, magnetic lineations run parallel to lines of crustal accretion and ensuing faulting. The near perpendicularity of lineations and faults in southern Afar signifies either that the lineations were developed in an earlier tectonic episode than the faults, or that if penecontemporaneous then each reflects a different structural level in the lithosphere, tectonism deriving from deeper than the magnetic lineations.

The problem is highlighted by the nonparallelism of magnetic lineations and axial volcanic ranges in northern Afar. Barberi & Varet (1975b, Figure 1) interpret the lineations there in terms of Pliocene-Quaternary sea-floor spreading, and identify anomalies 1, 2 (and 4) symmetrically disposed about the axial ranges themselves. NE-trending discontinuities in the magnetic pattern are identified with transform faults (Barberi & Varet 1975a). This stimulating interpretation of the magnetic pattern in northern and central Afar requires the following comments: first, the northern axial ranges trend at 375^g, the magnetic lineations at 340^g. Second, the interpretation demands that more than one-third the area of Afar was generated by sea-floor spreading during the last 3.5 m.y., which is not consistent with proportions in the Red Sea and Gulf of Aden (see below). Further, the Dubbi volcanic line (see above, and Figure 2), regarded by Barberi & Varet as marking a major transform from Red Sea spreading to Afar spreading, in fact manifests a succession of powerful but short lineations trending approximately east-west. This transverse orientation recalls the pattern in southwestern Afar.

It is unlikely that the obliquity of magnetic lineations and young surface faulting in Afar can be explained until the precise source of the lineations has been identified (see Cande & Kent 1976, Hall 1976), and the earlier history of Afar interpreted from other, geological information. The near-parallelism of magnetic, thermal, and gravity features with cross-rift trends in southwestern Afar and the Ethiopian rift valley (Mohr & Potter 1976) points to early transverse tectonism, preserved in these geophysical parameters, prior to the regional extensional tectonism that has barely reached the stage of crustal accretion. Furthermore, if attenuated continental crust is present under southern Afar, as Makris (1975) and Barberi & Varet (1975a) agree, then the magnetic lineations there cannot be due to mid-Tertiary sea-floor spreading, but might instead be analogous to the unexplained, oblique magnetic lineations of the Gregory rift of Kenya (Wohlenberg & Bhatt 1972).

EVOLUTION AND PLATE TECTONIC CONTEXT OF AFAR

The origin and nature of the Afar depression are tied to the evolution of the Red Sea and Gulf of Aden. After two decades of discussion, it is now generally accepted that sea-floor spreading has produced essentially the whole width of the Red Sea and Gulf of Aden (Laughton et al 1970, Coleman 1974, 1975, Girdler & Styles 1974, 1976). In terms of rigid plate tectonics, this signifies that the floor of Afar must be underlain by new crust. If attenuated continental crust should prove to be present under Afar, then a corresponding amount of extension must have been taken up in the bordering plateaus or else lithospheric horizontal decoupling must have occurred. Neither of these requirements appears likely.

There is good magnetic evidence for two-stage sea-floor spreading of the Gulf of Aden and Red Sea floors (Girdler & Styles 1974, 1976, 1978), though Gass (1977) prefers a single stage on geological evidence. It is clear that relatively rapid sea-floor spreading has been operating in the southern Red Sea for the last 4 m.y., at a half-spreading rate of 0.9 cm yr^{-1} (Searle & Ross 1975). The period of Girdler & Styles' first stage, however, is less sure: they originally identified it from short magnetic cross sections in the southwestern Red Sea as the 41–34-m.y. interval, but from additional data have revised this to the 30–15-m.y. interval. Geological evidence from the Red Sea coast of Arabia concurs that spreading began about 30 m.y. ago (Blank 1975).

How does the history of Afar fit with these data and concepts? The time of initiation of the Afar depression is presently obscure and controversial (Barberi et al 1975, Mohr 1975c). Profuse volcanism ringed the Afar region about 26 m.y. ago, but immediately to the west there is strong evidence that volcanism had begun 10 m.y. earlier than this (Zanettin & Justin-Visentin 1975, Jones 1976). The 26-m.y. date is taken by Barberi et al (1972, 1975) to mark the initiation of Afar. However, even neglecting the probable existence of earlier, mid-Oligocene basalts along the western margin of Afar, there seems no necessity for vertical and horizontal differential crustal movements to have been accompanied by volcanism. We recall the major development of the Gulf of Aden margins during the Upper Eocene, which lacked any associated magmatism. Black et al (1975) have demonstrated that, during the more recent history of eastern Afar, periods of extensional faulting have alternated rather than coincided with fissure volcanism (see Table 2). Blank (1975) considers that the ca 22-m.y.-old dike swarm along the Red Sea coast of southwestern Arabia (Coleman 1975) marks a *terminal* tensional episode in Red Sea development, which began some 30 m.y. ago. This is consistent with the thesis of Briden & Gass (1974) that a time lag of roughly 10 m.y. ensued from cessation of the African plate's drift over the mantle, ca 45–40 m.y. ago, to the first domal uplift and volcanism resulting from "hot spots" focussing on the base of the now static lithosphere.

The initiation of the Afar depression should have resulted in accumulation of contemporaneous sediments. Unfortunately, our knowledge of the Tertiary sediments of Afar is even poorer than for the early volcanics. In virtually all of central and southern Afar, the Pliocene stratoid basalts blanket and hide the nature of any basement under this pile. I estimate that ca 3×10^4 km^3 of rock has been denuded from the Afar plateau margins, at a mean rate of 35 m^3 km^{-2} yr^{-1} or a vertical lowering of .015 mm yr^{-1} (cf. McDougall et al 1975). This hypothesis requires a mean thickness of about 500 m of sediment over the entire floor of Afar, if it were evenly distributed and one excluded the axial volcanic ranges. Such a thickness may well be present in southern Afar, though the oldest exposed sediments there are only about 4 m.y. old (Taieb et al 1976). In central Afar, any significant thickness of sediments must lie under the stratoid pile, excepting the sediment-filled trough along the margin with the Ethiopian plateau, which is expressed in the gravity profile (Gouin & Mohr 1964, Figure 2v).

Accepting a two-stage history of the Red Sea and Gulf of Aden, then three-quarters to four-fifths of the present width of the Afar floor must have been formed during the first stage, especially in southern Afar where there has since been only very slow extension across the Wonji fault belt (Schaefer 1975, Mohr 1973, 1975c, Mohr et al 1978). This conclusion is irreconcilable with the thesis of Berckhemer et al (1975) and Makris et al (1975), which postulates a two- to fourfold thinning of plateau continental crust under Afar. This condition would allow Arabia to have occupied only half to three-quarters of the present area of Afar. Unless the Afar crust had been singularly "prethinned" by some process of subcrustal erosion, the Red Sea and Gulf of Aden likewise could have opened by plate separation for only half to three-quarters of *their* widths. One does not escape this problem by proposing oceanization of the Afar crust (Burek 1974) unless an equivalent process can be shown to have operated in the Red Sea and Gulf of Aden, and the evidence there is for sea-floor spreading (Girdler & Styles 1978).

Can plate tectonics be rigidly applied to Afar? As with Iceland, the need and desire for a plate model stems not from the geology of Afar itself, but from the neighboring, truly oceanic spreading zones. Plate tectonic analysis of these spreading zones, based on geometrical and earthquake-mechanism constraints, has led to simple single-stage models that have had a modicum of success regarding their gross implications for the African rift system, the third arm of the Afar triple junction (McKenzie et al 1970, Le Pichon et al 1973). But a detailed microplate story for Afar, at the locus of the triple junction, is less easy to decipher.

The first microplate model of Mohr (1970, 1972b), based on the idea of Laughton (1966), utilized a double-couple spreading in Afar and the Red Sea, separated by a rotating Danakil-Aisha superblock. The postulated superblock was an entity during the first stage of sea-floor spreading, but was split by the Gulf of Tajura during the second stage at which time the Aisha block remained anchored to the Somali plate. Paleomagnetic measurements confirm that the Danakil block, at least, has undergone an anticlockwise rotation of 20–30° since the mid-Tertiary (Burek 1974). And the proposed, second-stage separation of the Danakil and Aisha blocks is in excellent agreement (Mohr 1975c) with the spreading rates calculated by Girdler & Styles (1974) for the Red Sea at this latitude. Although Boucarut & Clin (1976) claim that the geology of Jibuti precludes any such separation, their evidence is not compelling. Greater geological problems arise from the requirement that the Aisha block slid east along the southern margin of Afar during the first spreading stage. There is no evidence for the required major fracture zone, but a specific study of the Aisha block–Somalian plateau contact has yet to be made. C. A. Wood (personal communication, 1977) suggests that the Aisha block is a part of the Somalian plateau that has been shifted northwards into Afar by sinistral transcurrent movement along the Marda fault zone (Black et al 1974, Purcell 1976). This intriguing idea raises problems of space and timing that cannot be enlarged on here.

Barberi & Varet (1975a) have developed the microplate model of Afar further, focussing on the second, Pliocene-Quaternary spreading episode. Their model is based on interpretation of magnetic anomalies, and requires that at least one-third the area of Afar was generated during this episode. They consider that during the Miocene, Afar developed only so far as a continental rift. The proportions of these two episodes of opening are difficult to reconcile with the spreading of the Red Sea and Gulf of Aden floors, and with the rotation of the Danakil block. Perhaps, as in Iceland (Pálmason & Saemundsson 1974), the magnetic anomaly pattern has been complicated by shifting spreading axes and by widespread flow of extruded lavas.

The Barberi-Varet microplate model proposes a southeastward bifurcation of the Erta-ali spreading line, the two branches reuniting again at the Asal graben in Jibuti. The two discontinuous lines of axial range basalts are indicated in Figure 2. Four NE–SW transform faults are required in internal Afar, plus two large transforms connecting Afar to the Red Sea spreading axis (Barberi & Varet 1975a, Figure 1B).

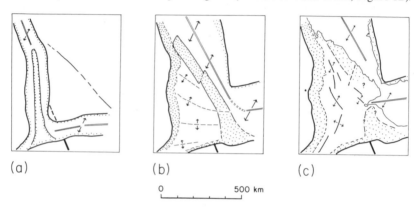

(a)   (b)   (c)

0   500 km

Figure 4   Schematic representation of the opening of Afar, based on ideas and information from Morton & Black (1975), Purcell (1976), and Mohr (1972b, 1974). The Ethiopian plateau is held fixed in this reference frame. Stippled areas mark attenuated continental crust.

(*a*) Configuration at start of first stage of Red Sea floor spreading (time not yet known, possibly early Oligocene). Note that previous to this configuration, there had been Mesozoic-early Tertiary movement along the proto-Red Sea line projecting SE right into the Somali plate as represented now by the Marda fault zone (thick trace); the geometry requires that this line also cut the nose of Yemen. Note also the symmetrical opening, at the southern end of the Red Sea, of the Ethiopian plateau margin and the Ataq fault zone of eastern Yemen: the former was ultimately utilized in making connection with the Gulf of Aden, perhaps because it more closely paralleled the regional Precambrian grain. Finally, the Danakil-Aisha superblock, surprisingly narrow for its length prior to crustal attenuation, became subject to this attenuation because of its location between or even above the loci of crustal spreading processes.

(*b*) Configuration at the end of first stage of Red Sea floor spreading (Middle Miocene). The extension directions shown within Afar do *not* stem from discrete lines of spreading in the positions marked, but are only indicative of regional extension. Note that rigid plate tectonics is not considered to have operated within Afar during this time, and such tectonism acting exteriorly in the Red Sea and Gulf of Aden was accomodated by regional deformation and magmatic processes in Afar itself. Note also that the greater part of Afar opened during the first stage, widening to proportions greater than simply a continental rift.

(*c*) Present-day configuration at the Afar triple junction, showing axial ranges and Wonji fault belt within Afar. Note that the second stage of Red Sea floor spreading, which began ca 4 m.y. ago, has split the Danakil and Aisha attenuated continental blocks along the Gulf of Tajura spreading zone.

The resulting Afar microplates are only some tens to one-hundred kilometers in size, with boundaries of comparable width. This is microplate tectonics with a vengeance! There is ongoing debate as to identification of transform fault zones in Afar (Fairhead & Girdler 1970, Tapponnier & Varet 1973; cf. Boucarut & Clin 1976, Gouin 1975). I would concur with the earlier convictions of Tazieff and co-workers (1972, CNR-CNRS Afar Team 1973) that no transform fault divides the Danakil block along the Dubbi volcanic line. The double-couple model for a united Danakil block appears adequate at the present state of knowledge.

Although the present areas of the Danakil and Aisha blocks can be reduced in the reconstruction of once contiguous Arabia and Africa, by a factor of 3 or 4 due to crustal attenuation (Morton & Black 1975), the presence of these blocks still remains a thorn in any plate tectonic picture. The stumbling block here is our ignorance of Afar evolution during the early spreading of the Red Sea and Gulf of Aden, especially southern Afar with its mysterious east-west magnetic lineations. However, ignorance makes bold, and Figure 4 summarizes schematically yet another proposed evolutionary model of Afar. Essentially, Afar has developed similarly to Iceland with the build-up of a terrestrial basalt pile, but a slower rate of spreading allied to a more dispersed and changing pattern of dike-feeder zones in a triple junction region has smudged out the mid-ocean ridge-rift process.

Nevertheless, the timing of Afar volcanic and tectonic episodes has a clear, if not one-to-one relationship with the spreading history of the Red Sea and Gulf of Aden as proposed by Girdler & Styles (1978). Volcanism and tectonism mark both the start (ca 26 m.y.) and cessation (ca 15 m.y.) of the possible first stage of spreading. In particular, profuse fissure basalts were erupted between 26 and 19 m.y. ago, to build a pile whose total thickness exceeds 3000 m at the margins of southern Afar. The cessation of spreading coincided with the first voluminous silicic volcanism in Afar, during the 15-11 m.y. interval. This is consistent with the notion of a large, spreading axis magma chamber left static 15 m.y. ago, so that massive fractionation could ensue. Tectonic development of the Afar margins some 9 m.y. ago, and the eruption of relatively small volumes of flood basalt during the following 2-3 million years, was not apparently accompanied by significant crustal spreading in the Red Sea and Gulf of Aden. Resumption of spreading about 4 m.y. ago was marked by a second voluminous basaltic volcanism, which however persisted for only ca 1 m.y. before the lava pile was broken by belts of intense faulting. Since about 3-2 m.y. ago, volcanism has been restricted to axial ranges with an accompanying narrowing of the zones of faulting.

If one may hazard a guess at the future development of Afar, it is that the underlying mantle plume/hot spot will maintain the region above sea level through excessive magmatism, despite continued separation of Arabia from Africa. The Ethiopian navy will never sail up to Addis Ababa.

ACKNOWLEDGMENTS

Mr. George Savard (University College of Addis Ababa) in 1958 gently and firmly informed me that the native name of Afar was more appropriate than the terms I had previously borrowed from the literature. Dr. Lucien Matte and Prof. Pierre Gouin gave me the impetus and the means to explore the reality beyond the name.

Literature Cited

Abbate, E., Sagri, M. 1969. Dati e considerazioni sul margine orientale dell' altopiano etiopico nelle province del Tigrai e del Wollo. *Boll. Soc. Geol. Ital.* 88:489–97

Baker, B. H., Williams, L. A. J., Miller, J. A., Fitch, F. J. 1971. Sequence and geochronology of the Kenya rift volcanics. *Tectonophysics* 11:191–215

Bannert, D., Brinckmann, J., Jordan, R., Kürsten, M., Ochse, G., Ries, H., Schmid, F. 1971. Beiträge zur Geologie der Danakil-Senke (NE-Athiopien). *Beih. Geol. Jahrb.* 116:1–199

Bannert, D., Brinckmann, J., Käding, K.-C., Knetsch, G., Kürsten, M., Mayrhofer, H. 1970. Zur Geologie der Danakil-Senke (N.E. Athiopien). *Geol. Rundsch.* 59:409–43

Barberi, F., Bonatti, E., Marinelli, G., Varet, J. 1974. Transverse tectonics during the split of a continent: data from the Afar rift. *Tectonophysics* 23:17–29

Barberi, F., Borsi, S., Ferrara, G., Marinelli, G., Santacroce, R., Tazieff, H., Varet, J. 1972. Evolution of the Danakil Depression (Afar, Ethiopia) in light of radiometric age determinations. *J. Geol.* 80:720–29

Barberi, F., Borsi, S., Ferrara, G., Marinelli, G., Varet, J. 1970. Relations between tectonics and magmatology in the northern Danakil Depression (Ethiopia). *Philos. Trans. R. Soc. London Ser. A* 267:293–311

Barberi, F., Ferrara, G., Santacroce, R., Varet, J. 1975. Structural evolution of the Afar triple junction. See Pilger & Rösler 1975, pp. 38–54

Barberi, F., Varet, J. 1970. The Erta Ale volcanic-range (Danakil Depression, Northern Afar, Ethiopia). *Bull. Volcanol.* 34:848–917

Barberi, F., Varet, J. 1974. Silicic peralkaline volcanic rocks of the Afar depression (Ethiopia). *Bull Volcanol.* 38:755–90

Barberi, F., Varet, J. 1975a. Recent volcanic units of Afar and their structural significance (A summary). See Pilger & Rösler 1975, pp. 174–78

Barberi, F., Varet, J. 1975b. Nature of the Afar crust: A discussion. See Pilger & Rösler 1975, pp. 375–78

Behle, A., Makris, J., Baier, B., Delibasis, N. 1975. Salt thickness near Dallol (Ethiopia) from seismic reflection measurements and gravity data. See Pilger & Rösler 1975, pp. 156–67

Berckhemer, H., Baier, B., Bartelsen, H., Behle, A., Burkhardt, H., Gebrande, H.,

Makris, J., Menzel, H., Miller, H., Vees, R. 1975. Deep seismic soundings in the Afar region and on the highland of Ethiopia. See Pilger & Rösler 1975, pp. 89–107

Berktold, A., Haak, V., Angenheister, G. 1975. Magnetotelluric measurements in the Afar area. See Pilger & Rösler 1975, pp. 66–79

Black, R., Morton, W. H., Rex, D. C. 1975. Block tilting and volcanism within the Afar in the light of recent K/Ar age data. See Pilger & Rösler 1975, pp. 296–300

Black, R., Morton, W. H., Rex, D. C., Shackleton, R. M. 1972a. Sur la découverte en Afar (Ethiopie) d'un granite hyperalcalin miocène: le massif de Limmo. *C.R. Acad. Sci. Ser. D* 274:1453–56

Black, R., Morton, W. H., Tsegaye Hailu 1974. Early structures around the Afar triple junction. *Nature* 248:496–97

Black, R., Morton, W. H., Varet, J. 1972b. New data on Afar tectonics. *Nature Phys. Sci.* 240:170–73

Blank, H. R. 1975. Aeromagnetic delineation of Miocene dikes and related structures in the Arabian margin of the Red Sea. *Trans. Am. Geophys. Union* 56 (6): 458 (Abstr.)

Boucarut, M., Clin, M. 1976. Conceptions nouvelles sur les relations Mer Rouge-Golfe d'Aden, à partir du Territoire des Afars et des Issas. *C.R. Acad. Sci.* 282:1145–48

Briden, J. C., Gass, I. G. 1974. Plate movement and continental magmatism. *Nature* 248:650–53

Brinckmann, J., Kürsten, M. 1971. Stratigraphie und Tektonik der Danakil-Senke (NE-Athiopien). *Beih. Geol. Jahrb.* 116:5–86

Brotzu, P., Morbidelli, L., Piccirillo, E. M., Traversa, G. 1974. Petrological features of the Boseti Mountains, a complex volcanic system in the axial portion of the main Ethiopian rift. *Bull. Volcanol.* 38:206–34

Burek, P. J. 1974. Plattentektonische Probleme in der weiteren Umgebung Arabiens sowie der Danakil-Afar-Senke. *Geotektonische Forsch.* 47:1–93

Cande, S. C., Kent, D. V. 1976. Constraints imposed by the shape of marine magnetic anomalies on the magnetic source. *J. Geophys. Res.* 81:4157–62

Chase, R. L. 1969. Basalt from the axial trough of the Red Sea. In *Hot Brines and Recent Heavy Metal Deposits in the Red Sea*, ed. E. T. Degens, D. A. Ross, pp. 122–28. New York: Springer. 600 pp.

Chessex, R., Delaloye, M., Müller, J., Weidmann, M. 1975. Evolution of the volcanic region of Ali Sabieh (T.F.A.I.), in the light of K-Ar age determinations. See Pilger & Rösler 1975, pp. 221–27

Christensen, N. I. 1974. The petrologic nature of the lower oceanic crust and upper mantle. In Geodynamics of Iceland and the North Atlantic area, ed. L. Kristjansson, pp. 165–76. Dordrecht: Reidel. 323 pp.

Christiansen, T. B., Schaefer, H.-U., Schönfeld, M. 1975. Geology of southern and central Afar. See Pilger & Rösler 1975, pp. 259–77

Civetta, L., De Fino, M., Gasparini, P., Ghiara, M. R., La Volpe, L., Lirer, L. 1975. Geology of central-eastern Afar (Ethiopia). See Pilger & Rösler 1975, pp. 201–06

Clague, D. A., Straley, P. F. 1977. Petrologic nature of the oceanic Moho. Geology 5: 133–36

CNR-CNRS Afar Team 1973. Geology of northern Afar (Ethiopia). Rev. Geogr. Phys. Geol. Dynam. (2) 15: 443–90

Cole, J. W. 1969. Gariboldi volcanic complex, Ethiopia. Bull. Volcanol. 33: 566–78

Coleman, R. G. 1974. Geologic background of the Red Sea. In Geology of Continental Margins, ed. C. A. Burk, C. L. Drake, pp. 743–51. New York: Springer. 1009 pp.

Coleman, R. G. 1975. A Miocene ophiolite on the Red Sea coastal plain. Trans. Am. Geophys. Union 56 (12): 1080 (Abstr.)

Comucci, P. 1933. Rocce dello Iemen raccolte dalla missione de S.E. Gasparini. Period. Mineral. 4: 89–131

d'Abbadie, A. 1873. Géodesie d'Ethiopie, Tome II. Paris

Dakin, F. M., Gouin, P., Searle, R. C. 1971. The 1969 earthquakes in Serdo (Ethiopia). Bull. Geophys. Obs. Addis Ababa 13: 19–56

De Fino, M., La Volpe, L., Lirer, L., Varet, J. 1973. Geology and petrology of Manda-Inakir range and Moussa Alli volcano, central eastern Afar (Ethiopia and T.F.A.I.). Rev. Geogr. Phys. Geol. Dynam. (2) 15: 373–86

Delibrias, G., Marinelli, G., Stieltjes, L. 1975. Spreading rate of the Asal Rift: a geological approach. See Pilger & Rösler 1975, pp. 214–21

DeWit, M. J., Stern, C. R. 1976. A model for ocean-floor metamorphism, seismic layering and magnetism. Nature 264: 615–19

Dow, D. B., Beyth, M., Tsegaye Hailu 1971. Palaeozoic glacial rocks recently discovered in northern Ethiopia. Geol. Mag. 108: 53–59

Drake, C. L., Girdler, R. W. 1964. A geophysical study of the Red Sea. Geophys. J. R. Astron. Soc. 8: 473–95

Dreyfuss, M. 1931. Etudes de géologie et de géographie physique sur la Côte Française des Somalis. Rev. Geogr. Phys. Geol. Dynam. (1) 4: 338–47

Fairhead, J. D., Girdler, R. W. 1970. The seismicity of the Red Sea, Gulf of Aden and Afar triangle. Philos. Trans. R. Soc. London Ser. A 267: 49–74

Filjak, R., Glumicic, N., Zagorac, Z. 1959. Oil Possibilities of the Red Sea Region in Ethiopia. Zagreb: Naftaplin. 104 pp.

Frazier, S. B. 1970. Adjacent structures of Ethiopia: that portion of the Red Sea coast including Dahlak Kebir Island and the Gulf of Zula. Philos. Trans. R. Soc. London Ser. A 267: 131–41

Gass, I. G. 1977. The age and extent of the Red Sea oceanic crust. Nature 265: 722–24

Gasse, F. 1975. Fluctuations of the Afar lake levels during the late Quaternary period. See Pilger & Rösler 1975, pp. 284–88

Gasse, F., Rognon, P. 1973. Le Quaternaire des bassins lacustres de l'Afar. Rev. Geogr. Phys. Geol. Dynam. (2) 15: 405–14

Gehlen, K., Forkel, W., Spies, O. 1975. Petrological interpretation of geophysical data from the Afar Depression, Ethiopia. See Pilger & Rösler 1975, pp. 151–55

Geukens, F. 1960. Contribution à la géologie du Yemen. Mem. Inst. Geol. Louvain 21: 117–80

Gibson, I. L. 1967. Preliminary account of the volcanic geology of Fantale, Shoa. Bull. Geophys. Obs. Addis Ababa 10: 59–67

Girdler, R. W., Hall, S. A. 1972. An aeromagnetic survey of the Afar triangle of Ethiopia. Tectonophysics 15: 53

Girdler, R. W., Styles, P. 1974. Two stage Red Sea floor spreading. Nature 247: 7–11

Girdler, R. W., Styles, P. 1976. The relevance of magnetic anomalies over the southern Red Sea and Gulf of Aden to Afar. In Afar between Continental and Oceanic Rifting, ed. A. Pilger & A. Rösler, pp. 156–70. Stuttgart: Schweizerbart. 216 pp.

Girdler, R. W., Styles, P. 1978. Sea floor spreading in the western Gulf of Aden. Nature. In press

Gouin, P. 1970. Seismic and gravity data from Afar in relation to surrounding areas. Philos. Trans. R. Soc. London Ser. A 267: 339–58

Gouin, P. 1975. Kara Kore and Serdo epicenters: relocation and tectonic implications. Bull Geophys. Obs. Addis Ababa 15: 15–25

Gouin, P. 1978. Earthquake history of Ethiopia. *Bull. Geophys. Obs. Addis Ababa* 16: In press

Gouin, P., Mohr, P. A. 1964. Gravity traverses in Ethiopia (interim report). *Bull. Geophys. Obs. Addis Ababa* 7: 185–239

Hall, J. M. 1976. Major problems regarding the magnetization of oceanic crustal layer. 2. *J. Geophys. Res.* 81: 4223–30

Harrison, C. G. A., Bonatti, E., Stieltjes, L. 1975. Tectonism of axial valleys in spreading centers: data from the Afar rift. See Pilger & Rösler 1975, pp. 178–98

Hermance, J. F., Grillot, L. R. 1970. Correlation of magnetotelluric, seismic and temperature data from southwest Iceland. *J. Geophys. Res.* 75: 6582–91

Hutchinson, R. W., Engels, G. G. 1970. Tectonic significance of regional geology and evaporite lithofacies in northeastern Ethiopia. *Philos. Trans. R. Soc. London Ser. A* 267: 313–29

Hutchison, R., Gass, I. G. 1971. Mafic and ultramafic inclusions associated with undersaturated basalt on Kod Ali island, southern Red Sea. *Contr. Mineral. Petrol.* 31: 94–101

Jones, P. W. 1976. Age of the lower flood basalts of the Ethiopian plateau. *Nature* 261: 567–69

Juch, D. 1975. Geology of the south-eastern escarpment of Ethiopia between 39° and 42° long. East. See Pilger & Rösler 1975, pp. 310–16

Justin-Visentin, E., Zanettin, B. 1974. Dike swarms, volcanism and tectonics of the western Afar margin along the Kombolcha-Eloa traverse (Ethiopia). *Bull. Volcanol.* 38: 187–205

Karrenberg, H. 1957. Junger magmatismus und vulkanismus in Südwestarabien (Jemen). *20th Int. Geol. Cong., Mexico City, 1956,* Vulcanologia del Cenozoico, 1: 171–85

Kazmin, V. 1975. Explanation of the geological map of Ethiopia. *Geol. Surv. Ethiopia Bull.* 1: 1–14

Knetsch, G. 1970. Danakil-Reconnaissance 1968. In *Graben Problems,* ed. J. H. Illies, St. Mueller, pp. 267–79. Stuttgart: Schweizerbart. 316 pp.

Kunz, K., Kreuzer, H., Müller, P. 1975. Potassium-argon age determinations of the Trap basalt of the south-eastern part of the Afar Rift. See Pilger & Rösler 1975, pp. 370–74

Kürsten, M. O. C. 1975. Stratigraphic units of northern Afar. See Pilger & Rösler 1975, pp. 168–70

Laughton, A. S. 1966. The Gulf of Aden, in relation to the Red Sea and the Afar depression of Ethiopia. *Geol. Surv. Canada Pap.* 66–14: 78–97

Laughton, A. S., Whitmarsh, R. B., Jones, M. T. 1970. The evolution of the Gulf of Aden. *Philos. Trans. R. Soc. London Ser. A* 267: 227–66

Lensen, G. J. 1958. Rationalized fault interpretation. *New Zealand J. Geol. Geophys.* 1: 307–17

Le Pichon, X., Francheteau, J., Bonnin, J. 1973. *Plate Tectonics–Developments in Geotectonics,* Vol. 6. Amsterdam: Elsevier. 300 pp.

Makris, J. 1975. Afar and Iceland—a geophysical comparison. See Pilger & Rösler 1975, pp. 379–90

Makris, J., Menzel, H., Zimmermann, J., Gouin, P. 1975. Gravity field and crustal structure of north Ethiopia. See Pilger & Rösler 1975. pp. 135–44

McDougall, I., Morton, W. H., Williams, M. A. J. 1975. Age and rates of denudation of Trap Series basalts at Blue Nile gorge, Ethiopia. *Nature* 254: 207–9

McKenzie, D. P., Davies, D., Molnar, P. 1970. Plate tectonics of the Red Sea and East Africa. *Nature* 226: 243–48

Megrue, G. H., Norton, E., Strangway, D. W. 1972. Tectonic history of the Ethiopian Rift as deduced by K-Ar ages and paleomagnetic measurements of basaltic dikes. *J. Geophys. Res.* 77: 5744–54

Merla, G., Minucci, E. 1938. *Missione geologica nel Tigrai.* Accad. Ital., Centro Studi A.O.I. 3: 1–363

Miller, J. A., Mohr, P. A., Rogers, A. S. 1967. Some new K/A age determinations of Basement rocks from Eritrea. *Bull. Geophys. Obs. Addis Ababa* 10: 53–7

Mohr, P. A. 1962. *The Geology of Ethiopia.* Addis Ababa: Univ. College. 268 pp.

Mohr, P. A. 1967. The Ethiopian rift system. *Bull. Geophys. Obs. Addis Ababa* 11: 1–65

Mohr, P. A. 1968. Annular faulting in the Ethiopian rift system. *Bull. Geophys. Obs. Addis Ababa* 12: 1–9

Mohr, P. A. 1970. The Afar triple junction and sea-floor spreading. *J. Geophys. Res.* 75: 7340–52

Mohr, P. A. 1971a. Ethiopian Tertiary dike swarms. *Smithsonian Astrophys. Obs. Spec. Rep.* 339: 1–53

Mohr, P. A. 1971b. Tectonics of the Dobi graben region, central Afar, Ethiopia. *Bull. Geophys. Obs. Addis Ababa* 13: 73–89

Mohr, P. A. 1972a. Regional significance of volcanic geochemistry in the Afar triple junction, Ethiopia. *Bull. Geol. Soc. Am.* 83: 213–22

Mohr, P. A. 1972b. Surface structure and

plate tectonics of Afar. *Tectonophysics* 15:3–18

Mohr, P. A. 1973. Crustal deformation rate and the evolution of the Ethiopian rift. In *Continental drift, sea floor spreading and plate tectonics: implications to the earth sciences*, ed. D. H. Tarling, S. K. Runcorn, pp. 767–76. London: Academic. 1184 pp.

Mohr, P. A. 1974. Mapping of the major structures of the African rift system. *Smithsonian Astrophys. Obs. Spec. Rep.* 361: 70 pp. & 15 maps

Mohr, P. A. 1975a. A new terminology for the Ethiopian volcanics, with especial reference to transitional basaltic and intermediate lavas and dikes. *Cent. Astrophys. Prepr. Ser.* 368: 39 pp.

Mohr, P. A. 1975b. Structural elements of the Afar margins: data from ERTS-1 imagery. *Bull. Geophys. Obs. Addis Ababa* 15:83–89

Mohr, P. A. 1975c. Structural setting and evolution of Afar. See Pilger & Rösler 1975, pp. 27–37

Mohr, P. A., Girnius, A., Rolff, J. 1978. Crustal strain rates at the northern end of the Ethiopian rift valley. *Tectonophysics.* 38: In press

Mohr, P. A., Gouin, P. 1967. Gravity traverses in Ethiopia (third interim report). *Bull. Geophys. Obs. Addis Ababa* 10:15–52

Mohr, P. A., Potter, E. C. 1976. The Sagatu Ridge dike swarm, Ethiopian rift margin. *J. Volcanol. Geotherm. Res.* 1:55–71

Mohr, P. A., Rogers, A. S. 1966. Gravity traverses in Ethiopia (second interim report). *Bull. Geophys. Obs. Addis Ababa* 9:7–58

Mohr, P. A., Wood, C. A. 1976. Volcano spacings and lithospheric attenuation in the Eastern Rift of Africa. *Earth Planet. Sci. Lett.* 33:126–44

Morton, W. H., Black, R. 1975. See Pilger & Rösler 1975, pp. 55–65

Needham, H. D., Choukroune, P., Cheminee, J. L., Le Pichon, X., Francheteau, J., Tapponnier, P. 1976. The accreting plate boundary: Ardoukôba rift (North-east Africa) and the oceanic rift valley. *Earth Planet. Sci. Lett.* 28:439–53

Ottonello, G., Vannucci, R., Bezzi, A., Piccardo, G. B. 1975. Genetic relationships between ultramafic xenoliths and enclosing alkali basalts in the Assab region (Afar, Ethiopia) based on their trace element geochemistry. See Pilger & Rösler 1975, pp. 206–14

Pálmason, G., Saemundsson, K. 1974. Iceland in relation to the mid-Atlantic Ridge. *Ann. Rev. Earth Planet. Sci.* 2:25–50

Pilger, A., Rösler, A., eds. 1975. *Afar Depression of Ethiopia, Inter-Union Comm. Geodyn. Sci. Rep. no. 14.* Stuttgart: Schweizerbart

Purcell, P. G. 1976. The Marda fault zone, Ethiopia. *Nature* 261:569–71

Rognon, P., Gasse, F. 1973. Dépôts lacustres quaternaires de la basse vallée de l'Awash (Afar, Ethiopie): leurs rapports avec la tectonique et le volcanisme sous-aquatique. *Rev. Geogr. Phys. Geol. Dynam.* (2) 15:295–316

Ruegg, J. C. 1975. Main results about the crustal and upper mantle structure of the Djibouti region (T.F.A.I.). See Pilger & Rösler 1975, pp. 120–34

Schaefer, H.-U. 1975. Investigations on crustal spreading in southern and central Afar (Ethiopia). See Pilger & Rösler 1975, pp. 289–96

Schilling, J.-G. 1969. Red Sea floor origin: rare earth evidence. *Science* 165:1357–59

Schilling, J.-G. 1973. Afar mantle plume: rare earth evidence. *Nature Phys. Sci.* 242:2–5

Searle, R. C., Gouin, P. 1971. An analysis of some local earthquake phases originating near the Afar triple junction. *Bull. Seismol. Soc. Am.* 61:1061–71

Searle, R. C., Ross, D. A. 1975. A geophysical study of the Red Sea axial trough between 20.5° and 22°N. *Geophys. J. R. Astron. Soc.* 43:555–72

Sickenberg, O., Schönfeld, M. 1975. The Chorora Formation—Lower Pliocene limnical sediments in the southern Afar (Ethiopia). See Pilger & Rösler 1975, pp. 277–84

Sleep, N. H. 1975. Formation of oceanic crust: some thermal constraints. *J. Geophys. Res.* 80:4037–42

Taieb, M., Johanson, D. C., Coppens, Y., Aronson, J. L. 1976. Geological and palaeontological background of the Hadar hominid site, Afar, Ethiopia. *Nature* 260:289–93

Tapponnier, P., Varet, J. 1973. La zone de Mak'arrasou en Afar: un équivalent émergé des "failles transformantes" océaniques. *C.R. Acad. Sci. Ser. D* 278:209–12

Tazieff, H. 1970. The Afar Triangle. *Sci. Am.* 222:32–40

Tazieff, H. 1971. Sur la tectonique de l'Afar central. *C.R. Acad. Sci. Ser. D* 272:1055–58

Tazieff, H., Marinelli, G., Barberi, F., Varet, J. 1969. Géologie de l'Afar septentrional—première expédition du CNRS-France et du CNR-Italie (decembre '67—fevrier '68). *Bull. Volcanol.* 33:1039–72

Tazieff, H., Varet, J., Barberi, F., Giglia, G. 1972. Tectonic significance of the Afar (or Danakil) Depression. *Nature* 235: 144–47

Tazieff, H., Varet, J. 1969. Signification tectonique et magmatique de l'Afar septentrional (Ethiopie). *Rev. Geogr. Phys. Geol. Dynam.* (2) 11: 429–50

United Nations, 1973. Geology, geochemistry and hydrology of hot springs of the East African rift system within Ethiopia. *U.N. Tech. Rep.* DP/SF/UN/116. New York: United Nations. 220 pp.

Voute, C. 1959. The Assab region: its geology and history during the Quaternary period. *Bull. Geophys. Obs. Addis Ababa* 2: 73–101

Wise, D. U. 1974. Linesmanship: guidelines for a thriving geological artform. In *Proc.*

First Int. Conf. New Basement Tectonics, ed. R. A. Hodgson, S. P. Gay, J. Y. Benjamins, pp. 635–36. Utah Geol. Assoc. Publ. 5

Wohlenberg, J., Bhatt, N. V. 1972. Report on airmagnetic surveys of two areas in the Kenya rift valley. *Tectonophysics* 15: 143–49

Zanettin, B., Gregnanin, A., Justin-Visentin, E., Nicoletti, M., Petrucciani, C., Piccirillo, E. M., Tolomeo, L. 1974. Migration of the Oligocene-Miocene ignimbritic volcanism in the central Ethiopian plateau. *Neues Jahrb. Geol. Paläontol.* Monatsh. 1974: 567–74

Zanettin, B., Justin-Visentin, E. 1975. Tectonical and volcanological evolution of the western Afar margin (Ethiopia). See Pilger & Rösler 1975, pp. 300–309

Ann. Rev. Earth Planet. Sci. 1978. 6: 173–204

THE INTERSTELLAR WIND AND ITS INFLUENCE ON THE INTERPLANETARY ENVIRONMENT

✳10091

Gary E. Thomas

Department of Astro-Geophysics and Laboratory for Atmospheric and Space Physics, University of Colorado, Boulder, Colorado 80309

INTRODUCTION

The existence of diffuse matter between the stars has been known for decades. However, only recently have we known of its presence in the immediate environs of the solar system. The discovery in 1970 of neutral interstellar gas pervading the solar system was a by-product of space exploration—specifically of the study of planetary atmospheres by ultraviolet photometry and spectroscopy. Because most of the work in this field was done by aeronomers using techniques borrowed from studies of the ultraviolet airglow, these new findings have not yet been widely disseminated in the astronomical community. However, studies of the interstellar medium have much to gain from these new results. To illustrate this point, we consider the hierarchy of sizes of neutral interstellar gas in numbers of parsecs (1 pc = 3×10^{18} cm): the galaxy (10^4), large interstellar clouds and spiral arms (10^2–10^3), small clouds (10^0), intercloud eddies (10^{-2}–10^{-1}), and the mean free path (10^{-3}–10^{-2}). Radio-astronomical techniques probe the largest scales (10^2–10^4), and absorption spectroscopy applies to smaller distances (10^1–10^3). In contrast, the ultraviolet back-scattering of solar resonance lines from the local gas constitutes literally a microscopic probe on a scale of 10^{-4} pc. It is as if a geologist were suddenly given the magnifying power to examine an individual atom of a rock sample, where before he was able to view only the individual grains. Unfortunately, we have only a single sample of the interstellar gas that we can study with such microscopic precision. Offsetting this disadvantage is the fact that for this "tiny" scale, the relevant laws of physics are simply those of single-particle motion (this is *not* true of the ionized component of the gas), and it is straightforward to infer the dynamical and chemical properties of the neutral gas. These properties can in turn be extrapolated to larger scales to infer the gas temperature and the chemical composition of at least the smallest irregularity in which the Sun is imbedded. This extrapolation is far safer

173

0084-6597/78/0515-0173$01.00

than the conventional extrapolation in astrophysics, which infers local properties from grossly averaged line-of-sight observations.

The relative motion of the Sun and the nearby gas causes an "interstellar wind" of velocity ~ 20 km sec^{-1}. As the gas approaches the Sun, it suffers a wide variety of processes, mainly due to the solar wind, which acts as a highly discriminating filter. It "neutralizes" the inflowing gas to a high degree by sweeping away the interstellar plasma and magnetic field to distances beyond ~ 75 (au). The resultant neutral gas flows virtually unimpeded by the solar wind to within a few au of the Sun. At that point the solar wind flux is sufficiently high to ionize the hydrogen component, whereupon the ionized fragments are swept away electromagnetically by the solar wind. This "de-hydrogenation" results in a helium-rich gas in the vicinity of the earth. Well inside Mercury's orbit, helium and other species are photoionized and subsequently swept away.

In this review we examine the details of the interaction of the Sun's emissions with the interstellar gas. First we discuss the general astronomical setting within which the solar system is imbedded. We next discuss the physical processes by which the interstellar gas and the Sun's emissions mutually interact. Within this framework we then describe a theoretical model of the neutral gas distribution and predict the backscattered intensity, which will then be compared to the measurements. Finally, we describe the measurements of the ultraviolet emission lines scattered from the two observed interstellar species, hydrogen and helium, and summarize the more recent observations in terms of what has been learned about the gas density, temperature, and flow velocity.

The scope of the review is deliberately narrow in some areas, partly because of space considerations, but more importantly because several aspects of the problem have already received full attention in a recent review by Holzer (1977b). In particular, he has reviewed the plasma interactions in greater detail. In addition, he has discussed alternative sources of neutral gas in interplanetary space, which is not covered here (outside 1 au, these sources are of minor importance). Finally, he describes some interesting consequences of the passage of the Sun through a dense interstellar cloud. Earlier reviews were given by Tinsley (1971), Axford (1972), Fahr (1974), and Thomas (1975). Recent reviews that have been helpful to me are those of Spitzer & Jenkins (1975) and Snow (1976), who describe recent results from Copernicus measurements, and Vidal-Madjar (1977) and Maloy et al (1978), who have reviewed the current status of the measurements of solar resonance lines.

THE GALACTIC SETTING

The Sun is located on the outskirts of our galaxy in the vicinity of a spiral arm, and is surrounded by a predominantly hydrogen gas with a temperature of $T \sim 10^4$ K and a density of $n_H \sim 0.1$ cm^{-3}. The properties of this local interstellar medium (LISM) are derived from interplanetary measurements and are distinct from the *average* properties of the interstellar medium (ISM) derived from astronomical measurements over enormously larger distance scales. The LISM gas flows past the Sun, with an apparent motion from the direction near the star χ Sco. It moves nearly in the galactic plane toward galactic longitude $l^{II} = 120°$ and somewhat faster than the local frame of rest, which is the center of motion of the nearest few hundred stars. While interplanetary measurements probe the nearby gas within a few tens of astronomical units (au), the techniques of stellar absorption spectroscopy (Bohlin 1975, Bohlin, Savage & Drake 1977) generally yield line-of-sight information over distances of 10^1–10^3 pc (1 pc = 2 × 10^5 au).

Only recently has information become available for distances r more closely approaching the scale of inhomogeneity of the gas. These analyses are based on the shape of the hydrogen Lyman alpha (Lα) and deuterium Lα absorption profiles superimposed on the chromospheric emission line of late-type stars. Estimates or upper limits for the hydrogen (H) column densities for $r < 30$ pc were reported by Moos et al (1974), McClintock et al (1975a,b), Dupree (1975), and Evans, Jordan & Wilson (1975) using the low-resolution (U2: ~0.2 Å) tube of the *Copernicus* satellite spectrometer. Some values were remarkably low. For example, Dupree gave $\langle n_H \rangle$ <0.01–0.02 cm^{-3} for the average hydrogen density over the 14-pc distance to α Aur. This small value was probably a consequence of unresolved geocoronal Lα contamination, and also an overestimate of the intrinsic central reversal of the stellar line. It was possible in more recent high-resolution measurements with the U1 (~0.06 A) tube on *Copernicus* to subtract the geocoronal influence and better resolve the interstellar line profiles. These improved data now yield a positive result for α Aur of 0.02–0.04 cm^{-3} for $\langle n_H \rangle$ (Dupree, Baliunas & Shipman 1977), a value that is typical for the 5 measured stars in the 11–21 pc range of distances.

There are now reports of measurements for the region inside 3.5 pc (which we call the *nearby* interstellar medium, or NISM). Analysis of data from ε Eri ($r = 3.3$ pc) yielded $\langle n_H \rangle$ ~0.1 cm^{-3}, $\langle v_r \rangle = 15.4 \pm 7.9$ km s^{-1}, and a velocity dispersion equivalent to a temperature range of $0.4 < T < 1.2 \times 10^4$ K (McClintock et al 1976). More recent high-resolution observations for ε Ind ($r = 3.5$ pc) and α CMi ($r = 3.5$ pc) are consistent with the above values (McClintock et al 1976, Anderson et al 1977, unpublished manuscripts). Dupree, Baliunas & Shipman (1977) have reported a value for $\langle n_H \rangle$ of 0.2 ± 0.5 cm^{-3} for the 1-pc distance to α Cen A. However, this result applies to a particular velocity dispersion. McClintock et al show that an equally good fit to the α Cen A *Copernicus* data for a somewhat-higher velocity dispersion yields $\langle n_H \rangle = 0.1$ cm^{-3}.

It appears that the interstellar medium within 3.5 pc of the Sun is more dense than that beyond 3.5 pc. There is indirect evidence of a *very* low average density in the direction of the white dwarf HZ 43 (Margon et al 1976). If the model of the extreme ultraviolet flux of this star by Auer & Shipman (1977) is correct, its visibility at wavelengths 100–300 Å implies the existence of a "tunnel" with $\langle n_H \rangle$ less than 0.01 cm^{-3} extending some 60 pc away from the galactic plane. Apparently, the Sun is imbedded in a condensation that nevertheless has a density somewhat smaller than the average intercloud density of 0.16 cm^{-3} (Bohlin, Savage & Drake 1977), and yet is surrounded by a still more tenuous gas.

From measurements of more distant hot stars, molecular hydrogen (H$_2$) is observed (Spitzer, Cochran & Hirschfeld 1975). H$_2$ is abundant (greater than 5% of H) in regions of large H density $\langle n_H \rangle$ ~0.7 cm^{-3}, and correlates better with $\langle n_H \rangle$ than with dust abundance as measured by reddening of starlight (Bohlin 1975). Recent evidence (York 1976) suggests that H_2 is present even along lines of sight to un-reddened stars.

At radio-astronomical wavelengths, 21-cm data yield $\langle n_H \rangle \sim 0.7$ cm^{-3}, 3–4 times larger than the average derived from Lα absorption measurements. Possibly the latter are subject to selection effects in that $\langle n_H \rangle$ applies to the line of sight to OB stars, over which the ionization of H may be greater than in the ISM gas as a whole (Bohlin 1975).

The presence of dust in the LISM has been examined by Bertaux & Blamont (1976). Using data from micrometeroid detectors in space, they find that the dust–gas ratio must be considerably smaller than the galactic mean ($\sim 1\%$ by mass) unless the particles are prevented from penetrating the inner solar system. An effective screening mechanism is the solar wind $\overline{V} \times \overline{B}$ force if the dust grains are electrically charged (Levy & Jokipii 1976). McClure & Crawford (1971) described evidence for a depletion of dust below the galactic average in the region $r \sim 50$ pc.

The best method of deducing electron density is through the dispersion of pulsar radio-frequency emission. A typical value for the electron density in the ISM is 0.03 cm^{-3}. However, Grewing (1975) has suggested that it is significantly higher (0.2 cm^{-3}) in the vicinity of the Sun.

Of particular importance is the scale of the local irregularities in comparing the interplanetary measurements and the NISM measurements. Often, 21-cm interfero-metric measurements show evidence of clouds of diameter 0.3–1 pc (Griesen 1976). However, much smaller scale sizes would have escaped detection unless the clouds were of very high density. If the irregularities are of comparable density to the sur-rounding medium, their scale sizes can be estimated to be 10–100 times the mean free path in the gas ($\sim 10^3$ au), or ~ 0.01–0.1 pc.

A large-scale correlation of the square of the magnetic field strength B^2 and total gas density ρ_H is usually adopted (Paul, Cassé & Cesarsky 1976), assuming the equipartition of magnetic field energy. If this holds for the LISM gas and if most of the H is in the neutral form, B is about one half of the 3 μG galactic average.

Evidence for $T \sim 10^4$ K in the ISM is weak, relying mainly on the large widths of 21-cm emission lines. These observations can also be explained by emission from small cold clouds with high-velocity dispersions (Greisen 1976).

A summary of the currently accepted values for the interstellar density, tempera-ture, and magnetic field is given in Table 1. The local properties inferred from reson-ance backscattering measurements are also shown for comparison. It is encouraging that the results from the two different methods (shown in the top two rows) are in good agreement.

It has become fashionable to cite the two-phase theory of the interstellar medium (Field, Goldsmith & Habing 1969) in accounting for the existence of a high tempera-ture in the solar neighborhood. In this picture, a hot tenuous "intercloud medium" is in pressure balance with cold dense interstellar clouds. However, it is now recog-nized that there are serious difficulties with this simple model. The identity of the heating mechanism is unknown. Recombination time scales are $\sim 10^7$ yr, much larger than the cooling time (Bottcher et al 1970). The state of excitation of the ISM could be a result of relic ionization from supernova bursts, or of previously hot stars

Table 1 Physical properties of the neutral interstellar medium

	r (pc)	n_H (cm^{-3})	n_e (cm^{-3})	T (K)	B (10^{-6} G)	$\langle v_r \rangle$ km sec^{-1}
Local Interstellar Medium (LISM)	$0.5\text{–}1 \times 10^{-5}$	0.05–0.1	?	1×10^4	1.4[a]	21 ± 3[b]
Nearby Interstellar Medium (NISM)	1–3.5	0.1[c]	0.2[d]	$0.4\text{–}1.2 \times 10^{4}$[c]	1.4[a]	15 ± 8[c]
Range of Values for Interstellar Medium (ISM), including cold clouds & spiral arms	$10^2\text{–}10^3$	$0.01\text{–}10^5$	0.03	$0.02\text{–}1 \times 10^{4}$[e]	3.6[f]	9[g]

[a] Estimated from the relationship $B^2 \propto \rho_H$ given in footnote f.
[b] Adams & Frisch (1977).
[c] McClintock et al (1977, unpublished manuscript).
[d] Estimated by Grewing (1975).
[e] Allen (1973).
[f] Paul, Cassé & Cesarsky (1976) and references therein.
[g] This value is the rms velocity along the line of sight for hydrogen clouds given by Allen (1973).

that have since left the main sequence and are no longer observable. Time-dependent models (Kafatos 1973) show that an initially very hot gas cools to 10^4 K in 3×10^4 yr, assuming a total density of 1 cm^{-3}, at which point the H and He densities are 0.07 and 0.1 cm^{-3} respectively. This large (He/H) ratio is ruled out for the LISM (see Table 3, p. 188). Steady-state heating theories relying on Lyman continuum or soft X-ray ionization also predict a large (He/H) ratio due to the larger ionization rate for H (Bergeron & Souffrin 1971, Grewing 1975, Blum & Fahr 1976). Cosmic-ray ionization rates are about the same for H and He (Field, Goldsmith & Habing 1969), and could maintain a high-temperature gas and a normal (H/He) ratio. However, the concept of a *uniform* cosmic-ray heating mechanism encounters serious difficulties (Salpeter 1976). Silk (1973) has mentioned various other possibilities for heat sources that produce no significant ionization.

An important question is whether the interplanetary measurements yield a true kinematic temperature for the nearby medium, or whether they represent only upper limits due to solar wind heating effects. Wallis (1973), Fahr & Lay (1974), Wallis (1975, 1976), and Fahr (1977, unpublished manuscript) have argued that elastic collisions of solar wind protons with the inflowing gas could significantly increase the temperature inside 3 au. Holzer (1977b) has argued that Wallis' cross sections for $H-H^+$ collisions are probably too high by a factor of ~ 4, and consequently that heating effects for H are negligible. The $H-H^+$ cross sections of Fahr & Lay (1974) and Fahr (1977) are also probably too large (Wallis 1975), since they used an incorrect Coulomb interaction for small impact parameters. This would yield an overestimate of the heating effect. The subject obviously requires further study.

INTERACTION OF SOLAR EMISSIONS AND THE NEARBY GAS

There are three important ways that the Sun influences the interstellar medium: (*a*) the expanding solar-wind plasma ionizes the neutral constituents and sweeps away the interstellar plasma and magnetic field; (b) solar gravity and solar Lα radiation pressure distort the interstellar density distribution; and (c) solar UV flux ionizes the neutral constituents. Only a brief sketch of the plasma processes (a) is given here. Detailed discussions are given by Blum & Fahr (1970), Holzer (1970), Fahr (1971), Wallis (1971), Axford (1972), Holzer (1972), Bhatnagar & Fahr (1972), Wallis (1973), Yu (1974), Fahr (1974), Wallis (1975), Axford (1976), and Holzer (1977a,b).

Plasma Interactions

In this section we describe a subject that is entirely theoretical. Our expectations are based on experience of solar wind interaction with various "obstacles"—the magnetospheres and ionospheres of planets and comets. However, we describe certain predicted effects of the interstellar medium on solar wind properties that may be observable.

The solar wind is a supersonic expansion of the solar corona, consisting mainly of protons and electrons. In interplanetary space the degree of ionization and the flow velocity are essentially constant, and the ion density decreases as r^{-2}. Thus, the dynamic ("ram") pressure of the solar wind at some large distance must become equal to the total interstellar "pressure." The latter is the sum of the thermal and magnetic pressures, and the opposing ram pressure of the inflowing gas. A small contribution is also provided by cosmic rays. For an interstellar electron density $\lesssim 0.1$ cm^{-3}, the interstellar magnetic pressure is dominant. At the point of pressure balance, a transition from supersonic to subsonic flow occurs abruptly as a shock discontinuity. [If the neutral density were significantly higher than 0.1 cm^{-3}, the transition may occur without the formation of a shock (Wallis 1971, Holzer 1972).] Beyond the shock is a region analogous to the earth's magnetosheath consisting of hot solar wind plasma (Figure 1). The "heliopause" is defined as the region dividing the solar wind plasma from the interstellar plasma and magnetic field B; its minimum

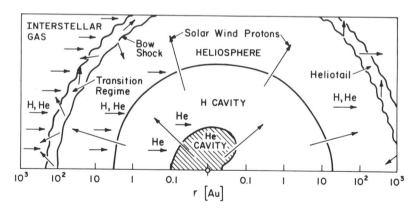

Figure 1 Schematic diagram illustrating the formation of the helium and hydrogen cavities, and the heliospheric shock and subsonic regions.

distance is estimated by Axford (1972) and Holzer (1977a) to be about 75 au, assuming $B = 3\ \mu G$. Along the heliospheric boundary layer (~ 50 au in thickness for a minimum heliopause distance of 100 au), the hot subsonic flow is redirected downwind. An elongated ionized wake is formed that becomes constricted downstream as the plasma cools. The interstellar magnetic pressure is expected to "squeeze off" the tail at $\sim 10^3$ au. The topology of the plasma flow in the wake region has been studied in detail by Yu (1974).

As shown by Blum & Fahr (1970), the neutral gas freely penetrates the heliosphere well into the inner solar system. Its influence on the solar wind is that of a "body force" in contrast to the "contact force" of the magnetic field (Wallis 1971, Holzer 1972). The primary interaction is resonant charge exchange, wherein a solar-wind proton picks up an electron from an interstellar H atom with negligible momentum exchange. Photoionization also occurs but at a rate only $\sim 15\%$ of the charge exchange rate. In either case, an ion is created within the solar wind at low heliocentric velocity, and is rapidly accelerated to solar wind speeds by $\overline{V} \times \overline{B}$ pickup. This acts as a frictional drag, reducing the solar-wind momentum and energy flux. The energy extracted from the bulk flow is partially converted to thermal energy of the ions and electrons, depending upon the effectiveness of some poorly understood plasma instabilities (Wu & Hartle 1974). The change in the flow speed and temperature in a complete thermalization scheme was described by Holzer & Leer (1973). For $n_H = 0.1$ cm^{-3}, the decrease in the speed is only about 15%; however, the heating of the solar wind in the outer solar system can be substantial. The neutral hydrogen and helium atoms formed by charge exchange move radially outward from the Sun with the solar-wind velocity.

A "splash" component of hot H atoms is produced in the turbulent boundary layer by charge exchange of hot solar-wind protons with interstellar H. A portion of these are injected into the inner solar system (Patterson, Johnson & Hanson 1963), but their contribution is small compared with the directly entering atoms (Fahr 1974, Holzer 1977b).

Since singly-ionized helium should be negligible in the solar wind, its presence is a direct indicator of interstellar helium (Holzer & Axford 1971). However, it has not yet been identified (Feldman et al 1974) because of observational constraints, and also possibly because the ions are not efficiently thermalized. Protons of interstellar origin might be discriminated from solar wind protons at ~ 10 au, provided the angular distribution of the ions in the forward hemisphere is measured (Vasyliunas & Siscoe 1976).

Effects of Ionization, Gravity, and Radiation Pressure

Whereas the collective interactions of the interstellar gas, magnetic field, cosmic rays, and solar wind plasma are very complex, the reciprocal influence of the solar wind on the neutral gas is a relatively simple matter. Almost all the important physics can be described by closed-form analytic solutions. Furthermore, observations of the resonance backscattering of solar emission lines from the interstellar gas represent *all* the information available so far. For these reasons, we describe in some detail the equations for a simplified model of the spatial distribution of the neutral gas density, and the equations for the backscattering intensity, to illustrate the various effects of ionization, radiation pressure, and gravity.

Since the mean free path of the interstellar neutrals is $\sim 10^3$ au, they execute collisionless Keplerian trajectories subject to an (approximately) inverse-square ionization loss process. An outward radial force, due to solar $L\alpha$ radiation pressure, is important for H, and diminishes the effective solar mass by the factor $(1-\mu)$ (μ denotes the ratio of the radiation pressure force to the Sun's gravitational force). For the present, we consider μ to be a constant for any given instant. However, since solar $L\alpha$ emission varies considerably with the solar activity cycle, the value of μ is ~ 0.5 at the minimum of solar activity, and ~ 1 at solar maximum. These values (which are uncertain to $\pm 30\%$) are based on the line-center solar flux measurements of Vidal-Madjar (1975). According to Meier & Mange (1973) and Keller & Thomas (1975), it is likely that periods of high solar $L\alpha$ emission exist for which the interstellar H atoms are *repelled* from the Sun ($\mu > 1$).

On the other hand, He atoms are unaffected by radiation pressure and are *attracted* to the Sun. Therefore, solar gravity tends to enhance He relative to H. Furthermore, solar wind charge exchange depletes H much more rapidly than He. As a result of these two influences, helium dominates hydrogen inside 1 au.

The relative motion of the Sun and gas causes an elongated ionization cavity to be created in the gas, as shown in Figure 1. The mathematical description of the distribution for a unidirectional flow (a "cold" gas) follows from a solution of the equation of continuity along a dynamical trajectory (Blum & Fahr 1970, Holzer 1970). [We ignore the contributions from the "splash" component and from the neutral H atoms moving radially outward with solar wind speeds, both of which are small (Fahr 1974).]

$$n_c(r, \theta) = n_\infty \sum_{j=1}^{2} \frac{p_j^2 \exp\left(-\beta_e r_e^2 \theta_j / |p_j|\right)}{u_0^2 \, r \sin \theta \, \{r^2 \sin^2 \theta - 2rr_p(0)(1-\cos\theta)\}^{1/2}} \tag{1}$$

In the equation above, n_c is the "cold atom" density at the radial distance r, and the angular distance θ from the upwind direction; β_e is the total ionization rate per atom at $r_e = 1$ au; u_0 is the initial flow speed of the "interstellar wind" (denoted v_r in Table 1); and n_∞ is the asymptotic interstellar gas density (denoted n_H for hydrogen in Table 1). With the assumption that the forces and ionization processes are spherically symmetric, n_c is cylindrically symmetric about the Sun-wind vector. The quantities $r_p(0)$ and p_j are defined by

$$r_p(0) = 2GM(\mu-1)/u_0^2, \tag{2}$$

$$p_{1,2} = \tfrac{1}{2}u_0 \{r \sin \theta \pm [r^2 \sin^2 \theta - 2rr_p(0)(1-\cos\theta)]^{1/2}\}, \tag{3}$$

where G is the gravitational constant and M is the solar mass. In Equation (3) the positive sign is associated with p_1 and the negative sign with p_2. Further, $\theta_j = \theta$ for $p_j > 0$ and $\theta_j = 2\pi - \theta$ for $p_j < 0$. The two terms in the sum correspond to the contributions from the two trajectories that intersect at (r,θ), having impact parameters p_1 and p_2. The exponential term is the net probability of survival.

For $\mu > 1$, solar radiation pressure exceeds gravity, and Equation (1) predicts a paraboloidal void inside the radius $r_p(\theta)$ given by

$$r_p(\theta) = 2r_p(0)/(1+\cos\theta). \tag{4}$$

Here $r_p(0)$ denotes the minimum (upwind) distance to this surface. For $\mu < 1$, no void exists and $r_p(0)$ is negative with no physical significance. These two cases are illustrated in Figure 2. For the special case of $\mu = 1$, Equation (1) reduces to

$$n_c(r, \theta) = n_\infty \exp\left[-r_i(\theta)/r\right], \tag{5}$$

where

$$r_i(\theta) = \beta_e r_e^2 \theta / u_0 \sin\theta. \tag{6}$$

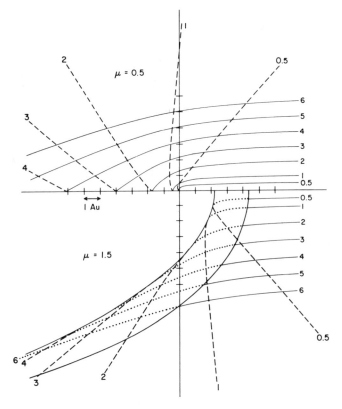

Figure 2 Trajectories of hydrogen atoms in two extreme cases, where μ is the ratio of the solar Lα radiation pressure to solar gravity. Each trajectory is identified by its impact parameter. The post-perihelion half of the trajectory is indicated by dashed lines. In the lower diagram, the dotted lines indicate the region where the probability of ionization is greater than $1-(1/e)$, assuming $\beta_e(\mathrm{H}) = 6.14 \times 10^{-7}\,\mathrm{sec}^{-1}$. Also shown are the locus of equal ionization probability (outer envelope) and the locus of the turning points (inner envelope).

Calculatons of the density distribution using Equation (1) have been published by Johnson (1972) and Axford (1972). The He ($\mu = 0$) and the H ($\mu < 1$) densities are infinite along the downstream axis ($\theta = \pi$), and are a result of the convergence of an infinite number of trajectories due to gravitational focussing. A density singularity also occurs at the surface $r_p(\theta)$ for $\mu > 1$ (Thomas 1973), similar to the infinite intensity occurring at a caustic surface in geometrical optics.

The mathematical form of Equation (1) is simplified by using a new independent variable $x^2 = (1 - r_p(\theta)/r)$ (Thomas 1973). For $\mu > 1$, $0 \leq x \leq 1$; and for $\mu < 1$,

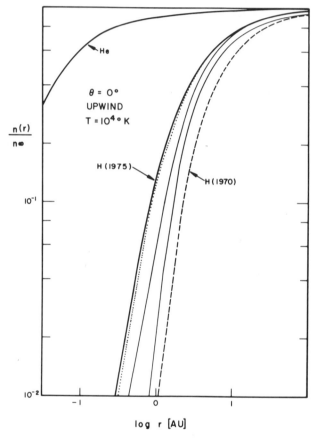

$$\frac{n(r)}{n_\infty}$$

$\theta = 0°$
UPWIND
$T = 10^{4} °K$

He

H (1975)

H (1970)

log r [AU]

Figure 3   Normalized upwind ($\theta = 0°$) radial density distributions for hydrogen and helium. The He density is for an ionization rate at 1 au of $\beta_e(\text{He}) = 6.67 \times 10^{-8} \text{ sec}^{-1}$. The H (1975) curve applies to $\mu = 0.5$ and $\beta_e(\text{H}) = 4.167 \times 10^{-7} \text{ sec}^{-1}$. The other two solid curves are for $\beta_e(\text{H}) = 4.167 \times 10^{-7} \text{ sec}^{-1}$ and for $\mu = 1$ and $\mu = 1.5$. The dashed curve applies to $\mu = 1$ and $\beta_e(\text{H}) = 7.48 \times 10^{-7} \text{ sec}^{-1}$. The dotted curve is computed from the "cold-gas" formula, Equation (1) or (7), for the same parameters as the H (1975) curve.

$1 \leqq x \leqq \infty$. Equation (1) takes the form

$$n_c(x) = (n_\infty/4x)[(1+x)^2 e^{-k_1(\theta)|x-1|} + (1-x)^2 e^{-k_2(\theta)(x+1)}], \tag{7}$$

where $k_1(\theta) = 2r_i(\theta)/|r_p(\theta)|$, and $k_2(\theta) = k_1(\theta) (\mu > 1)$ or $k_2(\theta) = k_1(\theta) (2\pi - \theta)/\theta$ ($\mu < 1$). The various singularities and limiting values are now explicit; Equation (7) shows that when $x = 1$ $(r = \infty)$, $n_c = n_\infty$; and when $x = \infty$ $(r = 0$ for $\mu < 1)$ or when $x = 0$ $[r = r_p(\theta)$ for $\mu > 1]$, $n_c \to \infty$.

The dotted curve in Figure 3 shows the "cold" upwind hydrogen density plotted versus radial distance for $\mu = 0.5$. The $(1/e)$ cavity radius depends mainly upon the ionization rate and less upon the solar Lα radiation pressure. As a result of the lower ionization rate for He ($\beta_e = 0.4 - 1.2 \times 10^{-7} s^{-1}$), He atoms penetrate to within ~ 0.1 au in the upstream direction.

Influence of a Finite Temperature

Since the interstellar temperature is 10^4 K, the situation described above must be drastically modified. At this temperature, the rms thermal speed is 16 km sec^{-1}, sufficient for many of the atoms to be incident at large angles from the upwind direction. Instead of a point in space receiving contributions from only two trajectories (excepting the degenerate case $\theta = \pi$), it actually receives contributions from a large portion of velocity space. This problem has been considered by Fahr (1971), Thomas (1972), Feldman et al (1972), Bertaux, Ammar & Blamont (1972), Fahr (1974), Blum, Pfleiderer & Wulf-Mathies (1975), and Meier (1977).

Most authors assume a Maxwell-Boltzmann distribution of velocities which, in polar coordinates, can be written (in solar rest frame) as

$$f(u, \delta) = (m/2\pi kT)^{3/2} u^2 \exp[-(m/2kT)\{u^2 \sin^2 \delta - (u \cos \delta - u_0)^2\}], \tag{8}$$

where m is the mass, k is Boltzmann's constant, u is the speed, and δ is the polar angle measured from the upwind direction; f is normalised by

$$\int_{4\pi} d\Omega \int_0^\infty du f(u, \delta) = 1,$$

where $d\Omega$ is the element of solid angle in velocity space $= \sin \delta d\delta d\phi$, and ϕ is the azimuthal angle. The contribution to the density from the element in velocity space $du d\Omega$ is

$$dn(r, \theta) = n_c(r, \theta; u, \delta, \phi) f(u, \delta) du d\Omega,$$

where the dependence of n_c on u, δ, and ϕ follows by replacing u_0 with u, and θ with the angle $\cos^{-1} (\cos \theta \cos \delta + \sin \theta \sin \delta \cos \phi)$. The expression for the density of the "hot" model n_h follows immediately:

$$n_h(r, \theta) = (2\pi)^{-3/2} \int_0^\infty dv v^2 \int_0^\pi d\delta \sin \delta \exp\{-v^2 \sin^2 \delta - (v \cos \delta - \gamma)^2\}$$
$$\int_0^\pi d\phi n_c(r, \theta; v, \delta, \phi). \tag{9}$$

In the above, v denotes the dimensionless variable $u/(2kT/m)^{1/2}$. In general, this three-dimensional integral must be evaluated numerically. However, Fahr (1971) and Blum, Pfleiderer & Wulf-Mathies (1975) have shown that in the upwind direction, and for all angles θ that differ appreciably from π, temperature has a minor influence on the density, and the "cold" density expression (Equation 1 or 7) is accurate for most purposes (see Figure 3). In the downwind direction, defined as lying with a cone of half-angle $\tan^{-1}(1/\gamma)$, where $\gamma = u_0/(2kT/m)^{1/2}$, the gas temperature has the effect of removing the singularity for $\mu < 1$. For $\mu > 1$ it removes the singular surface and causes the density everywhere within the paraboloidal cavity to be finite.

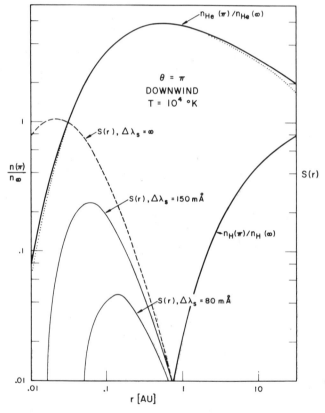

Figure 4 Normalized downwind ($\theta = \pi$) radial density distributions for hydrogen and helium and the 584-Å source functions for several solar line widths (full width at half maximum). The hydrogen density is for $\mu = 1$, $\beta_e(H) = 4.167 \times 10^{-7}$ sec^{-1}. The helium density is for $\beta_e(He) = 6.67 \times 10^{-8}$ sec^{-1}. The dotted curve is the helium density calculated from Equation (10). Note that $n(He)/n_\infty(He)$ can be > 1, illustrating the gravitational focussing effect.

Feldman, Lange & Scherb (1972) have derived an analytic expression for $\mu < 1$ for the downwind density in the limit of small temperature. Their result is

$$n_h(r, \pi) \cong n_\infty \, \gamma y \exp \left[-(\pi^{1/2}/2)k_1(0)y \right], \tag{10}$$

where $y = \left[\pi \lvert r_p(0) \rvert / r \right]^{1/2}$. The validity of Equation (10) depends upon the gas being cool ($\gamma^2 \gg 1$). For H, the value of γ is ~ 1.6; for He, $\gamma \sim 3.2$. It would appear that Equation (10) is invalid for H and possibly of limited validity for He. Blum, Pfleiderer & Wulf-Mathies (1975) showed that it provides an accurate approximation for the 584 Å source function (see Equation 11) inside 10 au, where the bulk of the resonance scattering occurs. Figure 4 shows a comparison of the He density calculated from Equation (10) and from an accurate numerical integration of Equation (9). Equation (10) correctly predicts that the density at $\theta = \pi$ is independent of u_0, in contrast to the upstream density, which varies strongly with u_0. This is a consequence of the fact that the "filling in" of the density on the downwind axis depends only on the perpendicular component of the wind velocity, which is proportional to γ^{-1} or $T^{1/2}$ (see Wallis 1976). Model interplanetary H and He densities were calculated from Equation (9) and are shown in Table 2.

RESONANCE SCATTERING FROM INTERPLANETARY ATOMS

So far the only method of determining the interplanetary gas distribution is through observations of the backscattering of two solar emission lines, H I Lα 1216 Å and He I 584 Å. All lines scattered from other interstellar species (such as O, C, Ne, etc) will be extremely weak (Wulf-Mathies & Blum 1975). In this section we consider the theoretically expected spatial pattern of the sky emission, and later describe the observations and how they are used in deducing properties of the interstellar gas.

Analytic Expressions for the Scattered Intensity

Meier (1977) has given a detailed description of the spectral distribution of the scattered intensity, taking into account the shape of the solar emission line and the distortion of the velocity distribution by solar gravity and ionization processes. The former is particularly important for helium since acceleration of the atoms can Doppler-shift the absorption frequency into the wings of the narrow 584 Å solar line. However, the basic properties of the brightness distribution are revealed by considering a much simpler model in which: (a) the solar emissions (the ionizing UV, solar wind, and solar emission lines) are spherically symmetric and time-independent, (b) the solar emission lines are spectrally flat and broad compared to the absorption profile of the interplanetary gas, (c) the scattering is isotropic and given by the optically thin approximation, and (d) the observer is located near the Sun. Assumption (d) is unnecessary but made here to simplify the geometry to allow analytic solutions. This assumption is realistic for Lα if the observer is at $\lesssim 1.5$ au as can be seen from Figure 3. It does not give the correct perspective for the 584 Å emission.

With the above assumptions, the backscattered intensity I at $r = 0$, arriving from

Table 2a Hydrogen densities for $n_\infty = 1$ cm^{-3}, $u_0 = 20$ km sec^{-1}, $\beta_e = 5 \times 10^{-7}$ sec^{-1}, $\mu = 0.75$, and $T = 10^4$ K. Estimated accuracy $\pm 0.2\%$. [4.76(−7) indicates 4.76×10^{-7}]

$\theta(°)$ \ r(au)	0.005	0.01	0.02	0.04	0.07	0.1	0.2	0.4	0.7	1	2	4	7	10	20	50	100
0						4.76(−7)	1.19(−4)	4.29(−3)	2.87(−2)	6.90(−2)	2.21(−1)	4.37	6.05	6.96	8.26	9.92	9.57
30						3.64(−7)	9.79(−5)	3.73(−3)	2.59(−2)	6.35(−2)	2.10(−1)	4.23	5.93	6.85	8.19	9.18	9.56
60						1.61(−7)	5.32(−5)	2.41(−3)	1.88(−2)	4.89(−2)	1.78(−1)	3.82	5.54	6.52	7.97	9.07	9.50
90						4.01(−8)	1.84(−5)	1.10(−3)	1.04(−2)	3.00(−2)	1.28(−1)	3.11	4.83	5.87	7.51	8.84	9.37
120						5.72(−9)	3.92(−6)	3.37(−4)	4.11(−3)	1.37(−2)	7.45(−2)	2.17(−1)	3.76	4.83	6.69	8.37	9.10
150						5.89(−10)	5.86(−7)	7.34(−5)	1.19(−3)	4.69(−3)	3.39(−2)	1.25(−1)	2.52	3.50	5.46	7.55	8.57
180						1.45(−10)	1.69(−7)	2.62(−5)	5.07(−4)	2.24(−3)	1.97(−2)	8.53(−2)	1.91(−1)	2.80	4.74	7.03	8.24

Table 2b Helium densities for $n_\infty = 1$ cm^{-3}, $u_0 = 20$ km sec^{-1}, $\beta_e = 7.5 \times 10^{-8}$ sec^{-1}, $\mu = 0$, $T = 10^4$ K.

r(au) θ(°)	0.005	0.01	0.02	0.04	0.07	0.1	0.2	0.4	0.7	1	2	4	7	10	20	50	100
0	5.59 (−3)	3.67 (−2)	1.28 (−1)	2.86	4.37	5.29	6.79	7.82	8.36	8.61	9.00	9.31	9.52	9.63	9.80	9.92	9.97
30	5.28 (−3)	3.55 (−2)	1.26 (−1)	2.84	4.36	5.30	6.83	7.86	8.40	8.65	9.02	9.32	9.52	9.63	9.79	9.92	9.97
60	4.43 (−3)	3.20 (−2)	1.19 (−1)	2.79	4.36	5.35	6.96	8.04	8.59	8.83	9.14	9.38	9.53	9.63	9.78	9.91	9.96
90	3.29 (−3)	2.70 (−2)	1.10 (−1)	2.74	4.45	5.56	7.43	8.70	9.28	9.48	9.63	9.65	9.68	9.71	9.79	9.89	9.95
120	2.16 (−3)	2.17 (−2)	1.03 (−1)	2.87	5.01	6.49	9.12	1.09 (0)	1.15	1.16	1.13	1.08	1.04	1.02	1.00	9.94 (−1)	9.95
150	1.42 (−3)	1.98 (−2)	1.21 (−1)	4.11	8.00	1.09 (0)	1.62	1.96	2.06	2.04	1.89	1.67	1.49	1.38	1.22	1.09	1.04
180	1.71 (−3)	3.27 (−2)	2.45 (−1)	9.36	1.92 (0)	2.67	4.06	4.95	5.16	5.09	4.62	3.92	3.33	2.97	2.37	1.77	1.47

Table 3   Summary of density/temperature determinations for the local interstellar medium

Spacecraft and Date	Experiment	References	Derived Density (cm^{-3})	Derived Temp. ($\times 10^4$ K)	Comments
OGO-5 1969–1971	Lα photo-meter (3° FOV)	Thomas & Krassa (1971, 1974), Thomas (1972)	$n_H = 0.12$	0.35	"Hot" model
OGO-5 1969–1971	Lα photo-meter 40 arc min FOV)	Bertaux & Blamont (1971), Bertaux, Ammar & Blamont (1972)	$n_H = 0.1$–0.2	0.1–1	"Hot" model
Rockets 1970, 1973	584 Å photo-meter	Paresce et al (1974a,b)	$n_{He} = 0.032$	0.4–0.6	Temperatures are lower limits
STP 72-1 1972–1973	584 Å photo-meter	Weller & Meier (1974), Meier (1977)	$n_{He} = 0.009$–0.024	0.25–1* 0.5–1**	* Uses modified cold model ** Uses Meier's more accur-ate model
Mars 7 1972–1973	Lα photo-meter plus H resonance cell	Bertaux et al (1976)		1.1–1.3	Value of tem-perature assumes galactic Lα intensity is zero
D2-A 1971	Lα photo-meter plus H resonance cell	Cazes & Emerich (1977)		0.8–1.2	Observed $\sim$10% inter-stellar polar-ization; verified that $\mu < 1$ in 1971
Mariner 10 Dec. 1973 & Jan. 1974	Spectrometer measured simultane-ously 1216 Å and 584 Å	Broadfoot & Kumar (1978), Ajello (1978)	$n_H = 0.02$–0.07 $n_{He} = 0.005$–0.011	1–2	He/H = $0.2^{+0.3}_{-0.13}$
Prognoz-5 1976	4-channel photometer includes 1216 Å, 584 Å, plus H resonance cell	Bertaux et al (1977)		0.78–0.98	

Table 3 (*continued*)

Spacecraft and Date	Experiment	References	Derived Density (cm^{-3})	Derived Temp. ($\times 10^4$ K)	Comments
Apollo–Soyuz 1975	EUV tele-scope & broad-band photometer	Freeman et al (1977)	$n_{He} = 0.002–0.006$		
Pioneer 10–11 1972–present	2-channel photometer 1216 Å and 584 Å	Suzuki, Carlson & Judge (1975)	$n_H = 0.06$ $n_{He} = 0.014$		He/H = 0.23 ± 0.1; $r = 1.5$ (1972) to 13 au (1977)
Black Brant IVB Rocket 1976	584 Å photo-meter and He resonance cell	Fahr, Lay & Wulf-Mathies (1977)	$n_{He} = 0.004–0.01$		

the direction θ, is

$$4\pi I(0, \theta) = g_e \int_0^\infty dr' n(r', \theta)(r_e/r')^2 \equiv \int_0^\infty dr' S(r', \theta), \tag{11}$$

where g_e is the number of solar photons scattered per atom per sec at 1 au evaluated at the center of the solar line. S is the source function, or the volume emissivity, and $g_e(L\alpha)$ and $g_e(584$ Å$)$ lie in the ranges $0.9 - 2.7 \times 10^{-3}$ sec^{-1} and $0.65 - 1.7 \times 10^{-5}$ sec^{-1} respectively. Using Equation (7) for the cold atom density in Equation (11), and noting that $r^{-2}dr = 2x\,dx/r_p(\theta)$, the integration can be performed analytically (Thomas 1973). The result is

$$4\pi I_c(0, \theta) = [g_e n_\infty r_e^2/r_p(\theta)] \times \left[\frac{2k_1^2(\theta) \pm 2k_1(\theta) + 1}{k_1^3(\theta)} \pm \frac{e^{-2k_2(\theta)}}{k_2^3(\theta)} \right], \tag{12}$$

where the $\pm$ sign refers to the sign of $1 - \mu$. For perfect balancing ($\mu = 1$) it can be shown that Equation (12) reduces to

$$4\pi I_c(0, \theta; \mu = 1) = g_e n_\infty r_e^2/r_i(\theta) = g_e n_\infty r_e^2 \sin \theta/r_i(0)\theta. \tag{13}$$

If the expression for the "hot model" density (Equation 9) is used in Equation (11), I have shown that by using Equation (12), the four-dimensional integral can be simplified to a sum of four one-dimensional integrals. Thus, the calculation of the backscattered intensity becomes an easy task for the computer and allows a large number of different cases to be run. In the following we show the results of calculations for different values of the various parameters.

The effect of interstellar temperature on the intensity $I(0, \theta)$ is illustrated in Figure

5 for both H I 1216 Å and He I 584 Å. The 1216 Å emission has its minimum value in the downstream direction and is gradually "filled in" as the temperature is increased. On the other hand, the 584 Å intensity is sharply peaked downwind, this feature being "smeared out" with increasing temperature. This behavior led to predictions (e.g., Axford 1972) that the maximum/minimum ratio and the half-width of the peak would be a useful "thermometer" for the interstellar medium. Unfortunately, this is true only if the remaining parameters (such as β_e and u_0) are accurately known. Also, the sensitivity to temperature of the 584 Å emission shown in Figure 5 is not necessarily realized in practice for reasons related to the actual observing geometries. The observing point is generally at ~ 1 au, outside the region of maximum downwind helium density (Figure 4), and is never exactly on the downwind axis (it comes within $\sim 7°$ on about December 15).

In this simplified geometry of a solar-based observer, the combined influences of

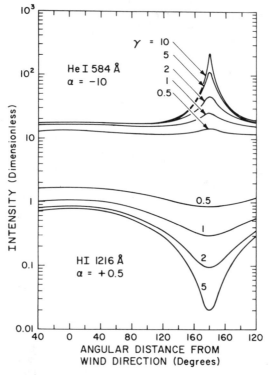

Figure 5   Theoretical sky background brightness variations with θ, the angle measured from the upwind direction. The upper curves apply to He 584 Å for which $\mu = 0$. The lower curves apply to the Lα intensity, for which the inflowing hydrogen atoms are repelled by the Sun ($\mu > 1$). The various curves apply to different ratios γ of the flow speed at infinity to the thermal speed, illustrating the effects of temperature on the spatial distribution.

ionization, velocity, and radiation pressure on the upwind/downwind intensity ratio are conveniently described by a dimensionless parameter α, where

$$\alpha = 2/k_1(0) = r_p(0)/r_i(0) = 2GM(\mu - 1)/r_e^2 u_0 \beta_e; \tag{14}$$

α is a measure of the importance of the effective gravitational field relative to ionization processes. For α large and negative (positive), gravitational focussing (repulsion) is dominant in shaping the cavity; and for $|\alpha| \ll 1$, ionization is dominant. Figure 6 illustrates this for the upwind/downwind intensity ratio. Note that for small temperatures (γ large) and for $\alpha < 0$, this ratio is <1, indicating that the intensity at the focussing peak is greater than the upwind intensity. As γ is decreased (temperature increased), the magnitude of the peak falls as also seen in Figure 5. For $\alpha > 0$, the upwind intensity always exceeds the downwind intensity. For high temperatures ($\gamma \leqq 1$), the effect of increasing α (decreasing the ionization rate and/or the velocity for fixed μ) is to increase the intensity ratio. For lower temperatures, the intensity ratio goes through a maximum at $\alpha > 0$. Eventually, as $\alpha \to +\infty$, all curves approach unity, with the effective cavity becoming very large and spherical in shape.

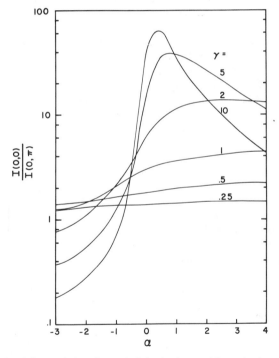

Figure 6   Ratio of the upwind to downwind sky background intensity for an observer at the Sun, plotted as a function of the parameter α, defined by Equation (14); α has the same sign as $\mu - 1$, and is a measure of the importance of the combined radiation pressure-gravity force to ionization.

As discussed in the next section, observational evidence indicates $T \sim 10^4$ K and $u_0 \cong 20$ km sec^{-1}, yielding $\alpha(H) \sim 1.6$. The observed range for the Lα intensity ratio is 3–6, indicating from Figure 6 that α must lie in the range -0.5 to $+0.5$. This can be immediately verified, since for $\mu = (0.5$ to $1.5)$, $\alpha(H) = (-1.1$ to $+1.1)$; for a "low" H ionization rate $\beta_e = 2.5 \times 10^{-7}$ sec^{-1}; for a "high" ionization rate $\beta_e = 7.1 \times 10^{-7}$ sec^{-1}, $\alpha = (-0.33$ to $+0.33)$.

Neglected Effects in the Simple Model

ANISOTROPIES   The assumption of spherical symmetry for the solar emissions is only an approximation. Active centers on the solar surface tend to congregate at moderate heliocentric latitudes. The effect on the interplanetary density of their longitudinal variations tends to be eliminated due to the fact that the 25-day sidereal period of rotation of the Sun is short compared to the effective flight time of atoms through the solar system (Fahr 1974). However, latitudinal variations would cause a latitudinal effect that would be revealed in the sky background maps (Fahr & Lay 1974). Since both UV and 584 Å fluxes tend to decrease toward the ecliptic pole, the former would cause the He density over the poles to *increase*; however, the smaller 584Å emission tends to cancel this increase. For H the spatial distribution of density is largely determined by the solar-wind (mass) flux. At the present time we have almost no information on the latitudinal variation of this quantity. Theoretical models have been constructed for which the solar wind flux takes on arbitrary latitudinal variations. Joselyn & Holzer (1975) have performed such studies, assuming that the flux both increases and decreases with latitude. Their isophotal maps for these nonsymmetrical ionization rates show a pronounced distortion of the nearly circular isophotes based on a symmetric solar wind. They did not consider the influence of a nonisotropic Lα flux, arguing that radiation pressure effects largely offset brightness increases.

Solar rotation also causes hydrogen to be illuminated by a continuously varying Lα flux. This can amount to $\pm 30\%$ for the entire solar line, and $\pm 50\%$ for the line-center flux (Vidal-Madjar 1977). This 27-day modulation was first observed by Mange & Meier (1970) in OGO-3 data, and later by Thomas & Bohlin (1972) in OGO-5 data, the latter work showing the highest correlation with solar activity (as measured by the Zurich sunspot number) in viewing directions opposite the Sun. This is expected if the primary source of variability is the Lα anisotropy. In this direction, all scattering points receive the same view of the Sun as a terrestrial observer, and are accordingly all in phase. In other directions, the line of sight receives different views of the Sun, and the correlation is reduced.

INFLUENCE OF SOLAR LINE SHAPE   Figure 7 shows a high-resolution spectrogram of the solar Lα line measured in July, 1975, from the OSO-8 spacecraft. As can be seen from the velocity scale, all H atoms with radial speeds $u_r < 25$ km sec^{-1} will experience very nearly the same solar Lα flux, and the assumption of a flat profile is completely valid. To reach $u_r = 25$ km sec^{-1}, an H atom must penetrate to within 1.6 au, even at solar minimum when the inward radial acceleration is maximum. However, more than 80% of the H atoms will have been ionized at this distance.

We conclude that the velocity-dependence of μ (and also that of the g-factor) can be safely ignored.

On the other hand, the 584 Å solar line has no influence on the He trajectories; however, the backscattering efficiency of the He atoms can be strongly influenced by the solar line shape for three reasons: (a) The solar 584 Å line is much narrower than the solar 1216 Å line (80–150 mÅ compared with ~ 1200 mÅ). The reference by Maloy et al (1978) contains a good review of all recent rocket measurements. (b) He atoms are accelerated to higher radial speeds than H. (c) The 584 Å source function peaks deeply within the inner solar system due to the comparitive immunity of He atoms to ionization. Consequently, the atoms are moving at considerably higher speeds.

By convolving the solar-line profile (assumed to be Gaussian) with a Doppler-shifted absorption profile, Wulf-Mathies & Blum (1977, unpublished manuscript) have derived an analytic expression for the source function:

$$S(r) = g_e(u_r = 0)n(r, \theta)(r_e/r)^2 \exp\{-(\lambda_0 u_r/c\Delta\lambda_e)^2\}, \tag{15}$$

where $\Delta\lambda_e^2 = \Delta\lambda_s^2 + \Delta\lambda_D^2$, λ_0 is the wavelength at line center, c is the speed of light, $\Delta\lambda_s$ is the solar line width, and $\Delta\lambda_D$ is the Doppler width of the scattering gas.

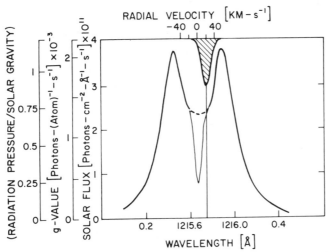

Figure 7 Solar Lα profile measured by the University of Colorado High Resolution Ultraviolet Spectrometer on OSO-8 (White & Lemaire 1976, Bruner 1978). The slit size was 1×900 arc sec and the spectral resolution was 0.03 Å. This spectrum was taken on July 22, 1975 and was calibrated on an absolute basis through the rocket data of G. Rottman (1975), who measured the total line intensity to be 2.47×10^{11} photons cm^{-2} sec^{-1} on July 28, 1975. The g-value applies to 1 au. The observed dip at line center is telluric hydrogen absorption. The dashed line segment is a guess for the chromospheric emission above the earth's atmosphere. The shaded feature at the top is a schematic drawing of the width and position of an absorption line due to a uniform volume of interstellar gas of 10^4 K moving away from the observer at 20 km sec^{-1}.

Wulf-Mathies & Blum assumed that the velocity distribution of the gas is Maxwellian everywhere, with a center of gravity of the distribution defined by the trajectory of the cold atoms. (For a critique of this modified cold-gas model, see Meier 1977.)

The effect of two different line widths on the downwind 584 Å source function line is shown in Figure 4. A full-width at half-maximum line width of 80 mÅ (the lowest measured value) changes the source function dramatically inside 0.3 au. The largest measured value of 150 mÅ causes an important change only at ≲0.1 au. Since the closest observation of the line of sight to the Sun is 0.61 au looking *across* the downwind cone (Broadfoot & Kumar 1978), this has so far turned out to be a minor correction.

TIME-DEPENDENT EFFECTS   I have recently shown (Thomas 1976) that it is incorrect to evaluate the quantity μ using the current Lα solar flux. This is because of the long flight time of H atoms through the effective scattering region (defined somewhat arbitrarily as enclosing 90% of the contribution to the observed backscattered intensity). For the upstream direction, this region extends to less than 2 au for 584 Å; however, for Lα it extends to ~35–40 au, corresponding to a flight time of ~8.5 yr. Consequently, the previous history of ionization and illumination of an incoming H atom must be considered in calculating the density. A crude estimate of this effect leads to a variation with θ of the effective value μ and β_e. Evidence of time-dependent effects appears to be present in the available Lα observations (Thomas 1976).

GALACTIC Lα EMISSION   Until recently, modeling efforts were frustrated by the possibility that galactic Lα emission may be contaminating the interplanetary observations. However, a number of recent observations have shown that no more than ~50 Rayleighs of an isotropic galactic background can exist (Bertaux et al 1976, 1977). R. Carlson (private communication) estimates that no more than 40 Rayleighs can be present in the Pioneer data. Thomas & Blamont (1976) have used the results of galactic Hα measurements to estimate an upper limit of ~10 Rayleighs.

MULTIPLE SCATTERING   Wallis (1974) suggested that the presence of higher-order scatterings complicates the interpretation of sky background measurements. However, the radiative transfer problem has not yet been solved (even for the simplified spherically symmetric problem). It appears that the effect may be important for Lα, but unimportant for 584 Å (work in progress by the author and H. U. Keller 1977).

Observations of Backscattered Solar Radiation

We now turn to the observational evidence of the presence of interstellar gas within the solar system. The existence of an extraterrestrial Lα sky background had been known since the pioneering rocket experiment of Morton & Purcell (1962). However, without a detailed spatial distribution, it was impossible to decide whether the radiation was of local or galactic origin.

1216-Å AND 584-Å SKY MAPS   The first firm evidence of nearby interstellar neutral hydrogen came in early 1970 from observations from two different Lα photometers on board the OGO-5 satellite. In a specially planned maneuver, the normally stabilized spacecraft was made to spin while it was near apogee (~15 earth radii), and presumably well above the influences of scattering from the terrestrial H geocorona. A series of Lα brightness maps revealed a large-scale spatial inhomogeniety of the emission, and a systematic seasonal variability of the spatial pattern. Such a non-uniform pattern was difficult to understand if the radiation were of galactic origin. The very large optical depths in the galaxy would impose a high degree of isotropy to the sky brightness. The observed seasonal variations completely ruled out a galactic source as the major contributor. As was borne out in subsequent observations, the interpretation by the OGO-5 group (Thomas & Krassa 1971, 1974, Bertaux & Blamont 1971) was correct—the seasonal variations were a result of parallactic shifts of the spatial pattern as the earth moved in its orbit. The large seasonal shift of the Lα maximum (38–50°) showed that the source of the emission lay within the solar system. That the source of the H atoms was interstellar had been anticipated by Blum & Fahr (1970) and Holzer (1970), who theoretically predicted many of the essential characteristics of the problem.

One important difference that could not have been anticipated was that the relative motion of the Sun and the gas is *not* directed from the classical "apex" of motion of the earth, but was instead nearly in the ecliptic plane, about 50° away from the apex. This simply meant that the nearby stars and gas have their own peculiar motions.

Also not anticipated was the very high temperature of the gas. As shown by Thomas (1972), if the galactic Lα background is negligible, the significant downwind Lα intensity implies a gas of temperature 10^3–10^4 K to fill in what would otherwise be a downwind void. Independent evidence of high temperatures and low galactic background was forthcoming from more recent 584-Å measurements and from resonance absorption cell measurements.

It was suggested by Fahr (1971), Axford (1972), Johnson (1972), Paresce & Bowyer (1973), and Fahr & Lay (1973) that a measurement of interstellar helium is observationally feasible if it has its cosmic abundance of ~0.1 H. Furthermore, they argued that the absence of radiation pressure on He atoms and its low ionization rate would cause an appreciable enhancement of the He density in the downwind direction. The first measurements of extraterrestrial 584-Å emission were made from rockets by Paresce, Bowyer & Kumar (1974a,b). However, the first reliable all-sky map that clearly revealed the presence of the downwind cone was made by Weller & Meier (1974) from broad-band measurements of 584 Å on the STP 72-1 satellite. This work clearly verified the predictions based on the Lα maps. The upwind direction was more accurately determined from the comparitively sharp downwind 584-Å peak than from the broad upwind Lα peak to be at right ascension $RA = 252°$ and declination $\delta = -15°$. This direction has prevailed until recently, being verified by measurements from Pioneer 10 and 11 (Suzuki, Carlson & Judge 1975), Mars 7 (Bertaux et al 1976), and Mariner 10 (Broadfoot & Kumar 1978, Ajello 1978). A revision of the wind direction has been recently suggested by Bertaux et al (1977) on the basis of a tentative analysis of data from the Prognoz-5 satellite that is in good agreement with the original OGO-5 measurements, $RA = 265 \pm 5°$. However, until the final analysis of the above data is made available, I would continue to recommend the Weller & Meier wind direction as being compatible with the largest number of sky background measurements.

Turning to more recent developments, perhaps the most dramatic has been the publication of the beautiful all-sky maps in *both* the Lα and 584-Å lines observed by Mariner 10 in 1973 and 1974 (Broadfoot & Kumar 1978). These interplanetary observations were made by a UV spectrometer at ~0.75 au. The importance of these new results is their excellent quality and the fact that they are the first all-sky maps constructed from data taken in interplanetary space where there is no possibility of atmospheric contamination.

Figure 8 shows the Mariner 10 Lα data taken on January 28, 1974. A deep (80 Rayleigh) minimum is present near the downwind direction. In addition, there is a smooth variation of intensity from the minimum to the maximum (not actually observed since the upwind direction was near the Sun). This behavior is consistent with the OGO-5 maps, and also with the data taken in interplanetary space from Mariners 5 and 6 (Barth 1970a,b) and Mariner 9 (Bohlin 1973). However, the deep minimum is more consistent with the value of 130 Rayleighs from Mariner 9 than the 215–240 Rayleighs measured by the OGO-5 instruments. It is now fairly certain that the OGO-5 maps were contaminated by geocoronal scattering in the region of

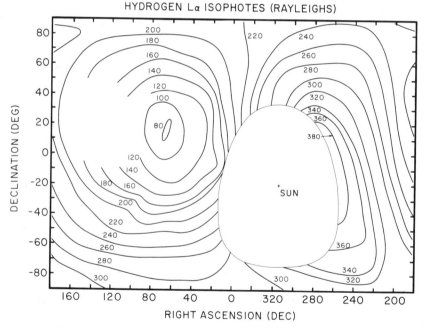

Figure 8   Hydrogen Lα isophotes measured by the Mariner 10 UV Spectrometer (Broadfoot & Kumar 1978). The data were collected on January 28, 1974 while the spacecraft was at a heliocentric distance of 0.757 au and ~60° from the downwind axis. The units are Rayleighs.

the minimum (Thomas & Krassa 1974). Furthermore, time-dependent effects from solar cycle variations probably play an important role in determining the Lα minimum intensity (Thomas 1976).

Ajello (1978) has recently performed a detailed analysis of the Mariner 10 data using a simple "cold" model to interpret the Lα measurements. Their model was capable of representing most of the data to an RMS fit of 18%. The effective ionization rate of the H atoms was found to be $\beta_e \sim 4.2 \times 10^{-7}$ sec^{-1} for a velocity $u_0 = 20 \pm 5$ km sec^{-1}.

Figure 9 shows a very different picture for the distribution of 584-Å emission. The 584-Å intensity does *not* peak in the direction of the Lα minimum as one might naively expect from Figure 6. In fact, the focussing cone is viewed at a large angle, and therefore is seen over a large portion of the sky. The 584-Å peak occurs closest to the Sun, and the isophotes trace out an elongated ridge, extending back to, and finally coinciding with the downwind direction. The upwind hemisphere is characterized by a nearly constant ~ 4 Rayleigh intensity, and is the behavior expected of an integration along directions over which the He density is nearly constant.

The width of the helium cone in directions perpendicular to the downwind axis seen by Mariner 10 was used by Ajello (1978) to derive a temperature from a modified cold-gas model (described earlier) and an approximation based on Equations (10) and (15). The theory predicts that the broader the cone, the larger is the tempera-

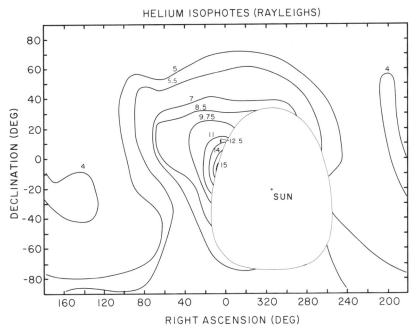

Figure 9 He 584 Å isophotes measured by the Mariner 10 UV Spectrometer for the same conditions given in Figure 8.

ture. Their best fit was for a temperature of 1.5×10^4 K; however, the uncertainty is $\pm 0.8 \times 10^4$ K, and is therefore consistent with more precise determinations to be discussed shortly. They also show that the value of the half-width of the He cone (35° as measured from the Sun) is constant between 0.61 au and 13 au, indicating an isothermal interstellar gas in the solar system, i.e. a minor if not negligible solar-wind heating of He atoms (Ajello 1978).

Of equal importance to the Mariner 10 results is the recent high resolution measurement of the interplanetary Lα line using the Copernicus spacecraft (Adams & Frisch 1977). For the first time, the Doppler-shift of the Lα signal has been measured, thus allowing a direct determination of the relative velocity of the gas. Assuming the direction of the wind to be given by the 584-Å measurement, they derived a velocity of $u_0 = 21.6 \pm 2.8$ km sec^{-1}. This result is significant since it removes a very serious uncertainty from theoretical models (the remaining principal uncertainty is now the ionization rate for each species).

ATOMIC RESONANCE ABSORPTION   A technique only recently applied to the study of the interplanetary medium is atomic absorption-cell spectroscopy. In this method a gas cell is placed in the light path of a conventional wide-band photometer. The cell contains the resonantly absorbing atomic species of interest whose column density in the cell can be controlled. For 1216 Å this requires a sealed H_2 sample that is partially thermally dissociated into H by means of a hot filament (Bertaux 1974, Blamont, Cazes & Emerich 1975). For 584 Å the helium column density within the cell is controlled by a pressure valve and a venting system (Freeman et al 1977). When the cell is "off" (transparent), the total brightness is measured; when the cell is "on" (partially absorbing), the signal is reduced, depending upon the line width and the Doppler shift of the sky emission. Thus both the emission temperature and the wind velocity can be inferred from measurements made in suitably chosen directions.

The CNRS group in Paris has recently reported analyses of observations using H absorption cells. From interplanetary measurements on the Soviet spacecraft Mars 7, Bertaux et al (1976) derived a gas temperature of $1.2 \pm 0.1 \times 10^4$ K, and a wind speed of 19.5 ± 1.5 km sec^{-1}. Similar measurements from the earth-orbiting D2-A satellite indicate a temperature of $1 \pm 0.2 \times 10^4$ K, and also evidence in April 1971 of a net focussing of H atoms towards the Sun (Cazes & Emerich 1977). A recent experiment on board the earth-orbiting Soviet spacecraft Prognoz-5 has yielded a value of the temperature of $0.88 \pm 0.10 \times 10^4$ K. This error estimate quoted by Bertaux et al (1977) is probably too small, since their analysis assumed $\mu = 1$.

Helium absorption cells flown in rockets are a more recent development. Freeman et al (1977) derived a lower limit of 10–15 km sec^{-1} for the wind velocity from the lack of an absorption in a He cell for a viewing direction $\sim 28°$ from the downwind direction. From an analysis of a recent rocket experiment Fahr, Lay & Wulf-Mathies (1977) derived a helium density in the range 0.0035–0.01 cm^{-3}.

SIZE OF THE H CAVITY: SOLAR CYCLE VARIATIONS   For a hot gas and for $\mu \neq 1$, Equations (5) and (13) can be generalized by replacing the quantity $r_i(\theta)$ by an effective cavity radius $r_c(\theta; \mu, T)$. It is found from numerical evaluation of Equation (9) that the upwind value $r_c(0° ; \mu, 10^4$ K) varies linearly with μ, according to the approximate formula

$$r_c(0° ; \mu, 10^4 \text{ K}) \cong 1.04 \, r_i(0°, \beta_e, u_0) + 1.185 \, (\mu - 1.28). \tag{16}$$

This behavior is also illustrated in Figure 3. As mentioned earlier, μ varies between ~ 1.5 at solar maximum and ~ 0.5 at solar minimum. The resulting change in r_c is 20–30% (depending upon the values of β_e and μ_0).

However, the observations indicate a considerably stronger variation with solar activity. Thomas (1972), Bertaux, Ammar & Blamont (1972), and Thomas & Krassa (1974) estimated $r_c(0) \sim 5.0$–6.2 au for the period 1969–1971. These values were determined from the magnitude of the seasonal parallactic displacements of the Lα maximum in the OGO-5 data. From the Pioneer 10 data (1972–1974), R. C. Carlson (private communication) evaluated the upwind cavity radius to be 3.8 ± 0.7 au from the slope of the curve of Lα intensity versus heliocentric distance (1.5 au to beyond 5 au). A value for $r_c(0)$ can be derived from the analysis of the Mariner 10 data by Ajello (1978). Assuming $u_0 = 20$ km sec^{-1}, $r_c(0) \sim 2.5$ au for 1974–1975. Thus, over the five-year period, the hydrogen cavity drew progressively closer to the Sun, from 5.5 to 3.8 to 2.5 au from 1970 to early 1975. This effect also shows up in the upwind/downwind Lα intensity ratio R. From a crude time-dependent model, I have shown that the downwind density remains nearly constant over the entire period 1970–1975 (Thomas 1976). It turns out that the density in the downwind direction in 1974 is controlled by the Sun's activity during the flat part of the previous maximum (1967–1971). As a result, the ratio R is controlled primarily by the upwind intensity, or equivalently the upwind cavity radius. The OGO-5 data indicated $R = 4 \pm 1$, whereas the Mariner 10 data yielded 5.7, consistent with a smaller upwind cavity radius in 1974–1975.

Since the variation in solar Lα radiation pressure fails by a large factor in explaining the observed change in the cavity dimensions, *it is likely that the H ionization rate decreased with decreasing solar activity between 1970 and early 1975*. Only a small decrease would result from a decreased photoionization rate. The required change in β_e is from $\sim 7.8 \times 10^{-7}$ sec^{-1} in 1970 to 4.3×10^{-7} sec^{-1} in early 1975. This implies that the solar wind mass flux decreased in this same period by nearly the same factor ($\sim 45\%$). This result is somewhat surprising in view of the evidence that the solar wind *speed* increased during this same period by about 30% (Bridge 1976, Bame et al 1976). Either the solar wind proton density decreased by a still larger amount ($\sim 60\%$), or else the solar wind variations in the ecliptic plane are not representative of the large-scale three-dimensional variations that are in fact more relevant for the Lα observations.

HYDROGEN AND HELIUM DENSITY IN THE LOCAL INTERSTELLAR MEDIUM   As can be seen from Equation (13), the asymptotic hydrogen density n_∞(H) is determined from a knowledge of the upwind intensity, the solar Lα "g-factor," and the effective cavity radius $r_c(\theta)$. Since $r_c(\theta) \ll 1$ for He, it is easily shown from evaluating Equation (11) with the lower limit of the integration set equal to 1 au, that $r_c(\theta) = 1$ au. Thus, to evaluate n_∞(He) requires only a knowledge of the 584-Å intensity and g(584-Å). This advantage is offset by the larger absolute calibration uncertainties at 584-Å. I estimate that the He and H densities can be determined with present techniques to an accuracy of a factor of $\sim 2-3$. In this regard we cite the recent work of Freeman et al (1977), who have used 584-Å measurements from the Apollo–Soyuz spacecraft to determine n_∞(He) to be 0.004 ± 0.002 cm^{-3}. As shown in Table 2, this value is somewhat smaller than earlier results, and can be traced to a different value for the upwind 584-Å intensity (0.5 Rayleighs). It compares with the results from STP 72–1 (1–2 Rayleighs), Pioneer 10 (3–5 Rayleighs), and Mariner 10 (4 Rayleighs). However, these differences may not be a fault of absolute calibration errors. It is interesting to note that a similar large spread in measurements of the 584-Å solar line also exists (see Figure 4 of Maloy et al 1977). Solar variability must again be invoked.

Table 3 summarizes all available results for both interstellar density and temperature. Within the considerable uncertainties, the He/H density ratio is consistent with cosmic abundances (Allen 1973); however, a wide range of ratios is permitted (1 : 2 to 1 : 20).

SUMMARY

In this review I stressed those properties of the local interstellar gas that have been measured, as opposed to those that are mainly of theoretical interest. With the large body of observations now at hand, it is clear that our conception of the "interstellar wind" phenomenon is well-founded. Furthermore, we now have more reliable values for the three important parameters of the interstellar gas—the density, temperature and flow velocity—than we had just a few years ago.

Important questions that still need to be answered are the following: (a) Are the nearby stars responsible for the high temperature of the local gas? A theoretical time-dependent study using the known positions and velocities of the nearby stars could answer this question. (b) What is the size of the "cell" in which the Sun is imbedded, and are the properties of this cell typical of the general intercloud medium? High-resolution Lα absorption measurements from nearby stars are now providing some of the answers to these questions. (c) How does the gas affect the velocity and temperature of the solar wind? Special mass spectrometer instrumentation flown to the outer solar system may be necessary. (d) Are there latitudinal gradients in the solar wind? Lα sky background maps may give a clue, but a definitive answer may come only from a space probe launched out of the ecliptic plane. (e) Are there solar cycle variations in the solar wind flux, and are these out of phase at high and low heliocentric latitudes? Again, measurements outside of the plane of the ecliptic may be required. (f) Does the solar wind significantly heat the interstellar gas? A critical review of the relevant inelastic collisions is important. However, the best way to answer this is to observe along the downwind He cone with a resonance absorption cell to determine whether the temperature decreases outward as predicted by theory. (g) What are the changes in the theoretical models when solar cycle effects are included, and do these produce a better fit to the observations? This requires constructing a time-dependent model, including long-term variations of solar wind flux, photoionization, and Lα radiation pressure.

New sky mappings in both 1216 Å and 584 Å are expected in late 1977 and 1978 from the Voyager UV Spectrometer Experiment (L. Broadfoot, private communication). New observing opportunities will arise when ultraviolet instruments are flown on forthcoming missions, such as the International Ultraviolet Explorer, Pioneer–Venus, and the Space Telescope; and on possible new deep-space explorations, such as the Solar Polar and Jupiter Orbiter-Probe Missions.

ACKNOWLEDGMENTS

I acknowledge with appreciation interesting discussions with T. E. Holzer, W. McClintock, J. Linsky, and R. R. Meier. I particularly want to thank T. E. Holzer for his careful reading of the manuscript, and C. Morgenstern for programming the "hot" model density calculation. This work was supported by NSF under grant number ATM76-23812.

Literature Cited

Adams, T. F., Frisch, P. C. 1977. High resolution observations of the Lyman Alpha sky background. *Astrophys. J.* 212:300–8

Ajello, J. M. 1978. An interpretation of Mariner 10 He (584 Å) and H (1216 Å) interplanetary emission observations. *Astrophys. J.* In press

Allen, C. W. 1973. *Astrophysical Quantities.* London: Athlone Press. 310 pp.

Auer, L. H., Shipman, H. L. 1977. A self-consistent model atmosphere analysis of the EUV white dwarf HZ 43. *Astrophys. J.* 211:L103–5

Axford, W. I. 1972. The interaction of the solar wind with the interstellar medium. In *The Solar Wind, NASA Spec. Publ. 308,* ed. C. P. Sonnett, P. J. Coleman, Jr., J. M. Wilcox, pp. 609–60

Axford, W. I. 1976. Flow of mass and energy in the solar system. In *Physics of Solar Planetary Environments,* ed. D. J. Williams, pp. 270–85. Washington, D.C.: Am. Geophys. Union. 1038 pp.

Bame, S. J., Asbridge, J. R., Feldman, W. C., Gosling, J. T. 1976. Solar cycle evolution of high-speed solar wind streams. *Astrophys. J.* 207:977–80

Barth, C. A. 1970a. Mariner 5 measurements of ultraviolet emission from the galaxy. In *Ultraviolet Stellar Spectra and Ground-Based Observations,* ed. L. Houziaux, H. E. Butler, pp. 28–35. Dordrecht: Reidel

Barth, C. A. 1970b. Mariner 6 measurements of the Lyman Alpha sky background. *Astrophys. J.* 161:L181–84

Bergeron, J., Souffrin, S. 1971. Heating of HI regions by hard UV radiation: Physical state of the interstellar matter. *Astron. Astrophys.* 11:40–52

Bertaux, J. L. 1974. *Atomic hydrogen in the terrestrial exosphere: measurements of intensity and line width of the Lyman-Alpha emission on board OGO-5 and their interpretation.* PhD thesis. Univ. Paris. 315 pp.

Bertaux, J. L., Ammar, A., Blamont, J. E. 1972. OGO-5 determination of the local interstellar wind parameters. In *Space Research,* ed. S. A. Bowhill, L. D. Jaffe, M. J. Rycroft, 12:1559–67. Berlin: Akademie-Verlag. 1198 pp.

Bertaux, J. L., Blamont, J. E. 1971. Evidence for a source of an extraterrestrial hydrogen Lyman Alpha emission: The interstellar wind. *Astron. Astrophys.* 11:200–17

Bertaux, J. L., Blamont, J. E. 1976. Possible evidence for penetration of interstellar dust into the solar system. *Nature.* 262:263–66

Bertaux, J. L., Blamont, J. E., Mironova, I., Kurt, V. G., Bourgin, M. C. 1977. Preliminary results from the H and He experiment on-board Prognoz-5. Presented at 20th Annual COSPAR Meeting, Tel Aviv, Israel.

Bertaux, J. L., Blamont, J. E., Tabarie, N., Kurt, V. G., Bourgin, M. C., Smirnov, A. S., Dementeva, N. N. 1976. Interstellar medium in the vicinity of the sun: A temperature measurement obtained with Mars-7 Interplanetary Probe. *Astron. Astrophys.* 46:19–29

Bhatnagar, V. P., Fahr, H. J. 1972. Solar wind expansion beyond the heliosphere. *Planet. Space Sci.* 20:445–60

Blamont, J. E., Cazes, S., Emerich, C. 1975. Direct measurement of hydrogen density at exobase level and exospheric temperature from Lyman α line shape and polarization. 1. Physical background and first results on dayside. *J. Geophys. Res.* 80: 2247–65

Blum, P. W., Fahr, H. J. 1970. Interaction between interstellar hydrogen and the solar wind. *Astron Astrophys.* 4: 280–90

Blum, P. W., Fahr, H. J. 1976. Revised interstellar neutral helium/hydrogen density ratios and the interstellar UV-radiation field. *Astrophys. Space Sci.* 39: 321–34

Blum, P. W., Pfleiderer, J., Wulf-Mathies, C. 1975. Neutral gases of interstellar origin in interplanetary space. *Planet. Space Sci.* 23: 93–105

Bohlin, R. C. 1973. Mariner 9 Ultraviolet Spectrometer Experiment: Measurements of the Lyman Alpha sky background. *Astron Astrophys.* 28: 323–26

Bohlin, R. C. 1975. Copernicus observations of interstellar absorption at Lyman Alpha. *Astrophys. J.* 200: 402–14

Bohlin, R. C., Savage, B. D., Drake, J. F. 1977. A survey of interstellar HI from Lα absorption measurements, II. *Goddard Space Flight Center Preprint X-681-77-255*, Greenbelt, Maryland

Bottcher C., McCray, R. A., Jura, M. 1970. Time-dependent model of the interstellar medium. *Astrophys. Lett.* 6: 237–41

Bridge, H. S. 1976. Solar cycle manifestations in the interplanetary medium. See Axford 1976, pp. 47–62

Broadfoot, A. L., Kumar, S. 1978. The interstellar wind: Mariner 10 measurements of H (1216 Å) and He (584 Å) interplanetary emission. *Astrophys. J.* In press

Bruner, E. C., Jr. 1978. OSO-8 High Resolution Ultraviolet Spectrometer Experiment: Introduction and optical design. *Space Sci. Instrumentation.* In press

Cazes, S., Emerich, C. 1977. Interstellar medium Ly α emission: Line profile, temperature and polarization measurements deduced from its geocoronal absorption. *Astron. Astrophys.* 59: 59–68

Dupree, A. K. 1975. Ultraviolet observations of Alpha Auriga from Copernicus. *Astrophys. J.* 200: L27–31

Dupree, A. K., Baliunas, S. L., Shipman, H. L. 1977. Deuterium and hydrogen in the local interstellar medium. Center for Astrophysics Prepr. No. 753, Harvard Univ., Cambridge, Massachusetts

Evans, R. G., Jordan, C., Wilson, R. 1975. Observations of chromospheric and coronal emission lines in F stars. *MNRAS* 172: 585–602

Fahr, H. J. 1971. The interplanetary hydrogen cone and its solar cycle variations. *Astron Astrophys.* 14: 263–74

Fahr, H. J. 1974. The extraterrestrial UV-background and the nearby interstellar medium. *Space Sci. Rev.* 15: 483–540

Fahr H. J., Lay, G. 1973. Interplanetary HeI 584 Å background radiation. In *Space Research* 13, ed. M. J. Rycroft, S. K. Runcorn, pp. 844–47; Berlin, Akademie-Verlag, 800 pp.

Fahr, H. J., Lay, G. 1974. Solar radiation asymmetries and heliospheric gas heating influencing extraterrestrial UV data. In *Space Research* 14, ed. M. J. Rycroft, R. D. Reasenberg, pp. 567–73; Berlin, Akademie-Verlag, 737 pp.

Fahr, H. J., Lay, G., Wulf-Mathies, C. 1977. Derivation of interstellar helium gas parameters from an EUV rocket observation. Presented at Annual COSPAR Meeting, 20th, Tel Aviv, Israel

Feldman, W. C., Asbridge, J. R., Bame, S. J., Kearney, P. D. 1974. Upper limits for the solar wind He$^+$ content at 1 A.U. *J. Geophys. Res.* 79: 1808–12

Feldman, W. C., Lange, J. J., Scherb, F. 1972. Interstellar helium in interplanetary space. See Axford 1972, pp. 684–97

Field, G. B., Goldsmith, D. W., Habing, H. J. 1969. Cosmic ray heating of the interstellar gas. *Astrophys. J.* 155: L149–54

Freeman, J., Paresce, F., Bowyer, S., Lampton, M., Stern, R., Margon, B. 1977. The local interstellar helium density. *Astrophys. J.* 215: L83–86

Grewing, M. 1975. The nearby interstellar radiation field between 1750 Å and 504 Å. *Astron Astrophys.* 38: 391–96

Greisen, E. W. 1976. The small-scale structure of interstellar hydrogen. *Astrophys. J.* 203: 371–77

Holzer, T. E. 1970. *Stellar winds and related flows.* PhD thesis. Univ. of Calif., San Diego. 308 pp.

Holzer, T. E. 1972. Interaction of the solar wind with the neutral component of the interstellar gas. *J. Geophys. Res.* 77: 5407–31

Holzer, T. E. 1977a. Interaction between the solar wind and the interstellar medium. In *Exploration of the Outer Solar System*, ed. E. M. Greenstadt, M. Dryer, D. S. Intrilligator, pp. 21–41. New York: AIAA. 237 pp.

Holzer, T. E. 1977b. Neutral hydrogen in interplanetary space *Rev. Geophys. Space Phys.* 15: 467–90

Holzer, T. E., Axford, W. I. 1971. Interactions between interstellar helium and the solar wind. *J. Geophys. Res.* 76: 6965–70

Holzer, T. E., Leer, E. 1973. Solar wind heating beyond 1 AU. *Astrophys. Space Sci.* 24: 335–47

Johnson, H. E. 1972. Backscatter of solar radiation-I. *Planet. Space Sci.* 20: 829–40

Joselyn, J. A., Holzer, T. E. 1975. The effect

of asymmetric solar wind on the Lyman α sky background. *J. Geophys. Res.* 80: 903–7

Kafatos, M. 1973. Time-dependent radiative cooling of a hot low-density cosmic gas. *Astrophys. J.* 182: 433–47

Keller, H. U., Thomas, G. E. 1975. A cometary hydrogen model: Comparison with OGO-5 measurements of Comet Bennett (1970II). *Astron Astrophys.* 39: 7–19

Levy, E. H., Jokipii, J. R. 1976. Penetration of interstellar dust into the solar system. *Nature* 264: 423–24

Maloy, J. O., Carlson, R. W., Hartmann, U. G., Judge, D. L. 1978. Measurement of the profile and intensity of the solar HeI λ584 Å resonance line. *J. Geophys. Res.* In press

Mange, P., Meier, R. R. 1970. OGO-3 observations of the Lyman-α intensity and the hydrogen concentrations beyond 5 earth radii. *J. Geophys. Res.* 75: 1837–47

Margon, B., Liebert, J., Gatewood, G., Lampton, M., Spinrad, H., Bowyer, S. 1976. An extrasolar extreme-ultraviolet object. II. The nature of HZ 43. *Astrophys. J.* 209: 525–35

McClintock, W., Henry, R. C., Moos, H. W., Linsky, J. L. 1975a. Detection of interstellar Deuterium Lyman Alpha. *Bull. Am. Astron. Soc.* 7: 547

McClintock, W., Linsky, J. L., Henry, R. C., Moos, H. W., Gerola, H. 1975b. Ultraviolet observations of cool stars. III. Chromospheric and coronal lines in α Tauri, β Geminorum, and α Bootes. *Astrophys. J.* 202: 165–82

McClintock, W., Henry, R. C., Moos, H. W., Linsky, J. L. 1976. Ultraviolet observations of cool stars. V. The local density of interstellar matter. *Astrophys. J.* 204: L103–6

McClure, R. D., Crawford, D. L. 1971. The density distribution, color excess, and matallicity of K Giants at the North Galactic Pole. *Astron. J.* 76: 31–39

Meier, R. R. 1977. Some optical and kinetic properties of the nearby interstellar gas. *Astron. Astrophys.* 55: 211–19

Meier, R. R., Mange, P. 1973. Spatial and temporal variations of the Lyman-alpha airglow and related atomic hydrogen distributions. *Planet. Space Sci.* 21: 309–27

Moos, H. W., Linsky, J. L., Henry, R. C., McClintock, W. 1974. High-spectral-resolution measurements of the HI λ1216 and MgII λ28000 emissions from Arcturus. *Astrophys. J.* 188: L93–95

Morton, D. C., Purcell, J. D. 1962. Observations of the extreme ultraviolet radiation in the night sky using an atomic hydrogen filter. *Planet. Space Sci.* 9: 455–8

Paresce, F., Bowyer, S. 1973. Resonance scattering from interstellar and interplanetary helium. *Astron. Astrophys.* 27: 399–406

Paresce, F., Bowyer, S., Kumar, S. 1974a. Observations of HeI 584 Å nightime radiation: Evidence for an interstellar source of neutral helium. *Astrophys. J.* 187: 633–39

Paresce, F., Bowyer, S., Kumar, S. 1974b. Further evidence for an interstellar source of nightime HeI 584 Å radiation. *Astrophys. J.* 188: L71–73

Patterson, T. N. L., Johnson, F. S., Hanson, W. B. 1963. The distribution of interplanetary hydrogen. *Planet. Space Sci.* 11: 767–78

Paul, J., Cassé, M., Cesarsky, C. J. 1976. Distribution of gas, magnetic fields, and cosmic rays in the galaxy. *Astrophys. J.* 207: 62–77

Salpeter, E. E. 1976. Planetary nebulae, supernova remnants, and the interstellar medium. *Astrophys. J.* 206: 673–78

Silk, J. 1973. Heat and ionization sources in the interstellar medium. *Publ. Astron. Soc. Pac.* 85: 704–13

Snow, T. P., Jr. 1976. A review of ultraviolet astronomical research with the Copernicus satellite. *Earth Extraterr. Sci.* 3: 1–25

Spitzer, L., Cochran, W. D., Hirschfeld, A. 1975. Column densities of interstellar molecular hydrogen. *Astrophys. J. Suppl.* No. 266. 28: 373–89

Spitzer, L., Jr., Jenkins, E. B. 1975. Ultraviolet studies of the interstellar gas. *Ann. Rev. Astron. Astrophys.* 13: 133–64

Suzuki, K., Carlson, R. W., Judge, D. L. 1975. *Pioneer 10 and 11 ultraviolet photometer observations of the interplanetary hydrogen and helium glow: The radial dependence.* Presented at Fall Ann. Meet. Am. Geophys. Union, San Francisco, California.

Thomas, G. E. 1972. Properties of nearby interstellar hydrogen deduced from Ly-α sky background measurements. See Axford 1972, pp. 668–83

Thomas, G. E. 1973. *The neutral interplanetary medium.* Presented at AIAA/AGU Conf. Explor. Outer Solar Syst., Denver, Colorado

Thomas, G. E. 1975. Interplanetary gas of nonsolar origin. *Rev. Geophys. Space Phys.* 13: 1063–65, 1081–83

Thomas, G. E. 1976. *Interaction of interstellar neutral hydrogen and the solar system.* Presented at AGU Int. Symp. Solar-Terr. Phys., Boulder, Colorado

Thomas, G. E., Blamont, J. E. 1976. Galactic Lyman Alpha emission in the solar vicinity. *Astron. Astrophys.* 51: 283–88

Thomas, G. E., Bohlin, R. C. 1972. Lyman Alpha measurements of neutral hydrogen in the outer geocorona and in interplanetary space. *J. Geophys. Res.* 77: 2752–61

Thomas, G. E., Krassa, R. F. 1971. OGO-5 measurements of the Lyman Alpha sky background. *Astron. Astrophys.* 11: 218–33

Thomas, G. E., Krassa, R. F. 1974. OGO-5 measurements of the Lyman Alpha sky background in 1970 and 1971. *Astron. Astrophys.* 30: 223–32

Tinsley, B. A. 1971. Extraterrestrial Lyman Alpha. *Rev. Geophys. Space Phys.* 9: 89–102

Vasyliunas, V. M., Siscoe, G. L. 1976. On the flux and spectra of interstellar ions. *J. Geophys. Res.* 81: 1247–52

Vidal-Madjar, A. 1975. Evolution of the solar Lyman Alpha flux during four consecutive years. *Solar Phys.* 40: 69–86

Vidal-Madjar, A. 1977. Measurements of the solar spectral irradiance at Lyman Alpha (1216 Å). In *The Solar Output and its Variation*, ed. O. R. White. pp. 213–36. Boulder, Colorado: Colorado Assoc. Univ. Press. 526 pp.

Wallis, M. 1971. Shock-free deceleration of the solar wind? *Nature* 233: 23–25

Wallis, M. 1973. Interaction between the interstellar medium and solar wind plasma. *Astrophys. Space Sci.* 20: 3–18

Wallis, M. 1974. Local hydrogen gas and the background Lyman Alpha pattern. *MNRAS* 167: 103–19

Wallis, M. 1975. Collisional heating of interplanetary gas: Fokker-Planck treatment. *Planet. Space Sci.* 23: 419–30

Wallis, M. 1976. Stochastic scattering and other contributions to the sun's wake in the local hydrogen cloud. *Celestial Mech.* 13: 65–74

Weller, C. S., Meier, R. R. 1974. Observations of helium in the interplanetary/interstellar wind: The solar-wake effect. *Astrophys. J.* 193: 471–76

White, O. R., Lemaire, P. 1976. A summary of scientific results from the sail experiments of OSO-8 during the first year of operation. *LASP OSO Rep. No. 2.* Boulder, Colorado: Univ. Colorado

Wu, C. S., Hartle, R. E. 1974. Further remarks on plasma instabilities produced by ions born in the solar wind. *J. Geophys. Res.* 79: 283–5

Wulf-Mathies, C., Blum, P. 1975. *Interplanetary abundance of neutral interstellar heavy elements.* Presented at Ann. COSPAR Meet., 18th, Varnz, Bulgaria

York, D. G. 1976. On the existence of molecular hydrogen along lines of sight with low reddening. *Astrophys. J.* 204: 750–58

Yu, G. 1974. The interstellar wake of the solar wind. *Astrophys. J.* 194: 187–202

Ann. Rev. Earth Planet. Sci. 1978. 6: 205–28

GLACIAL INCEPTION AND DISINTEGRATION DURING THE LAST GLACIATION

✻10092

J. T. Andrews

Institute of Arctic and Alpine Research and Department of Geological Sciences, University of Colorado, Boulder, Colorado 80309

R. G. Barry

Institute of Arctic and Alpine Research and Department of Geography, University of Colorado, Boulder, Colorado 80309

INTRODUCTION

The causes of the growth and collapse of the large Pleistocene ice sheets is a topic of intense scientific debate. The amount of new data and ideas has been prodigious during the last ten years. This review concentrates on the shifting patterns of glacial advance and shrinkage during the last Glaciation. We also examine reconstructions of conditions during the maximum of this Glaciation about 18,000 BP, and the mechanisms that might have caused the rapid collapse of many of the world's ice sheets between 18,000 and 8000 BP. Our paper focuses on the North Atlantic sector because it is there that the major Pleistocene ice sheets grew, developed, and retreated.

CLIMATIC CONTROLS OF THE GLACIAL CYCLE

Time Scales of Climatic Controls

Glaciation could have occurred according to three possible time scales. These are approximately 10^6–10^8, 10^4–10^5, and 10^3 years; we are concerned specifically with the last two.

The 10^4–10^5-year time scale involves changes in the Earth's orbital parameters (Milankovitch 1941). This external forcing mechanism has been established recently as a primary control of glacial/interglacial oscillations, at least during the late Pleistocene. Hays, Imbrie & Shackleton (1976) demonstrate the existence of spectral peaks in ocean-core records at about 100,000, 42,000, and 23,000 years. These intervals match within 5% the frequencies calculated for the astronomical factors that are due, respectively, to the eccentricity of the Earth's orbit about the Sun

205

0084-6597/78/0515-0205$01.00

(93,000 years), the obliquity of the ecliptic (41,000 years), and the precession of the equinoxes (21,000 years) (Vernekar 1972). The climatic variance within the oceanographic records corresponding to the three peaks is 50, 25 and 10%, respectively. The dominant 100,000-year component, unlike the other factors, appears to involve a nonlinear response of the climate system to the orbital forcing (Wigley 1976). The variations of solar radiation caused by orbital variations seem to be large enough to account for major changes in ice sheets on the northern continents, given high accumulation and ablation rates on the southern half of the ice sheets (Weertman 1976). However, the mechanism of the atmsopheric-cryospheric linkages with these variations of solar radiation remains to be shown in detail. It generally has been assumed that summer conditions over high middle latitudes would be most critical, although Kukla (1975) indicates that the highest sensitivity to solar radiation occurs in early autumn. At this season, values were low 17, 36, 51, 67, 89, and 111 thousand years ago and high 6, 28, 42, 55, 77, 101, and 122 thousand years ago.

The 10^3-year time scale of glacial variability is less well-defined. According to Bray (1970) and Denton & Karlén (1973), the record of Neoglaciation displays a recurrence interval of approximately 2500–2600 years, but the statement can be challenged. In this time range there is no generally accepted causal mechanism.

In addition to these time scales, there is also evidence of rapid shifts of climatic state over intervals of a century or less (Bryson 1974). These may represent the regional response of a proxy variable that was close to a particular critical threshold state, or the atmospheric circulation may switch its circulation mode. A sufficient cause of such a sharp transition may be a specific external event, or it could be the gradual trend of solar radiation due to the orbital variations.

Climatic Responses

If the prime climatic forcing is via a reduction or solar radiation, whether through orbital variations alone or in combination with an increase in atmospheric turbidity, then the next question concerns its effect on the atmospheric circulation and global climate. Modeling results indicate that a reduction in the solar constant of 2–5% would be sufficient to initiate an ice age (Sellers 1974). However, according to Berger (1977), the orbital effects are much more effective in modifying latitudinal temperature gradients than changes in solar constant. Simulations of the temperature effects of variations in the orbital parameters indicate changes 1–2 K at 65°N in a zonally averaged circulation model (Saltzman & Vernekar 1971) whereas the average ocean surface temperature was about 2.3 K lower than today at 18,000 BP (CLIMAP Project Members 1976).

These results suggest that orbital factors initiated a cooling process that was amplified and accelerated by feedbacks in the atmosphere-ocean-cryosphere system. A possible feedback mechanism in the cryosphere involves ice-sheet dynamics. The climatic implications of ice-sheet surges in Antarctica were recognized by Wilson (1964), and Hollin (1977) reports some geological evidence for a sharp rise in sea level (to about +16 m) at the end of the last interglacial (roughly 115,000 BP). This rise is related to a hypothetical Antarctic surge. Flohn (1974) and Hughes (1975) have both developed geophysical models of northern hemisphere glacial inception involving the response of Atlantic ocean and atmospheric circulations to such a surge, but in the absence of field evidence these speculations cannot be assessed.

The role of oceanic circulation appears to be a major one. During the inception and growth phases, it seems essential that there be warm ocean surfaces to provide the moisture (Barry 1966) and the "heat of glaciation" [released during condensation (88%) and freezing (12%)], which balances the energy budget over the ice sheets (Adam 1975). The ocean-core records indicate that, indeed, ice-sheet growth around 115,000 BP and 75,000 BP coincided with warm oceans (W. Ruddiman, personal communication 1977). Meridional temperature gradients would be enhanced during this phase, acting to increase poleward energy transport by the atmosphere (Newell, 1974). As the glacial cycle progresses, the North Atlantic ocean circulation becomes more zonal with the southward displacement of the Gulf Stream (Ruddiman & McIntyre 1976). In the model of Johnson & McClure (1976), the weakening of the subpolar gyre and the cessation of the West Greenland Current eliminate accumulation on the northern Laurentide Ice Sheet. Storm tracks across the Atlantic also remain well south of the European ice sheets (cf Williams & Barry 1975), giving a similar accumulation pattern there. This atmospheric circulation mode may still allow ice-sheet equilibrium, however. The shift to an interglacial regime seems to be dependent on the reduced formation of bottom water in the northern North Atlantic, due to the ice cover on the sea. (Weyl 1968, Newell 1974). This reduction leads to less upwelling of cold water giving warmer equatorial surface waters. Confirmation of this mechanism is provided by faunal evidence from the eastern equatorial Pacific (Pisias, Heath & Moore 1975). Decreases in bottom water production are identified 2600 years *prior* to changes in surface waters and 5600 years *prior* to ice-sheet recession.

THE NORTH ATLANTIC SECTOR—PRESENT CONDITIONS

The distinctive distribution of ice masses on land during the last Glaciation and the close similarity of earlier glaciations has been recognized by many writers (Tanner 1955, Flohn 1974, Adam 1973, Hughes 1975). The largest and most temporally variable ice sheets—the Laurentide, the Fennoscandian, the (postulated) Barents Sea—were all located around the margins of the Atlantic Ocean. Studies of air-sea interaction in the North Atlantic (Bjerknes 1964) demonstrate the close coupling of the atmosphere and ocean systems. The most significant aspect of the ocean circulation is the clockwise, wind-driven surface gyre in the Sargasso Sea and the concentration of flow along its western margin in the Gulf Stream. Continuation of the latter as the North Atlantic Drift and its offshoots transports warm saline water northeastward into the Norwegian and Greenland seas (Figure 1). Southward flows of water, primarily from the Arctic along the east coast of Greenland and from western Baffin Bay, are associated with extensive sea ice off East Greenland and along the Labrador coast in spring.

The atmospheric circulation in winter is dominated by the subtropical high pressure in the Azores and the subpolar low pressure in Iceland. This mean low is indicative of an area of frequent depression activity due to systems that form along the east coast of North America and then move generally northeastward, transporting moist, relatively warm air towards Iceland and northwestern Europe. In contrast, the eastern Canadian Arctic is anomalously cold for its latitude. The Azores anticyclone expands in summer while the subpolar low weakens and splits into centers over Iceland and Baffin Island.

Figure 1 is a map of the Glaciation Threshold or level (GT) (Østrem 1964). Low GTs occur around the Arctic Basin and steep gradients occur between the coasts and the more continental interiors, whereas the GT is higher in northern Alaska and northwestern Canada. The importance of Figure 1 is that it makes

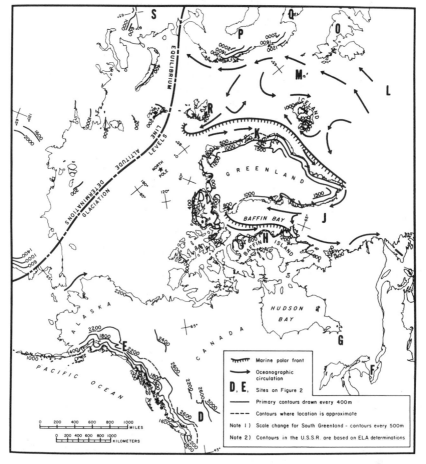

Figure 1   Map of North Atlantic area showing the elevation of the Glaciation Threshold (GT), major surface currents, position of the boundary between Polar (arctic) and Subpolar (subarctic) Water Masses, and location of the time/distance diagrams of Figure 2. Modified from an unpublished map compiled by J. T. Andrews and G. H. Miller, 1972.

possible assessments of potential climatic change in specific areas. The effect of climatic changes on the GT is strongly influenced by local topography. A drop in GT elevation of 200 m in mountains is much less effective in causing glaciation than a similar decrease across an elevated tilted plateau. This is the essence of the argument for "instantaneous glacierization" of the plateaus of the eastern Canadian Arctic.

WISCONSINAN STAGE–GLOBAL TRENDS AND NOMENCLATURE

An understanding of the last Glaciation involves differentiating between global and regional events. Changes in global sea level over the last 130,000 years have been documented from Barbados and New Guinea (Steinen, Harrison & Matthews 1973, Bloom et al 1974). Figure 2A suggests that there were six periods of glacial development during the last Glaciation. We refer to these as "Glacial Buildup Periods" (GBP). Did each period affect the same regions and in the same manner? The trend of ever-deepening glaciation was dramatically ended shortly after 18,000 BP. Each GBP resulted in a lowering of the sea level by 30 to 60 m at the rate of 6 to 2 m $(1000 \text{ yr})^{-1}$. An independent check on Figure 2A is available through $\delta^{18}O$ changes of planktonic foraminifera (e.g. Shackleton & Opdyke 1973, 1976) (Figure 2B). Regions of fast sedimentation indicate that the fall of sea level estimated from the $\delta^{18}O$ exceeded 1 m $(1000 \text{ yr})^{-1}$ between marine isotope stages 5e and 5d (Figure 2B; Shackleton 1976).

The onset of the Wisconsin Glaciation is placed between 120,000 and 70,000 BP (Suggate 1974). Such a time span is not surprising because by *definition* (e.g. Flint 1971) the boundaries of glacial stratigraphic units transgress time. Accordingly, a nomenclature is required that has global applicability. We suggest that the marine $\delta^{18}O$ record be the control for a global chronostratigraphy (Shackleton 1969, 1976).

LAST (WISCONSIN) GLACIATION–REGIONAL CHRONOLOGIES

What is meant by "glacial inception"? Judging from the diagrams in Figure 2 such a question has to be prefaced by "where" and "when". The sea-level evidence (Figures 2A and B) indicates that the initial GBP caused only partial ice storage. For paleoclimatic reconstructions we need to know where these early ice sheets were located, but many of the time-distance diagrams on Figure 2 are "floating" prior to 50,000 BP (30,000 BP on shells). We attempt to link the diagrams by using the deep-sea records as a master stratigraphy (e.g. Wijmstra & van der Hammen 1974).

Last (Eemian) Interglacial

Paleobiological evidence indicates that the last Interglacial was warmer than present by up to 6°–7°C in Russian and central Siberia (Frenzel 1973). In the western North Atlantic, sea-surface temperatures were 1°C higher than now (Sancetta,

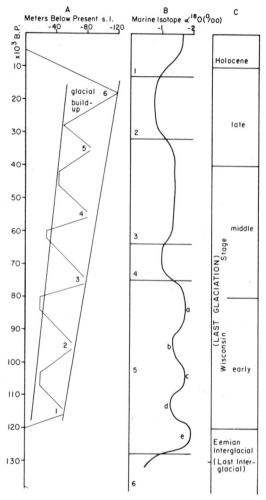

Figure 2 Time/distance diagrams from various geologic-climate records: global scale
(*A* and *B*) and regional stratigraphies (*D* onwards). Sites located on Figure 1. *A*, global
sea level (after Bloom et al 1974); *B*, marine isotopic record and stages (Shackleton &
Opdyke 1973); *C*, nomenclature of this paper; *D*, south-central British Columbia (Fulton
1976); *E*, Yukon and Alaskan Mountains (Denton 1974); *F*, Great Lakes/St. Lawrence
Valley (Dreimanis 1976); *G*, Hudson Bay (Skinner 1974); *H*, Baffin Island (Miller et al 1977);
I, Maritimes, Canada (Grant 1976); *J*, Labrador Sea sediment (Fillon 1976 and personal
communication); *K*, East Greenland (Funder & Hjort 1973); *L*, North Atlantic (Ruddiman
& McIntyre 1976); *M*, Greenland/Norwegian seas (Kellogg 1976); *N*, Northern Europe
vegetation (van der Hammen et al 1971); *O*, Great Britain (Worsley 1976); *P*, Fennoscandia
(Mörner 1974); *Q*, Skaerumhede II borehole, Denmark (Bahnson et al 1973); *R*, Spitzbergen
(Boulton & Rhodes 1975); *S*, USSR (Kind 1976).

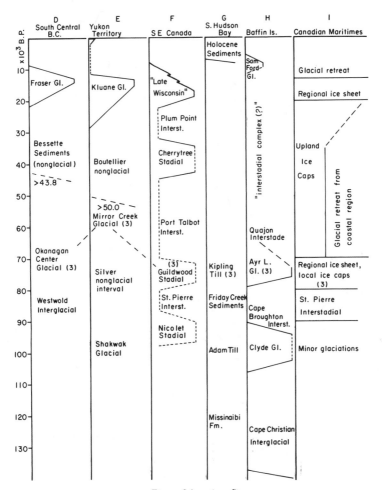

Figure 2 (continued)

Imbrie & Kipp 1973). In the Canadian Arctic, summer conditions were also warmer than they are now (Blake 1974). Macrofossils and pollen from sediments of the Flitaway Interglacial on Baffin Island (Terasmae, Webber & Andrews 1966) indicated that July temperatures were 2 to 4°C higher than present. Thus, in the Canadian Arctic, a key area for glacial inception, the change to glacial conditions was dramatic because the last Interglacial was significantly warmer than the present Interglacial.

Early Wisconsinan Stage (120,000 to 80,000 BP)

The early stadials and interstadials of the last Glaciation include global changes in ice storage represented by marine isotope stages 5*a*, *b*, *c*, and *d*. A critical

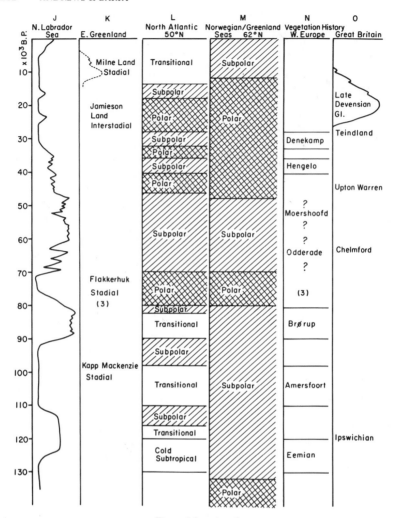

Figure 2 (continued)

question is the rate of sea-level lowering during GBP 1 and 2 (see Figure 2*A*). Marine isotope data suggest a rate of about 1 m $(1000 \text{ yr})^{-1}$. Numerical modeling of the growth of the Laurentide Ice Sheet indicates rates of sea-level lowering of up to 4 m $(1000 \text{ yr})^{-1}$ and a 20-m fall of the global sea level (Barry, Andrews & Mahaffy 1975, Andrews & Mahaffy 1976). This is far short of the 70-m fall suggested by Bloom et al (1974), which must be considered with scepticism even allowing for ice-sheet growth in Antarctica (Hughes 1975) and elsewhere.

We now examine the evidence for climatic change during the early part of the last Glaciation. The sparsity of absolute dating control prior to 50,000 BP must be

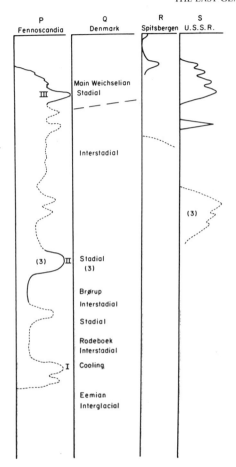

Figure 2 (continued)

re-emphasized. Correlations older than this are built around the assumption that the major glacial advance of GBP 3 (Ruddiman 1977a,b) can be recognized in many records.

Many stratigraphies for the early Wisconsinan indicate two periods of cooling and two interstadials. Evidence from the North Atlantic shows that two pulses of subpolar water crossed latitude 50°N (Ruddiman & McIntyre 1976), whereas subpolar water dominated the Greenland/Norwegian seas between 130,000 and about 80,000 BP (Kellogg 1975, 1976). In southern Scandinavia an initial cooling of the inshore marine environment was followed by faunas of the Rodeboek Interstadial (Figure 2). Mörner (1976) suggests that the Würm I stade of Fennoscandia dates from about 115,000 BP. If the Brørup Interstadial dates from about 80,000 BP (van der Hammen, Wijmstra & Zagwijn 1971, Wijmstra & van der

Hammen 1974, Wijmstra 1975), then such an age assessment is reasonable. There is no firm evidence for glacial ice during GBP 1, but glaciation apparently affected Scandinavia during the pre-Brørup stadial. The Brørup Interstadial was a period of significant climatic amelioration with temperatures within a degree or so of present-day values (van der Hammen et al 1967).

In the northwestern North Atlantic and in North America there is firm evidence for only one glacial stade prior to the St. Pierre Interstadial (D to K, Figure 2). Critical records are those from southern Hudson Bay (G), the St. Lawrence Lowlands/Great Lakes (F) and Baffin Island (H). McDonald (1971) concluded that the Missinaibi Formation is correlative with the St. Pierre Interstadial, whereas Prest (1970) and Skinner (1974) argue that the Missinaibi Formation is of last Interglacial age. All ^{14}C dates on the Missinaibi Formation (McDonald, 1971) are "greater than" with the "oldest" > 54,000 BP (Geological Survey of Canada, 1185). We conclude that the Nicolet Stadial and the Adam Till are broadly correlative and mark the first buildup and advance of ice southward and southwestward from the Labrador ice center. Retreat during the St. Pierre Interstadial resulted in lakes in the southern Hudson Bay lowlands into which Friday Creek nonglacial sediments were deposited.

The Quaternary stratigraphy of Baffin Island bears on this problem (Figure 2H). Radiocarbon dates on marine shells indicate that sediments of the Quajon Interstadial are > 48,000 BP (probably a minimum age, M. Stuiver, personal communication, 1977). Uranium-series dates on marine shells from units with similar "warm" faunas and similar amino acid ratios (Andrews et al 1976) were between 48,000 and 69,000 BP, suggesting that the Ayr Lake stade is > 60,000 BP (Miller, Andrews & Short 1977). On the basis of amino acid ratios sediments of the Cape Broughton Interstadial (Feyling-Hanssen 1976) are significantly older. Figure 3 illustrates a time/distance diagram from Baffin Island southward to the St. Lawrence Lowlands. The suggested correlations are qualitatively supported by simulated ice-sheet studies (Andrews & Mahaffy 1976). Available evidence indicates that during the early Wisconsinan stage the Laurentide Ice Sheet was confined to Labrador and Baffin Island with an extension across onto northern Keewatin. The Great Lakes and the Canadian Prairies were not glaciated during GBP 1 or 2 (see Figure 4).

Curves of δ^{18}O from arctic ice sheets (Dansgaard et al 1971, Paterson 1977) should provide evidence on the early Wisconsinan climate, but interpretation is bedeviled by uncertainties about the age of the ice in the lower layers (Robin 1977, p.60). Paterson et al (1977) noted that global glaciation was in progress by 115,000 BP, implying that the high δ^{18}O values occurring during the early Wisconsinan reflect a different precipitation source, rather than "interglacial" temperatures. Andrews et al (1974) previously suggested that the early Wisconsinan glacial advances in East Greenland (Funder & Hjort 1973) and eastern Baffin Island were a response to mild, moist conditions.

Field evidence from the North Atlantic region leads to the following conclusions: 1. there is no unequivocal evidence of glaciation during GBP 1; 2. there is broad agreement that significant glaciation affected Scandinavia and the eastern sector of

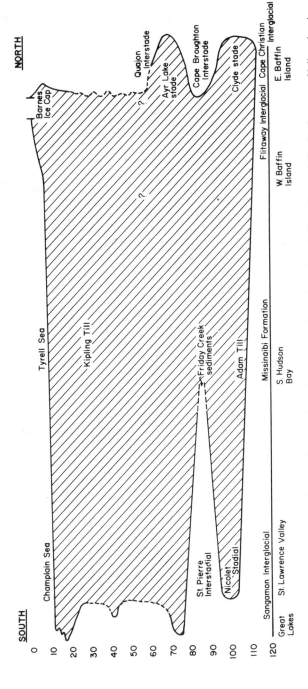

Figure 3   Time/distance diagram from Baffin Island southwestward and southward to Hudson Bay and the St. Lawrence Valley showing suggested correlations between three key areas (see Figure 1 and Figure 2, *F*, *G*, and *H*).

the Laurentide Ice Sheet immediately prior to the St. Pierre Interstadial; 3. significant deglaciation occurred during St. Pierre/Brørup times; probably centered about 85,000 BP. The argument for a global sea-level minimum of -70 m during both these periods of glacial buildup is difficult to reconcile with available evidence. The volume of the Laurentide Ice Sheet during the Nicolet/Clyde Stadial is equivalent to a 20 m fall in global sea level and, together with the growth of ice sheets elsewhere, could account for 35–40 m of global sea level stored within the early Wisconsin ice sheets. The first glacial nucleation occurred during marine isotope stage 5d when inital cooling of northern Europe was associated with the southward movement of Subpolar waters to latitude 44°N (Ruddiman & McIntyre 1976). Glacial buildup was interrupted by an interstadial, but the next southward penetration of Subpolar water into the North Atlantic triggered the expansion of the Laurentide Ice Sheet. Values of $\delta^{18}O$ from the ice sheets around Baffin Bay are close to present, suggesting the presence of considerable open water in Baffin Bay during GBP 2. Based on the above data, the Northern Hemisphere ice sheets during marine isotope stages 5b and 5d must have increased in volume at a rate equivalent to an average global sea level fall of 1.5 to 2 m $(1000 \text{ yr})^{-1}$.

During the St Pierre/Brørup Interstadial the Laurentide Ice Sheet was maintained over Hudson Bay, Labrador, and Baffin Island, whereas ice virtually may have disappeared from Fennoscandia.

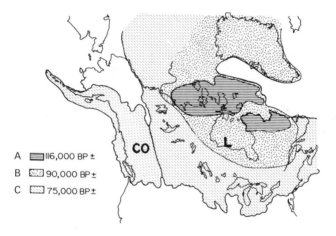

Figure 4   Schematic outline of the Laurentide (*L*) Ice Sheet. The area shown by *A* was affected by instantaneous glaciation about 116,000 BP. The core area shown by *B* was formed about 90,000 BP by glacial growth and flow from *A*. The ice extent was broadly similar during the Port Talbot Interstadial, 70,000–45,000 BP. *C* shows the unstable western and southwestern periphery where the ice advances and retreats were rapid and approximately synchronous with the growth and retreat of the Cordilleran (CO) Ice Sheet (75,000 BP).

Middle Wisconsinan stage (79,000 to 40,000 BP)

The *most* significant change in global environment during the last Glaciation occurred early in the middle Wisconsinan. During GBP 3 the Laurentide Ice Sheet expanded westward to the flanks of the Rocky Mountains (Fulton 1976, Klassen 1969) and southward into the Great Lakes and northern U.S.A. (cf. Dreimanis & Goldthwait, 1973) (see Figures 4 and 5).

The change in global sea level (Figure 2A) involved a drop of 40 to 60 m in 10,000 to 15,000 years. This rapid fall is commensurate with the greater change in $\delta^{18}O$ between marine isotope stages 4/5 compared with any other interstadial to glacial transition of the last Glaciation (Shackleton & Opdyke 1976, Table 3).

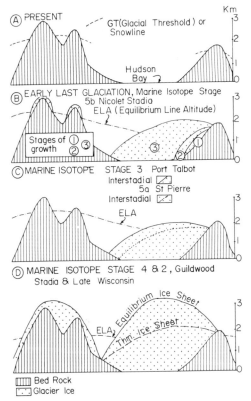

Figure 5 Schematic picture of changes in the Glaciation Threshold (GT) on west-east transect across North America at various times during the last Glaciation: *A*, present, *B*, early last Glaciation, marine isotope stage 5b; Nicolet Stadial, *C*, marine isotope stages 3 and 5a; Port Talbot and St. Pierre Interstadials, *D*, marine isotope stages 4 and 2; Guildwood Stadial and late Wisconsin.

However, the ice sheets that expanded during this period did so from considerable nuclei (Figures 3, 4 and 5). If our judgment of the age of the Friday Creek sediments is correct (Figure 3) and the margin of the Laurentide Ice Sheet advanced 1400–1800 km south and west in 10,000 to 15,000 years, then it is unlikely that the margin could have advanced by the lowering of the GT and instantaneous glacierization across the low continental interior. Ice must have flowed from accumulation areas over Labrador, Hudson Bay, and Foxe Basin. However, new glacial complexes developed in the Cordilleran Mountains of Canada and Alaska (Denton 1974, Fulton 1976, Easterbrook 1976), pointing to a significant and rapid fall in the GT across the western mountains (Figure 5).

A significant glacial retreat of several hundreds of kilometers in the western and southwestern sector of the Laurentide Ice Sheet (McDonald 1971, Dreimanis 1976, Fulton 1976) ushered in a long complex interstadial and/or nonglacial interval. The mountains of the Yukon territory and the Canadian Cordillera had less ice than now from about 50,000 BP to the start of GBP 6. Although records from the Great Lakes and northern Europe indicate glacial advances during GBP 5 these have not been recorded in the western United States (see, however, Easterbrook 1976).

The presence of Polar or Subpolar water in the North Atlantic between latitudes 50 to 62°N implies that northern Europe experienced a mainly cold, dry climate between 79,000 and 40,000 BP. The geological evidence suggests that a considerable part of the Fennoscandian Ice Sheet disappeared between 60,000 and 30,000 BP (Dreimanis & Raukas 1975). Climatic indicators from Middle Weichselian deposits around the Fennoscandian Ice Sheet indicate that, in continental areas, climates were as warm or nearly as warm as present. This represents a major area of difference between the oceanographic conditions in the North Atlantic and those inferred on land (M and N, Figure 2). In North America, conditions varied from similar to the present on the prairies, to 10° to 15°C cooler than present around the southern margin of the Laurentide Ice Sheet (Heusser 1972, Berti 1975, Dreimanis & Raukas 1975, Table 1; Fulton 1976).

In summary, the middle Wisconsinan stage comprises a major glacial advance dated between 70,000 and 80,000 BP, followed by a long and complex interstadial. A significant question is why the Laurentide Ice Sheet did not retreat across the St. Lawrence Valley and why it did not collapse over Hudson Bay (Figures 3, 4, and 5).

Late Wisconsinan stage (39,000 to 10,000 BP)

Reconstruction of the late Wisconsinan ice sheets is easier than for the periods just discussed. Radiocarbon dating is reliable, and postglacial isostatic deformation can be used to infer the geometry of former ice sheets (Schytt et al 1967, Peltier & Andrews 1976). However, the criteria adopted for reconstructing ice sheets from glacial isostatic deformation have not been applied consistently.

There is evidence of a climatic change about 40,000 BP during GBP 5. Global sea level fell to −80 m (Bloom et al 1974) and the southern margin of the Laurentide Ice Sheet readvanced during the Cherry-tree Stadial (Dreimanis 1976). There was a southward pulse of the polar front in the Greenland and Norwegian Seas

(Kellogg 1975, 1976), but conversely there is no recognized evidence for glaciation within the Cordilleran area (Fulton 1976; see, however, Easterbrook, 1976) nor the Alaska/Yukon mountains (Denton 1974).

The 18,000 BP world Global sea level and marine isotope curves (Figure 2) place the maximum Wisconsinan ice volume at 18,000 BP. By providing a reconstruction of global conditions at this time the CLIMAP group have made a fundamental contribution to Quaternary science (CLIMAP Project Members 1976, Cline & Hays 1976, Hughes, Denton & Grosswald 1977). The most pronounced oceanic changes occur in middle latitudes of the North Atlantic where August surface temperatures are ≥ 10 K lower than modern values. Even so, the global ocean surface was, on average, only 2.3 K cooler. The global albedo for August 18,000 BP is estimated to have been 22% compared with the modern value of 14% (Gates 1976a).

In considering the 18,000 BP world and subsequent events, there are three important points: 1. How synchronous was the 18,000 BP glacial maximum? 2. How accurate is the model of the final disintegration of the Laurentide Ice Sheet as sketched by Hughes, Denton & Grosswald (1977)? 3. What mechanism(s) caused the retreat of the Northern Hemisphere ice system?

SYNCHRONY OF THE 18,000 BP GLACIAL MAXIMUM AND ICE EXTENT AROUND NORTH ATLANTIC   The history of glacier advance and retreat over the last 18,000 years is diachronous. Anomalies exist throughout the glacial records of the last 18,000 years and they need to be recognized as important sources of information on local and regional paleoclimates.

Dates for maximum glacial advance close to 18,000 BP come primarily from the southern margins of the Laurentide, British, and Fennoscandian ice sheets (Worsley 1976). Especially in northern polar areas (Andrews 1974), evidence for an 18,000 BP glacial maximum is lacking—instead it occurred some 4000 to 10,000 years later (Figure 2).

The growth of the Cordilleran Ice Sheet (cf Armstrong et al 1965, Easterbrook 1969) is dated about 20,000 BP. The advance culminated close to 14,000 BP and was followed by rapid deglaciation. With the exception of northern Alaska (Hamilton & Porter 1975) and the northwestern Laurentide margin (Prest 1970), the dated glacial chronologies from northern polar areas (Figures 1 and 2) show a late Wisconsinan/early Holocene glacial maximum between 11,000 and 8000 years ago. In parts of Greenland, Spitsbergen, and the eastern Canadian Arctic, "young" glacial and marine deposits overlap "old" marine sequences dated > 30,000 BP (Funder & Hjort 1973, Miller & Dyke 1974, Boulton & Rhodes 1975, England 1976, Weidick 1976). In northern Russia, Grosswald (1974) presents evidence for a readvance of ice southward from the Barents Sea about 9800 BP.

Hughes, Denton & Grosswald (1977) portray the ice system at 18,000 BP on the basis of 1. an hypothesis that ice shelves are necessary for ice sheets; 2. on the distribution of erosional forms judged to be of glacial origin; and 3. on geological evidence for the position of the margin. They outline both a minimum and maximum cryosphere system, one of the principal differences between these being an unglaciated enclave along the coasts of the eastern Canadian Arctic from Bylot Island to the Torngat Mountains. Elsewhere, the continental ice sheets are shown as flowing across the continental shelves into ice shelves that cover considerable areas of the North Atlantic. Their proposal is a stimulating attempt to apply an "Antarctic model" to the Northern Hemisphere.

In our view, "Dry Valleys" were probably present within the 18,000 BP ice system in: northern Canada, the Canadian Maritimes (Grant, 1976), North Greenland, East Greenland, and Svalbard. Other data suggest that the ice did not extend across the North Sea and that the outer shelf of Norway might have been partly ice free. Not all evidence is compelling, but our reconstruction is a viable alternative to the proposal of Hughes, Denton & Grosswald (1977) (cf Paterson 1977).

Glacial isostatic studies have led to some intriguing data on the postulated Barents Ice Sheet and the Laurentide Ice Sheet. Peltier & Andrews (1976) introduced a restricted ice load consisting of an ice cap over the Spitsbergen Islands but with most of the shelf area left ice free. Isostatic rebound was predicted to be slightly *higher* than that observed, and hence incompatible with the presence of a major ice sheet across the shelf (Schytt et al 1967). Clarke (1977) attempted the "inverse" problem on the ice load over Hudson Bay (cf Peltier 1976) and derived an estimated maximum late Wisconsinan ice thickness of 2200 m.

In reconstructing ice extent at 18,000 BP it is still not known a priori whether we are dealing with ice sheets in a steady-state equilibrium, nor are we sure whether glacial isostatic equilibrium was achieved. All reconstructions of ice volume (Paterson 1972) make steady-state assumptions. Estimates of the drop of global sea level will be too *low* if these steady states were not achieved.

LATE WISCONSINAN/HOLOCENE DEGLACIATION

Since global sea level and marine isotope values are assumed to respond to a global change in ice volume, we may ask what caused global deglaciation to start shortly after the 18,000 BP maximum and to result in deglaciation of the northern continents within 11,000 years. Pollen data indicate that the temperature rise was limited until 12,500 BP (e.g. Webb & Bryson 1972, Whitehead 1973), although Coope (1975) demonstrates the uncertainties in climatic proxy data derived from pollen evidence. Nevertheless, cooling in northern Europe led to a significant glacial advance and glacier regrowth after 11,000 BP.

Andrews (1973) and Hare (1976) have been puzzled by the imbalance between the high rates of late glacial retreat and the inadequate incoming atmospheric energy for melting the Laurentide Ice Sheet. Two alternative solutions can be offered: 1. The calculations were based on steady-state ice profiles; faster rates of retreat would occur if the ice sheet were flatter (cf Mathews 1974). 2. The geological evidence for extensive glacial lacustrine and marine sediments within the margins of the Laurentide Ice Sheet suggests that wastage of the ice margin by calving into lakes and the sea was important (Andrews 1973). This suggestion ties in with the concept of the disintegration of marine-based ice sheets (Weertman 1974, Hughes,

Denton & Grosswald 1977). The theory of ice shelf and ice sheet dynamics has been discussed by Crary (1960), Mercer (1969), Weertman (1974), and Thomas & Bentley (1977). The critical control is the balance between the buoyancy forces operating at the margin of the ice shelf and the mass flux through the draining ice stream. An imbalance is inherently likely because the water depth at the margin will change due to: (*a*) slow glacial-isostatic deflection that lags the speed of build-up of the ice sheet; and (*b*) global changes of sea level caused by elastic and visco-elastic responses to shifts in the ocean water loads (Farrell & Clark 1976).

It is tempting to develop a hypothesis for deglaciation that is partly independent of climate (cf Emiliani & Geiss, 1957) in order to assist in explaining why deglaciation was well underway prior to paleobiological evidence of any substantial warming. A final answer probably lies in the interaction between ice disintegration and consequent changes in climate, superimposed on other climatic forcing functions. There is no reason to suspect that the cause(s) of deglaciation were simple. We propose that successive ice sheets collapsed due to a combination of isostatic and eustatic factors acting to deepen the water in the vicinity of the ice shelf and/or calving bay. The sequence of disintegration of the marine-based ice sheets: Barents Sea (?)→Fennoscandia→Laurentide→West Antarctic, suggests that successive collapses may induce surges in the rise of the global sea level that threaten the stability of other marine-based ice sheets. However, the size and position of the Laurentide Ice Sheet immediately prior to its collapse about 8000 years ago (Falconer et al 1965) was similar to that during the St. Pierre Interstade (Figure 5). It will be important to determine how the "core" of the Laurentide Ice Sheet, drained by the Hudson Strait ice stream, remained stable for about 80,000 years (Figures 4 and 5). This fact argues against a simple concept of unstable marine-based ice sheets.

Hughes, Denton & Grosswald (1977) outlined a mechanism for the deglaciation of Hudson Bay, based on the hypothesis that deglaciation occurred by the migration of a gigantic calving bay that was finally oriented east-west across Hudson Bay. However, their model of the final stage is contrary to the field evidence (Prest 1970, Andrews & Falconer 1969, Skinner 1974).

CLIMATIC MECHANISMS AFFECTING THE LAURENTIDE ICE SHEET

Inception and Growth Phase

Ives (1957) advanced a model of "instantaneous glacierization" by snowline lowering; this concept has been elaborated for the eastern Canadian Arctic and Fennoscandia (Ives, Andrews & Barry 1975). The identification of former Little Ice Age snow fields on Baffin Island provides a probable analog for glacial inception conditions, although the bias of a modern analog must be recognised (Barry, Andrews & Mahaffy 1975). Thin snow and ice cover increased over an area of 97,000 km^2 in north central Baffin Island from about 37% at present to 50% during some interval between 500–300 and 70–80 years ago [Locke & Locke (1977); also Koerner (1977), from Devon Island]. The estimated 250-m lowering of snowline associated with this event could be accounted for by a decrease of summer temperature of 1.3 K with the same precipitation input as at present (L. D. Williams 1977).

These concepts have been elaborated by modeling and analog studies. L. D. Williams (1975) shows that a summer cooling of 3 K, with solar radiation conditions representative of the 116,000 BP minimum, would permit snow cover, similar to that observed in August 1972, to persist over much of Baffin Island and the Canadian Arctic Archipelago. At the radiation minimum at 72,000 BP, an 8.5 K summer cooling would be required to extend this snow cover to northern Keewatin and Labrador-Ungava, which accords with Loewe's (1971) conclusion that, without increased snowfall, summer cooling would be insufficient to initiate glaciation in these plateau areas.

Circulation patterns causing significant snowfall over northeastern Canada have been examined by Barry (1966, see also Brinkmann & Barry 1972). Southeasterly flow components are essential for snowy winters, which are usually a degree or so milder than average, in Labrador-Ungava and Keewatin. The large-scale circulation is similar to the pattern hypothesized as favoring glacial inception by Flohn (1974). While such patterns explain individual snowy winters, it remains to be demonstrated that they can account for long-term snow accumulation by their persistence over several centuries.

Extensive snow cover, on a subcontinental scale, significantly modifies the atmospheric circulation through feedback effects. An upper trough develops over the snow-covered area with increased cyclonic activity along its southeastern margin (Lamb 1955, J. Williams 1975). Wiese (1924) showed analogous southward displacement of Atlantic cyclone tracks during years with more extensive sea ice. In view of such displacements, the means by which cyclonic systems continued to transport moisture into the eastern Canadian Arctic and Subarctic is problematical. This is especially so for Keewatin, which would tend to be dominated by anticyclonic situations given an amplified quasi-stationary trough to the east. An enhanced West Greenland Current would strengthen the tendency for low pressure systems to move into Baffin Bay (Johnson & McClure 1976). Such systems give heavy autumn snowfall at Cape Dyer today, but in north-central Baffin Island snowfall derives from lows over Foxe Basin (Andrews, Barry & Drapier 1970). Moreover, the effect of the North Water (an area of more or less permanently open water in Smith Sound and northern Baffin Bay) on autumn precipitation in south-eastern Ellesmere Island (Müller, Ohmura & Braithwaite 1976) seems limited to a local 33% increase compared with the regional "background" amounts. Therefore, while this general argument for warm current effects seems eminently reasonable, several questions remain to be answered.

Calculations with an ice-flow model, using inputs based on the foregoing consider-ations (Andrews & Mahaffy 1976), show that even given severe climatic conditions (including an accumulation rate up to three times modern values for most of Baffin Island), the computed contribution to the sea-level lowering rate only reached 20 m after 10,000 years. These results again illustrate the gap to be closed between the qualitative and quantitative reconstructions of conditions necessary for glacial inception over eastern North America. Since the calculated growth of the Laurentide and Fennoscandian ice sheets falls far short of the 65-m sea level lowering indicated by evidence from the New Guinea beaches (Bloom et al 1974), either ice growth rates over North America were more dramatic than yet envisaged or there was a substantial ice buildup elsewhere.

Westward Expansion

The expansion of the Laurentide Ice Sheet westward and southward, coincident with the growth of a Cordilleran Ice Sheet about 75,000 BP (Figure 4), has not been examined in detail from a climatic standpoint. Flint (1971, p. 467) envisages the coalescence of glaciers from the Coast Ranges and Rocky Mountains into piedmont lobes to form the Cordilleran ice complex and interprets the climatic pattern as essentially similar to today's but displaced southward. In the North Atlantic, Ruddiman (1977a) notes a major shift in the pattern of ice-rafted sand and ocean circulation, paralleling the isotopic evidence of ice build-up about 75,000 BP. Until similar data are available for the Pacific, no useful climatic reconstruction can be attempted. Based on modern analogs, mild snowy winters in western North America should tend to be associated with cold water in the central North Pacific (Dickson & Namias 1976). Cryospheric mechanisms must also be considered. The Laurentide "core" may have collapsed with surges to the west and southwest, assisted by heavier marginal accumulation due to forced ascent of warm air masses (Tanner 1965). This could explain the low gradient of the ice sheet (Mathews 1974) and Clark's (1977) estimate of a maximum of 2200 m of ice over Hudson Bay.

Climate 18,000 BP

The results of three simulations of global climate for July/August 18,000 BP using atmospheric general circulation models (GCMs) have been described (Gates 1976a,b, Manabe & Hahn 1977, Williams, Barry & Washington 1974, J. Williams 1974, Williams & Barry 1975). However, it must be emphasized that these experiments indicate only the broad global picture of a climatic state produced by the imposed boundary conditions. It is not yet possible to determine whether such a state is in thermal equilibrium with the boundary conditions or whether the ice sheet mass balance is in equilibrium with the calculated climate. This problem has an important bearing on the possible mechanisms for deglaciation.

CONCLUSION

This review has concentrated on the evidence for the temporal and spatial patterns of the ice sheets that developed around the North Atlantic during the last Glaciation. The major points are: 1. The Laurentide Ice Sheet spread out during the Nicolet Stadial from the plateaus of Labrador and Baffin Island; the ice extent was similar to that during the maximum recession of the Port Talbot Interstadial (Figures 4 and 5). The location and area of the Laurentide Ice Sheet during these two periods broadly resembled that about 9000 years ago. Environmental conditions in the northwestern Atlantic are critical to the build-up and maintenance of the stable core of the Laurentide Ice Sheet, although questions remain about the oceanic conditions during inception and growth phases. 2. The westward expansion of the Laurentide Ice Sheet into the Prairie Provinces appears to be associated in time with the growth of the Cordilleran Ice Sheet (Figure 5). The Keewatin margin is "unstable" on a time scale of 10^3 years. In Fennoscandia the major elements of a "stable" core and unstable periphery are the mirror image of North America with the most dramatic expansion occurring east of the Baltic Sea. 3. The most rapid fall in sea level (60 m in 7000 years) and the most extreme change in climate occurred between 80,000 and 70,000 BP (Figure 2, although a similar shift occurred during the final pronounced drop of sea level during marine isotope stage 2 (Figure 2A and B). These changes are difficult to account for with present climatic models. 4. Evidence is accumulating for extensive ice-free nunataks around the polar shores of the North Atlantic at 18,000 BP. 5. Evidence for late glacial advances of the northern ice sheet margins between 11,000 and 8000 BP may represent short-lived reactions to the break-up of sea ice cover and the northward ingress of storms into Baffin Bay and the Greenland and Norwegian seas. 6. The final collapse of the Northern Hemisphere ice sheets may not necessarily have been directly controlled by climatic factors.

ACKNOWLEDGMENTS

Our Quaternary research has been variously supported by the National Science Foundation (Geology, Geophysics and Meteorology Programs and the Division of Polar Programs). We are very grateful for this support. We also express our appreciation to our colleagues Drs. J. D. Ives and G. H. Miller for many discussions on Quaternary topics. L. D. Williams offered valuable comments on a first draft.

J. T. Andrews would like to thank the Institute for Quaternary Studies, University of Maine, Orono, for their hospitality and for numerous stimulating conversations with Drs. G. H. Denton, T. Hughes, R. Thomas, H. Borns Jr., and W. Karlén. He would also like to thank Dr. G. S. Boulton, University of East Anglia, Norwich, England, for discussions during the course of a University of Colorado Faculty Fellowship, 1976–77.

Finally, thanks are due to our typist Ardie Larson and to Marilyn Joel, John Adam, and Vicky Dow for drafting the illustrations.

Literature Cited

Adam, D. P. 1973. Ice Ages and the thermal equilibrium of the earth. *J. Res. U.S. Geol. Surv.* 1 : 587–96

Adam, D. P. 1975. Ice Ages and thermal equilibrium of the earth, II. *Quat. Res.* 5 : 161–71

Andrews, J. T. 1973. The Wisconsin Ice Sheet: dispersal centers, problems of rates of retreat, and climatic implications. *Arct. Alp. Res.* 5 : 185–99

Andrews, J. T. 1974. Cainozoic glaciations and crustal movements of the Arctic. *Arctic and Alpine Environments*, ed. J. D.

Ives, R. G. Barry, pp. 277–317. London: Methuen. 999 pp.

Andrews, J. T., Barry, R. G., Drapier, L. 1970. An inventory of the present and past glacierization of Home Bay and Okoa Bay, east Baffin Island, NWT, Canada. *J. Glaciol.* 9(57): 337–62

Andrews, J. T., Falconer, G. 1969. Late glacial and post glacial history and emergence of the Ottawa Islands, Hudson Bay, Northwest Territory: evidence on the deglaciation of Hudson Bay. *Can. J. Earth Sci.* 6 : 1263–76

Andrews, J. T., Feyling-Hanssen, R. W., Miller, G. H., Schlüchter, D., Stuiver, M., Szabo, J. B. 1976. Alternative models of early and middle Wisconsin events, Broughton Island, N.W.T. Canada: toward a Quaternary chronology. Int. Union. Geol. Sci/UNESCO, Int. Geol. Correl. Prog., Proj. 73/1/24, Bellingham, Prague, pp. 12–61

Andrews, J. T., Funder, S., Hjort, C., Imbrie, J. 1974. Comparison of the glacial chronology of eastern Baffin Island, East Greenland, and the Camp Century accumulation record. Geology 2:355–58

Andrews, J. T., Mahaffy, M. A. 1976. Growth rates of the Laurentide Ice Sheet and sea level lowering (with emphasis on the 115,000 B.P. sea level low). Quat. Res. 6:167–83

Armstrong, J. E., Crandell, D. R., Easterbrook, D. J., Noble, J. B. 1965. Late Pleistocene stratigraphy and chronology in southwestern British Columbia and northwestern Washington. Bull. Geol. Soc. Am. 76:321–30

Bahnson, H., Petersen, K. S., Konradi, P. B., Knudsen, K. L. 1973. Stratigraphy of Quaternary deposits in the Skaerumhede II boring: lithology, molluscs and foraminifera. Dan. Geol. Unders., Arbog 1973, pp. 27–62

Barry, R. G. 1966. Meteorological aspects of the glacial history of Labrador-Ungava with special reference to atmospheric vapour transport. Geogr. Bull. 8:319–40

Barry, R. G., Andrews, J. T., Mahaffy, M. A. 1975. Continental ice sheets: conditions for growth. Science 190:979–81

Berger, A. 1977. Power and limitation of an energy-balance climate model as applied to the astronomical theory of palaeoclimates. Palaeogeogr. Palaeoclimatol. Palaeoecol. 21:227–36

Berti, A. A. 1975. Paleobotany of Wisconsinan Interstadials, eastern Great Lakes region, North America. Quat. Res. 5:591–619

Bjerknes, J. 1964. Atlantic air-sea interaction. Adv. Geophys. 10:1–82

Blake, W., Jr. 1974. Studies of glacial history in Arctic Canada. II. Interglacial peat deposits on Bathurst Island. Can. J. Earth Sci. 11:1025–42

Bloom, A. L., Broecker, W. S., Chappell, J. M. A., Matthews, R. K., Mesolella, K. J. 1974. Quaternary sea level fluctuations on a tectonic coast: new ^{230}Th/^{234}U dates from the Huon Peninsula, New Guinea. Quat. Res. 4:185–205

Boulton, G. S., Rhodes, M. 1975. Isostatic uplift and glacial history in northern Spitsbergen. Geol. Mag. 111:481–500

Bray, J. R. 1970. Temporal patterning of post-Pleistocene glaciation. Nature 228:353

Brinkmann, W. A. R., Barry, R. G. 1972. Palaeoclimatological aspects of the synoptic climatology of Keewatin, Northwest Territories, Canada. Palaeogeogr. Palaeoclimatol. Palaeoecol. 11:77–91

Bryson, R. A. 1974. A perspective on climatic change. Science 184:753–60

Clark, J. A. 1977. Global sea level changes since the Last Glacial maximum and sea level constraints on the ice sheet disintegration history. PhD thesis, Univ. Colorado, Boulder

CLIMAP Project Members. 1976. The surface of the Ice-Age earth. Science 191:1131–37

Cline, R. M., Hays, J. D., eds. 1976. Investigation of late Quaternary paleoceanography and paleoclimatology. Geol. Soc. Am., Mem 145 464 pp.

Coope, R. J. 1975. Climatic fluctuations in northwest Europe since the Last Interglacial, indicated by fossil assemblages of Coleoptera. Ice Ages: Ancient and Modern, ed. A. E. Wright, F. Moseley, pp. 153–68. Liverpool: Seel House Press, 320 pp.

Crary, A. P. 1960. Arctic ice island and ice shelf studies, Part 2. Arctic 13:32–50

Dansgaard, W., Johnson, S. J., Clausen, H. B., Langway, C. C. 1971. Climatic record revealed by the Camp Century ice core. Late Cenozoic Ice Ages, ed. K. K. Turekian, pp. 37–56. New Haven: Yale Univ. Press. 606 pp.

Denton, G. H. 1974. Quaternary glaciations of the White River Valley, Alaska, with a regional synthesis for the northern St. Elias Mountains, Alaska and Yukon Territory. Bull. Geol. Soc. Am. 85:871–92

Denton, G. H., Karlén, W. 1973. Holocene climatic variations: their pattern and possible cause. Quat. Res. 3:155–205

Dickson, R. R., Namias, J. 1976. North American influences on the circulation and climate of the North Atlantic sector. Mon. Weather Rev. 104:1255–65

Dreimanis, A. 1976. Progress report on Late Pleistocene stratigraphy of southeastern Canada. See Andrews et al 1976, pp. 240–49

Dreimanis, A., Goldthwait, R. P. 1973. Wisconsin glaciation in the Huron, Erie, and Ontario lobes. The Wisconsinan Stage. Geol. Soc. Am. Mem. 136:71–106

Dreimanis, A., Raukas, A. 1975. Did Middle Wisconsin, Middle Weichselion and their equivalents represent an Interglacial or an Interstadial complex in the Northern Hemisphere. R. Soc. N.Z. Bull. 13:109–20

Easterbrook, D. J. 1969. Pleistocene chronology of the Puget lowland and San Juan Islands, Washington. *Bull. Geol. Soc. Am.* 80:2273–86

Easterbrook, D. J. 1976. Middle and early Wisconsin chronology in the Pacific Northwest. Andrews et al 1976, pp. 90–98

Emiliani, C., Geiss, J. 1957. On glaciations and their causes. *Geol. Rundsch.* 46:576–601

England, J. H. 1976. Late Quaternary glaciation of the eastern Queen Elizabeth Islands, N.W.T., Canada: alternative models. *Quat. Res.* 6:186–202

Falconer, G., Ives, J. D., Løken, O. H., Andrews, J. T. 1965. Major end moraines in eastern and central arctic Canada. *Geogr. Bull.* 7:137–53

Farrell, W. E., Clark, J. A. 1976. On postglacial sea level. *Geophys. J.* 46:647–68

Feyling-Hanssen, R. W. 1976. A mid-Wisconsinian interstadial on Broughton Island, Arctic Canada, and its foraminifera. *Arct. Alp. Res.* 8:161–82

Fillon, R. H. 1976. The Sangamonian/Wisconsinian transition in the Labrador Sea. *Geol. Soc. Am. Abstr. with Programs* 8:864

Flint, R. F. 1971. *Glacial and Quaternary Geology.* New York: John Wiley. 892 pp.

Flohn, H. 1974. Background of a geophysical model of the initiation of the next glaciation. *Quat. Res.* 4:385–404

Frenzel, B. 1973. *Climatic Fluctuations of the Ice Age.* Cleveland: Case Western Reserve Univ. Press. 306 pp.

Fulton, R. J. 1976. Quaternary history south-central British Columbia with adjacent areas. See Andrews et al 1976, pp. 62–89

Funder, S., Hjort, C. 1973. Aspects of Weichselian chronology in central East Greenland. *Boreas* 2:69–84

Gates, W. L. 1976a. Modelling the ice-age climate. *Science* 191:1131–44

Gates, W. L. 1976b. The numerical simulation of ice-age climate with a global general circulation model. *J. Atmos. Sci.* 33:1844–73

Grant, D. R. 1976. Glacial style and the Quaternary stratigraphic record in the Atlantic Provinces, Canada. *Quaternary Stratigraphy of North America,* ed. W. C. Mahaney, pp. 33–36. Stroudsburg, Pa.: Dowden, Huchinson & Ross. 512 pp.

Grosswald, M. 1974. On the probable role of glacier surges in the growth and decay of former ice sheets. *Akad. Nauk SSR Inst. Geogr. Mater. Glatsiologicheskikh Issled. Khron. Obsuzhdeniya.* 24:164–69

Hamilton, T. D., Porter, S. C. 1975. Itkillik Glaciation in the Brooks Range, Northern Alaska. *Quat. Res.* 5:471–97

Hare, F. K. 1976. Late Pleistocene and Holocene climates: some persistent problems. *Quat. Res.* 6:505–17

Hays, J. D., Imbrie, J., Shackleton, N. J. 1976. Variations in the earth's orbit: pacemaker of the Ice Ages. *Science* 194:1121–32

Heusser, C. J. 1972. Palynology and phytogeographical significance of a Late Pleistocene refugium near Kapaloch, Washington. *Quat. Res.* 2:189–201

Hollin, J. T. 1977. Thames interglacial sites, Ipswichian sea levels and Antarctic ice surges. *Boreas* 6:33–52

Hughes, T. 1975. The West Antarctic ice sheet: instability, disintegration, and initiation of ice ages. *Rev. Geophys. Space Phys.* 13:502–26

Hughes, T., Denton, G. H., Grosswald, M. G. 1977. Was there a late-Würm Arctic ice sheet? *Nature* 266:596–602

Ives, J. D. 1957. Glaciation of the Torngat Mountains, northern Labrador. *Arctic* 10:67–88

Ives, J. D., Andrews, J. T., Barry, R. G. 1975. Growth and decay of the Laurentide ice sheet and comparisons with Fenno-Scandinavia. *Naturwiss.* 62:118–25

Johnson, R. G., McClure, B. T. 1976. A model for northern hemisphere continental ice sheet variation. *Quat. Res.* 6:325–53

Kellogg, T. B. 1975. Late Quaternary climatic changes in the Norwegian and Greenland Seas. *Climate of the Arctic,* eds. G. S. Weller, S. A. Bowling, pp. 3–36. Fairbanks: Univ. of Alaska Press. 436 pp.

Kellogg, T. B. 1976. Late Quaternary climatic changes: evidence from deep-sea cores of Norwegian and Greenland Seas. See Cline and Hays 1976, pp. 77–110

Kind, N. V. 1976. Late Quaternary geochronology according to isotopes data. *Tr. Geol. Inst., Akad. Nauk SSR* 257. 255 pp.

Klassen, R. W. 1969. Quaternary stratigraphy and radiocarbon chronology in southwestern Manitoba. *Geol. Surv. Canada, Pap.,* 69–27 19 pp.

Koerner, R. M. 1977. Devon Island ice cap: core stratigraphy and paleoclimate. *Science* 196:15–18

Kukla, G. J. 1975. Missing link between Milankovitch and climate. *Nature* 253:600–3

Lamb, H. H. 1955. Two-way relationships between the snow or ice limit and 1000–500 mb thicknesses in the overlying atmosphere. *Quart. J. R. Meteorol. Soc.* 81:172–89

Locke, C. W., Locke, W. W., III. 1977.

Little Ice Age snow-cover extent and paleoglaciation thresholds: north-central Baffin Island, N.W.T. Canada. *Arct. Alp. Res.* 9:291–300

Loewe, F. 1971. Considerations regarding the origin of the Quaternary ice sheet of North America. *Arct. Alp. Res.* 3:331–44

Manabe, S., Hahn, D. G. 1977. Simulation of the tropical climate of an ice age. *J. Geophys. Res.* 82:3889–911

Mathews, W. H. 1974. Surface profiles of the Laurentide Ice Sheet in its marginal areas. *J. Glaciol.* 13:37–43

McDonald, B. C. 1971. Late Quaternary stratigraphy and deglaciation in eastern Canada. See Dansgaard et al 1971, pp. 331–56

Mercer, J. H. 1969. The Allerød oscillation: A European climatic anomaly *Arct. Alp Res.* 1:227–34

Milankovitch, M. 1941. *Canon of insolation and the ice age problem.* Transl. 1969, Isr. Prog. Sci. Transl., Jerusalem (from German). Springfield, Va.: U.S. Dept. Commer. 484 pp.

Miller, G. H., Dyke, A. S. 1974. Proposed extent of late Wisconsin Laurentide ice on eastern Baffin Island. *Geology* 2:125–30

Miller, G. H., Andrews, J. T., Short, S. 1977. The last interglacial/glacial cycle, Clyde Foreland, Baffin Island, N.W.T.: Stratigraphy, Biostratigraphy, and Chronology. *Can. J. Earth Sci.*, 14: In press

Mörner, N. A. 1974. Ocean paleotemperatures and continental glaciations. *Colloq. Int. CNRS* 219:43–49

Mörner, N. A. 1976. Global correlations and Weichselian chrono-stratigraphy. See Andrews et al 1976, pp. 327–38

Müller, F., Ohmura, A., Braithwaite, R. 1976. On the climatic influence of North Water. *Int. Geogr. Congr. Symp. on the Geogr. of Polar Countries*, 12th (extended summaries), pp. 55–58. Leningrad: Hydrometr. Publ. House

Newell, R. E. 1974. Changes in the poleward energy flux by the atmosphere and ocean as a possible cause for Ice Ages. *Quat. Res.* 4:117–27

Østrem, G. 1964. Ice-cored moraines in Scandinavia. *Geogr. Ann.* 46:282–337

Paterson, W. S. B. 1972. Laurentide Ice Sheet: estimated volumes during the late Wisconsin. *Rev. Geophys. Space Phys.* 10:885–917

Paterson, W. S. B. 1977. Extent of the Late-Wisconsin glaciation in Northwest Greenland and northern Ellesmere Island: A review of the glaciological and geological evidence. *Quat. Res.* 8:180–90

Paterson, W. S. B., Koerner, R. M., Fisher,

D., Johnson, S. J., Clausen, H. B., Dansgaard, W., Bucher, P., Oeschger, H. 1977. An oxygen-isotope climatic record from the Devon Island ice cap, Arctic Canada. *Nature* 266:508–11

Peltier, W. R. 1976. Glacial-isostatic adjustment—II. The inverse problem. *Geophys. J.* 46:669–706

Peltier, W. R., Andrews, J. T. 1976. Glacial-isostatic adjustment—I. The forward problem. *Geophys. J.* 46:605–46

Pisias, N. G., Heath, G. R., Moore, T. C., Jr. 1975. Lag times for oceanic responses to climatic change. *Nature* 256:716–17

Prest, V. K. 1970. Quaternary geology in Canada. *Geology and Economic Minerals in Canada*, ed. R. J. Douglas, pp. 676–764. Ottawa: Dept. Energy, Mines and Resources. 5th ed.

Robin, G. de Q. 1977a. Ice cores and climatic change, *Phil. Trans. R. Soc. London Ser. B.* 128:41–66

Ruddiman, W. F. 1977b. North Atlantic ice rafting: a major change at 75,000 years BP. *Science* 196:1208–11

Ruddiman, W. F. 1977. Late Quaternary deposition of ice-rafted sand in the Sub-polar North Atlantic (40°–65°N). *Bull. Geol. Soc. Am.* 88:1813–27

Ruddiman, W. F., McIntyre, A. 1976. Northeast Atlantic paleoclimatic changes over the past 600,000 years. See Cline & Hays 1976, pp. 111–46

Saltzman, B., Vernekar, A. D. 1971. Note on the effect of earth orbital variations on climate. *J. Geophys. Res.* 76:4195–97

Sancetta, C., Imbrie, J., Kipp, N. G. 1973. Climatic record of the past 130,000 years in the North Atlantic deep-sea core V23–82; correlation with the terrestrial record. *Quat. Res.* 3:110–16

Schytt, V. A., Hoppe, G., Blake, W., Jr., Grosswald, M. G. 1967. The extent of Würm Glaciation in the European Arctic. *Int. Assoc. Hydrol. Sci. Publ.* 79:207–16

Sellers, W. D. 1974. Climate models and variations in the solar constant. *Geofis. Int.* 14:303–15

Shackleton, N. J. 1969. The last interglacial in the marine terrestrial records. *Proc. R. Soc. London Ser. B.* 174:135–54

Shackleton, N. J. 1976. Oxygen-isotope evidence relating to the end of the Last Interglacial at substage 5e to 5d transition about 115,000 years ago. *Geol. Soc. Am. Abstr. with Programs* 8:1099–1100

Shackleton, N. J., Opdyke, N. D. 1973. Oxygen isotope and paleomagnetic stratigraphy of equatorial Pacific core V28–238: oxygen isotope temperatures and ice volumes on a 10^5 and 10^6 year scale. *Quat. Res.* 3:39–55

Shackleton, N. J., Opdyke, N. D. 1976. Oxygen isotope and paleomagnetic stratigraphy of Pacific core V23–239: late Pliocene to latest Pleistocene. See Cline & Hays 1976, pp. 449–64

Skinner, R. G. 1974. Quaternary stratigraphy of the Moose River Basin, Ontario. Geol. Surv. Can. Bull. 225, 77 pp.

Steinen, R. P., Harrison, R. S., Matthews, R. K. 1973. Eustatic lowstand of sea level between 125,000 and 105,000 years B.P.: evidence from the subsurface of Barbados, West Indies. Bull. Geol. Soc. Am. 84:63–70

Suggate, R. P. 1974. When did the Last Interglacial end? Quat. Res. 4:246–52

Tanner, W. F. 1955. North-south asymmetry of the Pleistocene ice sheet. Science 122: 642–43

Tanner, W. F. 1965. Cause and development of an ice age. J. Geol. 73:413–30

Terasmae, J., Webber, P. J., Andrews, J. T. 1966. A study of late-Quaternary plant-bearing beds in north-central Baffin Island, Canada. Arctic 19:296–318

Thomas, R. H., Bentley, C. R. 1977. Past decay and present growth of the West Antarctic ice sheet. X INQUA Cong., Birmingham, U.K., Abstr.:465

van der Hammen, T., Maarleveld, G. C., Vogel, J. C., Zagwijn, W. H. 1967. Stratigraphy, climatic succession and radiocarbon dating of the last glacial in the Netherlands. Geol. Mijnbouwkd. 45:79

van der Hammen, T., Wijmstra, T. A., Zagwijn, W. H. 1971. The floral record of the Late Cenozoic of Europe. See Dansgaard et al, pp. 391–424

Vernekar, A. D. 1972. Long-period global variations of incoming solar radiation. Meteorol. Monogr. 12(34), 21 pp.

Webb, T. III, Bryson, R. A. 1972. Late- and postglacial climatic change in the northern Midwest, U.S.A.: quantitative estimates derived from fossil pollen spectra by multivariate statistical analysis. Quat. Res. 2:70–115

Weertman, J. 1974. Stability of the junction between an ice sheet and an ice shelf. J. Glaciol. 13:3–11

Weertman, J. 1976. Milankovitch solar radiation variations on ice age ice sheet sizes. Nature 261:17–20

Weidick, A. 1976. Glaciations of Northern Greenland—New Evidence. Polarforschung 46:26–33

Weyl, P. K. 1968. The role of the oceans in climatic change: a theory of the Ice Ages. Causes of Climatic Change, ed. J. M. Mitchell, Jr., Materol. Monogr. 8(30): 37–62

Whitehead, D. R. 1973. Late Wisconsin vegetational changes in unglaciated eastern North America. Quat. Res. 3: 621–31

Wiese, W. 1924. Polareis und atmosphärische Schwankungen. Geogr. Ann. 6:271–99

Wigley, T. M. L. 1976. Spectral analysis and the astronomical theory of climatic change. Nature 264:629–31

Wijmstra, T. A. 1975. Palynology and paleoclimatology of the last 100,000 years. Proc. WMO/AMAP Symp. on Long-Term Fluctuations, W. M. O. No. 421, pp. 5–20

Wijmstra, T. A., van der Hammen, T. 1974. The Last Interglacial-Glacial cycle: state of affairs of correlation between the data obtained from the land and from the ocean. Geol. Mijnbouwkd. 53:386–92

Williams, J. 1974. Simulation of the atmospheric circulation model using the NCAR global circulation model with present day and glacial period boundary conditions. Univ. Colo. Boulder Inst. Arct. Alp. Res. Occas. Pap. 10, 328 pp.

Williams, J. 1975. The influence of snow-cover on the atmospheric circulation and its role in climatic change: an analysis based on results from the NCAR global circulation model. J. App. Meteorol. 14: 137–52

Williams, J., Barry, R. G. 1975. Ice Age experiments with the NCAR general circulation model: conditions in the vicinity of the northern continental ice sheets. See Kellogg 1975, pp. 143–49

Williams, J., Barry, R. G., Washington, W. M. 1974. Simulation of the atmospheric circulation using the NCAR global circulation model with ice age boundary conditions. J. Appl. Meteorol. 13(3):305–17

Williams, L. D. 1975. Effect of insolation changes on late summer snow cover in Northern Canada. See Wijmstra 1975, pp. 287–92

Williams, L. D. 1977. The late neoglacial (Little Ice Age) glaciation level on Baffin Island and its significance for ice sheet inception. Paleogeogr. Paleoclimatol. Paleoecol. In press.

Wilson, A. T. 1964. Origin of Ice Ages: an ice shelf theory for Pleistocene glaciation. Nature 201:147–49

Worsley, P. 1976. Correlation of the Last Glaciation glacial maximum and the extra-glacial fluvial terraces in Great Britain—a case study. See Dreimanis 1976, pp. 274–84

Ann. Rev. Earth Planet. Sci. 1978. 6 : 229–50

HYDROTHERMAL ALTERATION IN ACTIVE GEOTHERMAL FIELDS

�;10093

P. R. L. Browne

New Zealand Geological Survey, Box 30368, Lower Hutt, New Zealand

INTRODUCTION

In the fall of 1929 the Carnegie Institution drilled a 124-m well in the Upper Geyser Basin of Yellowstone National Park. Cores from this well and a subsequent well at Norris Basin were studied by Fenner (1934, 1936), who published the first detailed accounts of subsurface hydrothermal alteration in an active geothermal system. Of course, it had long been known from studies of hot springs and ore deposits that thermal fluids can react with the rocks they contact so that both change their compositions. Active geothermal systems, however, are places where these reactions are occurring now and where it is often possible to make direct physical and chemical measurements. Thus they may be regarded as large-scale, uncontrolled, open-end natural experiments. But because there still usually remain several unknown variables—such as the duration of thermal activity or the composition of the fluid before it enters the system—studies of alteration in geothermal samples complement, rather than replace, low-temperature mineral-stability experiments.

This chapter is a survey .of hydrothermal alteration in several geothermal systems recently explored by drilling. It is selective in that it usually mentions only papers published since about 1968, and it excludes discussion of ore minerals since these have been described elsewhere (White 1967, Weissberg et al 1978). Also excluded are submarine thermal systems (e.g. Ridge 1973), geopressured sedimentary basins (e.g. Carpenter et al 1974), and accounts of alteration at Steamboat Springs, Nevada (Schoen & White 1965, 1968, Sigvaldason & White 1961, 1962, White 1968), Pauzhetsk, Kamchatka (Naboko 1970, 1976), Tahuangtsui, Taiwan (Chen 1966, 1967), Wairakei (Steiner 1953, 1955, 1968, 1970), and Waiotapu (Steiner 1963). Many aspects of geothermal systems are discussed in three massive volumes forming the proceedings of a United Nations conference (1976), and an extensive bibliography is also now available (ERDA 1976).

229

0084-6597/78/0515-0229$01.00

HYDROTHERMAL MINERALS

A very wide range of hydrothermal minerals has been recognised in active geothermal systems (Table 1). Many occur in low-grade metamorphic rocks and hydrothermal ore deposits, but some are rare [e.g. buddingtonite (Erd et al 1964)] or perhaps unexpected at low temperatures [e.g. aegirine and lepidolite (Honda & Muffler 1970, Bargar et al 1973)]. Wairakite, a calcium zeolite, is the best known mineral first found in cores from an active geothermal field (Steiner 1955, Coombs 1955).

Several factors affect the formation of hydrothermal minerals and these vary in

Table 1   Some hydrothermal minerals in selected geothermal fields[a]

	Imperial Valley, California[a]	Yellowstone, Wyoming	The Geysers, California	Pauzhetsk, Kamchatka	Matsukawa, Japan	Otake, Japan	Tongonan, Philippines	Kawah Kamojang, Java	N.Z. Volcanic Zone	El Tatio, Chile	Low temp. Iceland	High temp. Iceland	Larderello, Italy
Allophane					x								
Quartz	x	x	x	x	x	x	x	x	x	x	r?	x	x
Cristobalite		x		x	x	x	x	x	x				
Kaolin group	d	x	x	x	x	x	x	x	x	x			
Montmorillonite	d	x		x	x	x	x	x	x	x	x	x	
Interlayered illite-mont.	x			x	x	x	x	x	x		x	x	
Illite	x	x	x	x	x	x	x	x	x	x			
Biotite	x				x				x				
Chlorite	x	x	x	x	x	x	x	x	x	x	?	x	x
Celadonite		x		x							x	x	
Alunite				x	x	x	x	x		x			
Anhydrite	x			x	x	x	x	x	x	x		x	x
Sulfur				x	x	x		x		x			
Pyrophyllite					r	x	r	r					
Talc	x						x						
Diaspore						x	x		x				
Calcite	x	x	x	x	x	x	x	x	x	x	x	x	x
Aragonite	x[b]						x[b]		x[b]				
Siderite				x	x				x		x	x	
Ankerite	x				x							x	
Dolomite	d												
Analcime		x			x					x	x	x	
Wairakite	x			x	x		x	x	x	x		x	x
Gmelinite											x		
Gismondine											x		

Table 1 (*continued*)

	Imperial Valley, California[a]	Yellowstone, Wyoming	The Geysers, California	Pauzhetsk, Kamchatka	Matsukawa, Japan	Otake, Japan	Tongonan, Philippines	Kawah Kamojang, Java	N.Z. Volcanic Zone	El Tatio, Chile	Low temp. Iceland	High temp. Iceland	Larderello, Italy
Erionite		x											
Laumontite		x	x	x	x				x	x	x	x	
Phillipsite			x										
Scolecite			x								x		
Chabazite			x								x		
Thomsonite			x								x		
Clinoptilolite		x					x						
Heulandite		x	x			x	x		x		x	x	
Stilbite											x	x	
Mordenite		x	x						x		x	x	
Prehnite	x		x						x	r?		x	
Amphibole	x		x				x					x	
Garnet	x		?									r	
Epidote	x		x			x	x	x	x		r	x	x
Clinozoisite									x				
Pectolite		x							x[b]				
Sphene	x		x					x	x	x			
Adularia	x	x	x	x		x	x	x	x	x			x
Albite	x		x			x	x	x	x			x	
Rutile					x	x	x						
Leucoxene				x	x	x			x		x		
Magnetite									x				
Hematite	x	x		x					x	x	x	x	
Pyrite	x	x	x	x	x	x		x	x	x	x	x	x
Pyrrhotite	x								x		x		
Marcasite									x		x		
Base-metal sulfides	x								x		x		
Fluorite		x						x					

References: Imperial Valley: Skinner et al 1967, Muffler & White 1969, Bird & Elders 1976, Kendall 1976b, Reed 1976, Browne & Elders 1976. Yellowstone: Fenner 1936, Honda & Muffler 1970, Bargar et al 1973, Keith & Muffler 1978, Keith et al 1978. The Geysers: Steiner 1958, McNitt 1964. Pauzhetsk: Naboko 1970. Matsukawa: Sumi 1968. Otake: Hayashi 1973. Tongonan: C. P. Wood, unpublished observations. Kawah Kamojang: Browne, unpublished observations. N.Z. Volcanic Zone: Steiner 1953, 1963, 1968, Browne & Ellis 1970, Browne, unpublished observations. El Tatio: Browne, unpublished observations. Iceland: Sigvaldason 1963, Tómasson & Kristmannsdóttir 1972, 1976, Kristmannsdóttir & Tómasson 1976a, b, c, Kristmannsdóttir 1976. Larderello: Marinelli 1969.

Note: d = detrital, r = relict.

[a] includes Cerro Prieto, Baja California, Mexico.

[b] deposited in discharge pipes and channels.

relative importance from field to field. Some are so intimately related that it is often impossible to separate one factor from another. The factors are: (a) temperature. (b) pressure, (c) rock type, (d) permeability, (e) fluid composition, and (f) duration of activity. This list is similar to several suggested from work on ore deposits, but geothermal studies have demonstrated that permeability and fluid composition are usually at least as important as temperature.

Temperature

Most liquid-dominated systems explored so far have reservoir temperatures below about 280°C, but several people have reported maxima above 300°C [including the Salton Sea, California at 360°C (Palmer 1975), Cerro Prieto, Mexico at 388°C (Mercado 1969) or 371°C (Mercado 1976), and Tongonan, Philippines at 324°C (C. P. Wood, personal communication)]. Vapor-dominated systems have maximum temperatures after exploitation of 236°C (White et al 1971).

Except for epidote, ortho, ring, and chain silicates are uncommon alteration minerals in active geothermal systems, but where present, usually occur at high temperatures. Hydrothermal garnet and tremolite are known from the Salton Sea (Muffler & White 1969, Kendall 1976b), and tremolite also occurs at Tongonan (C. P. Wood, personal communication). An unspecified amphibole is present at one of the Icelandic fields (Kristmannsdóttir 1976), and all its occurrences are where well temperatures exceed 280°C, but garnet is present only above 320°C. Epidote occurs at several fields (Table 1), usually above 240–260°C but in some (Seki 1972), such as at Reykjavik (Sigvaldason 1963), it apparently occurs down to 120°C. Possibly here, however, it is a relict formed when temperatures were higher.

Pyrophyllite and talc also form at high temperatures, but the former is considered to be relict at Kawah Kamojang, Indonesia and in several of the Japanese fields (Sumi 1968). The distribution of clay minerals is also temperature dependent (Browne & Ellis 1970, Muffler & White 1969, Steiner 1968). In the New Zealand fields surface kaolinite, formed at low pH, does not persist above about 60°C, although dickite is known in one well where measured temperatures are from 140 to 150°C. With increasing depth and temperature, the usually dominant clay, Ca-montmorillonite, becomes increasingly interstratified with illite, to become inter-layered illite-montmorillonite, but above about 220°C, illite plus chlorite is the typical clay-mineral assemblage. Of course, the apparent temperature dependence of the clays is aided by the near-uniform composition of fluids throughout these fields (Browne & Ellis 1970, Mahon & Finlayson 1972)—a condition not always met elsewhere.

As expected, the distribution of zeolites is strongly temperature dependent, and this is well illustrated by Icelandic and Japanese work (summarized later), although the greatest range of zeolites is at Pauzhetsk (Naboko 1970). Typical zonation in the New Zealand fields is mordenite-laumontite-wairakite, with mordenite forming near 50°C and wairakite usually above 215°C, but occasionally as low as 140°C (Browne & Ellis 1970, Steiner 1953, 1968). Cristobalite and siderite frequently occur below 100°C, but several minerals, including pyrite, calcite, and chlorite, form readily at both low and high temperatures.

Pressure

Fluid pressures in geothermal areas are low and seldom exceed 200 bars. In liquid-dominated fields they are usually close to, or slightly above, hot hydrostatic, but occasionally, where self-sealing has occurred—for example at Yellowstone—they must have exceeded lithostatic pressure (Muffler et al 1971). In vapor-dominated systems, fluid pressures are typically well below hydrostatic (White et al 1971). Over the drilled depths pressure has little *direct* affect on hydrothermal alteration, other than to influence the induration and lithification of sediments. However, *change* of fluid pressure can affect fluid composition. This is most obvious where boiling occurs and CO_2 is lost; zones of subsurface boiling are often characterized by hydrothermal quartz, K-feldspar and bladed calcite (Browne & Ellis 1970, Keith & Muffler 1978, Keith et al 1978). The effect of pressure on the K-mica–K-feldspar equilibrium results in a given solution being more alkaline at high pressure. Fluids released from high-pressure control, therefore, have a tendency to deposit K-feldspar (Ellis & McFadden 1972).

Rock Type

The parent rock influences hydrothermal alteration mainly through the control of permeability by texture and porosity. The initial mineralogy of the reservoir rocks seems to have little affect on equilibrium alteration assemblages above about 280°C. For example, albite, K-feldspar, chlorite, Fe-epidote, calcite, quartz, illite, and pyrite are the typical stable assemblage in basalts of Iceland, sandstones of the Imperial Valley, rhyolites of New Zealand, and andesites of Indonesia. At lower temperatures, however, the nature of the parent material clearly influences the alteration product. High-silica zeolites, such as mordenite, are common in rhyolitic fields at Yellowstone (Honda & Muffler 1970) and New Zealand (Browne & Ellis 1970, Steiner 1968), whereas lower silica zeolites, e.g. chabazite, thomsonite, scolecite, occur in the basalts of Iceland and andesites of Kamchatka (Kristmannsdóttir & Tómasson 1976c, Naboko 1970).

Permeability

Studies on alteration in geothermal fields have clearly recognized the important control of permeability on hydrothermal mineral deposition. Mineral reactions are seldom isochemical and extensive alteration and hydration needs more than pore water to proceed. In many cases, at least carbonate and sulfide species must be added to rocks from solutions.

In rocks of low permeability, equilibrium between rocks and the reservoir fluid is seldom achieved and primary minerals or glass can persist to high temperatures. For example, dense welded tuffs at Broadlands, Wairakei, and Yellowstone have locally remained little changed, even at high temperatures, because fluid access is too difficult (Browne & Ellis 1970, Keith et al 1978). The marked zoning of hydrothermal minerals about fluid channels in several Japanese geothermal fields clearly shows structural control of alteration and demonstrates how intensity and type of alteration reflect permeability (Sumi & Takashima 1976). Permeable fissure channels at Wairakei, Broadlands, Waiotapu, Kawerau (all in New Zealand),

and Tongonan, Philippines are characterized by vein adularia, usually together with quartz and calcite (Steiner 1968, Browne 1970). Browne (1970) showed that at Broadlands there is an approximate relationship between the nature of feldspars in aquifer rocks and the measured well permeabilities; in order of increasing permeability the feldspars are: primary andesine, albite, albite + adularia, adularia. Thus rocks from highly permeable zones are often increased three- to fourfold in K_2O.

Isotopic analysis of hydrothermal minerals also gives information about reservoir hydrology (Blattner 1975, Coplen et al 1975, Clayton et al 1968, Clayton & Steiner 1975, Eslinger & Savin 1973, Kendall 1976a, b, Olson 1976). Kendall's (1976b) study, for example, shows that the carbon and oxygen isotopic composition of minerals from the Salton Sea reflects variations in reservoir permeability, and she was able to relate depletion of O^{18} in calcite with zones of high permeability as indicated by Saraband logs.

Fluid Composition

This was recognized by Fenner (1934) as a factor in hydrothermal alteration at Yellowstone. However, only in the last few years has it been realized how close, in fact, is the relationship between fluid chemistry and alteration mineralogy. Successful use of the silica geothermometer to estimate deep temperatures depends on silica in the fluids being in equilibrium with a silica mineral, usually quartz

Table 2 Concentration (mg/kg) in waters separated from discharge at atmospheric pressure

Drillhole No.	Depth (m)	Source Temp (°C)	pH (15–25°)	Li	Na	K	Rb	Cs	Mg	Ca
Salton Sea 11D No. 2*	1776	332	—	210	53,000	16,500	70	20	10	28,800
Cerro Prieto* M.8	1300	305	5.43	10.9	4730	1180	—	—	0.2	272
Yellowstone Y-3	157	194	8.1	3.5	270	11	—	—	0.02	1.26
Matsukawa T1	162	100	2.8	—	—	—	—	—	2.1	12.0
Otake No. 9	550	220–240	8.1	5.15	936	131	—	—	0.2	12.3
Otake H-2	720	240–270	3.4	7.03	1186	228	—	—	0.47	15.0
Wairakei 44	695	260	8.4	14.2	1320	225	2.8	2.5	0.03	17
Broadlands 10	1092	260	8.6	9.5	910	143	1.1	1.4	0.05	1.1
Seltjarnarnes S-4	2050	114	—	—	362	10	—	—	0.09	132
Hveragerdi G-3	650	216	9.6	0.3	212	27	0.04	< 0.02	0.0	1.5
El Tatio 7	867	262	7.0	45.2	4840	830	0.6	17.4	0.16	211

[a] Helgeson 1968a
[b] Reed 1976
[c] Bargar et al 1973
[d] Sumi 1968
[e] Hayashi 1973
[f] Ellis 1967

(Mahon 1966, 1976, Fournier & Rowe 1966, Fournier 1973, Arnórsson 1975). Similarly, the Na-K and Na-K-Ca geothermometers require that the solutions be in equilibrium with albite and K-feldspar (Ellis & Mahon 1967, Fournier & Truesdell 1973, Truesdell 1976).

Several significant papers on experimental and thermodynamic work have greatly increased our understanding of hydrothermal alteration (Hemley et al 1969, Hemley & Jones 1964, Helgeson 1967, 1968b, 1970, 1971, Helgeson et al 1969a, b). For example, the experimentally derived activity diagram for equilibrium in the system K_2O–Na_2O–Al_2O_3–SiO_2–HCl–H_2O at 260°C (Figure 1), shows the relationships between several mineral phases and water of Broadlands composition (Browne & Ellis 1970); as expected from minerals in the cores, this water is in near equilibrium with K-feldspar (adularia), albite, and K-mica (illite).

By using petrographic studies of cores and analyses of well fluids and well temperatures, diagrams for other element combinations can be constructed. For example, the mineral stability diagram for calcium and potassium minerals (Figure 2) shows that the deep 260°C water at Broadlands has a composition close to equilibrium with K-mica, K-feldspar, wairakite, and calcite and emphasizes the critical role of CO_2 in determining the stability of calcium phases. Where CO_2 concentrations are high, as at Broadlands (0.15 moles), calcite tends to deposit at the expense of epidote (zoisite) or wairakite during steam separation. A close

for selected geothermal fluids

F	Cl	Br	I	SO_4	SiO_2	B	NH_3	CO_2	H_2S	Ref.
—	155,000	—	—	S = 30	400	390	—	500	—	a
—	9040	—	—	9	590	12	11.6	2580	624	b
30	278	—	—	19.1	—	3.6	—	HCO_3= 177	—	c
—	4.1	—	—	176	30	—	1.9	—	45.4	d
4.6	1474	3.4	0.3	136	665	32	—	40.7	—	e
1.5	1941	5.6	0.4	318	626	36	—	0.0	—	e
8.3	2260	6.0	0.3	36	690	29	0.15	19	1.0	f
6.2	1244	3.6	0.1	12	635	55	1.2	(CO_3 = —HCO_3—)553	—	g
0.95	554	—	—	209	101	—	—	17	—	h
1.9	197	0.45	0.0	61	480	0.6	0.1	55	7.3	f
3.0	8790	—	—	30	766	203	2.3	5.4(HCO_3=40)	—	i

g Browne & Ellis 1970
h Tómasson & Kristmannsdóttir 1976

i Cusicanqui et al 1976
* Corrected for steam loss

relationship between fluid composition and alteration (Tables 1 and 2) is seen in several Japanese fields including Otake. Here near-neutral pH waters form quartz, K-mica, heulandite, wairakite, albite, and adularia, whereas deep low pH fluids produce alunite (Hayashi 1973) in accord with the experimentally derived relations of Hemley et al (1969).

Little has so far been published on alteration in vapor-dominated fields, but the occurrence of wairakite at The Geysers (Steiner 1958) and epidote, adularia, quartz, calcite, and zeolites at Larderello (Marinelli 1969) indicates that a liquid was (or is) present in both fields. However, in rocks affected by steam, pyrrhotite may form in preference to pyrite (Browne & Ellis 1970), and clays, especially kaolinite or montmorillonite, may be more abundant under such conditions.

Duration

Much has yet to be learnt about the age and evolution of geothermal systems, but available evidence suggests they are long lived (Grindley 1965, Browne 1971, White 1974, Sumi & Takashima 1976). Steamboat Springs, for example, has been active, at least intermittently, for 1 million, and probably as long as 3 million years. During such long periods geothermal activity changes in both intensity and location, although many changes will be due to events not directly related to thermal

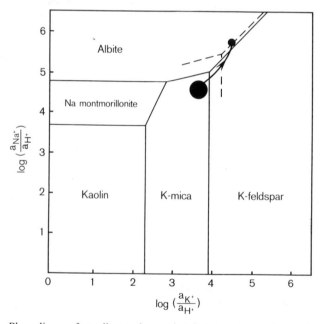

Figure 1   Phase diagram for sodium and potassium in terms of ion activity ratios at 260°C with quartz present. Solid circle plots Broadlands water composition and arrow shows trend with steam loss; dotted lines indicate positions of phase boundaries at 230°C (Browne & Ellis 1970).

activity, such as erosion, earth movement, or volcanic activity. Self-sealing, however, occurs at the tops and margins of fields where cooling waters deposit quartz, or less commonly zeolites (Bird 1975, Elders & Bird 1976, Keith et al 1978). Shifts in activity have long been recognized at Yellowstone from the migration of thermal features, and surficial activity has been extinguished at Matsukawa (Sumi & Takashima 1976) and, locally, Wairakei. Kristmannsdóttir & Tómasson (1976a) concluded that the Nesjavellir (Iceland) thermal field is shifting north and the minerals have not yet adjusted to the new higher temperatures. By contrast, cooling has been recognized at Kawah Kamojang and at Matsukawa (Sumi 1968), where pyrophyllite and diaspore are thought to be relicts from temperatures above 310°C, compared with 150–250°C at present.

Several proposed geothermometric methods have been tested on samples from geothermal areas. They include: aluminum-in-quartz (Hayashi 1973, Browne & Wodzicki 1977), sphalerite compositions (Browne & Lovering 1973), sulfur isotope fractionation between coexisting sulfide pairs (Browne et al 1975), oxygen isotope fractionation between hydrothermal carbonates and silicates (Blattner 1975, Eslinger & Savin 1973, Coplen et al 1975), and fluid inclusions (Browne et al 1976, Browne & Elders 1976, Hoagland 1976b). The last two methods appear to be the most useful in their applications to geothermal systems.

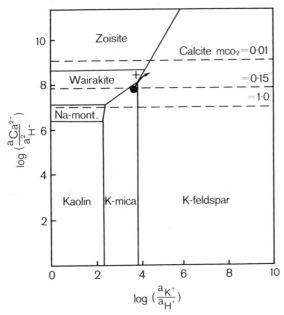

Figure 2 Phase diagram for potassium and calcium minerals in terms of ion activity at 260°C with quartz present; m = moles. Circle plots Broadlands and cross Wairakei water compositions; arrow shows trend with steam loss (Browne & Ellis 1970, Ellis & Mahon 1977).

The homogenization temperatures of 177 primary fluid inclusions in hydrothermal quartz and sphalerite crystals from seven Broadlands wells varied between 201° and 293°C; they ranged from 13°C below to 37°C above well temperatures but averaged 8°C above. One hundred and ninety-eight secondary or pseudosecondary inclusions, which had a wider temperature spread than the primary ones, gave filling temperatures that averaged 6°C below corresponding well temperatures. This close agreement between filling and measured temperatures indicates either that there has long been a stable thermal system at Broadlands or that the crystals and their inclusions are modern.

By contrast, fluid inclusions in calcite from Holtz No. 1 well at Heber, California appear to record two former thermal events; at 1525 m these were at $212 \pm 9°C$ and $235 \pm 5°C$ (Browne & Elders 1976), whereas present reservoir temperatures are probably about 190°C (Renner et al 1975). At well 6–1 East Mesa, homogenization temperatures show that the fluid inclusions formed 15–20°C above present well temperatures (Hoagland 1976b).

To date, there have been few measurements of the freezing temperatures of inclusions to attempt estimates of past fluid salinities, but such studies are likely to be extremely fruitful. In one study freezing temperatures were measured in 33 primary and 60 secondary (pseudosecondary) inclusions in eight hydrothermal quartz and one sphalerite crystal (Browne et al 1976) from Broadlands. The observed freezing point depressions (0.1 to 0.8°C) appear too large if only the dissolved salts in the deep waters (3000 ppm) are considered, but are in agreement with the present-day Broadlands water compositions when the range of downhole CO_2 compositions (0.1–0.3 moles) is included in the calculations.

HYDROTHERMAL ALTERATION IN SELECTED AREAS

Examples of hydrothermal alteration in geothermal fields recently studied show, in more detail, the variability of these usually interrelated factors.

Imperial and Mexicali Valleys, Western North America

Several geothermal fields have been recognized in the Imperial Valley of Southern California and its extension southward into the Mexicali Valley, Baja California. Drilling has so far taken place at the Salton Sea, Brawley, Heber, The Dunes, East Mesa, and Cerro Prieto. The main structural feature, the Salton Trough, is a depression filled with up to 7000 m of poorly sorted Colorado River detritus which form sandstones, shales, siltstones, and minor conglomerates (Biehler et al 1964, Muffler & Doe 1968). Five small young rhyolite domes occur at the Salton Sea, rhyodacite reaches the surface at Cerro Prieto, and minor intrusive basalts have been penetrated by wells at the Salton Sea (Robinson et al 1976) and Heber (Browne & Elders 1976).

The northernmost field, the Salton Sea, is the most famous, and here wells encountered brines with over 250,000 ppm dissolved solids (Table 2) at temperatures up to 360°C (Helgeson 1968a, Randall 1974, Palmer 1975). Salinities at the other fields are much lower and are about 50,000 ppm at Brawley, 2000 to 30,000 ppm at East Mesa, and about 17,000 ppm at Cerro Prieto. Copper- and silver-rich scales deposited from brines (Skinner et al 1967) discharged from the Salton Sea and their precipitation has been considered thermodynamically by Miller et al (1977). Alteration in cores and (mainly) cuttings from several drill holes here have been described by Muffler & White (1968, 1969), Keith et al (1968), Muffler & Doe (1968), and Kendall (1976b); mineral isotope studies have been made by Clayton et al (1968) and Kendall (1976a, b).

Muffler & White (1969) showed that the sediments, which contain detrital quartz, calcite, K-feldspar, plagioclase, montmorillonite, illite, dolomite, and kaolinite, gradually convert, in response to increasing temperatures, to an assemblage above 300°C of quartz, Fe-epidote, chlorite, K-feldspar, albite $\pm$ K-mica, pyrite, sphene, sphalerite, and hematite. Detrital montmorillonite below 100°C converts to illite-montmorillonite and then to K-mica below 210°C; ankerite can form as low as 120°C, and chlorite, calcite, and CO_2 are produced below 180°C in response to reaction between dolomite, ankerite, and kaolinite plus Fe^{2+} added from the brine. Fe-epidote and K-feldspar are abundant above 290°C, but calcite disappears at about this temperature. Temperature is thought to be the main control on the formation of this low-grade greenschist mineral assemblage, although lithostatic pressure influences induration and lithification so that rock densities increase from below 2.0 at the surface to 3.0 at 1500 m. Muffler & White (1968, 1969) considered that high CO_2 activity relative to H_2O precluded zeolite formation and favored carbonate precipitation. An interesting aspect of this is that the CO_2 at shallow levels has probably been released from deeper reactions between dolomite and kaolinite (150–200°C) and transformation of calcite into epidote (300–320°C). Kendall (1976b) has given a detailed description of alteration in three wells (Magmamax 2 and 3, Woolsey 1). As well as recognizing hydrothermal biotite, tremolite, garnet, and a high-temperature expanding smectite, she showed that there has been extensive oxygen and carbon isotope exchange between brines and carbonates down to 100–150°C; both detrital and authigenic quartz in sandstones have exchanged oxygen at 290°C generally confirming the conclusions of Clayton et al (1968). Kendall, however, attributes downwell variations in O^{18} to incomplete isotope exchange between fluid and rock caused by differences in reservoir permeability.

Hoagland (1976a, b) and Hoagland & Elders (1978) have described the alteration and Coplen (1976) measured the isotopic ratios of minerals from an 1830 m deep well at East Mesa, which has a maximum temperature of 188°C. The vertical extent of the reservoir can be defined by its alteration and, using thermochemical evaluation of aqueous species and petrographic observations, two distinct hydrothermal events were recognized. During the earlier event, quartz, pyrite, illite, Fe-chlorite, and rare adularia deposited in sandstone, as less porous siltstones and shales altered to illite and chlorite; however, detrital kaolin, montmorillonite, and interlayered clays persisted. Fluid inclusion measurements show that this occurred 15–20°C above present temperatures. Fluids responsible may have been similar to the sodium chloride brine with 26,900 ppm dissolved solids that discharged from a deeper (2200–2450 m) aquifer reached by a drill hole

400 m to the east. Later, cold water entered the system laterally and, on heating, partly dissolved the silicates and precipitated abundant calcite. Thermochemical calculations show that this may have been of the same composition as sodium bicarbonate–chloride fluid (2380 ppm dissolved solids) which discharges from the well and derives from 1660–1800 m. This fluid is in chemical, but apparently not isotopic (Coplen 1976), equilibrium with kaolinite, calcite, and montmorillonite rather than potassium silicates or dolomite.

A 612-m-deep well at the margin of the Dunes thermal area provides a good example of self-sealing (Bird 1975, Bird & Elders 1976, Coplen et al 1975, Elders & Bird 1976). In the upper 300 m, seven zones of dense quartzite have formed by episodic reaction between permeable sandstones and laterally moving sodium chloride fluids containing up to 3000 ppm dissolved solids. After the porosity of the sandstones was reduced, water flowed by way of fractures where quartz, adularia, pyrite, calcite, and hematite were deposited. Hydrothermal alteration has increased both the bulk densities of aquifer rocks (from 2.2 to 2.6 gm cm^{-3}) and their SiO_2 (35–50% above initial content) and K_2O (90–130%) relative to Al_2O_3. The temperature profile shows maxima of 110°C at 110 m and 104°C at 285 m, below which there is a reverse gradient. Further, the isotopic compositions of quartz-water, alkali feldspar-water and calcite-water pairs (Coplen et al 1975) show that temperatures here have never exceeded 130°C.

By contrast, temperatures as high as 388°C have been recorded at Cerro Prieto (Mercado 1969), although 370°C is more usual (Mercado 1976). More than 30 wells have been drilled to between 500 and 2630 m, but unlike the other very hot field, Salton Sea, deep fluids (Table 2) have salinities of only about 17,000 ppm (Mercado 1969). The two main factors affecting the formation and distribution of hydrothermal minerals here are temperature and fluid composition (Reed 1976). Kaolinite, montmorillonite, and illite-montmorillonite do not occur above 160°C, nor dolomite above 210°C, but chlorite and illite-chlorite form above 145°C, wairakite above 240°C, and epidote above 255°C. Authigenic orthoclase increases in abundance above 300°C, in accordance with thermodynamic calculations (Reed 1976) based on the composition of reservoir water, but is not easily distinguished in the cuttings from detrital orthoclase or microcline.

Yellowstone, U.S.A.

Because of environmental constraints, no deep drilling is normally allowed within Yellowstone National Park—one of the world's most famous and spectacular thermal areas. However, in 1967–1968 13 shallow diamond-core wells were drilled as part of a comprehensive scientific project by the U.S. Geological Survey. A major strength of this work, following the example of Fenner (1934, 1936) and Allen & Day (1935), is the close integration of geological, petrologic, and geochemical methods. Although the deepest well (Y-12) is only 329.6 m deep and most are near 150 m deep, both the drilling and the wells were closely monitored, so that an excellent physical record was obtained (White et al 1975). In addition, the near completeness of core recovery (usually above 90%) has made possible the most detailed work so far on hydrothermal alteration in an active geothermal system. Alteration studies have now been completed for wells Y-1 (Honda & Muffler 1970), Y-5 (Keith & Muffler 1978), Y-7, and Y-8 (Keith et al 1978). Less detailed information from Y-3 (Bargar et al 1973) and Y-11 (White et al 1971) is also available.

Wells in the Upper Geyser Basin (Y-1, Y-7, Y-8, and Fenner's C-1) penetrated sandstone, siltstone, and conglomerate mainly composed of rhyolitic detritus overlying rhyolite; Y-5, in Midway Geyser Basin, was drilled, for the most part, into densely welded rhyolitic ash flow tuff. Maximum well temperatures are: 171°C in Y-1, 170°C in Y-5, 143°C in Y-7, and 170°C in Y-8.

Several minerals not previously known in geothermal systems have been recognized. For example, aegirine occurs in Y-1 where the well temperature is 160°C and in Y-3 it also coexists, in the core from 23.8 to 28.5 m at temperatures of 130–140°C, with quartz, pectolite, montmorillonite, analcime, ?albite, and lepidolite (with 7.5% Li_2O). Fluids depositing these minerals (Table 2), although dilute, have high Li/K ratios (1.8) and low Cl/F ratios (about 5).

Factors affecting hydrothermal alteration at Yellowstone vary in importance from well to well and are not easily separated. The nature of the starting material is important in Y-1, Y-7, and Y-8 where detrital obsidian alters readily, but coexisting lithoidal rhyolite of essentially the same composition remains unchanged. The dense, welded ash-flow tuff in Y-5 shows only incipient alteration (to montmorillonite) as its devitrification products, quartz and sanidine, are more stable than glass so that hydrothermal minerals are here more abundant than in Y-5. Obsidian clasts in Y-7 and Y-8 alter, first nearly isochemically, to ?metastable potassic clinoptilolite (K_2O up to 5.7%) and then to analcime plus K-feldspar.

Temperature is important but its effect is often equivocal and difficult to distinguish from fluid composition or variations in permeability. Obsidian in Y-1 is completely replaced above 85°C, but persists to 170°C in Y-8 because of lower permeability in the vitrophyric flow breccia, which indicates that through-going solutions are needed here for even devitrification to occur. Opal occurs only below 43°C in Y-5 and near 104°C in Y-1, whereas in this well β-cristobalite is not present above 115°C. Erionite is present in cores from several wells where temperatures are below 110°C. Differences in fluid composition, especially SiO_2 activity, rather than temperature, evidently account for the observed distribution of clinoptilolite, mordenite, and α-cristobalite and the quartz-analcime association in Y-1. Subsurface boiling caused changes in fluid pH through loss of CO_2, thereby resulting in deposition of adularia in wells C-1 and Y-13 and bladed calcite in Y-8 and Y-5. The Yellowstone work, however, confirms the important effect of permeability in determining the extent and type of hydrothermal alteration in geothermal systems. For example, K-feldspar, quartz, mordenite, and celadonite have formed in permeable zones of Y-8, but in deeper impermeable parts of this well glass persists. Further, mineral deposition in pore spaces of a correlated unit in the 130 m between the sites of Y-7 and Y-8, the latter of which is thought to be closer to the upflow zone, has caused self sealing, so that there is now a pressure differential of .015 bar m^{-1} between the two wells.

These two wells also provide a good record of relationships among silica minerals in a geothermal field. X-ray—amorphous opal dominates in superficial sinter and also forms near-surface cement, whereas α- and β-cristobalite have irregular distributions at greater depth. In Y-7, β-cristobalite occurs mainly in veinlets and the α-form is in the groundmass, but cores from Y-8, the hotter well, contain quartz and chalcedony. This suggests that the distribution of silica minerals is controlled primarily by temperature. From a study of veinlets, Keith et al (1978) found that the silica forms, deposited from oldest to youngest, were quartz, chalcedony, α-cristobalite, β-cristobalite and opal; this is also the order of decreasing stability and increasing solubility with respect to water of constant temperature (Fournier 1973, Arnórsson 1975). Keith et al (1978) concluded that water from the wells is strongly supersaturated with respect to quartz, chalcedony, and α-cristobalite; is slightly undersaturated relative to β-cristobalite; and is undersaturated above 80°C with respect to opal. However, fluids in the reservoir itself are just saturated with quartz. Their suggested model is one in which ascending silica-rich water penetrates aquifers which, through self-sealing, become partially closed systems; the trapped pore water then becomes progressively lower in SiO_2 as more stable silica minerals are successively reconstituted from the earlier deposited phases.

Japan

Japan has about one hundred geothermal systems, excluding fumaroles on the craters of active volcanoes (Sumi & Takashima 1976). Several fields have been investigated by exploratory drilling and numerous accounts of their alteration have been published recently; space, however, permits mention of only a few.

Most Japanese geothermal systems, and indeed, many in the west Pacific, are associated with late Cenozoic andesite-dacite volcanic centers where there is usually a strong structural control on hydrology. A characteristic feature of their alteration is marked surface zonation; typically, between 3 and 10 alteration zones can be clearly recognized (Sumi & Takashima 1976). This, and the rarity of zone overlaps, shows that thermal episodes were commonly single events in contrast with many systems where mineral zoning, even when recognizable, is irregular. However, cross sections of alteration in several Japanese fields, such as Matsukawa (Sumi 1968), show more complicated zoning in the subsurface than is evident at the surface. Sulfates (particularly alunite) and kaolin are common, and pyrophyllite and zeolites occur in several fields (Hayashi & Fujino 1976, Hayashi & Yamasaki 1976, Sumi & Takashima 1976, Takashima 1971, Yamada 1976, Yoshida 1974, Yoshida et al 1976). An example of calcium zeolite formation is at Katayama, Onikobe where the zonation, with increasing depth and temperature, is mordenite-laumontite ($\pm$ yugawaralite, analcime)—wairakite (Seki et al 1969, Seki & Okumuru 1968). However, the transition temperatures between laumontite and wairakite (75°C to 175°C in four wells) seem to be below the usual formation temperatures of wairakite (about 220°C), possibly because water pressures at Katayama are low. However, it is not clear that the measured well temperatures are the same as those that prevailed before drilling. Yamada (1976) suggests that α-cristobalite occurs in the vapor zone at shallow depths, but zeolite alteration results from reaction between andesite and mildly alkaline solutions. Near the bottom of one 1200-m well (GO-10), however, quartz and kaolin have deposited from acid chloride solutions.

Another field where alteration shows a close dependence on fluid composition is at Otake, north-central Kyushu (Hayashi et al 1968, Hayashi 1973, Hayashi & Fujino 1976, Hayashi & Yamasaki 1976, Yamasaki & Hayashi 1976), where two chemically distinct fluids (Table 2) have deposited minerals zoned about fractures. In the Otake area itself deep waters are neutral to alkaline but wells at Hatchobaru, 3 km south, encountered fluids with a pH between 3.4 and 4.6. Hayashi (1973) classifies the alteration into five types (Table 3), based on the composition of the altering fluids, and subdivides further depending on measured well temperatures and fluid pressures. However, the temperatures in many wells were measured only a few days, or even hours, after drilling, so that they are unlikely to be quite the same as the temperatures of mineral deposition. Where acid fluids prevail, Type B alteration is conspicuous and Type D either occurs at shallow levels or where reaction is incomplete. By contrast, at Otake itself, mildly alkaline fluids have restricted Type B alteration but Types D and E are widespread. Major element analysis (Hayashi 1973) shows that massive amounts of SiO_2 were added to rocks with alteration of Type A, but during formation of Types B and C, SiO_2,

Table 3 Classification of hydrothermal alteration in the Otake geothermal area (after Hayashi 1973)

Type	Minerals[a]	Fluid	Temperature (°C)	$P_{fluid (bars)}$
A	crist	strong acid	< 100	< 15
	quartz	acid	100 to 230	30 to 50
B	alunite + crist	strong acid	< 100	< 15
	alunite + quartz	strong acid	100 to 230	15 to 50
C	kaol + crist	acid	< 100	< 15
	kaol + quartz	acid	100 to 200	< 30
	dickite + py + quartz	acid	150 to 250	< 60
	py + quartz	acid	> 230	> 50
D	mont + crist	weak acid	< 100	< 15
	mont S/M + quartz	weak acid	100 to 200	< 30
	ch + S/M + quartz	weak acid	150 to 250	< 60
	S + quartz	neutral	> 230	> 50
E	heul + crist	neutral	< 100	< 15
	laum + wair + quartz	neutral	100 to 200	< 30
	albite + quartz	weak alkaline	150 to 250	< 60
	adularia + quartz	weak alkaline	> 230	> 50

[a] Abbreviations are as follows: crist, cristobalite; kaol, kaolinite; py, pyrophyllite; mont, montmorillonite; ch, chlorite; S/M sericite-montmorillonite; S, sericite; heul, heulandite; laum, laumontite; wair, wairakite.

Al_2O_3, and TiO_2 are nearly immobile, as are Fe_2O_3, CaO, and Mgo during formation of Types D and E; however, Na_2O and K_2O are the most mobile constituents.

The Matsukawa field, northern Honshu, also has zoned surface alteration (Sumi 1968, 1969, Sumi & Takashima 1976) that extends over an area of 7 by 1.5 km. It is elongated about fissures so that structural control of fluid flow and alteration is obvious. From the margin inward four alteration zones are characterized by saponite plus chlorite, montmorillonite, kaolin, and alunite, but other hydrothermal minerals present include laumontite, calcite, and anhydrite in the chlorite zone, calcite, anhydrite, and quartz in the montmorillonite zone, and quartz in both the kaolin and alunite zones. A far smaller zone of pyrophyllite, usually with diaspore, zunyite, andalusite, quartz and anhydrite, which is overlapped by three other zones, is thought to be relict alteration formed above about 310°C, whereas present well temperature maxima are near 230 to 250°C (Sumi 1968, Truesdell 1976). When first opened, the wells typically discharged wet steam of low pH (about 5) and high sulfate contents (Table 2). Zoning from alunite-quartz at the center to outer kaolin is consistent with acid (at 200°C pH below 4) solutions moving outwards from channels and becoming more alkaline by reaction with reservoir rocks.

Iceland

Work on hydrothermal alteration in geothermal fields of Iceland has several important aspects: (a) it provides the best examples of reactions between thermal fluids and basalts, (b) there is extensive zeolitisation, especially in the low temperature ($< 150°C$) fields, (c) in the Reykjanes area the thermal fluid is modified sea water (Kristmannsdóttir 1975, 1976, Kristmannsdóttir & Tómasson 1976a, b, c, Sigvaldason 1963, Tómasson & Kristmannsdóttir 1972, 1976).

Reservoir rocks are basalt, basaltic hyaloclastites and minor dolerite. Glass readily alters first to opal, smectite, calcite, or a zeolite, and then to mixed-layer clays, although where permeability is low it persists to 200°C. Olivine is the most unstable mineral and it and pyroxene are often partly replaced by smectite-illite or chlorite. Complete replacement of plagioclase by epidote or albite is also rare, indicating that disequilibrium between fluids and rocks is usual. Factors controlling alteration are rock type, water composition, permeability, temperature, and the age of the system—although the first two are less important.

The high-temperature (up to $298 \pm 4°C$) fields occur in areas of active volcanism and, except for potash enrichment in some permeable zones at Reykjanes, alteration is near isochemical involving hydration and oxidation. Clays are abundant but poorly crystalline; iron-rich saponite, present below 200°C, is replaced by mixed-layer chlorite-smectite in the range 200° to 230°C and chlorite above 230°C. Zeolites are common below 230°C, but thermal gradients are so steep that a distinct temperature zoning cannot be readily detected. Epidote is abundant above 260°C and an amphibole (of unspecified composition) occurs locally where temperatures exceed 280°C. In the Nesjavellir field, prehnite is present above 250°C and calcite alternates in abundance with the zeolites. Well fluids are rich in carbonate (1355 ppm CO_2 equivalent) and sulfide (311 ppm), and consequently carbonate and sulfide minerals dominate the deep alteration. However, the observed hydrothermal minerals were not formed at the measured well temperatures because activity is shifting north and there has not been time for minerals to adjust to the higher temperature regime.

A high-temperature system in the Reykjanes area, southwest Iceland, is only one of several fields (Truesdell 1976) in which modified sea water is the thermal fluid, but it has the best record of alteration minerals formed from its reaction with reservoir rocks. Hydrothermal minerals are crudely zoned according to temperature (which reaches 300°C): (a) montmorillonite–zeolite–calcite zone; (b) mixed-layer clay–prehnite zone; (c) chlorite–epidote zone. These minerals occur in many fields in Iceland and elsewhere, although fluid compositions vary greatly, leading Tómasson & Kristmannsdóttir (1972) to conclude that permeability, porosity, and especially temperature are the main controls of hydrothermal alteration. Anhydrite, deposited when sea water heated, has an irregular distribution that may be due to about 20 to 30 sea water invasions which occurred when the usually impermeable boundaries of the field were ruptured by tectonic events.

Alteration in low-temperature (below 150°C at 1000 m) fields expands the smectite-zeolite zone of the hotter areas and confirms the control of temperature on zeolite distribution. Kristmannsdóttir (1976) distinguishes four zones character-ised by: (a) chabazite, opal, calcite ± levyne and stable to about 80°C; (b) mesolite-scolecite present from 80 to 90°C; (c) stilbite occurring between 100 and 120°C; and (d) laumontite above 120°C. Smectites occur at shallow depths typically coexisting with chlorite from about 180°C, although both "swelling" chlorite and random mixed-layered clays are also present. However, prehnite, epidote, quartz, and possibly chlorite in at least two low temperature fields are thought to be relicts from an earlier hotter regime.

THE FUTURE

No simple enumeration of factors affecting hydrothermal alteration can yet account for all the observed complications. With new or greatly expanded drilling programs in many parts of the world, however, much more will be learnt (and, hopefully published) so that it may soon be possible to isolate more confidently the effects of a single factor without distorting the general pattern.

Observations made so far confirm the important effect of temperature in the formation of hydrothermal mineral assemblages, but also demonstrate the critical role of permeability. Geochemical studies of thermal fluids have made a major contribution to knowledge of the intimate relationship between fluid composition and mineral deposition. However, few attempts have yet been made to interrelate the distribution of minor and trace elements in rocks, minerals, and fluids. Reasons why so many hydrothermal minerals occur in geothermal fields are still sought: possibly some phases are metastable, or stabilized by the substitution of minor or trace elements. Fluid inclusion measurements have great promise in helping unravel the evolution of geothermal systems, particularly when used in conjunction with a thermodynamic model for mineral deposition, such as that suggested by Helgeson

(1970). Experimental mineral syntheses and thermodynamic calculations of mineral equilibria should help clarify low-temperature phase relations, especially if the effects of solid solution are also considered.

Other rewarding approaches for some fields would be to evaluate quantitively mass transfer between fluid and solid phases; such studies, however, need careful, detailed petrologic and geochemical analysis of altered and unaltered rock samples.

Although an understanding of hydrothermal alteration in active geothermal systems will add greatly to our knowledge of low-grade metamorphism and hydrothermal ore deposition, much already known about these subjects could also be usefully applied to predict the evolution, and conditions, in geothermal reservoirs below drilling depth.

ACKNOWLEDGMENTS

I am grateful for the helpful comments of W. A. Elders, A. J. Ellis, J. H. Lowery, W. A. Watters, B. G. Weissberg, D. E. White and C. P. Wood, and I also thank Mrs. M. Ruthven for typing and Mrs. P. Williams for drawing the two figures.

Literature Cited

Allen, E. T., Day, A. L. 1935. The hot springs of the Yellowstone National Park. *Carnegie Inst. Washington Publ. 466*, 525 pp.

Arnórsson, S. 1975. Application of the silica geothermometer in low temperature hydrothermal areas in Iceland. *Am. J. Sci.* 725:763–84

Bargar, K. E., Beeson, M. H., Fournier, R. O., Muffler, L. J. P. 1973. Present-day deposition of lepidolite from thermal waters in Yellowstone National Park. *Am. Mineral.* 58:901–04

Biehler, S., Kovach, R. L., Allen, C. R. 1964. Marine geology of the Gulf of California. *Am. Assoc. Petrol. Geol. Mem.* 3:126–43

Bird, D. K. 1975. Geology and geochemistry of the Dunes hydrothermal system, Imperial Valley of California. *Inst. Geophys. Planet. Phys.—Univ. Calif. Riverside Rep. 75–2*. 123 pp.

Bird, D. K., Elders, W. A. 1976. *Proc. 2nd U.N. Symp. Dev. Use Geotherm. Resour. 1975.* 1:285–95. San Francisco: United Nations 844 pp.

Blattner, P. 1975. Oxygen isotopic composition of fissure-grown quartz, adularia, and calcite from Broadlands geothermal field, New Zealand. *Am. J. Sci.* 275:785–800

Browne, P. R. L. 1970. Hydrothermal alteration as an aid in investigating geothermal fields. *Geothermics.* Spec. Issue 2:564–70

Browne, P. R. L. 1971. Mineralization in the Broadlands geothermal field, Taupo

volcanic zone. *Soc. Min. Geol. Jap.* Spec. Issue 2:64–75

Browne, P. R. L., Elders, W. A. 1976. *Geol. Soc. Am. Boulder 793* (Abstr.)

Browne, P. R. L., Ellis, A. J. 1970. The Ohaki-Broadlands hydrothermal area, New Zealand: Mineralogy and related geochemistry. *Am. J. Sci.* 269:97–131

Browne, P. R. L., Lovering, J. F. 1973. Composition of sphalerites from the Broadlands geothermal field and their significance to sphalerite geobarometry and geobarometry. *Econ. Geol.* 68:381–87

Browne, P. R. L., Rafter, T. A., Robinson, B. W. 1975. Sulphur isotope ratios of sulphides from the Broadlands Geothermal Field, New Zealand. *N.Z. J. Sci.* 18:35–40

Browne, P. R. L., Roedder, E., Wodzicki, A. 1976. In *Proc. Int. Symp. Water-Rock Interaction 1974*, ed. J. Čadek, T. Pačes, pp. 140–9. Prague: Geol. Surv. 463 pp.

Browne, P. R. L., Wodzicki, A. 1977. The aluminium-in-quartz geothermometer. *N.Z. Dep. Sci. Ind. Res. Bull.* 218:35–36

Carpenter, A. B., Trout, M. L., Pickett, E. L. 1974. Preliminary report on the origins and chemical evolution of lead- and zinc-rich oil field brines in Central Mississippi. *Econ. Geol.* 69:1191–1206

Chen, C. H. 1966. Preliminary exploration of geothermal steam in Tatun volcanic region, Taiwan, Republic of China. *Ministry Economic Affairs.* 6 pp.

Chen, C. H. 1967. Exploration of geothermal

steam in Tahuangtsui thermal area, Tatun volcanic region, north Taiwan. *Ministry Economic Affairs.* 23 pp.

Clayton, R. N., Muffler, L. J. P., White, D. E. 1968. Oxygen isotope study of calcite and silicates of River Ranch No. 1 Well, Salton Sea geothermal field, California. *Am. J. Sci.* 266:968–79

Clayton, R. N., Steiner, A. 1975. Oxygen isotope studies of the geothermal system at Wairakei, New Zealand. *Geochim. Cosmochim. Acta* 39:1179–86

Coombs, D. S. 1955. X-ray observations on wairakite and non-cubic analcime. *Mineral. Mag.* 30:699–708

Coplen, T. B. 1976. Cooperative geochemical resource assessment of the Mesa geothermal system. *Inst. Geophys. Planet. Phys.—Univ. Calif. Riverside Rep. 76–1.* 97 pp.

Coplen, T. B., Kolesar, P., Taylor, R. E., Kendall, C., Mooser, C. 1975. Investigations of the Dunes geothermal anomaly, Imperial Valley, California, Part IV. *Inst. Geophys. Planet. Phys.—Univ. Calif. Riverside Rep. 75–20.* 42 pp.

Cusicanqui, H., Mahon, W. A. J., Ellis, A. J. 1976. See Bird & Elders 1976, pp. 703–11

Elders, W. A., Bird, D. K. 1976. See Browne, Roedder & Wodzicki 1976, pp. 150–57

Ellis, A. J. 1967. In *Geochemistry of Hydrothermal Ore Deposits,* ed. H. L. Barnes, pp. 465–514. New York: Holt, Rinehart & Winston. 670 pp.

Ellis, A. J., McFadden, I. M. 1972. Partial molal volumes of ions in hydrothermal solutions. *Geochim. Cosmochim. Acta* 36: 413–26

Ellis, A. J., Mahon, W. A. J. 1967. Natural hydrothermal systems and experimental hot water/rock interactions (Part II). *Geochim. Cosmochim. Acta* 31:519–38

Ellis, A. J., Mahon, W. A. J. 1977. *Chemistry and Geothermal Systems.* New York: Academic

Energy Research and Development Administration. 1976. *A Bibliography of Geothermal Resources—Exploration and Exploitation.* 617 pp.

Erd, R. C., White, D. E., Fahey, J. J., Lee, D. E. 1964. Buddingtonite, an ammonium feldspar with zeolitic water. *Am. Mineral.* 49:831–50

Eslinger, E. V., Savin, S. M. 1973. Mineralogy and oxygen isotope geochemistry of the hydrothermally altered rocks of the Ohaki-Broadlands, New Zealand geothermal area. *Am. J. Sci.* 273:240–67

Fenner, C. N. 1934. *Trans. Am. Geophys. Union, 15th Meet., Washington,* Pt. 1, pp. 240–43

Fenner, C. N. 1936. Bore-hole investigations in Yellowstone Park. *J. Geol.* 44:225–315

Fournier, R. O. 1973. *Proc. Symp. Hydrogeochem. Biogeochem.* 1:122–39

Fournier, R. O., Rowe, J. J. 1966. Estimation of underground temperatures from the silica content of water from hot springs and wet-steam wells. *Am. J. Sci.* 264: 685–97

Fournier, R. O., Truesdell, A. H. 1973. An empirical Na-K-Ca geothermometer for natural waters. *Geochim. Cosmochim. Acta* 37:1255–75

Grindley, G. W. 1965. The geology, structure, and exploitation of the Wairakei geothermal field, Taupo, New Zealand. *N.Z. Geol. Surv. Bull. 75.* 131 pp.

Hayashi, M. 1973. Hydrothermal alteration in the Otake geothermal area, Kyushu. *J. Japan Geotherm. Energy Assoc.* 10:9–46

Hayashi, M., Fujino, T. 1976. See Bird & Elders 1976, pp. 407–14

Hayashi, M., Yamasaki, T. 1976. See Browne, Roedder & Wodzicki 1976, pp. 158–69

Hayashi, M., Yamasaki, T., Matsumoto, Y. 1968. Secondary minerals of drilling cores from test borings T-1 and T-2 in the Otake geothermal area, Kujyu volcano group, Kyushu. *J. Japan Geotherm. Energy Assoc.* 17:93–8 (In Japanese with English abstr.)

Helgeson, H. C. 1967. In *Researches in Geochemistry,* ed. P. H. Abelson. 2:362–404 New York: Wiley. 663 pp.

Helgeson, H. C. 1968a. Geologic and thermodynamic characteristics of the Salton Sea geothermal system. *Am. J. Sci.* 266:129–66

Helgeson, H. C. 1968b. Evaluation of irreversible reactions in geochemical processes involving minerals and aqueous solutions—1. Thermodynamic relations. *Geochim. Cosmochim. Acta* 32:853–77

Helgeson, H. C. 1970. A chemical and thermodynamic model of ore deposition in hydrothermal systems. *Mineral. Soc. Am. Spec. Pub.* 3:155–86

Helgeson, H. C. 1971. Kinetics of mass transfer among silicates and aqueous solutions. *Geochim. Cosmochim. Acta* 35: 421–69

Helgeson, H. C., Garrels, R. M., Mackenzie, F. T. 1969a. Evaluation of irreversible reactions in geochemical processes involving minerals and aqueous solutions. Applications. *Geochim. Cosmochim. Acta* 33:455–81

Helgeson, H. C., Brown, J. H., Leeper, R. H. 1969b. *Handbook of Theoretical Activity Diagrams Depicting Chemical Equilibria in Geologic Systems Involving an Aqueous Phase at One Atm. and 0° to 300°C.* San Francisco: Freeman. 253 pp.

Hemley, J. J., Hostetler, P. B., Gude, A. J., Mountjoy, W. T. 1969. Some stability relations of alunite. *Econ. Geol.* 64: 599–612

Hemley, J. J., Jones, W. R. 1964. Chemical aspects of hydrothermal alteration with emphasis on hydrogen metasomatism. *Econ. Geol.* 59: 538–69

Hoagland, J. R. 1976a. *Geol. Soc. Am. Boulder 919* (Abstr.).

Hoagland, J. R. 1976b. *Petrology and geochemistry of hydrothermal alteration in borehole Mesa 6–2, East Mesa geothermal area, Imperial Valley, California.* M.S. thesis. Univ. Calif. Riverside. 90 pp.

Hoagland, J. R., Elders, W. A. 1978. The evolution of the East Mesa hydrothermal system, California, U.S.A. In *Proc. 2nd Int. Symp. Water-Rock Interaction, Strasbourg, 1977.* 10 pp. In press

Honda, S., Muffler, L. J. P. 1970. Hydrothermal alteration in core from research drill hole Y-1, Upper Geyser Basin, Yellowstone National Park, Wyoming. *Am. Mineral.* 55: 1714–37

Keith, T. E. C., Muffler, L. J. P., Cremer, M. 1968. Hydrothermal epidote formed in the Salton Sea geothermal system, California. *Am. Mineral.* 53: 1635–44

Keith, T. E. C., Muffler, L. J. P. 1978. Minerals produced during cooling and hydrothermal alteration of ash flow tuff from Yellowstone drillhole Y-5. Submitted to *J. Volcanol. Geotherm. Res.*

Keith, T. E C., White, D. E., Beeson, M. H. 1978. Hydrothermal alteration and self-sealing in Y-7 and Y-8 drillholes in northern part of Upper Geyser Basin, Yellowstone National Park, Wyoming. *USGS Prof. Pap. 1054A.* In press

Kendall, C. 1976a. *Geol. Soc. Am. Boulder 952* (Abstr.)

Kendall, C. 1976b. *Petrology and stable isotope geochemistry of three wells in the Buttes area of the Salton Sea geothermal field, Imperial Valley, California, U.S.A.* M.S. thesis. Univ. Calif. Riverside Rep. UCR/IGPP—76/17. 211 pp.

Kristmannsdóttir, H. 1975. Clay minerals formed by hydrothermal alteration of basaltic rocks in Icelandic geothermal fields. *Geol. Fören. Stockholm Förh.* 97: 289–92

Kristmannsdóttir, H. 1976. See Bird & Elders 1976, pp. 441–45

Kristmannsdóttir, H., Tómasson, J. 1976a. See Browne, Roedder & Wodzicki 1976, pp. 170–77

Kristmannsdóttir, H., Tómasson, J. 1976b. Hydrothermal alteration in Icelandic geothermal fields. *Soc. Sci. Islandica* 5: 167–75

Kristmannsdóttir, H., Tómasson, J. 1976c. *Zeolite Zones in Geothermal Areas in Iceland.* Presented at Zeolite Meet., 1976, Tucson.

Mahon, W. A. J. 1966. Silica in hot water discharged from drillholes at Wairakei, New Zealand. *N.Z. J. Sci.* 9: 135–44

Mahon, W. A. J. 1976. See Bird & Elders 1976, pp. 775–83

Mahon, W. A. J., Finlayson, J. B. 1972. The chemistry of the Broadlands geothermal area, New Zealand. *Am. J. Sci.* 272: 48–68

McNitt, J. R. 1964. Geology of the Geysers thermal area. *Proc. U.N. Conf. New Sources Energy 1961* 2: 292–301

Marinelli, G. 1969. Some geological data on the geothermal areas of Tuscany. *Bull. Volcanol.* 33: 319–34

Mercado, S. 1969. Chemical changes in geothermal well M-20, Cerro Prieto, Mexico. *Geol. Soc. Am. Bull.* 80: 2623–29

Mercado, S. 1976. See Bird & Elders 1976, pp. 487–95

Miller, D. G., Piwinskii, A. J., Yamauchi, R. 1977. The use of geochemical-equilibrium computer calculations to estimate precipitation from geothermal brines. *Univ. Calif. Livermore Rep. 52197.* 35 pp.

Muffler, L. J. P., Doe, B. R. 1968. Composition and mean age of detritus of the Colorado river delta in the Salton Trough, southeastern California. *J. Sediment. Petrol.* 38: 384–99

Muffler, L. J. P., White, D. E. 1968. Origin of CO_2 in the Salton Sea geothermal system, southeastern California, U.S.A. *23rd Int. Geol. Congr.* 17: 184–95

Muffler, L. J. P., White, D. E. 1969. Active metamorphism of Upper Cenozoic sediments in the Salton Sea geothermal field and the Salton Trough, southeastern California. *Geol. Soc. Am. Bull.* 80: 157–82

Muffler, L. J. P., White, D. E., Truesdell, A. H. 1971. Hydrothermal explosion craters in Yellowstone National Park. *Geol. Soc. Am. Bull.* 82: 723–40

Naboko, S. I. 1970. Facies of hydrothermally altered rocks of Kamchatka-Kurile volcanic arc. *Pac. Geol.* 2: 23–27

Naboko, S. I. 1976. See Browne, Roedder & Wodzicki 1976, pp. 184–95

Olson, E. R. 1976. *Geol. Soc. Am. Boulder 1036* (Abstr.)

Palmer, T. D. 1975. Characteristics of geothermal wells located in the Salton Sea geothermal field, Imperial County, California. *Univ. Calif. Livermore Rep. 51976.* 54 pp.

Randall, W. 1974. *An analysis of the subsurface structure and stratigraphy of the Salton Sea geothermal anomaly, Imperial Valley, California.* PhD thesis. Univ. Calif. Riverside. 92 pp.

Reed, M. J. 1976. See Bird & Elders 1976, pp. 539–47

Renner, J. L., White, D. E., Williams, D. L. 1975. *Assessment of geothermal resources of the United States—1975.* U.S. Geol. Surv. Circ. 726:5–57

Ridge, J. D. 1973. Volcanic exhalations and ore deposition in the vicinity of the sea floor. *Miner. Deposita* 8:332–48

Robinson, P. T., Elders, W. A., Muffler, L. J. P. 1976. Quaternary volcanism in the Salton Sea geothermal field, Imperial Valley, California. *Geol. Soc. Am. Bull.* 87:347–60

Schoen, R., White, D. E. 1965. Hydrothermal alteration in GS-3 and GS-4 drillholes, Main Terrace, Steamboat Springs, Nevada. *Econ. Geol.* 60:1411–21

Schoen, R., White, D. E. 1968. Hydrothermal alteration of basaltic andesite and other rocks in drillhole GS-6, Steamboat Springs, Nevada. *U.S. Geol. Surv. Prof. Pap.* 575–13:110–19

Seki, Y. 1972. Lower grade stability limit of epidote in the light of natural occurrences. *J. Geol. Soc. Jpn.* 78:405–13

Seki, Y., Okumuru, K. 1968. Yugawaralite from Onikobe active geothermal area, northeast Japan. *J. Jpn. Assoc. Miner. Petrol. Econ. Geol.* 60:27–33

Seki, Y., Onuki, H., Okumura, K., Takashima, I. 1969. Zeolite distribution in the Katayama geothermal area, Onikobe, Japan. *Jpn. J. Geol. Geogr.* 40:63–79

Sigvaldason, G. E. 1963. Epidote and related minerals in two geothermal drill holes, Reykjavik and Hveragerdi, Iceland. *U.S. Geol. Surv. Prof. Pap.* 450–E, pp. 77–9

Sigvaldason, G. E., White, D. E. 1961. Hydrothermal alteration of rocks in two drillholes at Steamboat Springs, Nevada. *U.S. Geol. Surv. Prof. Pap.* 424–D, pp. 116–22

Sigvaldason, G. E., White, D. E. 1962. Hydrothermal alteration in drillholes GS-5 and GS-7, Steamboat Springs, Nevada. *U.S. Geol. Surv. Prof. Pap.* 450-D, pp. 113–17

Skinner, B. J., White, D. E., Rose, H. J., Mays, R. E. 1967. Sulfides associated with the Salton Sea geothermal brine. *Econ. Geol.* 62:316–30

Steiner, A. 1953. Hydrothermal rock alteration at Wairakei, New Zealand. *Econ. Geol.* 48:1–13

Steiner, A. 1955. Wairakite, the calcium analogue of analcime, a new zeolite mineral. *Mineral. Mag.* 30:691–98

Steiner, A. 1958. Occurrence of wairakite at the Geysers, California. *Am. Mineral.* 43:781

Steiner, A. 1963. Waiotapu geothermal field. *N.Z. Dep. Sci. Ind. Res. Bull.* 155:26–34

Steiner, A. 1968. Clay minerals in hydrothermally altered rocks at Wairakei, New Zealand. *Clays Clay Miner.* 16:193–213

Steiner, A. 1970. Genesis of hydrothermal K-feldspar (adularia) in an active geothermal environment at Wairakei, New Zealand. *Mineral. Mag.* 37:916–22

Sumi, K. 1968. Hydrothermal rock alteration of the Matsakawa geothermal area, northeast Japan. *Geol. Surv. Jpn. Rep.* 225:42 pp.

Sumi, K. 1969. Zonal distribution of clay minerals in the Matsukawa geothermal area, Japan. *Int. Clay Conf. Tokyo.* pp. 501–12

Sumi, K., Takashima, I. 1976. See Bird & Elders 1976, pp. 625–34

Takashima, I. 1971. Hydrothermal rock alteration in Takenoyu geothermal area, Kumamoto Prefecture, Japan. *Jpn. Geol. Surv. Bull.* 23:26–32 (In Japanese with English abstr.)

Tómasson, J., Kristmannsdóttir, H. 1972. High temperature alteration minerals and thermal brines, Reykjanes, Iceland. *Contrib. Mineral. Petrol.* 36:123–34

Tómasson, J., Kristmannsdóttir, H. 1976. See Browne, Roedder & Wodzicki 1976, pp. 243–49

Truesdell, A. H. 1976. See Bird & Elders 1976, pp. liii–lxxix

United Nations 1976. *Proc. 2nd U.N. Symp. Dev. Use Geotherm. Res. 1975.* 3 vols. San Francisco: U.N. 2466 pp.

Weissberg, B. G., Browne, P. R. L., Seward, T. M. 1978. In *Geochemistry of Hydrothermal Ore Deposits*, ed. H. L. Barnes. New York: Wiley. 2nd ed.

White, D. E. 1967. See Ellis 1967, pp. 575–631

White, D. E. 1968. Hydrology, activity, and heat flow of the Steamboat Springs thermal system, Washoe County, Nevada. *U.S. Geol. Surv. Prof. Pap. 458-C.* 109 pp.

White, D. E. 1974. Diverse origins of hydrothermal ore fluids. *Econ. Geol.* 69: 954–73

White, D. E., Fournier, R. O., Muffler, L. J. P., Truesdell, A. H. 1975. Physical results of research drilling in thermal areas of Yellowstone National Park, Wyoming. *U.S. Geol. Surv. Prof. Pap.* *892.* 70 pp.

White, D. E., Muffler, L. J. P., Truesdell, A. H. 1971. Vapor-dominated hydrothermal systems compared with hot-water systems. *Econ. Geol.* 66:75–97

Yamada, E. 1976. See Bird & Elders 1976, pp. 665–72

Yamasaki, T., Hayashi, M. 1976. See Bird & Elders 1976, pp. 673–84

Yoshida, T. 1974. Alteration zones in the Kirishima geothermal area. *J. Jpn. Geotherm. Energy Assoc.* 11:35–40 (In Japanese with English abstr.)

Yoshida, T., Higuchi, K., Yahara, K. 1976. Alteration of rocks by geothermal activities in Satsuma Iwao-jima island, Kayoshima Prefecture, Japan. *J. Jpn Geotherm. Energy Assoc.* 13:9–19 (In Japanese with English abstr.)

Ann. Rev. Earth Planet Sci. 1978. 6 : 251–80

STRATIGRAPHY OF THE   ×10094
ATLANTIC CONTINENTAL SHELF
AND SLOPE OF THE
UNITED STATES

C. Wylie Poag

Paleontology and Stratigraphy Branch, U.S. Geological Survey, Woods Hole, Massachusetts 02543

INTRODUCTION

Purpose and Scope

As we await the first exploratory drilling on the Atlantic outer continental shelf of the United States, geologists are investigating the complex sedimentary framework and reconstructing the geological development of the region. Among the main concerns are the structure, age, sequence, composition, and origin of the layered sedimentary rocks that underlie the continental shelf and slope. Analysis and interpretation of these items are the major elements of stratigraphy. The purpose of this report is to present a current summary of what is known about these sedimentary rocks, and to point out where new knowledge is needed.

This summary includes a description of the geologic features of the area; a review of the postulated geologic development of the Atlantic Ocean; a chronological review of the major stratigraphic studies of the area; a description of studies presently underway; a description of the strata that fill the three major sedimentary basins along the U.S. Atlantic Margin; and lastly, a comparison of the stratigraphic record of the U.S. Margin with that of the Scotian Shelf (Canada) and the Grand Banks (Newfoundland).

Several comprehensive geologic summaries of the U.S. Atlantic Margin have been published or open-filed in the last few years, and much of the pre-existing information in this review comes from these summaries (Emery & Uchupi 1972, Keen 1974, Stehli 1974, Sheridan 1974, 1976, Dillon et al 1975, Mattick et al 1975, Schlee et al 1975, Scott & Cole 1975, Sheridan & Osburn 1975, Schlee et al 1976, Dillon et al 1978, Schlee et al 1977).

Besides the previously published data, this review presents new information derived from drill holes, core holes, dredges, and piston cores.

251

0084-6597/78/0515-0251$01.00

Description of the Atlantic Continental Shelf and Slope

The area described here stretches more than 2000 km from offshore Miami, Florida, to just north of Georges Bank, southeast of Cape Cod, Massachusetts (Figure 1). It includes the rocks that lie beneath the continental shelf and continental slope down to a water depth of about 2000 m, and those beneath the Blake Plateau. This area encompasses approximately 1,000,000 km².

Three major sedimentary basins contain most of the sedimentary rocks (largely Mesozoic and Tertiary strata; see Figure 1): the Blake Plateau trough, which is the southernmost basin; the Baltimore Canyon trough, which stretches beneath the shelf and slope from Cape Hatteras to Long Island; and the Georges Bank basin, which underlies Georges Bank (Figure 1).

GEOLOGIC DEVELOPMENT OF THE ATLANTIC OCEAN

Most geologists believe that the present Atlantic Ocean Basin began to form during the late Triassic and early Jurassic (about 200–180 m.y. ago) as a result of continental rifting and sea-floor spreading (Emery & Uchupi 1972, Schlee et al 1976). At that time, North America, Africa, and Europe were joined along what is now the U.S.–Canadian Atlantic continental slope. As the three continental plates began to move apart, the rocks along the rifted margins were faulted into large blocks that began to subside. The resulting basins began to fill with sediments eroded from the blocks and from the adjacent continents. As these basins subsided below sea level, they were invaded by sea water whose restricted circulation and high evaporation rates caused deposition of thick evaporitic deposits in some areas (Emery & Uchupi 1972, Meyerhoff & Hatten 1974, Tator & Hatfield 1975, Bhat et al 1975, Jacobs 1977).

In the late Jurassic and early Cretaceous, a vast shallow sea covered the eastern margin of North America. Along the seaward rim of the Bahama uplift, the southern Blake Plateau trough, the Baltimore Canyon trough, the Georges Bank basin, and even the Scotian basin off Newfoundland, an extensive series of reefs and carbonate platforms developed (Bryant et al 1969, Schlee et al 1976). Arched seismic reflectors of high acoustic velocity (attributed to dense carbonate rocks) show, for example, that this reef system is present 20–30 km seaward of the present shelf edge in the Baltimore Canyon trough area (Grow, Mattick & Schlee 1977, Mattick, Bayer & Scholle 1977). Emery & Uchupi (1972) have suggested that by mid-Cretaceous time the reef system may no longer have been an effective sediment dam; the depression behind it had filled with continental and shallow marine sediments, and the continental slope beyond it had developed to nearly its current configuration.

As the Atlantic Basin continued to widen during the Cenozoic the most effective controls on sedimentation along the U.S. Atlantic Margin were differential uplift and subsidence of the continent and margin, eustatic sea-level changes, and the activity of the Gulf Stream. Northward extension of warm tropical to subtropical shelf environments accompanied the Gulf Stream during the Paleocene, Eocene,

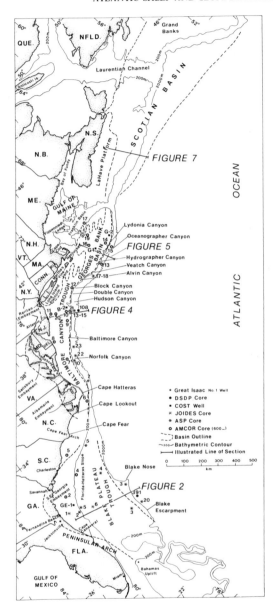

Figure 1 Location map showing general geography, bathymetry, major sedimentary basins and uplifts, sample sites (core holes and drill holes), and location of geologic cross sections illustrated in Figures 2, 4, 5, and 7 herein. (Note that AMCOR Site 6018 is the same as COST G-1.)

and Oligocene, resulting in carbonate-rich sediments all along the U.S. Margin. During the Miocene, however, the Gulf Stream shifted southward, or farther offshore, and the northern sediments became increasingly clastic. The three major basins continued to subside, and had received most of their sediments by the end of the Pliocene. During the Pleistocene and Holocene, most clastic sediments bypassed the shelf, to be deposited on the continental slope and rise. This process was accentuated during glacial sea levels when the shoreline moved close to the present shelf break.

HISTORY OF SIGNIFICANT INVESTIGATIONS

The earliest stratigraphic data from the U.S. Atlantic Margin were obtained from rock fragments dredged from sea-floor outcrops (Verrill 1878, Aggasiz 1888, Dall 1925, Cushman 1936, Stephenson 1936, Stetson 1936, 1949).

After World War II, petroleum exploration became moderately active on the Atlantic Coastal Plain, and as a result of numerous subsurface tests (as deep as 3000 to 4000 m), the coastal stratigraphy became known in some detail (Herrick & Vorhis 1963, Maher 1971, and especially Brown, Miller & Swain 1972). Belatedly, after a few shallow submarine foundation borings had been drilled (McClelland Engineers 1963, McCollum & Herrick 1964), the first systematic attempt at deeper stratigraphic coring on the shelf was undertaken by the Joint Oceanographic Institutions for Deep Earth Sampling (JOIDES) group (Bunce et al 1965, Charm, Nesteroff & Valdes 1970, Schlee 1977). Six core holes were drilled as deep as 320 m below the sea floor (Figure 1) in a transect east of Jacksonville, Florida, across the shelf and the Blake Plateau (J. Schlee, R. D. Gerard, unpublished results).

Two years later (1967), a consortium of oil companies (Exxon, Chevron, Gulf, and Mobil) conducted their Atlantic Slope Project (ASP). They drilled core holes at 14 sites along the base of the continental slope and in submarine canyons (Figure 1). Sediment penetration was limited to approximately 330 m. Weed et al (1974) have summarized part of the dredging, oceanographic coring, and ASP results on a map showing pre-Pleistocene geology from Cape Hatteras to Georges Bank. From the Blake Plateau, Kaneps (1970) selected several of the most complete oceanographic cores and synthesized the Miocene-Holocene stratigraphy of that region.

During the late 1960s and 1970s, the Deep Sea Drilling Project (DSDP) probed the deeper parts of the Atlantic Basin, but only one DSDP core hole was located on the U.S. continental slope (Figure 1; DSDP Leg 11, 1970), and none were drilled on the shelf (Hollister et al 1972). The first deep Continental Offshore Stratigraphic Test (COST) well, No. B-2, was drilled in 1975 to a depth of 16,043 ft (4862 m) in the Baltimore Canyon trough, 154 km east of Atlantic City, New Jersey (Figure 1). This was the first in a series of four deep stratigraphic tests (two additional wells on Georges Bank and one in the Southeast Georgia embayment) that penetrated the major subshelf sedimentary basins (Figure 1), but only data from the COST B-2 are currently available (Smith et al 1976, Scholle 1977).

Meisburger & Field (1976) and Meisburger (1977) summarized the stratigraphy of the nearshore sediments off northern Florida and central North Carolina as deduced from more than 200 short vibrocores and 4 deeper borings (maximum depth 38 m).

In 1976, the U.S. Geological Survey undertook the Atlantic Margin Coring Project (AMCOR), the first systematic stratigraphic coring of the continental shelf from Georgia to Georges Bank (Hathaway et al 1976). This project resulted in the coring of 19 sites (Figure 1). Analyses of the JOIDES, ASP, and AMCOR cores are the primary geologic basis for constructing a stratigraphic framework for the Cenozoic of the U.S. Atlantic Margin (Poag et al 1977).

These geologic data coupled with an abundance of excellent geophysical data allow extrapolation of cored Cenozoic strata into undrilled areas. Seismic reflection and refraction profiles allow generalized stratigraphic interpretations of the deeply buried (undrilled) Mesozoic rocks and basement. The first offshore geophysical studies were published by Ewing and his associates (1937), whose work continued up through the 1950s (Drake, Ewing & Sutton 1959). These seismic records are supplemented by gravity and magnetics investigations, which are summarized by Mayhew (1974). A much more detailed discussion is presented by Emery & Uchupi (1972).

In 1973, the U.S. Geological Survey began collecting high-resolution and multi-channel seismic reflection profiles across the shelf and slope (see, for example, Schlee et al 1976). These data coupled with results from oil drilling on the adjacent Scotian Shelf (e.g. Jansa & Wade 1975, Williams 1975) are improving stratigraphic interpretations significantly.

STRATIGRAPHY OF MAJOR BASINS

Blake Plateau Trough and Environs

PHYSIOGRAPHY AND STRUCTURE   The area, between latitudes 26° N and 35° N, encompasses three major physiographic features: the continental shelf off Florida, Georgia, and South Carolina; the Florida-Hatteras slope; and the Blake Plateau (Figure 1). The major geologic structure under this area is the Blake Plateau trough, which underlies the Blake Plateau. It is an elongate north-south trending depression, bounded to the north by the Cape Fear arch, to the south by the Bahama uplift, and to the west by the Peninsular arch of Florida. A northwestward extension of the Blake Plateau trough beneath the shelf and the Georgia Coastal Plain is known as the Southeast Georgia embayment (Figure 1; Dillon et al 1978).

More than 12 km of principally Triassic, Jurassic, and Cretaceous sedimentary strata fill the Southeast Georgia embayment-Blake Plateau trough (Dillon et al 1975). Tertiary sediments are as thick as 800 m beneath the shelf, but are only about half that thick on the Blake Plateau. A thin Pleistocene-Holocene veneer is draped over the entire region, except for extensively eroded areas on the inner Blake Plateau. A diagrammatic cross section illustrates the general configuration and age of the beds (Figure 2).

256   POAG

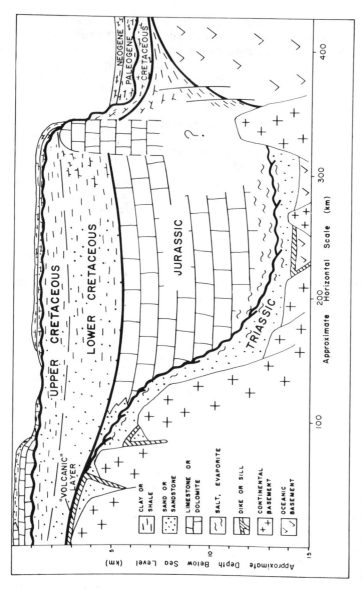

Figure 2   Schematic geologic cross section of the Southeast Georgia embayment and Blake Plateau trough. The Blake Plateau is the topographic surface between 85 and 330 km. Major seismic reflectors indicated by heavy black lines. General geometry of major rock units is approximately to scale, but the thicknesses *within* the Cenozoic section are not to scale. (This distortion is necessary to show the stratigraphic relationships.) Vertical exaggeration approximately 17:1 (composited from Dillon et al 1978, Sheridan 1976.)

BASEMENT   Basement rocks beneath the Coastal Plain are of igneous and metamorphic origin, and range in age from Precambrian to Devonian; some of the rocks presumably extend under the shelf and Blake Plateau. However, much of the Southeast Georgia embayment appears to be underlain by undeformed tuffaceous clastic rocks intermixed with basaltic and rhyolitic flows and ash (Dillon et al 1978).

TRIASSIC AND JURASSIC   The oldest sediments of the Blake Plateau trough are thought to be of Triassic age. Dillon et al (1978) have observed a strongly angular seismic unconformity that ranges from about 2 to 12 km below the sea surface under the Blake Plateau (Figure 2), below which are presumed Triassic continental deposits that fill faulted rift basins. Above this unconformity is a strong smooth seismic reflector, which corresponds to a horizon that also exhibits high refraction velocities (5.8 to 6.2 km sec^{-1}). It extends beneath the inner Blake Plateau and the continental shelf, between northern Florida and Cape Fear, North Carolina. This reflector may be correlated with basalt and rhyolite drilled in South Carolina and Georgia, which yielded a minimum radiometric age of Middle Cretaceous (Maher 1971, Gohn et al 1978). Dillon et al (1978) assigned the rocks of this "volcanic" layer to the Jurassic based on structural considerations (Figure 2). Northeast of Jacksonville, Florida, about 5 km of presumed Jurassic rocks lie between the "volcanic" layer and a seismic reflector that is correlated with a Tithonian (late Jurassic) event previously documented at DSDP Site 391 (Figure 1), seaward of the Blake Escarpment (Benson et al 1976). The bulk of the Jurassic rocks beneath the Blake Plateau is believed to be more than 7 km of dense carbonate deposits (interbedded limestones and dolomites, with anhydrite and salt layers; see Figure 2). This interpretation is based on extrapolation from the Great Isaac No. 1 well, which is located adjacent to the southern margin of the Blake Plateau trough (Tator & Hatfield 1975, Jacobs 1977, W. P. Dillon, personal communication 1977).

CRETACEOUS   Lower Cretaceous limestones are the oldest rocks that have been recovered from the Blake Plateau trough. Chalky calcilutite, oolitic calcarenite, and calcarenite containing abundant Foraminifera and algal fragments (Neocomian to Aptian age) have been recovered from the middle and base of the Blake Escarpment (Heezen & Sheridan 1966, Sheridan, Smith & Gardner 1969). These samples, in conjunction with geophysical evidence, indicate derivation from an extensive Lower Cretaceous (and possibly partly Upper Cretaceous) shelf-edge reef (Figure 2). Some sort of reef-bank system appears to have rimmed nearly the entire Gulf of Mexico–Atlantic Coast region, and extended as far north as the Grand Banks of Newfoundland (Bryant et al 1969, Emery & Uchupi 1972). Lower Cretaceous backreef and shelf strata have not yet been sampled, but geophysical evidence indicates their presence throughout the Blake Plateau trough and Southeast Georgia embayment. Just south of the trough (Great Isaac No. 1 well and south Florida; see Figure 1), more than 2000 m of limestone, dolomite, anhydrite, gypsum, and thin halite beds (upper part of Marquesas Supergroup of Meyerhoff & Hatten 1974) constitute the Lower Cretaceous rocks (Tator & Hatfield 1975, Jacobs 1977).

Upper Cretaceous sediments have been sampled in cores from the Blake

Escarpment and in AMCOR 6004. The Cretaceous/Tertiary contact in AMCOR 6004 is an unconformity with hard, grey, silty clay of Middle Paleocene age resting on slightly softer, grey, silty clay of Middle Maestrichtian age (late Cretaceous, equivalent to the Peedee Formation of the South Carolina Coastal Plain; see Figure 3). The erosional contact was recovered, and the bathyal microfaunas on either side indicate submarine erosion. Upper Cretaceous sediment from the Blake Escarpment is a deep-water calcilutite of Cenomanian Age.

Because AMCOR 6004 was drilled on seismic profile BT4 of Dillon et al (1978), the unconformity can be confidently traced under the shelf and Blake Plateau; this allows us to infer thick, flat-lying Cretaceous sediments throughout the trough. In fact, erosion on the northern part of the inner Plateau has exposed Cretaceous beds over wide areas of the sea floor (Uchupi 1970). The lithology of Upper Cretaceous rocks in the southern part of the trough may be extrapolated from the nearby Great Isaac No. 1 well. Tator & Hatfield (1975) reported approximately 1400 m of dolomitic carbonates and evaporites of the Pine Key and Card Sound Formations in the Upper Cretaceous interval. Upper Cretaceous rocks with increased terrigenous components are expected in the northern part of the trough, based on extrapolation of rocks beneath the Georgia Coastal Plain (Herrick & Vorhis 1963).

Uppermost Cretaceous sediments have not been firmly documented offshore or from the Atlantic Coastal Plain. [Olsson's 1960 and 1964 reports of this interval in New Jersey are in dispute (C. C. Smith, personal communication 1977)]. This lends some support to the concept of a global unconformity reported to have formed during this time (Vail, Mitchum & Todd 1975).

PALEOCENE   Tertiary rocks have been recovered in dredges and short cores from the Blake Plateau and Blake Escarpment, but most data come from boreholes: 2 U.S. Coast Guard test holes off Savannah, Georgia (McCollum & Herrick 1964), 6 core holes drilled by the JOIDES project (Bunce et al 1965), and 3 AMCOR core holes (Hathaway et al 1976). Tertiary rocks are no more than 1.5 km thick in the Southeast Georgia embayment, dip gently seaward from coastal outcrops (Emery & Uchupi 1972), and thin to about 0.5 km under the Blake Plateau. Most strata under the shelf are coarse grained and were deposited in shelf-like paleoenvironments. On the Blake Plateau, however, the entire Tertiary sequence is finer grained, rich in carbonates and contains bathyal microfossils. To the north of Cape Hatteras, it is diluted by terrigenous detritus from a more clastic regime.

Paleocene rocks are known from AMCOR 6004 and 6005, JOIDES 4, ASP 3 and 5, and from oceanographic cores on the Blake Plateau and Blake Escarpment (Emery & Uchupi 1972, Saito, Burckle & Hays 1974, Hathaway et al 1976, Schlee 1977). Paleocene strata range from shallow water calcareous clay and limestone (AMCOR 6005; inner continental shelf) to greenish gray, calcareous clay and shale containing abundant planktic and benthic microfossils of bathyal origin (ASP 3; Blake Nose). These strata reach more than 100 m in thickness.

EOCENE   Eocene rocks are highly calcareous like the Paleocene rocks, but are much thicker and more widely sampled offshore. The principal offshore sample

SERIES	FLORIDA	GEORGIA	SOUTH CAROLINA	NORTH CAROLINA	MARYLAND VIRGINIA	NEW JERSEY	LONG ISLAND	SAMPLED OFFSHORE
PLEISTOCENE	ANASTASIA MIAMI OOLITE KEY LARGO FORT THOMPSON UNIT A CALOOSA-HATCHEE		"PAMLICO" FLANNER BEACH WACCAMAW - CROATAN		KEMPS-VILLE NORFOLK		GARDINERS	
PLIOCENE	PINE CREST TAMIAMI		BEAR BLUFF DUPLIN		YORKTOWN	COHANSEY		
MIOCENE	HAWTHORN *			PUNGO RIVER	ST. MARYS CHOPTANK CALVERT *	KIRKWOOD		AMCOR 6002,6004 6007,6010 6011,6012 6016,6021 JOIDES 1,2 COST B-2
OLIGOCENE	SUWANNEE OCALA *	FLINT RIVER	Upper COOPER * Lower COOPER	SILVERDALE TRENT				
EOCENE	AVON PARK * TALLA-HASSEE LAKE CITY OLDSMAR	TWIGGS McBEAN	SANTEE Upper BLACK MINGO	CASTLE-HAYNE	NANJEMOY	SHARK RIVER MANASQUAN		JOIDES 1,2 AMCOR 6002
PALEOCENE	CEDAR KEYS		Lower * BLACK MINGO	BEAUFORT	MARLBORO AQUIA BRIGHT-SEAT	VINCEN-TOWN HORNERS-TOWN		AMCOR 6004 6005
UPPER CRETACEOUS	PINE KEY CARD SOUND	TUSCALOOSA	PEEDEE * BLACK CREEK MIDDENDORF CAPE FEAR		MONMOUTH GROUP * MATAWAN GROUP * MAGOTHY * RARITAN *			NANTUCKET MARTHA'S VINEYARD COST B-2 AMCOR 6004
LOWER CRETACEOUS	MARQUESAS SUPER GROUP				POTOMAC GROUP *			NANTUCKET COST B-2

Figure 3 Generalized stratigraphic chart of surface and subsurface formations of the US Atlantic Coastal Plain. For each series of rocks, the most widely used formation names are listed, but their precise lateral relationships are not accurately depicted. Additional unnamed rock sequences are known, but their presence is not indicated here. Asterisks indicate those formations whose lithologic and biostratigraphic equivalents have been sampled in rotary cores or drill cuttings from the continental shelf or slope (does not include dredgings and oceanographic cores).

points are AMCOR 6002, JOIDES 1–6, ASP 3, 5, and 20, several core holes near the South Carolina shoreline, and short cores and dredgings from the Blake Plateau. Beneath the continental shelf (JOIDES 1, 2; AMCOR 6002), the Upper and Middle Eocene rocks are shallow water, white-to-buff, massive, bioclastic limestone and dolomite and clayey sand. The total Eocene section here is as much as 500 m thick (according to seismic data). Larger benthic species dominate in nearly all the foraminiferal faunas and indicate deposition on a shallow carbonate shelf.

Gray calcareous clay was cored on the Florida-Hatteras slope (JOIDES 5), and light gray, silty, sandy calcilutite is prevalent beneath the Blake Plateau and on the Blake Escarpment (JOIDES 3, 4, 6; ASP 3, 5, 20), where the dominantly planktic microfossils suggest bathyal deposition. Maximum thickness here is about 70 m (in JOIDES 6).

OLIGOCENE   Oligocene strata are also highly calcarous shallow-water deposits, but are less extensive and much thinner than the Eocene beds. Principal sample points are JOIDES 1–6, AMCOR 6002, 6004, ASP 5, and dredges and short cores from the Blake Plateau and Escarpment.

Under the continental shelf and Blake Plateau, Oligocene strata are as much as 73 m thick (AMCOR 6002, JOIDES 3) and consist of olive-green to gray calcilutite and silty clay. Abundant benthic and planktic Foraminifera suggest deposition on the outer shelf and slope.

MIOCENE   Miocene sediments have been sampled in several 5-m vibrocores and 4 deeper borings (maximum penetration 38 m) just offshore from northern Florida (Meisburger & Field 1976) in AMCOR 6002 and 6004, JOIDES 1–4, and ASP 3, 5, and 20.

Miocene strata beneath the inner shelf (e.g. JOIDES 1) differ markedly from the carbonate-rich Paleogene sediments. Terrigenous quartzose sand and sandy and clayey silt predominate, with light to blackish-brown sand-sized phosphatic pellets sometimes constituting as much as 95% of the total sediment, especially in Lower Miocene beds. Much of the Miocene bioclastic debris in JOIDES 1 (79 m thick) has also been altered to phosphate.

Lower Miocene microfossils in JOIDES 1 are largely diatoms and radiolarians as compared to mainly Foraminifera and calcareous nannoplankton in the Paleogene. This assemblage is partly the result of diagenetic leaching and phosphatization of the calcareous fauna, but also it reflects the drastically different environment of deposition. Beneath the outer shelf (JOIDES 2), Miocene sediments are largely quartzose sandy clay about 75 m thick; phosphatic grains are still common in the lower 24 m.

Beneath the Florida-Hatteras slope and Blake Plateau, Miocene sediment becomes olive-gray to pale tan clay and foraminiferal calcarenite and calcilutite, reaching a maximum penetrated thickness of 130 m (ASP 5). The abundance of phosphatic grains is reduced compared to the shelf section, and these sediments compose the sea floor over wide stretches of the Slope.

PLIOCENE   Pliocene strata have been identified in nearshore cores off northern Florida (Meisburger & Field 1976) in AMCOR 6004, in ASP 5, and in short cores from the Blake Plateau. Pliocene strata thicken from about 6 m just off the northern Florida coast to about 15 m on the continental slope. The lithology changes from quartzose sand near shore to light olive-gray calcarenite on the slope.

Beneath the inner Blake Plateau, Pliocene beds are missing in extensive patches, but they have been documented on the outer plateau by piston cores. Kaneps (1970) described Pliocene strata there as unusually homogeneous, tan, foraminiferal calcilutites about 8 m thick.

PLEISTOCENE   Pleistocene sediments appear to be present at nearly every sampling location, but they have been lumped together with Pliocene and Holocene strata as undifferentiated post-Miocene beds in the JOIDES core holes.

On the inner shelf, Pleistocene strata are missing at several sites but, where sampled, consist of locally channeled quartzose sand of about 6-m maximum thickness (Meisburger & Field 1976). This sand extends to the shelf edge (JOIDES 2), where 49 m of post-Miocene quartzose sand and gravel (grains as large as 10–40 cm in diameter) were cored.

On the Florida-Hatteras slope, Pleistocene sediments change to yellowish or olive-gray calcarenite made up of foraminiferal and molluscan fragments with lesser amounts of silty clay (maximum thickness 15 m at ASP 5). Seismic profiles indicate that the thickest Pleistocene sediments are on the Florida-Hatteras slope.

Where present on the Blake Plateau and Escarpment, Pleistocene sediments are bathyally derived grayish-yellow or brownish foraminiferal calcarenite and calcilutite, 18 m thick or less where penetrated. Pleistocene beds are missing in extensive patches on the inner plateau and on the escarpment.

HOLOCENE   Near the Florida shoreline, fine to coarse-grained, shelly, phosphoritic, Holocene quartz sand is present, apparently derived from reworking of Tertiary strata. It is generally only a meter or so thick and is locally discontinuous (Meisburger & Field 1976).

With few exceptions, modern deposition does not occur on the middle and outer shelf (Milliman, Pilkey & Ross 1972), but, where present, Holocene surface sediments of the inner Blake Plateau are sands, similar to those of the slope north of Florida. The most striking sedimentary feature is an area of phosphorite and manganese (nodules and pavements) covering approximately 5000 km^2 of the inner plateau off South Carolina and Georgia. Elsewhere on the plateau, Holocene sediments are almost entirely calcareous, composed of coral debris in areas of irregular topography, and of foraminiferal calcarenite in smooth areas.

Baltimore Canyon Trough and Environs

PHYSIOGRAPHY AND STRUCTURE   This area, between latitudes 35° N and 40° N, extends eastward to about 72° W longitude, encompassing the continental shelf and slope from North Carolina to New Jersey (Figure 1). Geophysical data show that the shelf and slope are the physiographic expression of a seaward-thickening wedge of Mesozoic and Cenozoic sediment overlying crystalline basement. The basement is warped and faulted to form a series of basins and arches, the dominant feature of which is the Baltimore Canyon trough. The trough extends subparallel to the shoreline for more than 600 km between Long Island and Cape Hatteras; it is widest (200 km) off New Jersey, where its maximum sediment fill also occurs (> 15 km thick).

The seaward border of the Baltimore Canyon trough is a ridge of igneous basement or Mesozoic sedimentary rocks that underlies the present upper continental slope. The trough's southern margin is the Cape Fear arch, and its northern border is the Long Island platform. Beneath the Coastal Plain of eastern Virginia, Maryland, Delaware, and most of New Jersey, a westward extension of the trough is known as the Salisbury embayment (Figure 1). Another coastal embayment, the Albemarle embayment, plunges eastward beneath the North Carolina Coastal Plain and Cape Hatteras, and joins the southern end of the Baltimore Canyon trough. A third shallow embayment underlies the northern New Jersey Coastal Plain (Raritan embayment).

A diagrammatic cross section from coastal New Jersey through the deep COST B-2 well illustrates the stratigraphic framework of the Baltimore Canyon trough–Salisbury embayment and environs (Figure 4).

BASEMENT, TRIASSIC, AND JURASSIC   By analogy with basement rocks of the Coastal Plain, the basement offshore is inferred to be of Paleozoic and Precambrian (?) metamorphic rocks. Down-faulted basement grabens filled with Triassic arkosic sandstone, shale, basaltic lava flows, and diabase intrusions are recognized from offshore geophysical profiles and from outcrops and wells along the Coastal Plain.

The basal part of the sedimentary section may also include Paleozoic and marine Triassic rocks. Seismic velocities in the strata below about 6 km exceed 5 km sec^{-1}, which suggests that this is a thick section of carbonate rocks. By analogy with well data from the Scotian Shelf, these are interpreted as Jurassic carbonates and evaporites, overlain by a kilometer of Jurassic sands and shales (Figure 4; Scholle 1977; Schlee et al 1977). More than half a kilometer of Jurassic sandstones and shales was penetrated by the COST B-2 well (E. I. Robbins, personal communication 1977).

CRETACEOUS   Approximately 2 km of largely nonmarine Lower Cretaceous sandstone and shale with interbedded lignite and coal beds was penetrated by the COST B-2 well. The sandstones are fine, medium, and coarse grained; the shales are red, green, and gray.

Seaward, however, beneath the shelf edge and upper slope, seismic velocities increase to more than 4 km sec^{-1} in what appear to be Upper Jurassic and Lower Cretaceous beds. The increase of seismic velocity is interpreted to indicate limestone or dolomite that may be part of the shelf-fringing reef complex (discussed earlier) that rimmed much of the east coast during the late Jurassic and early Cretaceous.

Upper Cretaceous strata in the COST B-2 well are more marine; nearly 1000 m

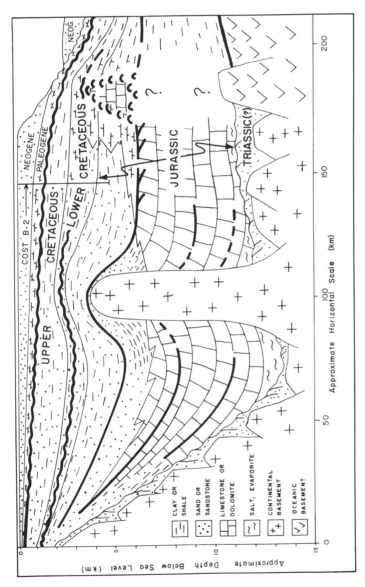

Figure 4 Schematic geological cross section of the Baltimore Canyon trough. Major seismic reflectors are indicated by heavy black lines. General geometry of major rock units is approximately to scale, but thicknesses *within* the Cenozoic strata are not to scale. (This distortion is necessary to show the stratigraphic relationships.) A large igneous body locally called the "Great Stone Dome" intrudes Jurassic sediments under the shelf. Vertical exaggeration approximately 8 : 1 (composited from Scholle 1977, Schlee et al 1976, Sheridan 1976.)

of dominantly shallow marine sandstone, limestone, and shale was penetrated. Seaward from COST B-2, there is no major seismic velocity change in the Upper Cretaceous beds, which extend with an increased dip under the slope. The slope sequence of Upper Cretaceous mudstone and foraminiferal ooze appears to have been deposited in bathyal conditions, as sampled in dredge hauls from the submarine canyons off New Jersey.

PALEOCENE   Paleocene rocks from beneath the shelf have been sampled only in the COST B-2 well. Approximately 10 m of relatively dense buff limestone contains microfossil assemblages deposited on the middle to outer continental shelf.

EOCENE   Eocene strata have been sampled in COST B-2, AMCOR 6011, ASP 15, 22, and several dredges and cores in Hudson Canyon and on the shelf and slope off North Carolina.

Beneath the inner shelf of New Jersey AMCOR 6011 penetrated approximately 20 m of blackish to gray-green silty clay and fine, silty, glauconitic sand of mid to early (?) Eocene age. The calcareous Foraminifera, which are mainly benthic species, indicate middle-shelf conditions during deposition. Eocene strata also are present near the sea floor on the inner North Carolina shelf. The beds there contain a bryozoan hash and shallow-water benthic Foraminifera (Meisburger 1977).

Near the axis of the Baltimore Canyon trough (COST B-2 well), Eocene rocks are 270 m thick and appear to represent the entire Eocene. They are composed of impermeable, buff, calcareous shale in the upper third, white chalky limestone in the middle third, and buff shale with interbedded cream limestone in the lower third. These rocks are rich in planktic micro- and nannofossils that accumulated in bathyal conditions, and they represent the maximum marine transgression known in the Baltimore Canyon trough area.

On the continental slope ASP 15, 22, DSDP 108, and Hudson Canyon dredgings have encountered middle Eocene, pale yellow-gray, siliceous calcilutite, which contains rich radiolarian assemblages in addition to abundant planktic Foraminifera and calcareous nannoplankton (bathyal deposition).

OLIGOCENE   Oligocene rocks are best known from the COST B-2 well, which penetrated approximately 150 m of Upper and Lower Oligocene strata, composed largely of buff to light gray calcareous shale and claystone, with impermeable limestone near the top of the section.

Between Cape Fear and Cape Lookout, North Carolina, shallow cores and seismic reflection profiles reveal a homogeneous fine sand, up to 80 m thick, which contains shallow-water benthic Foraminifera of Oligocene age (Meisburger 1977). The deepest-water Oligocene locality is on the Hatteras Slope, where calcareous glauconitic siltstone, sandstone, and tan chalk have been dredged.

MIOCENE   Miocene strata are known in this area in considerably more detail than the older rocks. They were penetrated by AMCOR 6007, 6009, 6010, 6011, 6012, and 6016; ASP 10, 14, 15, and 22; and COST B-2.

Near the New Jersey shore AMCOR 6011 drilled approximately 172 m of largely shelly, gray to grayish-brown, fine, medium, and coarse sand, with interspersed beds of gray clay and grayish-brown silty clay. Microfossils in this section are largely diatoms; radiolarians are also common, but calcareous nannoplankton and Foraminifera are few. These strata accumulated in shallow marginal marine conditions.

In the Baltimore Canyon trough (COST B-2), Miocene rocks thicken dramatically. More than 800 m of unconsolidated to semi-consolidated gray sand, gravel, and clay were penetrated. Diatoms are again prominent, and Foraminifera are largely benthic species indicative of deltaic or near-deltaic conditions. Thick sequences of foreset bedding seen on geophysical profiles support the interpretation that the biota are deltaic. Along the outer shelf from Virginia to Massachusetts AMCOR 6007, 6009, 6010, and 6012 encountered Upper (?) Miocene sediments of similar lithology, microfossils, and derivation (maximum thickness about 130 m in 6009).

On the New Jersey slope (ASP 14, 15), diatomaceous sediments probably of Miocene age are composed of dark olive-gray to yellowish brown glauconitic sand and silty clay (possibly as thick as 83 m in ASP 14). Outer shelf conditions of deposition are inferred for these strata.

In the Norfolk Canyon area off Virginia (ASP 10, 22), Middle Miocene diatomaceous beds were cored. The thickest section (110 m at ASP 22) is composed of olive-gray to yellowish brown clay, sandy clay, and clayey sand, without conspicuous glauconite except in the lower few meters. The diatoms are benthic and planktic marine species and suggest inner-shelf deposition. Slightly upslope, at ASP 10, the sediments and fossils are similar, but the recovered section is slightly thinner; it also contains more abundant freshwater diatoms, a factor suggesting that fluvial-marine deposition occurred nearby.

PLIOCENE   Pliocene sediments in this region are known in AMCOR 6007, 6009, and 6010; in ASP 7, 10, and 22, and in scattered piston cores. Beneath the inner New Jersey shelf, about 65 m of gray-to-white sand and silty clay is barren of fossils; some of it may be Pliocene. Beneath the middle and outer New Jersey shelf (AMCOR 6010 and COST B-2), a 66-m interval of similar strata contains sparse inner-shelf benthic Foraminifera that suggest Pliocene age.

To the south, AMCOR 6007 and 6009 and ASP 22 encountered gray clay and silty clay containing marine and freshwater diatoms and shallow shelf assemblages of Pliocene Foraminifera.

On the Hatteras Slope (ASP 7) and in Hudson Canyon, olive-gray, highly calcareous, clayey sand and sandy clay contains abundant Upper Pliocene planktic Foraminifera, and calcareous nannoplankton.

PLEISTOCENE   Pleistocene sediments have been recovered in almost every core that has been taken in this area. Beneath the inner shelf (AMCOR 6008 and 6011), up to 120 m of unconsolidated gravel and gray, shelly-to-barren sand and silty clay have been penetrated. These strata constitute marginal marine, lagoonal, and probably fluvial channel-fill deposits. Beneath the outer shelf, Pleistocene sediments thicken and change facies. At AMCOR 6007, 6009, and 6010, Pleistocene beds are dark gray to olive-gray clayey and silty sand (often shelly) and silty clay. The maximum thickness is around 170 m at AMCOR 6010. Complex vertical facies are formed of alternating lagoonal, inner-shelf, and middle-to-outer shelf strata; glacial/interglacial cycles have not yet been clearly defined.

The COST B-2 well penetrated nearly 250 m of white-to-gray, unconsolidated, shelly-to-barren gravel, sand, and clay of Pleistocene age. The vertical facies changes were somewhat masked by the drilling methods, but shallow marine to nonmarine strata appear to dominate.

On the continental slope, surprisingly thick Pleistocene sequences of gray to dark gray, gassy, silty, and sandy clays were found in AMCOR 6012 and 6021 and in ASP 7 and 23. Thicknesses range from around 285 m in AMCOR 6012 and ASP 7 to more than 300 m in AMCOR 6021 and ASP 23, where pre-Pleistocene strata were not reached. Planktic Foraminifera of temperate water origin are scattered throughout the strata, and some intervals are rich in diatoms, middle-to-outer-shelf benthic Foraminifera, and dark organic particles. This association suggests deposition in anoxic depressions on the shelf. With the exception of a thin interval of slope deposits in the bottom of AMCOR 6021, all Pleistocene strata sampled in AMCOR and ASP cores were derived from the shelf.

More than 30 piston cores from the Hudson Canyon area (slope and rise) were studied by Ericson et al (1961), who recognized among them an abundance of Pleistocene turbidites. High-resolution seismic profiles indicate that Pleistocene turbidites and submarine slumps are common all along the incised continental slope of this region.

HOLOCENE   Emery & Uchupi (1972) have concluded that about 73% of the surface sediment on the middle and outer Atlantic Shelf is relict (early Holocene). The strata are largely coarse yellow-brown sand (iron-stained quartz and feldspar grains), often containing peat, remains of land vertebrates, and oyster shells. The only area on the shelf that seems to be accumulating modern sediments is a narrow strip (a few kilometers wide) next to the shoreline, where gray-to-white sand appears to be prograding over the relict iron-stained beds.

Modern sediments on the continental slope are mainly green silty clay and clayey silts, whose carbonate content (mainly planktic Foraminifera) increases with depth, and which are subject to frequent mass movement downslope.

Georges Bank Basin and Environs

PHYSIOGRAPHY AND STRUCTURE   This area, bounded by latitudes 74° W and 66° W and longitudes 40° to 43° N, encompasses the continental shelf and slope off Long Island and New England (including Georges Bank; see Figure 1). The major physiographic feature here is Georges Bank, a broad, shallow, topographic high, 150 km wide, extending about 350 km east-southeast from Cape Cod, Massachusetts.

Multichannel seismic profiles across Georges Bank reveal a thick sedimentary wedge (more than 8 km) occupying a broad northeast-southwest trending depression over irregularly flexed and faulted basement rocks. The southern edge of the basin

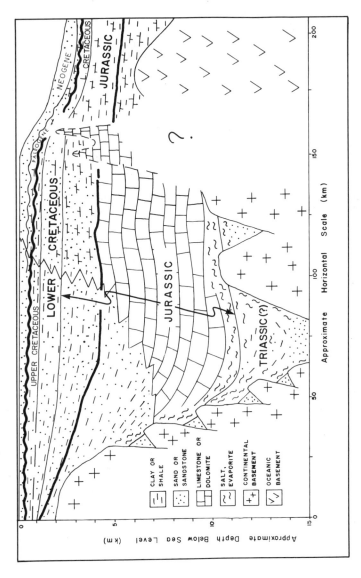

Figure 5 Schematic geologic cross section of the Georges Bank basin. Major seismic reflectors are indicated by heavy black lines. General geometry of major rock units is approximately to scale, but thicknesses *within* the Cenozoic strata are not to scale. (This distortion is necessary to show the stratigraphic relationships.) Vertical exaggeration approximately 8:1 (composited from Schlee et al 1976, 1977, Sheridan 1976.)

is underlain by a faint irregular ridge that is double-humped (about 20 km wide) along some seismic reflection profiles (Figure 5). Under the shelf, the ridge is about 3 km below the sea surface, but it rises to within 1 km under the continental slope. The northeastern end of the basin is bounded by a block-faulted structural high known as the LaHave platform (Figure 1; some authors have called the southern tip of the LaHave platform the "Yarmouth arch"). A part of this platform appears to extend into Georges Bank basin, dividing the northeastern end into two sub-basins. The southwestern boundary of Georges Bank basin is another block-faulted high—the Long Island platform. Figure 5 shows a diagrammatic cross section of the Georges Bank basin.

BASEMENT   Basement rocks have not been penetrated in Georges Bank basin. Schultz & Grover (1974) assumed, however, that the rocks would be similar to basement beneath the Scotian basin where Cambro-Ordovican slate, quartzite, argillite, schist, and gneiss and Devonian granite have been encountered.

TRIASSIC (?)   The nature of pre-Tertiary rocks filling the Georges Bank basin is inferred almost entirely from geophysical data and by extrapolation of rock units drilled on the Scotian Shelf, for as yet, no data have been released from two deep stratigraphic tests drilled in 1976–1977 on Georges Bank (COST G-1 and G-2). Schlee et al (1976) interpreted a gently undulating major seismic reflector around 8 km below sea level to be the top of Triassic or pre-Triassic rocks. More recent interpretations of additional seismic profiles place the Triassic (?) top near 10 km (J. Schlee, personal communication 1977). The presumed Triassic/pre-Triassic rocks are thought to be basaltic volcanics, evaporites, and continental sandstones and shales similar to those filling the Triassic grabens of the northeastern United States and the Bay of Fundy (Klein 1962, Sanders 1963).

JURASSIC   Presumed Jurassic rocks thicken from less than 1 km on the northern edge of the basin to more than 8 km on the southern margin. High acoustic velocities within these rocks are probably produced by dense carbonates and shales.

CRETACEOUS   Inferred Cretaceous rocks also thicken southward across the basin from less than 1 km to nearly 2 km. Acoustic reflectors are anastomosing and discontinuous to the north but become continuous and regular southward. This probably is associated with facies change from shallow-water marine shale and terrestrial sandstone in the north to more evenly bedded marine shale in the south (Schlee et al 1976). The lower rocks of this unit appear to intersect the ridge that forms the southern edge of the basin. Schlee et al (1976) tentatively suggested that this ridge is a reef or structureless carbonate mass that could be part of the same system of Upper Jurassic-Lower Cretaceous reefs bordering the Blake Plateau and Baltimore Canyon trough.

Rocks of late Cretaceous age have been recovered in dredges and cores from Alvin, Block, Oceanographer, Lydonia, and Veatch Canyons. In addition, ASP core holes 17 and 18 penetrated around 400 m of late Cretaceous sediments in Veatch Canyon (Coniacian? through Maestrichtian). The Coniacian-Santonian section is around 300 m thick, consisting largely of dark to light olive-gray, calcareous to noncalcareous clay and claystone with occasional silty and sandy layers. The remaining 100 m of Campanian and Maestrichtian strata is principally dark olive-gray to brownish gray, laminated and massive silt, with minor amounts or dark olive-gray sand and medium-gray clay. Microfossils and nannofossils indicate bathyal deposition.

Two U.S. Geological Survey core holes on Martha's Vineyard and Nantucket encountered Upper and Lower (?) Cretaceous sediments (maximum thickness of 350 m beneath Nantucket). These beds are principally unconsolidated, light gray, fine-to-coarse clayey sand with interbedded light gray, red, and yellow, micaceous, silty, and sandy clay of terrestrial origin.

PALEOCENE   Seismic profiles show presumed Tertiary strata containing flat-lying, weak, discontinuous reflectors; the beds thicken southward across the basin from less than 0.5 km to about 1 km.

Paleocene rocks are known from Nantucket Island and Martha's Vineyard, and are composed of clayey silt containing a few planktic and benthic Foraminifera of outer shelf origin. A dredge sample from Oceanographer Canyon contains bathyal brown silty clay, with abundant planktic Foraminifera.

EOCENE   Eocene strata have been sampled in or near Block Canyon, Alvin Canyon, and Veatch Canyon, from Fippennies Ledge in the Gulf of Maine, in AMCOR 6019 on northern Georges Bank, and from the core holes beneath Nantucket and Martha's Vineyard.

Beneath Nantucket and Martha's Vineyard, Eocene rocks are sparsely fossiliferous clayey silt and greensand, no more than 20 m thick. However, only a part of the Eocene is present here (Lower Eocene ?). A shallow mid-shelf environment is postulated.

On the northern edge of Georges Bank, AMCOR 6019 bottomed in about 15 m of light to dark green, glauconitic clay, and hard gray limestone of Middle Eocene age. The rich foraminiferal and calcareous nannofossil assemblages indicate deposition on the outer shelf and upper slope.

Lower, Middle, and Upper Eocene strata have been sampled in the canyons of the continental slope. The sediments include greenish gray silty clay, yellow clay and chalk, and white chalk with rich radiolarian and diatom assemblages along with bathyal Foraminifera and calcareous nannofossils.

At the northernmost locality, Fippennies Ledge, pebbles and cobbles of hard, light gray, fossiliferous, calcareous, opaline porcellanite of mid to late Eocene age have been dredged (Schlee & Cheetham 1967).

OLIGOCENE   The only Oligocene rocks sampled to date come from Oceanographer Canyon. Two samples—one a buff calcareous clay, the other a brown glauconitic sandy clay—contain planktic Foraminifera.

MIOCENE   Miocene rocks have been recovered from a well on Nantucket Island, AMCOR 6016, Lydonia, Double, and Hydrographer Canyons, and four dredge stations on the northern edge of Georges Bank.

At least 9 m of clayey silt in a core hole on Nantucket (USGS 6001) contain benthic Foraminifera of Miocene-Pliocene age deposited in a middle-shelf environment. The basal section of AMCOR 6016 contains about 10 m of stiff, dark gray, micaceous, clayey silt with mid-shelf benthic Foraminifera and a few planktic Foraminifera. On the northern edge of Georges Bank, dredged blocks of indurated pebble sandstone yielded mainly Miocene molluscan remains, but a few inner-shelf benthic Foraminifera also were recovered.

The submarine canyon samples are brown silty clay, green clayey silt, and green sandstone containing sparse benthic and planktic Foraminifera probably of shelf origin (Gibson, Hazel & Mello 1968).

PLIOCENE   Definite Pliocene sediments are unknown in this region, and although a quartzose greensand recovered from Lydonia Canyon has been reported to be of Pliocene age (Stetson 1949), confirmation is needed.

PLEISTOCENE   Detailed high-resolution seismic profiles across Georges Bank (Lewis & Sylwester 1977) reveal an eastern and a western wedge of inferred Pleistocene strata lying on either side of a Tertiary "mid-bank divide." The western wedge thickens westward to at least 175 m. Seismic reflectors within it appear to represent three sets of truncated foreset beds. The eastern wedge resembles a massive deltaic sequence that progrades eastward and thickens to at least 200 m. The youngest sedimentary unit is generally 20–50 m thick and is devoid of internal reflectors. It overlies a nearly horizontal unconformity that truncates all pre-existing bank structures and is thought to be of late Pleistocene age.

The most important Pleistocene sample locations are AMCOR 6012–6019 and ASP 18. AMCOR 6014, 6016, and 6018, drilled on top of Georges Bank, penetrated the western wedge of Pleistocene sediments. To the north, 6016 encountered 60 m of gray sand and gravel, and in the south, 6014 drilled 102 m of shelly, glauconitic, olive or gray, silty sand and clay. In between 6018 sampled 40 m of coarse shelly sand containing a few richly organic layers. These strata were deposited in inner shelf, lagoonal, near-delta, and possibly fluvial environments.

On the northern flank of the Bank (6019), sediments are finer grained. Fifty-five m were cored, which consisted of dark, olive-gray, sandy or silty clay with a few gravel layers. The upper 18 m is rich in organic detritus and diatoms. Lithoclasts derived from underlying Eocene rocks are common in the lower beds, which were deposited on the inner and middle shelf.

At AMCOR 6017, in the Franklin Basin just north of Georges Bank (Figure 1), 90 m of soft, dark olive-gray, sandy, and silty clay with scattered large pebbles (5 cm in diameter) compose the Pleistocene section. Miocene and, especially, Eocene lithoclasts and microfossils are abundantly reworked into the inner and middle shelf Pleistocene at this locality.

On the continental slope south of the Bank (AMCOR 6013 and 6015), the Pleistocene section contains at least 305 m of dark gray or olive, glauconitic, silty sand and silty clay with diatom-rich intervals. Most of these beds also originated on the inner and middle shelf.

HOLOCENE   Little modern sediment is accumulating except along the shoreline, especially in estuaries and near river mouths and stream channels. Surficial sediments of this region are largely sand, gravel, and boulders of glacial origin, winnowed during the last post-glacial rise in sea level.

RELATIONSHIPS TO SCOTIAN SHELF AND GRAND BANKS

Several investigators (McIver 1972, King 1975, Jansa & Wade 1975, Gradstein et al 1975, King & MacLean 1976, Given 1977) have shown that the sedimentary rocks offshore from Nova Scotia and Newfoundland are largely a continuation of the Mesozoic and Cenozoic succession underlying the Atlantic Margin of the United States. Jansa & Wade (1975) have shown that the Scotian basin is a basement depression beneath the eastern Scotian Shelf, the Laurentian Channel, the western Grand Banks, and the adjacent continental slope and rise (Figure 1). This shift in locus of deposition to the rise is a major difference from the U. S. basins, which are largely confined to the shelf and slope

Geological data from more than 100 deep exploratory wells, coupled with extensive seismic reflection profiling, have revealed a sequence of mappable sub-surface formations (Figure 6), whose stratigraphic framework and terminology were originally set up by McIver (1972) and Amoco Canada Petroleum Co., Ltd. and Imperial Oil Ltd. (1974), and were modified by Jansa & Wade (1975), Williams (1975), Parsons (1975), and Given (1977). Figure 7 shows a diagrammatic cross section of the Scotian basin, and Figure 8 shows the relationships between the U.S. and Canadian margins.

In general, Mesozoic and Cenozoic rocks of the Scotian basin have thicknesses, lithologies, and depositional environments similar to their presumed equivalents farther south, but more details and older strata are known as a result of the numerous wells.

Some investigators have concluded that the Triassic through Pleistocene sections of the Scotian basin are free of regional unconformities (Jansa & Wade 1975), although numerous local diastems have been recognized (King, MacLean & Fader 1973). In contrast, several regional unconformities have been detected off the U.S. Margin, although much more drilling is needed to determine their precise nature. Perhaps the most significant difference in the known stratigraphic sequences of the two areas is the proven presence of thick Jurassic salt that forms numerous diapirs in the Scotian basin. Some authors expect deep drilling in Georges Bank basin to reveal a southward extension of the salt (Schultz & Grover 1974), and recently obtained multichannel seismic reflection profiles (Schlee et al 1977) have revealed a few diapirs of possible salt composition in the eastern part of the basin. Also, Grow & Markl (1977) and Grow, Dillon & Sheridan (1977) have observed 8 diapirs (salt ?) in the Baltimore Canyon trough off New Jersey and under the continental slope off North Carolina.

More direct evidence of deep-seated salt beds in the Baltimore Canyon trough and Georges Bank basin comes from measurements of interstitial salinity in ASP 15

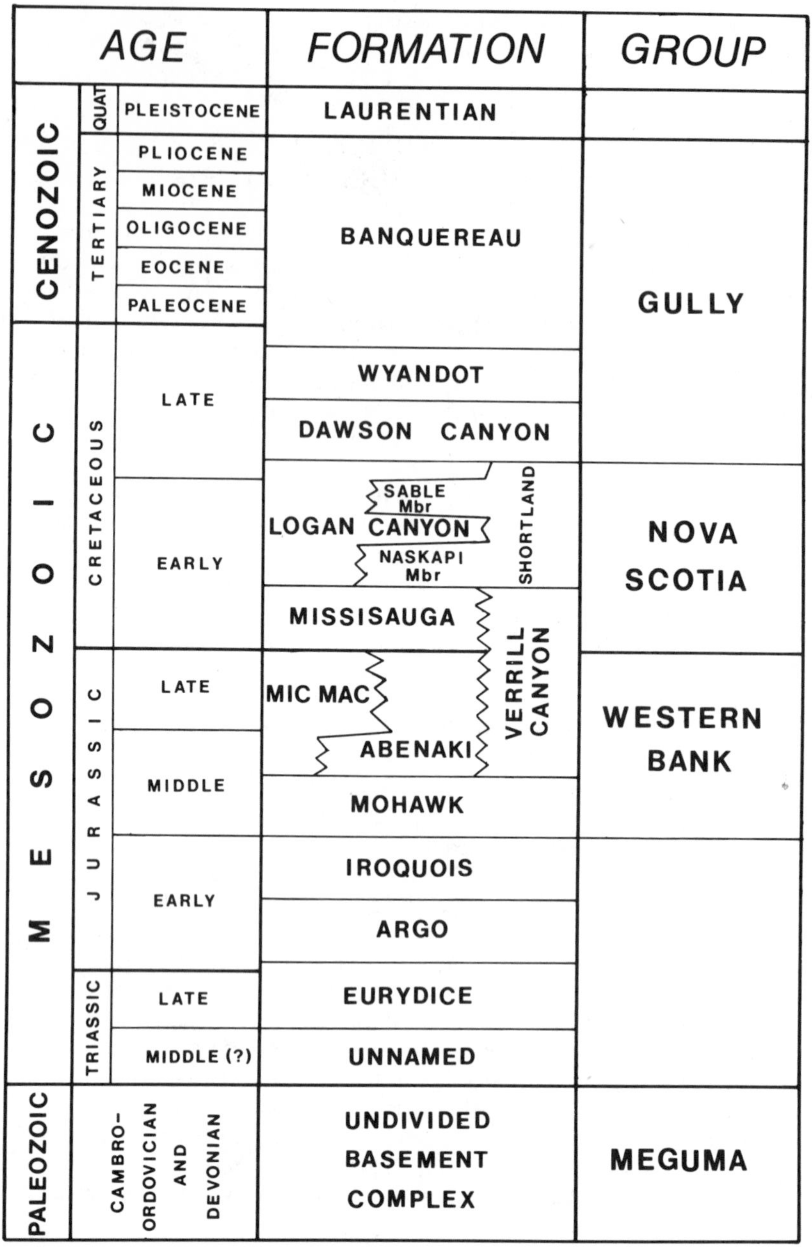

AGE			FORMATION		GROUP
CENOZOIC	QUAT	PLEISTOCENE	LAURENTIAN		
CENOZOIC	TERTIARY	PLIOCENE	BANQUEREAU		GULLY
CENOZOIC	TERTIARY	MIOCENE	BANQUEREAU		GULLY
CENOZOIC	TERTIARY	OLIGOCENE	BANQUEREAU		GULLY
CENOZOIC	TERTIARY	EOCENE	BANQUEREAU		GULLY
CENOZOIC	TERTIARY	PALEOCENE	BANQUEREAU		GULLY
MESOZOIC	CRETACEOUS	LATE	WYANDOT		GULLY
MESOZOIC	CRETACEOUS	LATE	DAWSON CANYON		GULLY
MESOZOIC	CRETACEOUS	EARLY	LOGAN CANYON — SABLE Mbr / NASKAPI Mbr	SHORTLAND	NOVA SCOTIA
MESOZOIC	CRETACEOUS	EARLY	MISSISAUGA	VERRILL CANYON	NOVA SCOTIA
MESOZOIC	JURASSIC	LATE	MIC MAC	VERRILL CANYON	WESTERN BANK
MESOZOIC	JURASSIC	LATE	ABENAKI	VERRILL CANYON	WESTERN BANK
MESOZOIC	JURASSIC	MIDDLE	MOHAWK		WESTERN BANK
MESOZOIC	JURASSIC	EARLY	IROQUOIS		WESTERN BANK
MESOZOIC	JURASSIC	EARLY	ARGO		WESTERN BANK
MESOZOIC	TRIASSIC	LATE	EURYDICE		WESTERN BANK
MESOZOIC	TRIASSIC	MIDDLE (?)	UNNAMED		WESTERN BANK
PALEOZOIC	CAMBRO-ORDOVICIAN AND DEVONIAN		UNDIVIDED BASEMENT COMPLEX		MEGUMA

Figure 6 Stratigraphic chart of formations recognized in the Scotian basin. (After McIver 1972, Jansa & Wade 1975, Given 1977.)

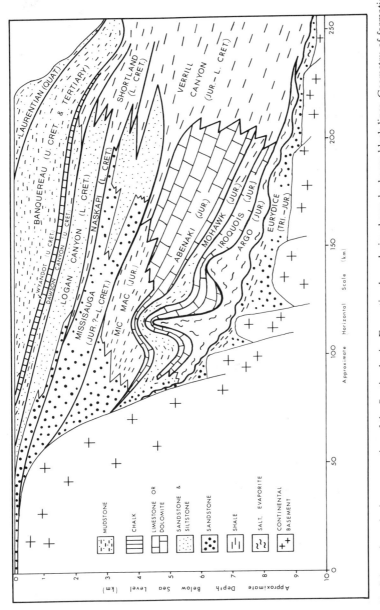

Figure 7 Schematic geologic cross section of the Scotian basin. Formation boundaries shown by heavy black lines. Geometry of formations is approximately to scale. Vertical exaggeration approximately 15:1 (composited from Sheridan 1974, Jansa & Wade 1975, Parsons 1975, Given 1977.)

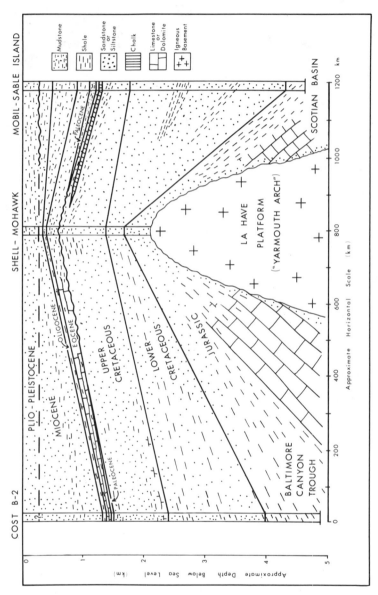

Figure 8 Schematic geologic cross section showing stratigraphic relationships of drilled strata between Baltimore Canyon trough and Scotian basin. Major series boundaries shown by heavy black lines. Approximately to scale. Vertical exaggeration approximately 174:1. (After Jansa & Wade 1975, Williams 1975, Smith et al 1976.)

and 17 (Figure 1). Manheim & Hall (1976) concluded that the increase in salinity toward the bottom of these core holes (as high as $53\,\mathrm{g\,kg^{-1}}$) represents diffusion from buried Jurassic evaporites.

SUMMARY AND CONCLUSIONS

The Atlantic Margin of the United States is underlain by a series of elongate sedimentary basins and coastward embayments that are filled with as much as 15 km of Mesozoic and Cenozoic rocks. The older rocks (Triassic and Jurassic) are almost completely unsampled, their characteristics having been interpreted from seismic reflection and refraction profiles and extrapolation from deep wells nearby. The oldest strata are believed to be Triassic continental redbeds that overlie downfaulted crustal blocks formed during the initial separation of North America and Africa. Salt and evaporites may have accumulated in all the U.S. Atlantic Margin basins during the early Jurassic (e.g. diapirs on seismic profiles in Georges Bank basin, Baltimore Canyon trough, and beneath North Carolina slope), but the only firm geologic evidence of this comes from wells on the adjacent Bahamas uplift (southern edge of Blake Plateau trough) and Scotian basin and from high interstitial salinities in two offshore core holes. Upper Jurassic and Lower Cretaceous rocks are largely thick, nonmarine sands and silty shales under the inner parts of the present shelves, but there are probably thick marine carbonates in the seaward parts of the basins. At the Upper Jurassic–Lower Cretaceous shelf edge, dense limestones formed a discontinuous series of sediment dams (reefs and carbonate platforms) that spanned much of the coastline. During the Middle and Upper Cretaceous, increasingly marine calcareous sands and shales filled the backreef basins, eventually burying the reefs and spilling across them into the oceanic basin. Faunal evidence points to a wide latitudinal distribution of nearly homogeneous microfaunas during this time.

Marine conditions continued during the early Tertiary. Highly calcareous sediments of Paleocene and Eocene age are known the length of the margin, but latitudinally distinctive lithofacies and biofacies began to develop. Thick dolomitic limestone characterized by abundant larger benthic Foraminifera occupy the Southeast Georgia embayment, but give way to smaller types in the more clastic environments of the northerly basins. The thick limestones of the southern shelf change seaward to thin calcilutites of bathyal origin beneath the continental slope and Blake Plateau. Rich radiolarian faunas dilute the biogenic carbonate in Eocene rocks all along the lower Continental Slope.

Oligocene beds are also widespread and rich in carbonate and larger Foraminifera under the southern shelf, but their northern extent is restricted. Oligocene rocks are currently not known on the U.S. Atlantic Margin north of the COST B-2 well. Their absence on much of the northern Coastal Plain and from the shallow water facies southward suggest a major regressive cycle of deposition.

Clastic deposition increased dramatically during the Miocene all along the Margin, and it continued, although at a diminished rate, through the Pleistocene. Calcareous, often phosphatic, open-shelf facies predominate south of Cape Hatteras, but are replaced by diatomaceous deltaic facies off Maryland and New Jersey. Southern facies are also characterized by warmer water planktic organisms than those north of Cape Hatteras.

Pliocene depositional regimes were similar; mixed carbonate-clastic shelf facies south of Cape Hatteras become dominantly clastic to the north. A major delta complex, characterized by freshwater diatom assemblages, developed off Virginia and Maryland.

Pleistocene sands and silts are widespread and they also become increasingly clastic to the north, culminating in abundant glacial sediments on Georges Bank. Pleistocene strata are especially thick on the northern continental slope; density and turbidity flows have transported large volumes of shelf sediments onto the slope and rise through shelf-edge canyons.

Holocene sediments are virtually absent from the middle and outer continental shelf. The only significant modern accumulations are gray and white sands and silts along the coastline and green silty clays on the continental slope.

Similar sequences of Mesozoic and Cenozoic strata occupy the adjacent Scotian basin under the eastern Scotian shelf, slope, and rise, the Laurentian Channel, and the western Grand Banks.

The stratigraphic record as summarized above is based mainly on indirect geophysical methods (i.e. seismic reflection and refraction studies that have been concentrated on the continental shelf), and on geological extrapolations from the Bahama uplift, the Canadian Margin, and the Atlantic Coastal Plain. New technology (multichannel recording) and new methods of interpretation (seismic stratigraphy) have greatly improved the reliability of geophysical techniques, but with respect to direct analysis of actual rock samples, the U.S. Atlantic Margin has hardly been scratched. However, during the next few years, many new rock samples should become available when data from the three additional COST wells are released, and numerous exploratory wells are drilled on the shelf, slope and Blake Plateau. Geologic data from each new exploratory well are to become public two years after each well is drilled, according to new Federal regulations. Continuing programs of seismic, gravitational, and magnetic profiling will allow extrapolation into undrilled areas, especially on the continental slope and rise, that will eventually fill many of the gaps in the present stratigraphic framework.

Nevertheless, in spite of this expected abundance of wells, much critical geological evidence will be left in the ground because the strata will not be continuously sampled (i.e. the upper 300 m are generally not sampled and, below that, rotary cuttings are recovered only about every 3 to 10 m). We desperately need strategically located deep boreholes that are cored continuosuly, if we expect ever to completely understand the geology of this vast region. Stratigraphers must immediately devote their best efforts to achieving this goal.

ACKNOWLEDGMENTS

I should like to thank John Schlee, William P. Dillon, John A. Grow, and Thomas G. Gibson for their editorial and technical advice in preparing this summary.

Page C. Valentine, Raymond E. Hall, Charles C. Smith, William H. Abbott, Paul F. Huddlestun, Norman Frederiksen, Raymond A. Christopher, and Richard N. Benson contributed biostratigraphic and paleoecologic interpretations. Peggy Hempenius, David Towner, and Diane Johnson prepared the typescript and illustrations.

Literature Cited

Agassiz, A. 1888. *Three Cruises of the United States Coast and Geodetic Survey Steamer "Blake" in the Gulf of Mexico, in the Caribbean Sea, and Along the Atlantic Coast of the United States, from 1877 to 1880.* Boston: Houghton Mifflin. 2 Vols.

Amoco Canada Petroleum Company Ltd., Imperial Oil Ltd. 1974. Regional geology of Grand Banks. *Am. Assoc. Petrol. Geol. Bull.* 58:1109–23

Benson, W. E., Sheridan, R. E., Enos, P., Freeman, T., Gradstein, F., Murdmaa, I. O., Pastouret, L., Schmidt, R. R., Stuermer, D. H., Weaver, F. M., Worstell, P. 1976. Deep-sea drilling in the North Atlantic. *Geotimes* 21(2):23–26

Bhat, H., McMillan, N. J., Aubert, J., Porthault, B., Surin, M. 1975. North American and African drift—The record in Mesozoic coastal plain rocks, Nova Scotia and Morocco. In *Canada's Continental Margins*, ed. C. J. Yorath, E. R. Parker, D. J. Glass, pp. 375–89. Calgary: Can. Soc. Petrol. Geol., Mem. 4. 889 pp.

Brown, P. M., Miller, J. A., Swain, F. M. 1972. Structural and stratigraphic framework and spatial distribution of permeability of the Atlantic Coastal Plain, North Carolina to New York. *U.S. Geol. Surv. Prof. Pap. 796.* 70 pp.

Bryant, W. R., Meyerhoff A. A., Brown, N. K., Jr., Furer, M. A., Pyle, T. E., Antoine, J. W. 1969. Escarpments, reef trends, and diapiric structures, eastern Gulf of Mexico. *Am. Assoc. Petrol. Geol. Bull.* 53:2506–42

Bunce, E. T., Emery, K. O., Gerard, R. D., Knott, S. T., Lidz, L., Saito, T., Schlee, J. 1965. Ocean drilling on the continental margin. *Science* 150:709–16

Charm, W. B., Nesteroff, W. D., Valdes, S. 1970. Drilling on the continental margin off Florida: detailed stratigraphic description of the JOIDES cores on the continental margin off Florida. *U.S. Geol. Surv. Prof. Pap. 581-D.* 13 pp.

Cushman, J. A. 1936. Geology and paleontology of the Georges Bank canyons, Part IV, Cretaceous and Late Tertiary Foraminifers. *Geol. Soc. Am. Bull.* 47: 413–40

Dall, W. H. 1925. Tertiary fossils dredged off the northeastern coast of North America. *Am. J. Sci.*, 5th Ser. 10:213–18

Dillon, W. P., Girard, O. W., Weed, E. G. A., Sheridan, R. E., Dolton, G., Sable, E., Krivoy, H. L., Grim, M., Robbins, E., Rhodehamel, E. C., Amato, R., Foley, N. 1975. Sediments, structural framework, petroleum potential, environmental conditions, and operational considerations of the United States South Atlantic Outer Continental Shelf. *U.S. Geol. Surv. Open File Rep. 75–411.* 262 pp.

Dillon, W. P., Paull, C. K., Buffler, R. T., Fail, J-P. 1978. Structure and development of the Southeast Georgia embayment and northern Blake Plateau: preliminary analysis. *Am. Assoc. Petrol. Geol. Mem.* In press

Drake, C. L., Ewing, M., Sutton, G. H. 1959. Continental margins and geosynclines: the east coast of North America north of Cape Hatteras. In *Physics and Chemistry of the Earth*, ed. L. H. Ahrens et al, 3:110–98. New York: Pergamon. 464 pp.

Emery, K. O., Uchupi, E. 1972. *Western North Atlantic Ocean: Topography Rocks, Structure, Water, Life, and Sediments.* Tulsa: Am. Assoc. Petrol. Geol., Mem. 17. 532 pp.

Ericson, D. B., Ewing, M., Wollin, G., Heezen, B. C. 1961. Atlantic deep-sea sediment cores. *Geol. Soc. Am. Bull.* 72: 193–286

Ewing, M., Crary, A. P., Rutherford, H. M. 1937. Geophysical investigations in the emerged and submerged Atlantic Coastal Plain. I. Methods and results. *Geol. Soc. Am. Bull.* 48:753–801

Gibson, T. G., Hazel, J. E., Mello, J. E. 1968. Fossiliferous rocks from submarine canyons of northeastern United States. *U.S. Geol. Surv. Prof. Pap. 600-D.* pp. 222–30

Given, M. M. 1977. Mesozoic and early Cenozoic geology of offshore Nova Scotia. *Can. Petrol. Geol. Bull.* 25:63–91

Gohn, G. S., Higgins, B. B., Smith, C. C., Owens, J. P. 1978. Preliminary report on the lithostratigraphy of the deep core hole (Clubhouse Crossroads Corehole

#1) near Charleston, South Carolina. *U.S. Geol. Surv. Prof. Pap. 1082.* In press

Gradstein, F. M., Williams, G. L., Jenkins, W. A. M., Ascoli, P. 1975. Mesozoic and Cenozoic stratigraphy of the Atlantic Continental Margin, Eastern Canada. See Bhat et al 1975, pp. 103–31

Grow, J. A., Markl, R. G. 1977. IPOD-USGS multichannel seismic reflection profile from Cape Hatteras to the Mid-Atlantic Ridge. *Geology* 5(10):625

Grow, J. A., Dillon, W. P., Sheridan, R. E. 1977. Diapirs along the continental slope off Cape Hatteras. Presented at Ann. Int. Meet. Soc. Explor. Geophys., 47th, Calgary

Grow, J., Mattick, R. E., Schlee, J. 1977. Depth conversion of multichannel seismic-reflection profiles over Atlantic Outer Continental Shelf and Upper Continental Slope between Cape Hatteras and Georges Bank. *Am. Assoc. Petrol. Geol. Bull.* 61(5):790–91

Hathaway, J. C., Schlee, J., Poag., C. W., Valentine, P. C., Weed, E. G. A., Bothner, M. H., Kohout, F. A., Manheim, F. T., Schoen, R., Miller, R. E., Schultz, D. M. 1976. Preliminary summary of the 1976 Atlantic Margin Coring Project of the U.S. Geological Survey. *U.S. Geol. Surv. Open File Rep.*, 76–844. 217 pp.

Heezen, B. C., Sheridan, R. E. 1966. Lower Cretaceous rocks (Neocomian-Albian) dredged from Blake Escarpment. *Science* 154:1644–47

Herrick, S. M., Vorhis, R. C. 1963. Subsurface geology of the Georgia Coastal Plain. *Georgia State Div. Conserv., Geol. Surv., Inf. Circ. 25.* 78 pp.

Hollister, C. D., Ewing, J. I., Habib, D., Hathaway, J. C., Lancelot, Y., Luterbacher, H., Paulus, F. J., Poag, C. W., Wilcoxon, J. A., Worstell, P. 1972. Site 108—Continental Slope. *Initial Rep. Deep-Sea Drilling Proj.*, 11:357–64. Washington, D. C.:GPO.

Jacobs, C. 1977. Jurassic lithology in Great Isaac 1 well, Bahamas: discussion. *Am. Assoc. Petrol. Geol. Bull.* 61(3):443

Jansa, L. F., Wade, J. A. 1975. Geology of the continental margin off Nova Scotia and Newfoundland. *Can. Geol. Surv. Pap. 74–30,* pp. 51–105

Kaneps, A. G. 1970. *Late Neogene biostratigraphy (planktonic foraminifera), biogeography, and depositional history.* PhD thesis. Columbia Univ., New York. 185 pp.

Keen, M. J. 1974. The continental margin of eastern North America, Florida to Newfoundland. In *The Ocean Basins and Margins,* ed. A. E. M. Nairn, F. G.

Stehli, 2:41–78. New York: Plenum. 598 pp.

King, L. H. 1975. Geosynclinal development on the continental margin south of Nova Scotia and Newfoundland. See Jansa & Wade 1975, pp. 199–206

King, L. H., MacLean, B. 1976. Geology of the Scotian Shelf. *Can. Geol. Surv. Pap. 74–31.* 31 pp.

King, L. H., MacLean, B., Fader, G. B. 1973. Unconformities on the Scotian Shelf. *Can. J. Earth Sci.* 11(1):89–100

Klein, G. de V. 1962. Triassic sedimentation, Maritime Provinces, Canada. *Geol. Soc. Am. Bull.* 73:1127–46

Lewis, R. S., Sylwester, R. E. 1977. Shallow sedimentary framework of Georges Bank. See Grow, Mattick & Schlee 1977, p. 808

Maher, J. C. 1971. Geologic framework and petroleum potential of the Atlantic Coastal Plain and Continental Shelf. *U.S. Geol. Survey Prof. Pap. 659.* 98 pp.

Manheim, F. T., Hall, R. E. 1976. Deep evaporitic strata off New York and New Jersey—evidence from interstitial water chemistry of drill holes. *J. Res. U.S. Geol. Survey* 4:497–502.

Mattick, R. E., Bayer, K. C., Scholle, P. A. 1977. Petroleum potential of possible Lower Cretaceous reef trend beneath U.S. Atlantic Continental Slope. See Grow, Mattick & Schlee 1977, p. 811

Mattick, R. E., Perry, W. J., Robbins, E., Rhodehamel, E. C., Weed, E. G. A., Taylor, D. J., Krivoy, H. L., Bayer, K. C., Lees, J. A., Clifford, C. P. 1975. Sediments, structural framework, petroleum potential, environmental conditions, and operational considerations of the United States mid-Atlantic Outer Continental Shelf. *U.S. Geol. Surv. Open File Rep. 75–61.* 143 pp.

Mayhew, M. A. 1974. Geophysics of Atlantic North America. In *Geology of Continental Margins,* ed. C. A. Burke, C. L. Drake, pp. 409–27 New York: Springer. 1009 pp.

McClelland Engineers 1963. *Fathometer survey and foundation investigation: Frying Pan Shoals offshore structure, Cape Fear, North Carolina.* Rep. to Commandant, U.S. Coast Guard, Washington, D.C. by McClelland Engineers, Inc., Soil and Foundation Consultants, Houston, Texas. 12 pp.

McCollum, M. J., Herrick, S. M. 1964. Offshore extension of the upper Eocene to Recent stratigraphic sequence in southeastern Georgia. *U.S. Geol. Surv. Prof. Pap. 501-C,* pp. 61–63

McIver, N. L. 1972. Cenozoic and Mesozoic stratigraphy of the Nova Scotia Shelf. *Can. J. Earth Sci.* 9:54–70

Meisburger, E. P. 1977. Shallow framework of Inner Continental Shelf of Cape Fear region, North Carolina. See Grow, Mattick & Schlee 1977, pp. 812–13

Meisburger, E. P., Field, M. E. 1976. Neogene sediments of Atlantic Inner Continental Shelf off northern Florida. *Am. Assoc. Petrol. Geol. Bull.* 60:2019–37

Meyerhoff, A. A., Hatten, C. W. 1974. Bahamas salient of North America: tectonic framework, stratigraphy and petroleum potential. *Am. Assoc. Petrol. Geol. Bull.* 58:1201–39

Milliman, J. D., Pilkey, O. H., Ross, D. A. 1972. Sediments of the continental margin off the eastern United States. *Geol. Soc. Am. Bull.* 83:1315–34

Olsson, R. K. 1960. Foraminifera of latest Cretaceous and earliest Tertiary age in the New Jersey Coastal Plain. *J. Paleont.* 34:1–58

Olsson, R. K. 1964. Late Cretaceous planktonic foraminifera from New Jersey and Delaware. *Micropaleontology* 10:157–88

Parsons, M. G. 1975. The geology of the Laurentian Fan and the Scotian Rise. See Bhat et al 1975, pp. 155–67

Poag, C. W., Valentine, P. C., Smith, C. C., Hall, R. E., Abbott, W. H., Huddlestun, P. 1977. Preliminary biostratigraphy of the U.S. Atlantic Continental Margin. See Grow, Mattick & Schlee 1977, pp. 820–21

Saito, T., Burckle, L. H., Hays, J. D. 1974. Implications of some pre-Quaternary sediment cores and dredgings. In *Studies in Paleo-Oceanography*, ed. W. W. Hay, pp. 6–36. Tulsa: Soc. Econ. Paleontol. Mineral, Spec. Pub. 20. 218 pp.

Sanders, J. E. 1963. Late Triassic tectonic history of northeastern United States. *Am. J. Sci.* 261:501–24

Schlee, J. 1977. Stratigraphy and Tertiary development of the continental margin east of Florida. *U.S. Geol. Surv. Prof. Pap. 581-F.* 25 pp.

Schlee, J., Behrendt, J. C., Grow, J. A., Robb, J. M., Mattick, R. E., Taylor, P. T., Lawson, B. J. 1976. Regional geologic framework off northeastern United States. *Am. Assoc. Petrol. Geol. Bull.* 69:926–51

Schlee, J., Cheetham, A. H. 1967. Rocks of Eocene age on Fippennies Ledge, Gulf of Maine. *Geol. Soc. Am. Bull.* 78:681–84

Schlee, J., Martin, R. G., Mattick, R. E., Dillon, W. P., Ball, M. M. 1977. Petroleum geology on the U.S. Atlantic-Gulf of Mexico margins. *Proc. Southwest Legal Found.* 15:47–93

Schlee, J., Mattick, R. E., Taylor, D. J., Girard, O. W., Grow, J. A., Rhodehamel, E. C., Perry, W. J., Bayer, K. C., Furbush, M., Clifford, C. P., Lees, J. A. 1975. Sediments, structural framework, petroleum potential, environmental conditions, and operational considerations of the United States North Atlantic Outer Continental Shelf. *U.S. Geol. Surv. Open File Rep. 75–353.* 179 pp.

Scholle, P. A., ed. 1977. Geological studies on the COST No. B-2 well, U.S. mid-Atlantic Outer Continental Shelf area. *U.S. Geol. Surv. Circ. 750.* 71 pp.

Schultz, L. K., Grover, R. L. 1974. Geology of Georges Bank Basin. See Meyerhoff & Hatten 1974, pp. 1159–68

Scott, K. R., Cole, J. M. 1975. Geology of the U. S. Continental Margin from Maine to Florida—A resumé. See Bhat et al 1975, pp. 33–43

Sheridan, R. E. 1974. Atlantic continental margin of North America. See Mayhew 1974, pp. 391–407

Sheridan, R. E. 1976. Sedimentary basins of the Atlantic margin of North America. *Tectonophysics.* 36:113–32

Sheridan, R. E., Osburn, W. L. 1975. Marine geological and geophysical studies of the Florida-Blake Plateau-Bahamas area. See Bhat et al 1975, pp. 9–32

Sheridan, R. E., Smith, J. D., Gardner, J. 1969. Rock dredges from near Great Abaco Canyon. *Am. Assoc. Petrol. Geol. Bull.* 53:2551–58

Smith, M. A., Amato, R. V., Furbush, M. A., Pert, D. M., Nelson, M. E., Hendrix, J. S., Tamm, L. C., Wood, G. Jr., Shaw, D. R. 1976. Geological and operational summary, COST No. B-2 well, Baltimore Canyon trough, Mid-Atlantic OCS. *U.S. Geol. Surv. Open File Rep. 76–774.* 79 pp.

Stehli, F. G. 1974. Geology of the Bahama-Blake Plateau region. See Keen 1974, pp. 15–39

Stephenson, L. W. 1936. Geology and paleontology of the Georges Bank canyons. Part II. Upper Cretaceous fossils from Georges Bank (including species from Banquereau, Nova Scotia). See Cushman 1936, pp. 367–410

Stetson, H. C. 1936. Geology and paleontology of the Georges Bank canyons. Part I. Geology. See Cushman 1936, pp. 339–66

Stetson, H. C. 1949. The sediments and stratigraphy of the east coast continental margin—Georges Bank to Norfolk Canyon. *Mass. Inst. Technol., Woods Hole Oceanog. Inst. Pap. Phys. Oceanogr. Meterol.* 11:1–60

Tator, B. A., Hatfield, L. E. 1975. Bahamas present complex geology. *Oil Gas J.* 73(43):172–76; 73(44):120–22

Uchupi, E. 1970. Atlantic Continental Shelf and Slope of the United States: shallow structure. *U.S. Geol. Surv. Prof. Pap. 529-I.* 44 pp.

Vail, P. R., Mitchum, R. M., Todd, R. G. 1975. *Global Unconformities from Seismic Sequence Interpretation.* Presented at Ann. Meet. Soc. Explor. Geophys., 45th, Denver

Verrill, A. E. 1878. Occurrence of fossiliferous Tertiary rocks on the Grand Bank and Georges Bank. *Am. J. Sci.*, 3rd Ser. 16:323–24

Weed, E. G. A., Minard, J. P., Perry, W. J., Rhodehamel, E. C., Robbins, E. I. 1974. Generalized pre-Pleistocene geologic map of the northern United States Atlantic continental margin. *U.S. Geol. Surv. Misc. Geol. Invest. Ser. Map I-861* (scale 1:1,000,000)

Williams, G. L. 1975. Dinoflagellate and spore stratigraphy of the Mesozoic-Cenozoic, offshore eastern Canada. See Jansa & Wade 1975, pp. 107–61

Ann. Rev. Earth Planet. Sci. 1978. 6 : 281–303

CHEMICAL EXCHANGE ACROSS SEDIMENT-WATER INTERFACE

×10095

A. Lerman
EAWAG and Institute of Aquatic Sciences, Swiss Federal Institute of Technology, Dübendorf-Zürich, Switzerland[1]

INTRODUCTION

The sediment-water interface separates a mixture of solid sediment and interstitial water from an overlying body of water. Wherever sediment accumulates, growth of the sediment pile is achieved by sedimentation of solid particles and inclusion of water in the pore spaces among the particles. Growth or erosion of sediments on the bottom of a body of water results in a rise or fall of the sediment-water interface relative to an observer on shore. If, however, an imaginary observer were positioned at the sediment-water interface, he would observe the sediment particles and water moving downward past him when the sediment column grows, or moving upward when the sediment is being eroded. In considering various chemical, physical, and biological processes taking place near the sediment-water interface, it is often more convenient to regard the interface as a plane of reference, as if one were an imaginary observer who sees sediment particles and water flow by him while he balances himself on the interface.

Arrival of sediment particles and inclusion of water in the interparticle pores space are the two major fluxes of materials across the sediment-water interface. The rates of sediment deposition vary from the low values of millimeters per 1000 years in the pelagic ocean up to 1 centimeter per 1 year in lakes and near-shore oceanic areas. Sediments near the interface commonly contain 70–90 vol % of water, but upon compaction the volume fraction of water (called the sediment porosity) usually decreases to 40–60% within a few tens of centimeters to a few meters below the interface. Representative values of the mass fluxes of solids and dissolved materials across the sediment-water interface can be estimated for areas of low and high deposition rates as follows. The slower sedimentation rates in the ocean are of the order of 5×10^{-4} cm yr^{-1} (5 mm 1000 yr^{-1}), and the higher rates in lakes and coastal sections are

---

[1] Current address: Department of Geological Sciences, Northwestern University, Evanston, Illinois 60201

0084-6597/78/0515-0281$01.00

5×10^{-1} cm yr^{-1}; for a sediment made of equal volumes of water and solid particles of density 2.5 g cm^{-3} (that is, sediment volume fraction of 0.5), the mass flux of solids is in the range

$$6 \times 10^{-4} \quad \text{to} \quad 6 \times 10^{-1} \text{ g cm}^{-2} \text{ yr}^{-1}.$$

Ocean water contains 35 g l^{-1} dissolved materials and for fresh water lakes a typical figure is 0.1 g l^{-1}. For the same range of sedimentation rates and sediment porosity as used in the computation of solid mass flux, the flux of dissolved materials across the sediment-water interface is in the range

$$0.1 \times 10^{-4} \quad \text{to} \quad 0.1 \times 10^{-1} \text{ g cm}^{-2} \text{ yr}^{-1}.$$

In fresh-water lakes, the fluxes of dissolved material into sediments are closer to the lower values of 0.1×10^{-4} g cm^{-2} yr^{-1}.

Because the flux of dissolved materials is only of the order of 1% of the flux of solids to the sediment, the dissolved material in pore water cannot add much to the mass of solids. The opposite, however, is commonly observed: the chemical composition of pore waters changes owing to reactions with solid phases in the sediment.

In some environments, the sediment-water interface is a quiet zone where settling materials come to rest on the bottom. In other environments, the zone of the interface is subject to a variety of disturbances: benthic organisms mix and rework the sediment within several centimeters of the interface; fluctuations in near-bottom water currents resuspend and shift the sediment; and sediments accumulating on inclined surfaces creep and slide, altering the local topographic bottom relief to a greater or lesser extent.

The settling residence times of solid particles in the water column of lakes and oceans are short in comparison to the rate of sediment column growth, such that the materials arriving at the sediment surface may continue to react either inorganically or through bacterially mediated processes. Transport of chemical species across the sediment-water interface is caused by the advective processes of sediment-column growth, as discussed above, as well as by diffusional migration in pore waters, which is caused by differences in concentration (or thermodynamic activity) that develop between the two sides of the interface. Biogenic skeletal material made of silica and calcium carbonate phases, organic matter, and some inorganic mineral phases that are out of a thermodynamic equilibrium with pore-water solutions react and cause changes in concentrations of dissolved species. Examples of such processes include decomposition of organic matter, reduction of sulfate to sulfide, production of methane, ammonia, and carbon dioxide, and also dissolution of mineral phases of inorganic and biogenic origin, caused by changes in pore-water chemistry. An increase or decrease in concentration of a dissolved species in sediment pore water relative to the overlying water establishes a condition for the diffusional flux that can be realized if pore water is in contact with the bottom water above the sediment-water interface. Those chemical species whose concentrations increase in pore water are at least in part removed by diffusional fluxes up, across the sediment-water interface, or down the sediment pore-water column. Some of the chemical species (for example, phosphate and ammonia) generated in sediments are not completely removed by diffusional fluxes: for these species, uptake by solid sediment particles is an additional sink.

The preceding paragraph summarizes very concisely a large body of information from the literature dealing with the geochemistry of waters and sediments. Some of the more recent publications dealing specifically with the problems of chemical exchange across the sediment-water interface are referenced in the subsequent sections of this paper.

A reverse process, desorption from sediment, is of a potential significance to the recovery of some chemical substances stored in sediments. Heavy metals, bound to sediment particles via organic and inorganic ligands, can be desorbed by acids or chelating agents in solution. In the sediments of lakes, inland seas, and oceanic sections of industrialized regions, concentrations of such heavy metals as lead, copper, zinc, cadmium, nickel, and cobalt have been increasing during industrial times, especially during the last several decades (for example, Bruland et al 1974, Erlenkeuser et al 1974, Iskandar & Keeney 1974). A concomitant phenomenon of industrial times is an increasing content of acids and a variety of organic substances in atmospheric precipitation and surface waters. Continuing or accelerated inputs of acids and chelating substances may lead to some removal of heavy metals from sedimentary substrates to solution, with all the potentially hazardous consequences to the environment and its inhabitants.

The mechanisms of transport of chemical species across the sediment-water interface can be grouped into the following categories:

1. Sedimentation flux of solids (mineral, skeletal, and organic materials)
2. Dissolved material and water flux into sediment, owing to the growth of the sediment column
3. Dissolved material and pore water flow upward, caused by hydrostatic pressure gradients of ground water in aquifers on land
4. Molecular diffusional fluxes in pore water
5. Mixing of sediment and water at the interface (bioturbation and water turbulence)

Processes 1–3 are advective fluxes, to distinguish them from molecular diffusion responsible for process 4. The significance of process 3, upward flow of pore water, depends to a large extent on a local configuration of a body of water and ground-water aquifers. Flow of ground water into and out of lakes can amount to some fraction of the total water budget, depending on the geological conditions within the drainage basin. For the ocean, an amount of ground-water inflow has been estimated as 5% of the river and glacier melt inflow, 1600 vs 32,000 km^3 yr^{-1} (Nace 1967, Menard 1974).

GENERAL ROLE OF THE SEDIMENT-WATER INTERFACE

The quantitative role played by the fluxes at the sediment-water interface within a bigger biogeochemical cycle of materials in the ocean is demonstrated in this section for two elements, calcium and silicon, both components of biogenic and inorganic solids. The diagram in Figure 1 shows the major fluxes of $CaCO_3$ and SiO_2 involved in the biological productivity and chemical regeneration cycle in the ocean. For both chemical species the model diagram in Figure 1 represents only global average conditions and it does not apply to the local sections of excessively high blooms of calcareous and siliceous organisms, where $CaCO_3$ or SiO_2 dominate the sedimentary fluxes.

Some of the calcium carbonate and silica fixed biologically within the euphotic zone of the ocean are being regenerated to solution in surface ocean water. A remaining fraction settles into the deeper water, as indicated by flux A in Figure 1. This flux includes all of the suspended $CaCO_3$ or SiO_2 material of biogenic origin, that is, material settling as individual particles, aggregates, or fecal pellets. Some of the skeletal material that dissolves in deep ocean water can dissolve during settling, and this dissolution flux is shown by B in Figure 1. Solid particles reaching the bottom remain in contact with overlying water as long as they are not covered by new sediment, and at the sedimentation rate of about 1 cm $1000 \ yr^{-1}$ this time is relatively long. Continuing dissolution at the sediment-water interface returns dissolved material to ocean water, as indicated by flux C. The material that is buried and becomes part of the sedimentary record (flux D) continues to react with pore water, and its dissolution results in flux E, upward across the interface.

With the amount of biogenic $CaCO_3$ or SiO_2 that enters the deeper ocean at 1 km depth with flux A taken as 100%, the estimated proportions of the other dissolution and sedimentation fluxes are listed in Table 1. About 60% of $CaCO_3$ entering the deep ocean dissolves at the sediment-water interface, and up to 1% dissolves deeper in the sediments. The magnitude of the flux E out of pore water is absorbed by the plus-or-minus uncertainty margin on the estimates of the other fluxes. The relatively large fraction of $CaCO_3$ that dissolves at the sediment-water interface and during settling to the bottom is the result of undersaturation of deep ocean water with respect to the $CaCO_3$ minerals calcite and aragonite, and the faster dissolution rate of calcite in the deep ocean.

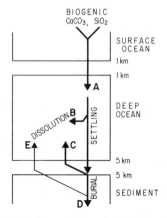

Figure 1   Dissolution and settling fluxes of biogenic $CaCO_3$ and SiO_2 in the ocean. Flux C denotes dissolution at the sediment–water interface. Flux values are in Table 1.

Table 1 Sedimentation and dissolution fluxes of biogenic $CaCO_3$ and SiO_2 in the ocean, shown schematically in Figure 1 (recalculated from data in Lerman & Lal 1977)

Flux (Figure 1)	$CaCO_3$ % of flux A	$CaCO_3$ μmole cm^{-2} yr^{-1}	SiO_2 % of flux A	SiO_2 μmole cm^{-2} yr^{-1}
A	100	35	100	47 ± 12
B	35 ± 17	12 ± 6	50 ± 10	23 ± 4
C	53	18	46	22
D	12	4	4	2
E	1 ± 1	0.5 ± 0.5	2 ± 2	1 ± 1

For biogenic SiO_2, the estimated fraction that dissolves at the sediment-water interface is about 50%.

The sum total of the fluxes $B + C + D$ (Figure 1) is the rate of regeneration of $CaCO_3$ or SiO_2 in the deep ocean. The small value of the flux E carrying material out of the sediment in comparison to the dissolution flux C at the sediment-water interface (Table 1) can be attributed to the following factors: 1, molecular diffusion responsible for the fluxes of dissolved species in pore water is generally a slow process; 2, concentrations of Ca and SiO_2 in pore waters are closer to saturation with respect to the dissolving phases, such that the rates of dissolution are slower; 3, newly formed mineral phases precipitate on the dissolving solids reducing their total reactive area.

MODELS OF THE SEDIMENT-WATER INTERFACE

Three conceptual models of the sediment-water interface are shown in Figure 2, all assume the interface to be a horizontal plane. In the simplest model of Figure 2A, the water above the bottom is a well-mixed homogeneous solution (no concentra-

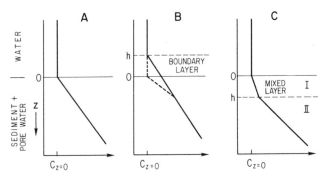

Figure 2 Three models of the sediment-water interface. The vertical axis is the distance coordinate z, and the horizontal axis is concentration of a dissolved species in pore water and overlying water. $C_{z=0}$ is concentration at the interface.

tion gradients of dissolved species). Below the interface, concentration gradients can exist because of the slower transport rates in pore water. At the interface, concentrations in the bottom water and pore water are equal (but the concentration gradients are not equal in the geometric representation of Figure 2A).

A sediment-water interface with a water boundary layer above it is shown in Figure 2B. Laminar flow and molecular diffusion take place within the boundary layer, as opposed to the eddy turbulent conditions above it. In the deep ocean, the thickness of the laminar boundary layer is measurable in centimeters (Wimbush & Munk 1971). Existence of such boundary layers with concentration gradients within them, as drawn in Figure 2B, is potentially significant to computation of the diffusional fluxes at the sediment-water interface (Morse 1974). The diffusional flux across the interface $F_{z=0}$ is

$$F_{z=0} = -\phi D(\Delta C/\Delta z) \qquad (\text{g cm}^{-2} \text{ yr}^{-1}), \tag{1}$$

where ϕ is the sediment porosity (fraction), D (cm^2 yr^{-1}) is the molecular diffusion coefficient of a dissolved species in pore water, and $\Delta C/\Delta z$ (g cm^{-4}) is the concentration gradient at the interface. Sampling methods of sediment pore waters are such that reliable sample withdrawals and analyses are often not possible in the uppermost 1–3 cm of the pore-water column. Therefore a concentration gradient $\Delta C/\Delta z$ is often estimated from the concentration at the highest point sampled (C at some distance z below the interface) and from the concentration in the overlying water near the bottom, taken as the concentration at the interface $C_{z=0}$: $\Delta C = C - C_{z=0}$. The linear distance Δz is taken as the distance between the smallest depth sampled and the interface at $z = 0$. If Δz is of the order of 10^0 cm, then the presence of a water boundary layer of a comparable thickness increases the Δz value and lowers the estimate of the flux computed using Equation (1). Note that if concentration at the interface $C_{z=0}$ can be determined independently of the near-bottom concentration in overlying water outside the boundary layer, then the flux value based on (1) would be more correct.

Figure 2C is a diagram of the interface zone, where a sediment layer immediately below the interface is characterized by a faster transport than the sediment below it. In near-shore carbonate sediments of Bermuda and in sediments of Long Island Sound, which are rich in organic matter, seasonal variations in concentrations of dissolved species have been observed within the upper few to a few tens of centimeters of sediments (Thorstenson & Mackenzie 1974, Goldhaber et al 1977). Sediment zones of comparable thickness, showing the effects of mixing and reworking, have also been reported in the Caribbean deep-sea sediments and in coastal sediments of the North Sea (Guinasso & Schink 1975, Vanderborght et al 1977). The reasons behind the formation of a mixed sediment layer are the burrowing activity of benthic organisms (bioturbation) and sporadic or continuous penetration of water turbulence near the bottom into the sediment pore space. If the vertical transport within the mixed sediment layer is regarded as analogous to the diffusional flux model of Equation (1), then conservation of the flux at the boundary $z = h$ between two layers I and II (Figure 2C) requires the following condition to be obeyed for dissolved species in pore water

$$[D(dC/dz)]_I = [D(dC/dz)]_{II}. \tag{2}$$

The coefficient D_I for the fast-transport layer can be computed from (2), by using the observed values of the concentration gradients in the mixed layer $(dC/dz)_I$ and in undisturbed sediment below it $(dC/dz)_{II}$, and by using the molecular diffusion coefficient D_{II} for the undisturbed layer pore water. For near-shore environments, values of D_I have been reported (Goldhaber et al 1977, Vanderborght et al 1977) in the range

$D_I/D_{II} = 5$ to 100,

where the molecular diffusion coefficient D_{II} is of the order of 10^{-6} cm^2 sec^{-1}.

For two- and three-layer models, mathematical relationships for concentrations and fluxes in pore waters, in the presence of diffusion, pore water advection, and dissolution and precipitation reactions, have been given in Lerman (1977).

FLUXES AT THE SEDIMENT-WATER INTERFACE

General Flux Equation

The total flux across the sediment-water interface $F_{z=0}$ due to the diffusion in pore water and advective transport of a dissolved species in pore water and on settling sediment particles is

$$F_{z=0} = -\phi D(dC/dz) + \phi UC + \phi U_s C_s \big|_{z=0} \quad \text{(g cm}^{-2}\text{ yr}^{-1}\text{)}, \tag{3}$$

where the term $-\phi D(dC/dz)$ represents the molecular diffusional flux defined by Fick's first law of diffusion in Equation (1); U is the rate of pore water advection; U_s is the rate of sediment deposition relative to the sediment water interface (cm yr^{-1}); C is concentration in pore water; and C_s is concentration in solids, both in units of mass per 1 cm^3 of pore water. The subscript $z = 0$ denotes that all the parameters are taken at their values at the sediment-water interface.

The relative magnitudes of the first two terms in Equation (3) determine the relative importance of diffusional and advective transport in pore water. If the concentration gradient is positive $(dC/dz > 0)$, then concentration increases with depth, indicating that the diffusional flux $-\phi D(dC/dz)$ is directed upward (the value is negative). In a growing sediment column, U is always positive, and if the two fluxes are of opposite sign then diffusion tends to counteract the effects of transport into sediment. When C and ΔC in Equation (3) are of comparable magnitude, then the ratio $D/\Delta z$ and U, both having the dimensions of velocity, can be compared in their relative importance. Taking $z \simeq 5$ cm and $D \simeq 5 \times 10^{-6}$ cm^2 sec^{-1}, the ratio is $D/\Delta z \simeq 10^{-6}$ cm sec$^{-1} = 30$ cm yr^{-1}; sedimentation rates U are of the order of <1 cm yr^{-1}. Thus we have $D/\Delta z > U$, and this indicates that the diffusional flux $D(dC/dz)$ term in the general flux Equation (3) is always significant in comparison to the advective flux UC in pore water.

Diffusion in Sediment Pore Water

The molecular diffusion coefficient D of dissolved species in water-filled sediments depends to some extent on the texture of the solid framework through its porosity and tortuosity. In general, smaller pore sizes and greater tortuosity retard diffusional fluxes through the pore space. If pore sizes are small they impose constraints on the diffusional paths of ions and molecules in solution; and tortuosity means that a molecule has to travel on a long winding path between any two points within a porous material, in comparison to the straight-line distance between them. The smaller values of the flux $F = \phi D \, (dC/dz)$ for a given concentration gradient across a section of sediment pore water are reflected in the smaller values of D for smaller values of porosity ϕ.

The chloride ion Cl^- is one of the least interacting dissolved species with solids. Diffusion coefficients of Cl^- (as NaCl) in different sediments are shown as a function of sediment porosity in Figure 3. The diffusion coefficient is a decreasing power function of porosity

$$D = D_0 \, \phi^n \qquad (cm^2 \, sec^{-1}),$$ (4)

where D_0 is the diffusion coefficient in bulk solution ($cm^2 \, sec^{-1}$), ϕ is porosity (fraction), and n is an empirical constant. The values of D plotted in Figure 3 correspond to n between 1.2 and 2.8, with a mean near $n \simeq 2$. For dissolved species in pore waters, the relationship $D = D_0 \, \phi^2$ has been used by a number of investi-

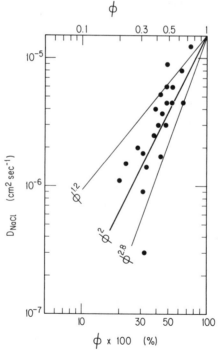

Figure 3 Molecular diffusion coefficient of chloride-ion (as NaCl) as a function of porosity ϕ in various sediments. Data from Manheim (1970).

gators (Lerman 1975, Stiller et al 1975) to estimate diffusional coefficients from bulk solution data. Because of the varied nature and sources of data plotted in Figure 3, caution should be exercised in using the relationship $D = D_0 \, \phi^n$ to estimate diffusional fluxes $-\phi D(dC/dz)$. A simple experimental method for the determination of diffusion coefficients in pore water is based on the relationship

$$D = \frac{F}{\phi \, dC/dz} \quad (\text{cm}^2 \text{ sec}^{-1}), \tag{5}$$

where the concentration gradient dC/dz across a slice of sediment, the flux F at the low concentration end, and sediment porosity ϕ can be measured. An experimental diffusion coefficient can also be computed as some effective diffusion coefficient D_{eff} for a given sediment

$$D_{\text{eff}} = \phi D = \frac{F}{dC/dz} \quad (\text{cm}^2 \text{ sec}^{-1}). \tag{6}$$

If some of the diffusion coefficients for pore waters reported in the literature were computed according to Equation (6), then from the data in Figure 3 and Equation (4) the following relationship follows

$$D_{\text{eff}} = D_0 \, \phi^n \tag{7}$$

or

$$D = D_0 \, \phi^{n-1}. \tag{8}$$

If the value of the power exponent n is about $n \simeq 2$, then the diffusion coefficients D computed from either Equation (4) or (8) differ by a factor of ϕ. However, near the sediment-water interface the values of ϕ are in the range $0.7 < \phi < 0.9$, and the difference between ϕ- or ϕ^2-dependence is not great.

A number of experimentally determined values of D for ionic species in sediment-water mixtures, approximating sediment pore waters, are listed in Table 2. The differences between the self-diffusion coefficients in pure water and sea water are very small. In sediment-water mixtures, the ionic diffusion coefficients are by a factor of two lower than in bulk sea water. The D values for sediments fall within the order of magnitude of 10^{-6} cm^2 sec^{-1}, in the temperature range between 5 and 25°C. Thus an approximate value of $D \simeq 3 \times 10^{-6}$ cm^2 sec^{-1} may be used as an approximation for ionic species in sediment-pore waters.

Fluxes on Solids

If concentration of a chemical species in water is small in comparison to its concentration on solid sediment particles ($C \ll C_S$), then the flux at the sediment-water interface is, from Equation (3),

$$F_{z=0} = \phi U_s C_s \quad (\text{g cm}^{-2} \text{ yr}^{-1}), \tag{9}$$

where C_s is concentration on the solid, in units of mass per 1 cm^3 of pore water. A more convenient unit of concentration is mass per unit mass of solid. Denoting C_g concentration on solid in units of g g^{-1}, and solid density ρ_s (g cm^{-3}), the

Table 2 Molecular diffusion coefficients of some dissolved species in bulk sea water and sediment–sea-water mixtures approximating pore waters ($\phi \simeq 0.7$)

Aqueous species	Temperature (°C)	Sea water	Pore water
		$D\ (10^{-6}\ \text{cm}^2\ \text{sec}^{-1})$	
Na^+ [a]	5	8.0	3.5
	24	13.4	5.8
K^+ [a]	5	11.4	
	24	17.9	
Ca^{2+} [a]	5	5.0	8.5
	24	7.5	15.5
Cl^- [a]	5	11.5	5.9
	24	18.6	10.2
SO_4^{2-} [a]	5	5.8	3.3
	24	9.8	5.3
	7		2.0–2.8 [b]
	19		3.7–3.9 [b]
SiO_2	5		3.3 [c]
	25	10.0 [d]	

[a] Li & Gregory (1974)
[b] Goldhaber et al (1977)
[c] Fanning & Pilson (1974)
[d] Wollast & Garrels (1971)

relationship between the two concentrations becomes $C_s = C_g \rho_s (1 - \phi)/\phi$, and the flux due to deposition of sediment is

$$F_{z=0} = U_s C_g \rho_s (1 - \phi) \quad (\text{g cm}^{-2}\ \text{yr}^{-1}). \tag{10}$$

Equation (10) can be generalized for lakes and near-continent oceanic sediments where the sedimentation rates U_s are of the order of 10^{-1} cm yr^{-1}. If we use $\phi \simeq 0.7$ for the sediment porosity near the sediment-water interface and the solid density $\rho_s \simeq 2.5$ g cm^{-3}, the flux into the sediment is

$$F_{z=0} \simeq 10^{-1} C_g \quad (\text{g cm}^{-2}\ \text{yr}^{-1}). \tag{11}$$

Concentrations of some heavy metals on sediment particles are usually measurable in ppm to hundreds of ppm (10^{-6} to 10^{-4} g g^{-1}), and this produces fluxes of the order of 0.1 to 10 μg cm^{-2} yr^{-1}. In Table 3 are summarized concentrations of four heavy metals in settling material and in solution in a fresh-water lake. The fluxes of Pb, Cd, Cu, and Zn to the sediment are almost exclusively due to deposition of these metals with settling solids. Potential concentrations of the metals in lake water, if they were completely solubilized from suspended materials, are of the order of 1 μg/liter (more if surface sediments were also contributing). Such potential concentration levels in lake water should be viewed against the currently acceptable health

Table 3   Heavy metals in water and settling sediments in Greifensee, fresh-water lake near Zurich, Switzerland. Mean sedimentation rate $U_s = 0.85$ cm yr^{-1}; solid sediment density $\rho_s = 2.0$ g cm^{-3}; sediment porosity $\phi = 0.7$; suspended matter concentration (as dry solids) 15 ± 5 mg l^{-1} (J. Tschopp, personal communication, 1977)

Metal	Concentration in		Flux to sediments
	Lake water C (μg l^{-1})	Settling material C_g (μg g^{-1})	$F_{z=0}$, Eq. (10) (μg cm^{-2} yr^{-1})
Cd	0.06	4	2
Cu	0.95	34	17
Pb	0.6	77	39
Zn	4.9	134	68

hazard limits for heavy metal concentrations in natural waters, of the order of 10^0 to 10^1 μg/liter.

Exchange Equilibrium and Adsorption

If an exchange equilibrium exists between a chemical species in solution and on solid sediment particles, then concentration of the species taken up by the solids is some function of its concentration in solution. The Freundlich adsorption relationship $C_s = KC^n$ is often used to describe concentration on the solid C_s as a function of concentration in solution C, with K and n as empirically determined constants. (Experimental determination requires measurements of C and C_s over some range of values, and K and n can be evaluated from a graph of log C_s plotted against log C). In those cases where experimental data show that the power exponent n can be taken as approximately $n \simeq 1$ without a great sacrifice of accuracy, the equilibrium exchange relationship reduces to a simple form of $C_s = KC$. Substitution of this relationship for C_s in the general flux Equation (3), with $U = U_s$, gives

$$F_{z=0} = -\phi D \, dC/dz + \phi U(K+1)C \big|_{z=0}. \tag{12}$$

The condition $U = U_s$ is valid when the porosity of the sediment column does not change with depth, and the solid sediment and pore water move downward relative to the sediment-water interface at a rate equal to the mean rate of sedimentation. When adsorption is strong ($K \gg 1$), then Equation (12) reduces to the form of Equation (9) describing the flux due to deposition of sediment only. The role of adsorption in migration of chemical species in sediments has been discussed by Berner (1976) and Lerman (1977).

The distribution factor K in the relationship $C_s = KC$ can be a dimensionless or a dimensioned quantity, depending on the concentration units used for the aqueous and solid phases. Here K is dimensionless if C_s and C are in units of mass per 1 cm^3 of solution. If concentration on the solid C_g is in units of g g^{-1}, then the distribution

Table 4 Distribution factor $K = C_s/C$ for some chemical species in oceanic and fresh-water sediments[a]

Chemical species	K	Notes and Sources
Ca	1.6	Sea water and clays (Li & Gregory 1974)
Cs	3,500–10,000	[137]Cs in fresh-water sediments (Lerman & Lietzke 1975)
Mg	10	Deep-ocean sediments (Lerman & Lietzke 1977)
Na	0.3	Sea water and clays (Li & Gregory 1974)
	2	Fresh-water sediments (Lerman & Weiler 1970)
NH_4^+–NH_3	3	Anoxic marine sediments (Berner 1974)
Phosphate	8	Sea water and clays (Berner 1974)
Ra	2000	[226]Ra in deep-ocean sediments (Lerman 1977)
Sr	6–12	[90]Sr in Ligurian Sea sediments (Cerrai et al 1969)
	45–120	[90]Sr in fresh-water sediments (Lerman & Lietzke 1975)

[a] K is dimensionless; concentration in solids C_s and in solution C are in units of mass per unit volume of solution.

equation is $C_g = K'C$, where K' has the dimensions of $cm^3\ g^{-1}$. The relationship between the two distribution factors K and K' is

$$K' = \frac{\phi K}{(1-\phi)\rho_s} \quad (cm^3\ g^{-1}), \tag{13}$$

where ρ_s (g cm^{-3}) is the solid particles density.

Some values of the distribution coefficient K for a number of chemical species in different sediment and water environments are listed in Table 4. The differences between the K-values for different species, and the differences between adsorption in sea water and fresh water (Na, Sr) are large. Additional data on K-values for radionuclides have been given by Aston & Duursma (1973).

Fluxes of Reactive Species

Bacterially mediated decomposition of organic matter in sediments leads to changes in concentrations of some of the chemical species in pore water, and inorganic precipitation, adsorption, or dissolution reactions involving the same chemical species may take place concomitantly with the biologically controlled processes. A model that takes into account one-dimensional (vertical) changes in concentrations of dissolved species in pore waters in the presence of (a) molecular diffusion, (b) pore-water advection, (c) precipitation, (d) dissolution, and (e) organic production or consumption can be written in the form of the following equation for the steady state

$$D\frac{d^2C}{dz^2} - U\frac{dC}{dz} + k_1(C_{s,1}-C) + k_2(C_{s,2}-C) \pm P(z) = 0 \quad (g\ cm^{-3}\ yr^{-1}), \tag{14}$$

where the diffusion coefficient D and sedimentation rate U are taken as constant and independent of the vertical distance coordinate z (cm). $P(z)$ is either the rate of production $(+)$ or consumption $(-)$ of a chemical species, in units of mass per 1 cm^3 of pore water. The term $k_1(C_{s,1} - C)$ represents a first-order precipitation reaction, where the solubility of the precipitating phase is $C_{s,1}$ (g cm^{-3}), the reaction rate constant is k_1 (yr^{-1}), and the condition $C_{s,1} < C$ applies. For a dissolution reaction, the term $k_2(C_{s,2} - C)$ has a similar significance, with $C_{s,2} > C$. Equation (14) can be simplified by setting

$$k = k_1 + k_2 \text{ (yr}^{-1}),\tag{15}$$

$$J = k_1 C_{s,1} + k_2 C_{s,2} \qquad \text{(g cm}^{-3}\text{ yr}^{-1}),\tag{16}$$

which reduces the number of constants, giving

$$D\frac{d^2C}{dz^2} - U\frac{dC}{dz} - kC + J \pm P(z) = 0.\tag{17}$$

Solutions of Equation (17) and computed values of k, J, and $P(z)$ for such chemical species as sulfate, ammonia, and phosphate in pore waters of pelagic and near-shore sediments in the ocean have been given by Toth & Lerman (1977). When sulfate is being bacterially reduced to sulfide in pore water, and carbon dioxide, ammonia, and phosphate are being produced, there is a decrease in the organic matter content of sediment, or at least a decrease in the biologically consumable fraction of organics. The decrease in the consumable organic matter with depth has been described (Berner 1974) as an exponential relationship of the type $C_{org}^0 \exp(-z\kappa_i/U)$ (mol cm^{-3}), where C_{org}^0 is the reactive organic matter concentration at the sediment-water interface ($z = 0$), and κ_i (yr^{-1}) is a first-order reaction rate constant corresponding to the rate of production of a certain chemical species from organic matter (i stands for either H$_2$S, CO$_2$, NH$_3$, or H$_3$PO$_4$ as stoichiometric species). The rate of production or consumption of each of the species in pore water is, therefore,

$$P(z) = \kappa_i \alpha_i C_{org}^0 \exp(-z\kappa_i/U) \qquad \text{(mol cm}^{-3}\text{ yr}^{-1}),\tag{18}$$

where α_i is a stoichiometric coefficient relating the number of moles of each species consumed or produced when 1 mol of organic matter is being oxidized. In a simpler notation, Equation (18) can be written as

$$P(z) = J_* e^{-\beta z} \qquad \text{(mol cm}^{-3}\text{ yr}^{-1}),\tag{19}$$

where $\beta = \kappa_i/U$ cm^{-1} and $J_* = \kappa_i \alpha_i C_{org}^0$ (mol cm^{-3} yr^{-1}).

Computed values of the consumption or production rate constant κ_i for SO$_4^{2-}$, NH$_4^+$, and PO$_4^{3-}$ in oceanic sediments show a positive correlation with the square of the sedimentation rate U

$$\kappa_i \propto U^2,\tag{20}$$

and explicit relationships for the three species are (Toth & Lerman 1977):

$$\kappa_S = 0.057 \; U^{1.94 \pm 0.22},$$

$$\kappa_P = 0.041 \; U^{1.83 \pm 0.51},$$

$$\kappa_N = 0.010 \; U^{1.83 \pm 0.43},$$

where κ_i is in units of yr^{-1} and U is in cm yr^{-1}.

The proportionality relationship $\kappa_i \propto U^2$ reflects the residence time of reactive organic materials at the sediment-water interface. When the sedimentation rate U is small, less organic material settles on the bottom and the settled material remains exposed to ocean water for longer periods of time, such that as a result only the more refractory components of organic matter are buried. In the areas of faster sedimentation rates, relatively more of reactive organics is buried in sediments, such that the decomposition rates are higher for the material in the sediment. The values of the rate constants given above apply to decomposition reactions of organic matter in oceanic sediments, and they should not be confused with the rates of organic matter decomposition in the course of settling through ocean water.

Fluxes $F_{z=0}$ in pore water across the sediment-water interface can be computed from Equation (12) $(K = 0)$, using the explicit relationships for concentration C and its gradient dC/dz from the solutions of Equation (17). Such explicit relationships for the sediment-water interface fluxes will be given in the next section, whereas here only some of the results will be shown. For sulfate, ammonia, phosphorus, calcium, magnesium, and bicarbonate, the computed fluxes $F_{z=0}$ are plotted against the sedimentation rate for a number of near-shore and pelagic sediments in Figure 4. There is an approximately linear relationship between the flux and the sedimentation rate

$$F_{z=0} \propto U, \tag{21}$$

and for the six species shown in Figure 4 the relationships are:

$$F_{S,z=0} = 81 \, U^{0.98 \pm 0.15}$$

$$F_{P,z=0} = -0.31 \, U^{1.29 \pm 0.58}$$

$$F_{N,z=0} = -18 \, U^{1.32 \pm 0.54}$$

$$|F|_{Ca,z=0} = 88 \, U^{1.27 \pm 0.17}$$

$$F_{Mg,z=0} = 48 \, U^{0.86 \pm 0.09}$$

$$F_{HCO_3,z=0} = -310 \, U^{1.22 \pm 0.52}$$

where U is in cm yr^{-1} and the fluxes $F_{i,z=0}$ are in units of μmol cm^{-2} yr^{-1}. Negative values of the flux, such as for phosphorus, ammonia, and bicarbonate, indicate flux out of the sediment into the overlying water. Positive fluxes indicate transport of material from the ocean to the sediment. For calcium in pore waters, both positive and negative values are encountered in different sediments, suggesting that the fluxes of calcium are to a considerable degree dependent on the mineral–pore-water relations and the mineralogical composition of the sediment (Lerman & Lietzke 1977).

In rapidly deposited sediments (large U), the rates of decomposition of organic matter and the corresponding rates of production or consumption (κ_i) of phosphorus, nitrogen, carbonate, and sulfate are high, as explained under Equation (20). The higher production rates generally produce stronger concentration gradients, which are responsible for the stronger diffusional fluxes. The building up of the sediment–

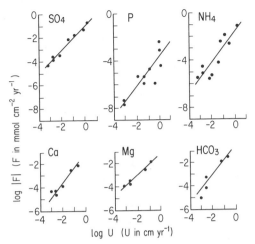

Figure 4 Fluxes at the sediment-water interface (F) of near-shore and pelagic sediments as a function of the sedimentation rate (U). Equations for the straight lines $F \propto U$ are given in the text. As F can be either <0 or >0, the plot shows $|F|$ against U. From Toth & Lerman (1977) and Toth (1977).

pore-water column at a faster rate also enhances the flux across the sediment-water interface through the term ϕUC in Equations (3) and (12), although the latter may be of opposite direction to the diffusional flux.

The sediment-water interface fluxes plotted in Figure 4, if extrapolated to the deep-ocean bottom area, give residence times in ocean water—(mass dissolved in ocean)/$F_{z=0}$—of the order of 10^7 to 10^8 yr. These residence times are 10^1 to 10^3 times longer than the residence times of the same chemical species with respect to input in river inflow. Alternatively, the fluxes across the sediment-water interface are weaker than the river input flux. As the computed fluxes $F_{z=0}$ shown in Figure 4 are based on concentration against depth profiles in long sections of sediment pore waters, the flux values represent the effects of diagenetic reactions throughout the entire length of a sediment section. In the notation of Figure 1, these are the fluxes labeled E.

DISCUSSION OF THE FLUX EQUATION

Explicit flux equations for dissolved species, based on the relationship

$$F_{z=0} = -\phi D(dC/dz) + \phi UC|_{z=0}, \tag{22}$$

are discussed in this section with reference to the sediment-water interface model shown in Figure 2A. The model describes a semi-infinite layer of sediment pore water in contact with the overlying water, and the processes far below within the sediment column do not affect the fluxes near the sediment-water interface.

From the steady-state model given in Equation (17), with the boundary condition of constant concentration $C_{z=0}$ at the sediment-water interface ($z = 0$), the flux $F_{z=0}$ is

$$F_{z=0} = \frac{-\phi D(\pm J_*)}{D\kappa_i^2/U^2 + \kappa_i - k}\left(\frac{\kappa_i}{U} + \frac{U}{2D} - R\right) - \phi D\left(C_{z=0} - \frac{J}{k}\right)\left(\frac{U}{2D} - R\right) + \phi U C_{z=0},$$

(23)

where the parameters are as defined in Equations (17)–(19), and

$$R = \left(\frac{U^2}{4D^2} + \frac{k}{D}\right)^{1/2} = \frac{U}{2D}\left(1 + \frac{4kD}{U^2}\right)^{1/2} \quad (\text{cm}^{-1}).$$

(24)

The term $+J_*$ denotes production of a species in pore water, and $-J_*$ denotes consumption or removal by bacterially mediated reactions.

The direct proportionality relationship between the sediment-water interface flux and sedimentation rate, as given in Equation (21), is a consequence of the correlation between the reaction rate constant κ_i and U^2 shown in Equation (20). First, the term $\phi U C_{z=0}$ in Equation (22) shows that the flux is to some extent proportional to the sedimentation rate U. Additional contribution to the proportionality relationship $F_{z=0} \propto U$ arises from the following. Parameter R in Equation (24) is approximately proportional to the first power of U, $R \propto U$. If $constant \times U$ is substituted for R, and $constant \times U^2$ is substituted for κ_i in Equation (23), then it can be verified by inspection that $F_{z=0} \propto U$ is approximately valid, as also suggested by the data plotted in Figure 4.

Equation (23) can be simplified for a number of particular cases, where some of the chemical or transport parameters can be neglected.

1. $k \ll U^2/4D$. This is the case when the dissolution or precipitation reaction rates, given by the value of the rate constant k (yr^{-1}) in Equation (14), are slow in comparison to the rates of removal of the dissolved species by molecular diffusion and pore-water advection. In this case we have $U/2D - R \simeq 0$, and Equation (23) becomes

$$F_{z=0} = -\phi U\left(\frac{D(\pm J_*)}{D\kappa_i + U^2} - C_{z=0}\right).$$

(25)

Equation (25) also shows that the inorganic dissolution and precipitation reactions (k) are insignificant in comparison to the organic matter reactions (κ_i and J_*).

2. $\kappa_i = 0$. In this case there are no organic matter decomposition reactions, and dissolved species are involved in inorganic reactions only. For this case Equation (23) becomes

$$F_{z=0} = -\phi D\left(C_{z=0} - \frac{J}{k}\right)\left(\frac{U}{2D} - R\right) + \phi U C_{z=0}.$$

(26)

Equation (26) is the flux equation for a semi-infinite pore-water layer within which concentration approaches a steady value $C_{z=\infty} = J/k$ at depth, and concentration gradient tends to zero. Concentration profiles of this type are shown in Figures 5–7.

2a. $U = 0$. This is the case of zero sedimentation rate and zero of pore-water advection relative to the sediment-water interface. With $U = 0$, Equation (26) or

(23) reduces to

$$F_{z=0} = \phi(kD)^{1/2}(C_{z=0} - J/k).$$ (27)

2b. $k = 0$. Because of the definitions of parameters k and J in Equations (15) and (16), the case of $k = 0$ formally corresponds to a constant dissolution rate of a solid, $J = k_1(C_{S,1} - C_{S,2})(\text{g cm}^{-3}\,\text{yr}^{-1})$, or precipitation rate if $J < 0$. When k tends to 0, then it follows from Equation (24) that $U/2D - R \simeq -k/U$. Consequently, the flux defined in Equation (26) becomes

$$F_{z=0} = -\phi DJ/U + \phi U C_{z=0}.$$ (28)

2c. $J = 0, k = 0$. This is a case of no inorganic reactions between dissolved species and solid sediment. In the absence of any chemical reactions, there are no concentration gradients in pore water and the flux across the sediment-water interface is due only to the growth of the sediment–pore-water column

$$F_{z=0} = \phi U C_{z=0}.$$ (29)

Example: Phosphorus Flux

In Figure 5 are shown dissolved phosphorus concentrations in pore waters of a sediment core taken in the Timor Trough in the southwestern Pacific Ocean (Cook 1974). The concentration maximum near a depth of 50 m in sediment suggests that decomposition of organic matter may be responsible for the higher phosphorus concentration. Concentration gradients indicate that phosphorus may be transported in pore water by molecular diffusion, upward and downward from the concentration peak depth. For the upward flux, the sink is ocean water. For the downward flux, the sink is likely to be the solid particles in the sediment. The curve drawn

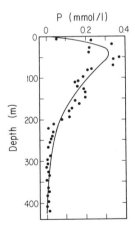

Figure 5 Phosphorus in sediment pore waters. Deep-Sea Drilling Project Site 262. The dots are concentrations reported by Cook (1974), and the computed curve is from Toth & Lerman (1977): $C = 0.746\exp(-0.011\,z) - 0.726\exp(-0.042\,z)\,\text{mmol l}^{-1}$, where z is in meters.

through the points was computed from the model based on Equations (17)–(19) using the following values of the transport and chemical kinetic parameters:

$D = 1 \times 10^{-6}$ cm^2 sec$^{-1} = 32$ cm^2 yr^{-1}

$U = 13$ cm 1000 yr$^{-1} = 1.3 \times 10^{-2}$ cm yr^{-1}

$\phi_{z=0} = 0.7$

$C_{z=0} \simeq 0.02$ μmol cm^{-3}

$J_* = 6.5 \times 10^{-6}$ μmol cm^{-3} yr^{-1}

$J = 1.6 \times 10^{-11}$ μmol cm^{-3} yr^{-1}

$k = 1.9 \times 10^{-6}$ yr^{-1}

$\kappa_P = 5.4 \times 10^{-6}$ yr^{-1}

In the above list of numerical values, the rate of sedimentation U is from Heirtzler et al (1974), and the values of the other parameters, as well as the method of computation of the concentration profile, have been given in Toth & Lerman (1977). The porosity at the sediment-water interface (ϕ) and diffusion coefficient of dissolved phosphorus in pore water (D) are assumed values. The remaining chemical parameters were evaluated from the concentration against depth profile.

The kinetic parameters κ_P and J_* represent the production of dissolved phosphorus through decomposing organic matter in sediment, whereas the parameters k and J represent the mineral sinks. Concentration in pore water at "infinite" depth in the sediment column is virtually zero or, in terms of the model, $J/k = 8.4 \times 10^{-6}$ μmol cm^{-3}. The rate constants for the organic production (κ_P) and mineral sink reactions (k) are comparable one to another; the values of k and the quotient $U^2/4D$ are also similar, suggesting that none of the processes (that is, organic production, removal into solids, diffusion in pore water, and transport by pore water advection) included in the model should be excluded for the sake of simplification without considering first the complete flux model given in Equation (23).

Using the listed values of the kinetic and transport parameters successively in the flux Equations (23) through (29), the following estimates of the flux at the sediment-water interface are obtained:

Equation (23) $F_{z=0} = -4.6 \times 10^{-3}$ μmol cm^{-2} yr^{-1}

Equation (25) $F_{z=0} = -5.4 \times 10^{-3}$

Equation (26) $F_{z=0} = +0.23 \times 10^{-3}$

Equation (27) $F_{z=0} = +0.11 \times 10^{-3}$

Equation (28) $F_{z=0} = +0.18 \times 10^{-3}$

Equation (29) $F_{z=0} = +0.18 \times 10^{-3}$

In the above tabulation of the flux values, Equation (23) gives the most complete estimate that takes into account all the processes included in the model. The flux of dissolved phosphorus owing to the growth of the sediment column is given by Equation (29), and it amounts to less than 5% of the flux out of the sediment, -4.6×10^{-3} μmol cm^{-2} yr^{-1}. The estimates of $F_{z=0}$ based on Equations (26)–(28) that take into account only inorganic reactions but ignore the production of phosphorus from organic matter are erroneous. Equation (25) ignores the role of the mineral sinks for dissolved phosphorus, and the flux estimate based on it is therefore somewhat higher than the estimate based on the more complete Equation (23).

SILICA IN THE INTERFACE ZONE

Deep Sea

Concentration profiles of silica in pore water of the eastern Caribbean Sea sediments, within several tens of centimeters below the sediment-water interface, are shown in Figure 6. The increase in concentration from the near-bottom value of 28 μmol SiO$_2$ l^{-1} is well pronounced. The magnitude of 100 to 400 μmol l^{-1} is characteristic of the solubilities of clay minerals and zeolites in sea water near 25°C and atmospheric pressure (Lerman et al 1975). The positive concentration gradients within the upper 5 cm of the sediment–pore-water column show that silica concentration increases six to tenfold over a short distance. From about 5 cm depth down, some of the profiles remain nearly constant with depth, and some show a slight decrease in concentration, suggesting the presence of mineral sinks in sediments. The shape of the profiles displaying a concentration maximum can be compared with the dissolved phosphorus profile, shown in Figure 5. Despite the large differences in the lengths of the sediment sections, the mechanisms responsible for

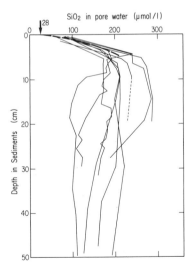

Figure 6 Silica in pore waters of sediments in eastern Caribbean Sea. SiO$_2$ concentration in near-bottom water shown by arrow. Heavy dashed line shows a generalized concentration gradient below the interface. (K. A. Fanning, personal communication, 1977).

the occurrence of either phosphorus or silica concentration maxima are in principle the same: production of the dissolved species, as represented by the production rate term J_* in Equations (17)–(19), which is probably due to decomposition of organic matter in the case of phosphorus, and dissolution of skeletal and, perhaps, mineral material in the case of silica.

The flux of silica across the sediment-water interface, according to Equation (22), is the sum of the diffusional and advective fluxes. Taking for the sediment porosity at the interface $\phi = 0.7$, the diffusion coefficient of SiO_2 in pore water $D = 3 \times 10^{-6}$ cm^2 sec^{-1} = 100 cm^2 yr^{-1}, and the concentration gradient from Figure 6, the molecular diffusional flux, is

$$-\phi D \, \Delta C / \Delta z = -0.7 \times 100 \times (200 - 30) \times 10^{-3}/5$$
$$= -2.4 \, \mu mol \, cm^{-2} \, yr^{-1}$$

The diffusional flux is out of the sediment and its magnitude is comparable to the value of the silica flux E listed in Table 1.

The advective flux in pore water into the sediment, owing to inclusion of dissolved silica in the growing pore-water column, at the rates of sedimentation between 1 and 20 cm 1000 yr^{-1}, is

$$\phi U C_{z=0} = 0.7 \times (1 \text{ to } 20) \times 10^{-3} \times 30 \times 10^{-3}$$
$$= 0.2 \times 10^{-4} \text{ to } 4 \times 10^{-4} \, \mu mol \, cm^{-2} \, yr^{-1}$$

In this case, the advective flux is negligibly small in comparison with the diffusional flux out of the sediment, such that an effective mean value for $F_{z=0}$ is essentially equivalent to the diffusional flux, about 2.4 μmol cm^{-2} yr^{-1}.

Near-Shore Sediments

Concentration profiles of dissolved silica in near-shore sediments similar in shape to those of the deep-sea pore waters (Figure 6) have been reported in the literature, and one example is shown in Figure 7A. The sediment is made of a mixture of biogenic carbonates and detrital alumino-silicates transported from soils and partly weathered volcanic rocks of the Island of Oahu, Hawaii. The sedimentation rate in the near-shore environment, from which the data shown in Figure 7A were obtained (Kaneohe Bay) is rapid, about 4.5 cm yr^{-1}. The net rate of supply of silica to pore water, as measured by the rate parameter J in Equation (17), is also fast, $J = 0.18$ μmol cm^{-3} yr^{-1}. Below 100 cm depth, concentration is near a steady value of about $J/k = 0.35$ μmol cm^{-3}, and this gives the reaction rate parameter $k = 0.5$ yr^{-1}. The latter values of J and k can be compared with the much lower values of these parameters computed from the deep-sea pore-water profile of phosphorus in Figure 5.

Mineralogical composition of the fine fraction of sediment (particle size $< 0.2 \, \mu$m, Figure 7B) shows that the alumino-silicate mineral nontronite appears at depth of about 20 cm and its relative abundance increases with an increasing depth. A possible reaction between kaolinite, dissolved silica, and other cations (in sea water or in solid phases) producing nontronite can be written as

$$Al_2Si_2O_5(OH)_4 + 12H_4SiO_4 + \text{cations} \rightarrow 2(\text{cations})Fe_4AlSi_7O_{20}(OH)_4.$$

kaolinite   nontronite

Combination of the data in Figures 7A and B suggests that nontronite grows at silica concentrations higher than about 250 μmol/liter and it constitutes a sink for silica in pore water. If the removal of silica from pore water at depths $z > 20$ cm is considered as a first-order precipitation reaction, then the removal rate is $k(C - C_{z=20})$ μmol cm^{-3} yr^{-1}. The length of time precipitation takes place is $(z-20)/U$ yr, where U is the sedimentation rate (cm yr^{-1}). Thus the amount of silica at any depth $z > 20$ cm removed from pore water into solids of density ρ_s (g cm^{-3}) is

$$\frac{\phi k(C - C_{z=20})(z-20)}{(1-\phi)\rho_s U} \quad \text{mol SiO}_2 \text{ per 1 g of solids.}$$

Silica concentration in pore water below 75-cm depth (Figure 7A) is nearly constant and the difference $C - C_{z=20}$ is therefore also constant with depth. If porosity does not change appreciably below 75 cm, then doubling of the distance $(z-20)$ cm in sediment corresponds to doubling of the amount of silica precipitated. If all the precipitated silica goes into the formation of nontronite, then the nontronite fraction in sediment should increase linearly with depth below 75 cm. This is nearly the case, as shown by the few points plotted in Figure 7C: below 75-cm depth, the abundance of nontronite increases approximately linearly with depth. The fraction of nontronite in sediment, 4% at $z = 75$ cm, increases by a factor of two, to 8% at $z = 130$ cm, and this is also an increase by a factor of two in the distance measured from the 20-cm depth level $(z-20)$.

The occurrence of nontronite in the very fine size fraction of the sediment and its low concentration in bulk sediment point to the general difficulties that are encountered in identification of the mineral sinks for dissolved chemical species in pore waters. The presence or absence of mineral sinks controls the magnitude of

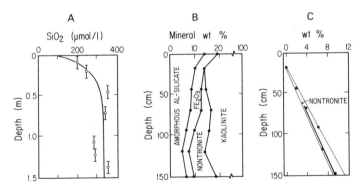

Figure 7 Dissolved silica and mineralogical composition of fine fraction (<0.2 μm) of sediments in a core from Kaneohe Bay, Island of Oahu, Hawaii (B. L. Ristvet, personal communication and Ph.D. dissertation, Northwestern University, 1977). (A) SiO$_2$ concentration in pore water, computed curve is $C = 0.34 - 0.24 \exp(-0.055 z)$ mmol l^{-1}, where z is in cm (Lerman, 1975). (B) wt % of minerals in the fine sediment fraction (<0.2 μm) amounting to $\sim$0.5 wt % of dry sediment. (C) wt % of nontronite in the fine sediment fraction. Two thin lines bracket nontronite weight percentages. Thicker line is a computed increase in nontronite abundance with depth, as explained in the text.

diffusional fluxes of reactive species in or out of sediments, even though the mass of the mineral sink can be very small in comparison with the bulk of solid phases making the sediment.

SUMMARY

Fluxes of chemical species across the sediment-water interface are generated by (a) deposition of sediments, (b) growth of the sediment–pore-water column taking place in the course of deposition, and (c) diffusion in pore waters. For different chemical species, the relative importance of each of the three fluxes depends on the nature and type of chemical reactions between solids and solution. When uptake or adsorption from solution is strong, the flux on solids usually predominates over the two other modes of transport. When the rate of sedimentation is slow, then the diffusional fluxes (provided concentration gradients exist) can be the main transport mechanism upward or downward across the sediment-water interface.

For those chemical species, concentrations of which in sediment pore waters are affected by biochemical reactions (sulfate, bicarbonate, phosphate, and nitrogen species), the fluxes into or out of sediments are approximately directly related to the sedimentation rates. The reasons for this relationship are probably in the faster rates of organic matter reactions in sediments that are deposited faster: in areas of slower deposition rate, less reactive organic material reaches the bottom.

A global geochemical role of the oceanic sediment-water interface is reflected in the amounts of materials, such as biogenic calcium carbonate and silica, that are being regenerated to ocean water during their residence time at the interface. Of the total of biogenic skeletal materials reaching the ocean bottom, significant fractions dissolve before burial in sediments. In general, regeneration of biogenic materials at the sediment-water interface can significantly affect the chemical composition of overlying water and it can be, in human terms, environmentally harmful when relatively large quantities of material dissolve, oxidize, or decay in a relatively small volume of water, such as in seasonally stratified eutrophic lakes.

ACKNOWLEDGMENTS

I thank Dr. K. A. Fanning (University of South Florida) and Mr. J. Tschopp (EAWAG) for permission to quote their unpublished data. Part of the material reported in this paper was obtained through work supported by the National Science Foundation grant OCE-75-13844. Support received from EAWAG and from the John Simon Guggenheim Memorial Foundation of New York is gratefully acknowledged.

Literature Cited

Aston, S. R., Duursma, E. K. 1973. Concentration effects on ^{137}Cs, ^{65}Zn, ^{60}Co, and ^{106}Ru sorption by marine sediments, with geochemical implications. *Neth. J. Sea Res.* 6:225–40

Berner, R. A. 1974. In *The Sea*, ed. E. D. Goldberg, 5:427–50. New York: Wiley
Berner, R. A. 1976. Inclusion of adsorption in the modelling of early diagenesis. *Earth Planet. Sci. Lett.* 29:333–40

Brúland, K. W., Bertine, K., Koide, M., Goldberg, E. D. 1974. History of metal pollution in Southern California coastal zone. *Environ. Sci. Technol.* 8:425–32

Cerrai, F., Mezzandri, M. G., Triulzi, C. 1969. Sorption experiments of strontium, cesium, promethium, and europium on marine sediment samples. *Energ. Nucl.* (*Milan*) 16:378–85

Cook, P. J. 1974. Geochemistry and diagenesis of interstitial fluids and associated calcareous oozes, Deep Sea Drilling Project. Leg 27, Site 262, Timor Trough. *Initial Rep. Deep Sea Drill. Proj.*, 27:463–80. Washington, D.C.: GPO

Erlenkeuser, H., Suess, E., Willkomm, H. 1974. Industrialization affects heavy metal and carbon isotope concentrations in recent Baltic Sea sediments. *Geochim. Cosmochim. Acta* 38:823–42

Fanning, K. A., Pilson, M. E. Q. 1974. The diffusion of dissolved silica out of deep-sea sediments. *J. Geophys. Res.* 79:1293–97

Goldhaber, M. B., Aller, R. C., Cochran, J. K., Rosenfeld, J. K., Martens, C. S., Berner, R. A. 1977. Sulfate reduction, diffusion, and bioturbation in Long Island Sound sediments. *Am. J. Sci.* 277:193–237

Guinasso, N. L., Jr., Schink, D. R. 1975. Quantitative estimates of biological mixing rates in abyssal sediments. *J. Geophys. Res.* 80:3032–43

Heirtzler, J. R., Veevers, J. J., Bolli, H. M., Carter, A. N., Cook, P. J., Krasheninnikov, V., McKnight, B. K., Proto-Decima, F., Renz, G. W., Robinson, P. T., Rocker, K., Jr., Thayer, P. A. 1974. Site 262. *Initial Repts. Deep Sea Drilling Project* 27:193–278. Washington, D.C.: GPO

Iskandar, I. K., Keeney, D. R. 1974. Concentration of heavy metals in sediment cores from selected Wisconsin lakes. *Environ. Sci. Technol.* 8:165–70

Lerman, A. 1975. Maintenance of steady state in oceanic sediments. *Am. J. Sci.* 275:609–35

Lerman, A. 1977. In *The Sea*, ed. E. D. Goldberg, 6:695–738. New York: Wiley

Lerman, A., Lal, D. 1977. Regeneration rates in the ocean. *Am. J. Sci.* 277:238–58

Lerman, A., Lietzke, T. A. 1975. Uptake and migration of tracers in lake sediments. *Limnol. Oceanogr.* 20:497–510

Lerman, A., Lietzke, T. A. 1977. Fluxes in a growing sediment layer. *Am. J. Sci.* 277:25–37

Lerman, A., Mackenzie, F. T., Bricker, O. P. 1975. Rates of dissolution of aluminosilicates in seawater. *Earth Planet. Sci. Lett.* 25:82–88

Lerman, A., Weiler, R. R. 1970. Diffusion and accumulation of chloride and sodium in Lake Ontario sediment. *Earth Planet Sci. Lett.* 10:150–56

Li, Y.-H., Gregory, S. 1974. Diffusion of ions in sea water and in deep-sea sediments. *Geochim. Cosmochim. Acta* 38:703–14

Manheim, F. T. 1970. The diffusion of ions in unconsolidated sediments. *Earth Planet. Sci. Lett.* 9:307–9

Menard, H. W. 1974. *Geology, Resources and Society*, p. 253. San Francisco: Freeman

Morse, J. W. 1974. Calculation of diffusive fluxes across the sediment-water interface. *J. Geophys. Res.* 79:5045–48

Nace, R. L. 1967. Water resources: a global problem with local roots. *Environ. Sci. Technol.* 1:550–60

Stiller, M., Carmi, I., Münnich, K. O. 1975. Water transport through Lake Kinneret sediments traced by tritium. *Earth Planet. Sci. Lett.* 25:297–304

Thorstenson, D. C., Mackenzie, F. T. 1974. Time variability of pore water chemistry in recent carbonate sediments, Devil's Hole, Harrington Sound, Bermuda. *Geochim. Cosmochim. Acta* 38:1–19

Toth, D. J. 1977. *Organic and inorganic reactions in near-shore and deep-sea sediments*, PhD thesis. Northwestern Univ., Evanston, Ill. 159 pp.

Toth, D. J., Lerman, A. 1977. Organic matter reactivity and sedimentation rates in the ocean. *Am. J. Sci.* 277:465–85

Vanderborght, J.-P., Wollast, R., Billen, G. 1977. Kinetic models of diagenesis in disturbed sediments. *Limnol. Oceanogr.* 22:787–803

Wimbush, M., Munk, W. 1971. In *The Sea*, ed. E. D. Goldberg, 4:731–58. New York: Wiley

Wollast, R., Garrels, R. M. 1971. Diffusion coefficient of silica in seawater. *Nature* (*Phys. Sci.*) 229:94–96

Ann. Rev. Earth Planet. Sci. 1978. 6:305–24
Copyright © 1978 by Annual Reviews Inc. All rights reserved

CRITICAL MINERAL RESOURCES

×10096

Charles F. Park, Jr.

Department of Applied Earth Sciences, Stanford University, Stanford, California 94305

About 2000 minerals are known in the crust of the earth, but of these fewer than 100 furnish the great bulk of materials that have enabled man to create modern civilization. While the future availability of organic or "energy" minerals, coal, petroleum, and natural gas is one of the most critical problems facing the world today, this subject has been discussed so much that little can be added to what has already been said (Park 1975). Therefore this review is limited to a discussion of the equally vital but less publicized inorganic minerals, the ores of the metals and the industrial minerals. Many articles have been published on this subject in the recent past, and much of the material is repetitive. However, I have tried to be selective in the bibliography, including only those references I consider most useful. Statistics concerning consumption and production of individual countries (especially in the third world) are not published regularly and usually are difficult to obtain. Very few countries—even the developed nations—publish anything comparable to the Minerals Yearbooks of the U.S. Bureau of Mines.

Owing to the vagaries of nature, minerals are irregularly distributed throughout the world in deposits of diverse sizes and widely varying mineral contents. No country is self-sufficient in all mineral resources. All must import some materials, and many of the lesser industrialized countries depend upon the export of their surpluses to keep their economies healthy. The largest consumers of raw materials are, of course, the heavily industrialized nations of western Europe, Russia, Japan, and the United States, but many of the developing countries are now establishing industrial complexes and are increasing their consumption of raw materials faster than are the more developed nations.

The demands for all raw materials have increased very rapidly during the past few decades because of the world's growing human population and the increasing pressure for generally higher living standards. Many developing nations depend almost entirely upon the export of raw materials for foreign exchange and trade. Understandably these countries wish to obtain maximum benefits from the sale of their products and are reluctant to deplete the raw materials they may need in the future. Almost without exception they have chronic unemployment, and to provide needed jobs they want their mineral products to be processed and fabricated in the country of origin. Greatly increased worldwide demands, fears of shortages,

305

and the success of the Arab embargo in 1973 on petroleum exports to the western nations have resulted in uncertainty and distrust among the industrial nations that depend upon imports for their survival. They are making widespread efforts to diversify sources of supply and to find new deposits. Supplies must be guaranteed for the future if standards of living are to be maintained or improved (MacGregor 1976).

A study of production statistics published by the United States Bureau of Mines (Annual Minerals Yearbook), combined with tables of population growth published by the United Nations, shows clearly that world wide, per capita consumption of raw materials is increasing at a rate faster than the population growth. Table 1 shows the growth rate in consumption of copper and lead between the periods 1960 and 1970 related to the human population growth rate. Such accelerating rates of consumption can be maintained for a few years, possibly to the end of this century, but it must be recognized that resources are finite and that many of them are rapidly being exhausted. Absolute exhaustion may never take place but a material might as well be used up if prices are driven up by scarcity so people cannot afford to buy what is needed. Population increases must be brought under control or all efforts to improve living standards will fail and hunger and existence at subsistence levels will result (Kesler 1976, Marsden 1975).

The United States is self-sufficient in only a few materials: boron, molybdenum, phosphorus, salt, silicon, sulfur, and some of the other industrial minerals such as limestone, dolomite, and a few others (U.S. Bureau of Mines 1976, Brobst &

Table 1   Per capita consumption of copper and lead, 1960 and 1970[a]

World 1960: population 3 billion people
       1970: population 3.3 billion people
U.S.A. 1960: population 180 million people
       1970: population 200 million people

Copper consumption
World 1960: 4,724,000 short tons = 3.15 pounds per person
       1970: 7,139,000 short tons = 4.32 pounds per person
U.S.A. 1960: 1,422,000 short tons = 15.0 pounds per person
       1970: 2,079,000 short tons = 20.8 pounds per person

Lead consumption
World 1960: 2,629,000 short tons = 1.75 pounds per person
       1970: 3,707,000 short tons = 2.25 pounds per person
U.S.A. 1960: 1,021,000 short tons = 11.4 pounds per person
       1970: 1,422,000 short tons = 14.22 pounds per person

On a worldwide basis, including the U.S.A., consumption over the period 1960–1970 has increased:
Copper: 1.17 pounds per person in world or 37%
        5.79 pounds per person in U.S.A. or 39%
Lead: 0.5 pounds per person in world or 29%
        2.8 pounds per person in U.S.A. or 25%

[a] Data from U.S. Bureau of Mines; reproduced in Park (1975).

Pratt 1973). Copper possibly should be added to this list inasmuch as the nation can meet its needs during times of emergency and when copper prices are high. Many copper deposits in the United States are low in grade and cannot compete with higher-grade and cheaper foreign ores. At present the United States imports nearly 100% of the ores of 9 materials that it requires—columbium-tantalum, amorphous graphite, industrial diamonds, quartz crystal, tin, manganese (without which no flaw free steels can be made), sheet mica, chromium (so essential in manufacturing stainless steel alloys), and the titanium mineral rutile. About 90% of others, the ores of aluminum, cobalt, beryllium, long-fiber asbestos, platinum metals, antimony, bismuth, and nickel, are imported. All other nonrenewable essential raw materials, including the ores of iron, lead, zinc, mercury, potassium, uranium, and others that are little known to the average person, such as fluorite, barite, tungsten minerals, and zirconium minerals, must be imported to a greater or lesser extent (U.S. Bureau of Mines 1977). Many can be obtained only from countries that have been characterized by political instability or unfriendly attitudes toward us. Our sources of supply for many commodities are insecure, especially during times of international crisis.

Australia, Canada, and South Africa are among the richest nations in the varieties and quantities of nonrenewable resources they possess. In the past they have furnished large quantities of raw materials to the industrial nations, and they have widely sought and welcomed foreign investments in the mineral industry. In this way they have used their resources to further their economic development and to improve the standards of living of their people while serving as steady, dependable suppliers of raw materials to industry throughout the world. In recent years, however, Australia and Canada appear to have become disenchanted with their roles and have become strongly nationalistic. They now insist upon national ownership and control and, by raising taxes and imposing restrictive legislation, they have discouraged foreign investors. The apartheid policies and racial unrest in South Africa have been equally discouraging to potential investors, although South Africa still produces more than 8% of the total value of nonenergy minerals of the non-Communist world (Rensburg & Pretorius 1977) and the South African government continues to offer attractive concessions to foreign entrepreneurs. As a result of these conditions, their former customers are searching for other sources of supply, and foreign venture capital no longer flows freely into these three countries. Development of their resources is lagging and almost certainly their standards of living will suffer.

Japan, Great Britain, and the nations of western Europe are minor producers of raw materials, but they are among the largest consumers. These nations generally maintain only small inventories and stockpiles and thus are dependent upon imports from many parts of the world. Without these steady imports of raw materials and the export of fabricated products their economies face serious trouble.

Growing needs for raw materials lead naturally to more intensive searches for substitutes and especially to greater exploration efforts. The growth of political and economic competition, including the rise of socialism and state ownership of raw materials and mineral deposits in many parts of the world, has become a dominant trend in recent years. The resulting uncertainties have curtailed some exploration efforts and have encouraged or forced industrial consumers to search in their own lands and to diversify their sources of supply as widely as possible. Private companies must, of necessity, cooperate with government-owned organizations and, at times, act as management consultants and sales agents. Examples of such government–private-industry cooperation are numerous and are growing. The governments of Gabon and Liberia retain part ownership of their iron ore deposits, which are operated by private industry, and the copper deposits of Iran are operated by the Anaconda company under contract to the Iranian Government. The search for new supplies and new sources of raw materials in politically stable countries, for example in the United States, has been given considerable impetus as a result of political activism elsewhere. A surprisingly large number of Canadian mining companies are changing their exploration efforts from Canada to the United States.

Mineral exploration is closely dependent upon resource management, and where governments control the management of resources they also control exploration. No longer can mining companies send geologists anywhere in the world and select their targets, because many areas are closed to them. Search is conducted under the auspices of the country of origin and in many developing nations trained personnel and financial resources to support such search are not available. Exploration must then depend upon agencies of the United Nations or a cooperative program organized with a friendly nation.

The extensive search for new mineral deposits has encouraged development of new methods of exploration. The use of geochemistry and geophysics is now standard practice in many places, and in addition such tools as aerial photography, airborne geophysical methods, and remote sensing have been improved and are finding increasing applications in reconnaissance selection of areas favorable for more detailed examinations. Lower grades of ore are being sought and larger tonnages are being mined as earth-moving equipment and transportation facilities improve. A few years ago ore containing 1% copper was considered to be low grade. Now, however, it can be mined nearly anywhere in the world if the body is large enough; in some districts 1% copper in ore is considered high grade. At places in western Canada and the United States, rock containing as little as 0.35% copper or 7 pounds per short ton is being mined.

Exploration groups are now constructing models for different types of ore deposits and such models are finding widespread use. Deposits such as the massive sulfide (Kuroko) ores of Japan, the porphyry or disseminated copper and molybdenum deposits of the western United States, and lead-zinc ores of the type found in the Mississippi Valley readily lend themselves to this kind of study (Hollister, Potter & Barker 1974). Criteria for the recognition of the various kinds of ore deposits may also be categorized and set up on a computer for ready use (Hart 1975).

Another area of increasing concern—and expense—to the minerals industry is the growing awareness of the need for a clean environment. Mine plants and particularly smelters are prime targets for environmental organizations, and mining companies are being forced to change many of their former practices. These changes, while long overdue, have been expensive and have diverted large blocks of capital from exploration and development. Increased costs, certainly in part resulting from environmental expenditures, are placing the United States mining companies at a disadvantage when competing with foreign companies that have no such costs, and few nations in the world have shown serious concern for the environment. Nevertheless, benefits have accrued to the companies in the form of improved smelter processes and metal recoveries. Another factor of great consequence to mineral exploration and development in the United States is the very large areas of government-owned lands that are being set aside for wilderness and recreational purposes in which these activities are prohibited. Some of these lands are among the best places in the country to search for the raw materials our economy so badly needs. A compromise will eventually have to be reached between mining and environmental objectives. Reasonable people believe it is possible to have both a clean and attractive environment and essential mining.

Because many minerals are in economic concentrations at only a few places in the world they readily lend themselves to the formation of cartel-like organizations and activities by both private companies and governments. Such organizations have been given a great boost by the successful activities of OPEC—The Organization of Petroleum Exporting Countries—in raising the price of petroleum products. Producers of other raw materials have been encouraged to emulate OPEC. Active and aggressive organizations of producers, or associations as they are frequently called, now exist in bauxite, copper, diamonds, iron ore, mercury, tin, and tungsten. The International Tin Council is unique in that it includes both producers and consumers and has as its purpose the stabilization of price and production of tin. The others are cartel-like organizations of producers and governments without participation of consumers. They are designed primarily to increase the amounts of money the producing nations obtain for their raw materials. To be effective a cartel must control more than 70% and as much more than that of the production of a commodity as possible. None of the associations named, other than the diamond cartel, yet controls this high a percentage of output, and to date they have been largely ineffective in dictating prices and rates of production. Moreover, most developing nations have economies too weak to enable them to keep their surplus raw materials off the international market for an extended period of time. They depend upon revenues from the sales of their raw materials for day-to-day operations of their governments. The requests for higher prices by developing nations in many instances appears to be justified: the industrial countries in the past have obtained many materials at unreasonably low prices.

The construction of ore deposits models and the recognition of broad types of ore deposits has led to considerable progress in the understanding of the genesis of the deposits. Of particular interest is the fact that many so-called insoluble elements can now be shown without doubt to travel in solutions in the form of complex ions (K. B. Krauskopf, personal communication 1977). While this fact may not result in the direct discovery of new ore deposits, it has resulted in a better understanding of the processes of ore transport and deposition; it may prove to be of great long term value. Geologists are still, however, unable to say why ores are localized in certain places whereas in other apparently identical geological environments the rocks are barren.

The theory of global or plate tectonics is of great interest to students of ore deposits, particularly to those seeking an explanation of metallogeny or the restriction of mineral deposits of specific types to certain segments or provinces of the earth's crust. One particularly fertile field for study is the relationships of ore deposits to different types of plate boundaries. For example, the massive sulfide deposits of Cyprus represent a type of ore associated with ophiolites and are thought to be related to an accreting plate margin such as the mid-Atlantic ocean ridge (Moores & Vine 1971). Other types of deposits, such as the porphyry coppers of Peru and Chile, are associated with calc-alkaline magmas formed by remelting along subduction zones and consuming plate margins. Still other types of ore may form along transform plate margins, margins of uncertain nature or location, and island arcs. Guild in particular has contributed to the understanding of these theories as applied on a broad scale (Guild 1972, 1974, 1976). Although discussions of the theory have helped clarify many problems of ore genesis, a great deal more work is needed before the theory of plate tectonics is generally applicable in detailed exploration.

Increased understanding of known deposits will undoubtedly aid in the timely application of theory to practice; e.g., Sillitoe & Sawkins attempt to relate the ore deposits of Peru and Chile to consuming plate margins and subduction zones (Sillitoe 1972a,b, 1975, Sawkins 1972). The ore deposits of Chile have long been recognized as being zoned parallel to the coast; hence, these are parallel to the zone of subduction that lies just offshore (Ruiz 1965). In the deformed metamorphic rocks close to the coast are the magnetite and hematite deposits of El Tofo, Romeral, Algarrobo, and many others. Progressing eastward away from the coast are bedded or stratabound copper deposits such as Bandurrias, Portezuelo, Grupo Avion, Santa Teresa, and Guayacan, and numerous bedded deposits of manganese oxides, including those of Corral Quemada, Romero, and Coquimbana. These ores of copper and manganese are associated with andesitic tuffs and flows, limestones, and clastics. In the interior valley are the formerly productive gold and silver veins at Inca del Oro, Huantacaya, and Chanarcillo. Farther east are the extensive and well-known disseminated copper deposits associated with Tertiary igneous bodies. These include Chuquicamata, El Salvador, Potrerillos, Andina, Disputada, and El Teniente. The ore deposits of Chile thus furnish an unusually good example of zoning parallel to a subduction feature, and their study helps delineate favorable areas for exploration of certain types of deposits. The knowledge obtained from this type of examination may also prove helpful in other localities where relationships are less clearly defined.

Ore deposits in the interiors of the continents, for example the copper deposits of southwestern United States and the lead-zinc deposits of the Mississippi Valley, are difficult to relate convincingly and directly to global tectonics and to plate boundaries. Many geologists are inclined to attribute their origin to other processes

(Lowell 1974, Sillitoe 1975). One suggestion that has been emphasized by several geologists is that ore deposits in the interiors of the continents are associated with mantle plumes, especially where these formed at triple junctions of linear features in the underlying rocks (Burke & Dewey 1973, Wyman 1976). Many other ideas, too numerous to go into here, have also been advanced to explain the origin and location of these interior ore deposits. P. M. Hurley (personal communication 1975) suggested that during migration and breakup of the continents, the lands did not move equally throughout their masses. The twisting motion that resulted may have caused the opening in the interior of deep-seated fractures that tapped underlying sources of magma and minerals.

Kuroko or black ores are generally fine-grained volcanogenic ores typical of many deposits in the green tuff series in Japan. They have been defined as: "strata-bound polymetallic sulfide-sulfate deposits genetically related to Miocene felsic volcanism" (Sato 1974). Excellent descriptions are available (Ishihara 1974, Tatsumi 1970). The better understanding of the distribution and causes of localization of these ores has resulted in more effective exploration and the discovery of new deposits. Directly contributing to the understanding of the Kuroko type ores has been the discovery and study of the layered sulfides and mineral brines of the Red Sea (Degens & Ross 1969) and the investigation of many other geothermal areas (White 1974). The sources of the metals and methods of emplacement of the massive sulfide deposits have been considerably clarified, as shown by a modern description of Kuroko ore.

The Kuroko ores are usually associated with volcanic piles of felsic, andesitic, and at places basaltic tuffs and flows, and especially where these piles contain domes of rhyolite or dacite. The mineral bearing fluids broke through the shallow oceanic crust from an underlying source. Their contents were deposited in irregular beds interspersed with tuffs and flows on the floor of the ocean. The passageways of the ore-bearing fluids are marked by downward decreasing amounts of sulfides with a change at depth from galena and sphalerite to chalcopyrite and an increase in silica. The ores on the ocean floor were quickly covered with volcanic debris that locally was altered to a minor degree by the nearly spent mineralizing solutions from the underlying rocks. Such alteration is useful in indicating the presence of ore below. Japanese Kuroko ores are generally of Miocene age and they are little deformed or metamorphosed. Knowledge gained from their study has been applied in many places, Canada and Australia for example, in older and deformed terranes. Volcanic rocks throughout the world are now being subjected to long overdue examinations and a surprisingly large number of Kuroko-type ore deposits are being found.

Recent work by Roberts in the massive sulfide deposits of Saudi Arabia indicates that not all such deposits are of volcanogenic origin, and even the genesis of some of the shallow-layered deposits may be open to question (Roberts 1976). Roberts thinks that the Arabian deposits were possibly affiliated with volcanogenic processes during the early stages of deposition, but that later more dominant processes were directly related to shallow magmatic or epithermal processes. The deposits thus have a dual origin; they are volcanogenic with later epigenetic activities masking the early stages. Roberts has proposed that they be called epigenetic-volcanogenic or epivolcanic deposits.

While the continuing growth in demands for all materials is creating real shortages of some commodities, existing shortages of others such as copper and aluminum result from man-imposed restrictions of exploration, development, mining, and trade. Sufficient amounts of these metals exist to supply the needs of the world for years to come.

Individual commodities are described in the following section in order to better understand the problems of future supplies, substitutions, and uses.

Silver

Silver is an example of a metal that is in short supply internationally. Silver consumption has long exceeded annual primary production, and since the United States Treasury demonitized silver and sold most of its stock at auctions beginning in 1967, this situation has persisted. The shortfall has been made up from existing inventories, old coins, and silver released from hoards in the Far East, particularly in India. Silver is no longer a major factor in the monetary system, but it is still bought by many investors as a hedge against inflation and for speculative purposes. The use of silver in industry has also increased, particularly in the manufacture of film, where no substitute is available, and in electronic devices, where inferior substitutes may be used. Though the use of silver in jewelry can hardly be considered essential, silver is commonly substituted for the scarcer and higher priced gold. As the price of gold increases, more silver is substituted. Table 2 shows that the production of silver has increased at a very modest rate during the past forty-five years.

Most silver is now obtained as a byproduct from the mining of other metals, particularly copper, lead, and zinc. Hence when production of these metals is depressed for any reason, the production of silver also falls, even though larger amounts could be used. The production of silver therefore cannot be readily expanded or contracted. Mines where silver is the dominant metal are few indeed and are mostly small. The so-called bonanza deposits of the past—Comstock Lode, Pioche, Goldfield, Tonopah, and many others in the mountainous belt from Chile to Alaska—are mainly of historical interest, for they produce little or no metal. They were formed within a few hundred or at most a few thousand feet of the surface of the earth, and commonly show wide conspicuous zones of iron-stained and brightly colored alteration products and hence were easily detected by early prospectors. Very few bonanzas have been discovered in recent years. Only in Mexico, at Real del Monte and particularly in the Veta Madre vein system in the Guanajuato district, have new discoveries resulted in a considerable increase in production.

The needs for silver in industry continue to increase, and with continued growth of the photographic and electronic industries, consumption of silver seems likely to accelerate. So long as the difference between production and consumption can be met by old coins and from the recycling of various stocks and hoards, the market for silver will remain reasonably quiet with a slow steady upward pressure on price. When, however, the shortfall cannot be met, the consumption of silver will have to be curtailed and the price could rise dramatically. The ability of the mining industry to expand production does not appear to be good; within a few years silver everywhere could be in very short supply.

Gold

Gold is in the same situation as silver in that industrial consumption equals or surpasses primary production. Industry has long prospered using only a bare minimum of gold. Nevertheless, there are a few industrial processes that use gold exclusively, and such uses have been growing, albeit slowly, in spite of the high price and uncertainty of a steady supply of the metal. The marketing and pricing of gold have been monopolies of governments for so long that consumers are reluctant to rely on a supply that may later be diverted for monetary or government uses. The future of gold is still a political issue. Partly successful efforts by the United States to have gold removed from the international monetary system and the sale of gold from the United States Treasury and from the International Monetary Fund stocks resulted in temporary surplus in the marketplace with a consequent drop in the price of the metal to a point so low that many mines were forced to close. Gold is again slowly advancing in price, but gold mining cannot regain its vigor until the overhanging monetary supply is exhausted, put to use, or the price increases as people purchase gold to hedge against worldwide growing inflation. Many gold producers would like to see gold removed from politics and a free market established—a situation not easy to accomplish because people accept gold as a symbol of both wealth and beauty, and the metal is eagerly sought by both individuals and governments. Especially during times of inflation or fears of inflation, people seek and hoard gold. The future of gold in the international monetary system remains unclear, although the United States is pushing for its complete removal. The backing of currencies by gold so far has been the only successful means by which the issuance of unrestrained quantities of paper money has been controlled and by which effective monetary discipline has resulted.

Most gold mines operate on low-grade ores, many as low as 0.20 ounces per ton or of much lower grade in surface mines. The Witwatersrand district of South Africa produces about 70% of the annual output of new gold in the non-Communist world. Its mines are among the deepest in the world and some are operating at depths in excess of 3355 meters (11,000 feet). They are scarcely profitable at prices of $120–$130 per ounce of gold. No large new discoveries have been reported in recent years in spite of searches involving the expenditure of a great deal of time and money. Gold ore reserves throughout the world are gradually being exhausted as existing mines become less profitable, and unless new discoveries are made, gold, like silver, will more and more become a byproduct of other mines, particularly of the large open pit copper properties. The price of gold will then increase. Whether gold will become just another minor metal with minor industrial uses or will take a place as a major material is unknown.

Copper

Copper is an essential commodity with many industrial uses that have been growing rapidly. However, during the worldwide industrial slowdown of recent years many consumers of copper were caught with large inventories. As a result the price of the metal has been depressed to a point where most mines in the United States curtailed production and a few have been closed. Most developing countries that produce copper have not curtailed production, and thus have contributed to the current glut of the metal. Nevertheless, as inventories are reduced and international trade recovers, probably within about five years, the price will gradually strengthen and copper mining will be restored to its pattern of steady growth as in the past. Table 2 shows clearly how spectacular this growth has been. The consumption of copper is notoriously cyclical.

No actual shortage of copper exists at present and none is anticipated before the end of this century. Existing problems concerning copper that must be satisfied are those of financing, development, distribution, and pricing. Many large deposits are known that are not currently in production. Peru alone is said to have at least forty known deposits of disseminated ores that are largely unevaluated (F. C. Kruger, personal communication 1977). Some are in remote sections of the Andes mountains where they are unprofitable to develop at present prices of copper, but others might prove to be economical. Large initial capital investments, on the order of $300 to $1000 million or more, are required for development of each property, but well-substantiated fears of expropriation discourage private investments of this magnitude. Most developing countries have neither the personnel nor the the the money to proceed on their own.

The United States is nearly self-sufficient in copper, but many of its deposits are low in grade and the metal is expensive to extract. Most deposits are mined by open-pit methods, but several new ore bodies are being developed in underground mines where the numerous problems of extraction are both difficult and challenging. The mining and hoisting of 25,000 to 50,000 tons of ore a day from depths of 610–915 meters (2000 to 3000 feet) present engineering problems of the first magnitude.

A very large potential supply of copper is present in the widely distributed and abundant manganese nodules on the floors of the deep ocean basins. As many as 50 billion tons of nodules are estimated to be present (Frazer & Arrhenius 1972, Holser 1976, Horn, Delach & Horn 1973). These nodules appear to contain an average of about 2.5% of combined copper, nickel, and cobalt, and mining plans are for a cutoff grade of about 2%. The technology to permit economic mining and recovery of these nodules has been developed by private companies and is said to be available now. Production cannot be started, however, until the difficult questions of ownership and insurance are resolved. The complex legal problems are being considered by a Commission of the United Nations but have become bogged down in discussions of ownership and royalties. Should the deposits be owned and operated by a company or companies belonging to the United Nations, which would then receive all profits, or should private industry, which has perfected the technology, be permitted to mine the nodules, paying royalties to whoever owns the deposits? Present investments in exploration and technology are so large and the potential benefits to civilization so great that production should not be long delayed.

Aluminum

The ore of aluminum, generally called bauxite, forms as a residual product on the surface of deeply weathered rocks that were originally high in aluminum or where aluminum-rich products of weathering have been transported and deposited in sedimentary basins. Such ores are abundant in the heavy rainfall belts of the tropics, but are scarce or absent in the temperate, more heavily industrialized nations. The United States, for example, has only a few small transported sedimentary bauxite deposits in the coastal plains near the Gulf of Mexico. The United States imports over 90% of its aluminum ores, mostly from Jamaica and Surinam. Russia depends upon its satellites, particularly Hungary, for a goodly part of its needs for high-grade ore.

The principal problem in the manufacture of aluminum is not the availability of ore but rather the very large amounts of energy required to convert the ore to metal. Aluminum reduction works have therefore been located where cheap, abundant water power is available—places such as Norway, Kitimat in western Canada, and the basin of the Columbia River. There is no reason, however, why aluminum should not be produced in the tropical countries where the ore is abundant and where large rivers could furnish ample water power. Such a development has taken place in Ghana with the establishment of the Volta River reduction works. Elaborate plans to dam the Congo River near the Stanley rapids and to use the power generated to manufacture aluminum from the abundant nearby ores were abandoned when Zaire became an independent nation. Many such possibilities exist, and if implemented would have definite advantages for both the industrialized and underdeveloped nations. It seems inappropriate that industrial, energy-short countries should dissipate their badly needed electric power on the manufacture of aluminum metal when such manufacturing can be done effectively and cheaply in underdeveloped countries near the source of the ore. When the United States built the large dams on the Columbia River the disposal of surplus power was a problem, and establishment of aluminum reduction works was encouraged by selling power at reduced rates. Surplus power no longer exists in this area and, although continuation of aluminum reduction may perhaps be justified on the basis of self-sufficiency of a vital resource, the ore must be imported.

Shortages of aluminum ore seem unlikely in the foreseeable future, although problems of reduction and distribution caused by human and political dissent cannot be discounted. Strong attempts have been made by aluminum ore producers to organize a cartel. If developing nations in the tropics are able to finance construction of the needed dams and reduction works, they could eventually take over most of the aluminum industry, thus creating jobs and generating capital that they badly need. At the same time they would be helping to relieve the critical energy shortages in the industrial countries. Nations such as Brazil and Venezuela, which have reasonably stable governments, would seem to be ideally situated to establish viable aluminum reduction and fabrication plants.

Large quantities of clays and rocks that contain appreciable amounts of aluminum are widely distributed, and aluminum has been recovered from them in pilot plant tests. Although these deposits are considered to be potential sources of aluminum, their reduction to the metal requires considerably more power than does reduction of ordinary bauxite. At present, clays are too expensive to reduce to metal and require too much energy to be exploited.

Lead-Zinc

Lead and zinc are two common base metals that have wide industrial applications. Although lead has been removed from gasoline and from pigments and glazes, this decrease has been more than offset by the rapid growth in use of lead for storage batteries, and demands for lead have increased slowly in recent years. The uses of zinc have expanded much more rapidly—in galvanizing, pigments, vulcanizing of rubber, manufacture of brass and bronze, and many other alloys and die castings.

The two metals are found together commonly, but not everywhere. They are widely distributed, with the principal sources being Australia, Canada, Russia, Mexico, Peru, the United States, and several countries in Europe. Minor amounts come from many other places in Asia, Africa, and South America. The wide

Table 2   World production of several metals[a]

Year	Silver (Troy ounces)	Copper (metric tons)	Lead (metric tons)	Zinc (metric tons)	Iron Ore (metric tons × 000)
1930	248,139,100	1,573,000	1,659,810	1,382,270	181,860
1939	264,957,000	2,182,000	1,572,738	1,482,945	204,700
1950	192,000,000	2,687,000	1,560,040	1,762,300	248,920
1960	239,500,000	4,489,650	2,294,710	2,920,540	515,200
1961	236,900,000	4,643,840	2,399,015	3,247,060	502,515
1962	244,700,000	4,843,380	2,385,410	3,405,785	507,640
1963	250,300,000	4,970,360	2,462,505	3,473,810	522,305
1964	248,545,000	5,253,090	2,556,165	3,692,380	582,624
1965	259,415,000	5,536,890	2,640,370	3,947,780	620,965
1966	266,731,000	5,558,910	2,745,025	4,097,915	635,810
1967	259,081,000	5,386,725	2,773,200	4,126,550	625,385
1968	275,264,000	5,488,095	2,944,940	4,626,570	678,830
1969	290,469,000	6,005,750	3,270,600	4,963,535	718,495
1970	301,745,000	6,225,600	3,366,600	4,904,265	766,365
1971	294,713,000	5,978,710	3,272,600	4,742,665	787,072
1972	301,510,000	6,657,025	3,410,200	5,120,910	777,712
1973	308,584,000	7,109,035	3,546,500	5,256,385	849,825
1974	295,562,000	7,390,665	3,584,600	5,413,140	893,485
1975	291,100,000		3,196,300		

[a] Data from U.S. Bureau of Mines, Silver Institute, and Lead Institute.

distribution and abundance of the ores has meant in the past that they have been readily available on the market and the prices have been reasonable. The largest deposits are replacement ore bodies in Precambrian schists at Broken Hill, Australia, and the Sullivan mine in Canada, and the bedded or stratabound deposits of the Mississippi Valley type in limestones and dolomites.

A few large new districts have been found in recent years in Tennessee and southeastern Missouri, and no shortage of either lead or zinc exists at present. However, reserves are not large when compared with the demands, and shortages could develop toward the end of the century if consumption continues to grow. Zinc appears to be more vulnerable to shortages than does lead. Table 2 shows the annual production of lead and zinc in the world from 1930 to 1975. Note especially the rapid growth in production of zinc in recent years.

Tin

Tin is a metal obtained from only a few places in the world. Southeast Asia, particularly Malaysia, Thailand, Indonesia, and China dominate the industry, but lesser amounts come from Australia, Bolivia, Nigeria, and in the past few years Brazil. With the exception of the Rondonia district in northwestern Brazil, which has been recognized for only about a decade, the other districts have been known and operated for many years. Malaysia produces about one third of the total world production, mostly from stream gravels and offshore sands. Tin is extracted from layers of pyrrhotite at the Rennison-Bell mine in Tasmania. Only in Bolivia and in several minor districts such as Cornwall, England, are vein mines being operated.

Substitutes are available for most uses of tin, and even though reserves of the metal gradually diminish and prices increase, shortages would cause inconvenience and fiscal difficulties in the producing countries but would not be of dire consequences for civilization.

Iron

Iron is one of the most abundant elements in the earth's crust and is also one of the most useful. Its ores are so plentiful and so widespread that no shortages of it are predicted for many years in the future. Exploration following World War II resulted in the discovery of many new large deposits, principally in Australia, Brazil, Canada, Venezuela, and west Africa—Angola, Gabon, Liberia, and Mauritania. Iron ores are so abundant and so widespread that nations possessing them are anxious to sell large quantities. Development is encouraged, competition is keen, and politically inspired difficulties are fewer than with many other minerals. Table 2 shows the world production of iron ores in recent years. By about 1990 the world will be using about one billion tons of iron ore a year.

Manganese

Manganese is one of the most useful of all metals. When added to steel (8 to 14 pounds of manganese to a ton of steel) it acts as a scavenger, collecting sulfur and oxygen into the slag, thus reducing bubbles and producing sound steel. In spite of extensive research, no good substitute for manganese in the manufacture of steel has been found. Manganese is thus essential to the maintenance of a heavy industrial complex.

The largest deposits of manganese are in Australia, Brazil, Gabon, India, Mexico, Russia, South Africa, and Zaire, although many other areas produce small quantities. All of the larger deposits apparently are of sedimentary or residual origin, and most, but not all, are closely associated with volcanic terranes, commonly andesitic. All industrial nations with the exception of Russia must import manganese. The United States presently is importing more than 98% of its needs.

A potentially large source of manganese and one that should forestall shortages for many years is the manganese nodules on the floors of the deep ocean basins. Estimates of as much as 50 billion tons of dry nodules have been made and appear to be well substantiated (Holser 1976). The grade of manganese in the nodules is far too low to be used without beneficiation (about 20% manganese in the nodules). Most commercial manganese ore at present averages 45% of manganese or more. The separation of manganese oxides from ore where they are intimately mixed with iron oxides, as they are in nodules, is difficult metallurgically, and most of the announced plans to treat nodules expect to recover the copper, nickel, and cobalt, but not the manganese or iron. However, some plans for manganese recovery do exist and as shortages develop, for whatever reason, these processes will be improved and their costs lowered. Shortages of manganese are not expected in the near future, and as ocean mining is advanced the possibility of such shortages will be moved even farther into the future.

Ferroalloy Metals

The group of metals known as the ferroalloy metals includes chromium, cobalt, molybdenum, nickel, tungsten, vanadium, and zirconium. The variable and diverse properties of steel depend in large part upon the presence of small amounts of these elements; they are essential to civilization. As offshore drilling and ocean mining are expanded, a great deal more stainless and corrosion-resistant steels will be needed. These steels cannot be made without the ferroalloy metals, and hence growth of the alloy industry seems to be assured.

Chromium is one of the best known and most widely used of the ferroalloy elements. Because of its pleasing color and resistance to oxidation and corrosion, it is a component in most stainless steels and in the superalloys used in aircraft. Three grades of chromium ores are recognized. The most widely used type is metallurgical ore, which contains about 46% of chromium. This is the material needed in the alloy industry. The second type of ore contains about 45% chromium and is used in the chemical industry and in pigments. The third type contains about 30 to 34% chromium and is used as a refractory, particularly to line furnaces where a nonreactive neutral refractory is needed. Most chromite deposits, independent of the type, are in layers in ultramafic intrusive bodies.

Chromium ore—chromite—is obtained from Russia, South Africa, Rhodesia, Philippine Islands, and Turkey. Minor amounts have also been obtained from eastern Europe, particularly from Albania, Yugoslavia, and a few other places. Very large reserves are said to exist in Russia, South Africa, and Rhodesia, but in other deposits the reserves are small and expensive to mine (Rensburg & Pretorius 1977). The ore is commonly present in layered mafic intrusive rocks such as the Bushveldt Complex in South Africa.

Because of the location of the principal reserves, the marketing of chromium ore has been beset with political problems that seem likely to continue for some years. None of the industrialized nations except Russia contains significant quantities of chromium ores.

Cobalt has many uses, particularly when alloyed with iron or carbon. Probably the most valuable is as a ferroalloy in the permanent magnets that enable electronic systems such as television, radio and many other devices to function. Cobalt is largely obtained as a byproduct from the stratabound copper ores of Zambia and to a lesser extent from Zaire, with smaller amounts coming from the mining of lateritic nickel ores. Supplies of cobalt are adequate at present and will be considerably augmented when laterite mining is expanded, and particularly when mining and recovery of deep-sea manganese nodules are implemented. Most industrial nations import all or part of the cobalt they require.

Nickel is another of the ferroalloy metals that is used in large quantities by the industrial nations. It is a common component of stainless steel alloys and as a native metal is used in coins and many other items that require a hard metal that is resistant to wear and corrosion.

Most nickel ore comes from Canada where, in the form of sulfides and arsenides, it is associated with mafic (noritic) igneous rocks at Sudbury, Ontario and Thompson Lake, Manitoba. Similar ores obtained from the Kambalda and other districts in western Australia are likewise associated with mafic igneous rocks. The laterites or oxide soils so widespread in the tropics also are becoming of increasing economic interest. Many serpentinites and other mafic rocks contain minor amounts of nickel. As these rocks weather, the nickel is carried down in the groundwater. It accumulates in the partly decomposed rock below the typical clay and pellet layers of the surficial laterites. Nickel in laterites is generally in the form of a silicate, and it is expensive and difficult to recover. However, the metallurgy is gradually being improved, and in many areas laterites are now being mined and the metal recovered profitably.

The laterite deposits of New Caledonia are well known, and at one time furnished most of the nickel used in industry. Similar deposits are now being worked in Australia, Brazil, Colombia, Cuba, Guatemala, Indonesia, and the Philippine Islands. Large accumulations of lateritic nickel ores are known in other areas, but because of difficulties of extraction these materials have been little used to date although more and more they are being sought and developed as the demands for nickel grow.

Still another potential source of very large amounts of nickel is the manganese nodules of the deep ocean basins. Larger and larger amounts of nickel will be used as the demands of the world grow, particularly in the increasing needs for stainless steels as more work is done in the oceans. Still it would appear as if sufficient quantities are available to satisfy the needs for many years provided that the processes for recovery from laterites and nodules can be perfected. Most industrial nations contain only small amounts of nickel ores, and they are dependent upon imports for their supplies.

Tungsten, when alloyed with iron, gives a hard, tough metal that will retain a cutting edge at high temperatures. For this reason it is indispensable as a cutting tool steel. When combined with carbon, tungsten also forms the very hard tungsten carbide, a material that is used as an edge in many cutting and drilling tools.

Tungsten comes principally from China, North and South Korea, and Russia, with lesser amounts from Portugal and Bolivia. The United States has one large tungsten mine near Lone Pine, California, and several other properties that generally are worked only when the price of the metal is high. The minerals that contain tungsten are those of the wolframite-huebnerite series, commonly found in veins, and scheelite, present in many zones of igneous or contact metamorphism. Search for new sources of tungsten have turned up numerous small deposits but none that will permit complacency. It seems as if shortages of tungsten are likely to become realities in a few years.

The United States is fortunate in having the world's largest deposits of molybdenum ores at the Climax and Henderson mines in Colorado. Molybdenum is also recovered as a byproduct during the concentration of many porphyry copper ores. Enough molybdenum ore is recovered in the United States to permit the export of considerable amounts annually. Molybdenum steel is hard and tough and is finding growing uses in the transportation industry. Most industrial nations other than the United States are dependent upon imports of this metal.

Two other widely used ferroalloy metals are vanadium and zirconium. Vanadium is obtained in the United States as a byproduct, largely from the flue dust in smelter stacks. Zirconium is extracted from the mineral zircon, a common constituent in many beach sands. Most zircon comes from Australia though large reserves are said to exist in the beach sands of India and Brazil (Park 1975). The governments of India and Brazil will not permit the mining of beach sands because of the content of radioactive thorium minerals they contain.

A glance at the distribution of the ferroalloy metals shows why the under-developed nations are so important to the welfare of the industrial countries. Industry is totally dependent upon the ferroalloy metals and, with the possible exception of Russia, all nations must import a large part of their needs from distant and, at times, unfriendly sources.

Uranium

The metal uranium has attained prominence in recent years because of its use in atomic armaments and nuclear power plants. Uranium, more than any other solid raw material, with the possible exception of gold, is a subject of controversy: environmentalists want to ban its use; power companies say that it must be used in nuclear power plants or shortages of power will occur within a few years; people are afraid of radiation leakage; and everyone is worried about the disposal of atomic waste.

The sources of uranium ore are principally Australia, Canada, Russia, South Africa and the United States. Other nations, such as Niger, Gabon, and

Czechoslovakia, have small but viable deposits. Uranium is present in many minerals, but probably the most common is uraninite or pitchblende, found as interstitial material in both layered sandstones and quartzites and in veins and replacement deposits of hydrothermal origin. Uranium is a byproduct of gold mining in South Africa where it is present in thucolite, a uraniferous carbon. Conglomerates similar to those in the Witwatersrand district of South Africa are present at Elliott Lake, Canada, and contain commerical quantities of uranium ores.

In recent years the price of uranium has jumped from about $6 per pound to $40 or more, and the search for additional deposits has correspondingly been stimulated. Not only the mining companies, but the oil companies, power generating companies, and makers of nuclear power equipment such as Westinghouse and General Electric have joined the search. Several new districts and extensions of known areas have been recognized, but large new districts capable of supplying the needs of the many nuclear power plants expected to be built in the next few years have not been found. The consensus of opinion among uranium exploration groups seems to be that sufficient uranium is available to provide for the energy needs of the next generation or so, but other sources of energy must be developed for the long-range future or the breeder reactor must become operative (Adams 1975, Atomic Industrial Forum 1975, Burnham et al 1974, Hanrahan 1975, Kroft 1976, Patterson 1975).

Fertilizer Minerals

Among the many nonmetallic minerals the fertilizer materials rank with the most valuable and most needed of all. The three common constituents in fertilizers are nitrogen, potassium, and phosphorus. Of these, nitrogen is obtained largely from natural gas and the atmosphere by the fixation process, but minor amounts come from the nitrate fields of the arid deserts of northern Chile and from the Guano deposits of Peru.

The salts of potassium are abundant in a few places. Saskatchewan, Canada, has very large reserves and is one of the major producers. In recent years the Saskatchewan Provincial Government has become strongly nationalistic and socialistic and, with compensation, has confiscated some of the deposits for operation by a state-owned company. Continued stringent governmental regulations and controls, and the well-founded fears of continued expropriation, have made Saskatchewan less suitable as a place for investment and development. Most companies are now looking elsewhere for deposits. Potassium salts are recovered in the United States in New Mexico and Utah and search is being conducted in North Dakota in an effort to find the southern continuation of the Canadian deposits at depth. Large amounts of potassium salts are recovered from saline deposits in East and West Germany, France, and Russia, but deposits elsewhere are minor. Potassium salts are generally interlayered with halite or ordinary salt, and their mining entails the removal of a large amount of this salt, which at times presents environmental problems.

Phosphate rock is obtained from only a few districts. The United States has considerable reserves in Florida, and in the shaley deposits of the western districts in Idaho and neighboring states. Much of the rest of the world's supply comes from Morocco, which recently annexed large additional reserves in what was formerly known as Spanish Sahara. Tunisia and Algeria also have smaller deposits. Russia obtains more than adequate phosphate rock from the apatite deposits in the Kola peninsula. Recent reports are that a large deposit of phosphate rock has been found in Australia and development is now underway to put this district into production.

Some of the largest and most populous nations of the world, India, China, Japan, Indonesia, and many nations of South America and Africa have neither phosphate rock nor potassium salt deposits in sufficient quantities to enable them to establish their own fertilizer industries based upon domestic raw materials. All of these nations are dependent upon imported fertilizers or fertilizer components to sustain their agriculture. In some nations, for example, India, the poorer farmers have insufficient cash resources to enable them to buy adequate fertilizers and the results show up in low yield of crops.

Fertilizers become more essential as the population of the world grows and the needs for food increase. Whereas temporary surpluses of fertilizer materials exist and lend a false sense of security and abundance, the reserves of phosphate rock, while large, give no cause for complacency. Morocco, which controls a large segment of the international market, has also shown a tendency to raise prices whenever possible and to attempt to create a monopoly. To prevent starvation the costs of fertilizers must be kept low enough so that the poorer nations can affford to buy what they need.

Sulfur

It has been said that the degree of industrialization of a country can be measured by its consumption of sulfur and sulfuric acid. Native sulfur (elemental sulfur) has many uses, in insecticides, fertilizers, chemicals, explosives, dye and coal tar products, paint and varnish, as a bleach in the manufacture of white paper, and in the processing and preserving of foods. As sulfuric acid the element is used in the manufacture of fertilizers, pigments, rayon and other cellulose products, industrial explosives, textiles, refining of petroleum, the chemical industry, and in the metallurgy of iron and steel and other metals.

The bulk of sulfur and sulfuric acid come from the United States, Canada, Poland, Mexico, France, and Russia, with minor amounts from the volcanic provinces of Italy, Japan, and South America. Sulfur is obtained from the tops of salt domes along the coast of the Gulf of Mexico and the Isthmus of Tejuantepec, Mexico, and from "sour" gas and petroleum products rich in sulfur. Large amounts of sulfuric acid and some elemental sulfur are recovered as byproducts from the smelting of metallic sulfide ores. In the future the washing of coal should also supply considerable sulfur. Sufficient sulfur exists to supply the needs of the world for many years in the future. Continued efforts to improve the environment and to obtain cleaner air mean that more and more sulfur will be removed from smelter gases, from petroleum and natural gas, and from coal. Sulfur and sulfuric acid are low-priced products and because of costs cannot be transported for long distances. As a result they may locally be in surplus supply. However, if demands continue to grow in the future as in the past such surpluses will be short lived.

Conclusions

Problems of supply of the nonrenewable resources are many and difficult. Shortages of a few metals such as gold and silver already exist, and supplies are insufficient for the needs of industry and government. With other metals such as lead, zinc, and tungsten, current supplies are adequate but serious shortages could develop around the turn of the century unless adequate conservation measures are taken or new supplies or substitutes are found. Other metals, iron, aluminum, and those in the manganese nodules of the deep-ocean basins are abundant and are sufficient to last for many years provided the problems of ownership, pricing, and trade are satisfactorily resolved. Clearly though, the economies of the world cannot continue to expand in the future as they have in the recent past because of serious shortages of several critical materials. The chances of the developing nations attaining standards of living equal to those of the major industrial nations are not good. It is doubtful that enough raw materials are available even to support the economies of present human populations at current living standards for long, and unless population growth is controlled, standards of living will deteriorate. Shortages and price increases of petroleum have already started the deterioration of some living standards, and shortages of essential minerals can only accelerate the process. Improved conservation, recycling, substitution of the more abundant materials for the scarcer ones, and research to better all phases of exploration, mining, and consumption are essential for the future if civilization is to prosper (National Committee on Materials Policy 1973).

Literature Cited

Adams, S. S. 1975. Problems in the conversion of uranium resources to uranium reserves. In *Mineral Resources and the Environment*, pp. 168–81. Comm. Min. Resour. Environ. Natl. Acad. Sci.

Atomic Industrial Forum 1975. Foreign resources of uranium. *Atomic Ind. Forum*, Dec. 1975, No. 89

Brobst, D. A., Pratt, W. P. 1973. United States mineral resources. *U.S. Geol. Surv. Prof. Pap. 820*

Burke, K., Dewey, J. F. 1973. Plume-generated triple junctions: key indicators in applying plate tectonics to old rocks. *J. Geol.* 81:406–33

Burnham, J. B., Brown, R. E., Enderlin, W. I., Hanson, M. S., Hartley, J. N., Hendrickson, P. L., Paasch, R. K., Rickard, W. H., Shreckhise, R. G., Watts, R. L., Gonser, B. W., Zegers, T. W. 1974. *Assessment of Uranium and Thorium Resources in the United States and the Effect of Policy Alternative*. Battelle Pac.

Northwest Lab. 236 pp.

Degens, E. T., Ross, D. A. eds. 1969. *Hot Brines and Recent Heavy Metal Deposits in the Red Sea*. New York: Springer

Frazer, L., Arrhenius, G. 1972. World-wide distribution of ferromanganese nodules and element concentrations in selected Pacific Ocean nodules. *Natl. Sci. Found. IDOE Tech. Rep. No. 2*

Guild, P. W. 1976. Discovery of natural resources. *Science* 191:709–13

Guild, P. W. 1974. *Application of global tectonic theory to metallogenic studies*. Paper presented at Meet. Int. Assoc. Genesis of Ore Deposits, Varna, Bulgaria, 1974

Guild, P. W. 1972. Metallogeny and the new global tectonics. *24th Int. Geol. Cong.*, Sect. 4, pp. 17–24

Hanrahan, E. J. 1975. Domestic and foreign uranium requirements. *ERDA Uranium Indust. Semin.*, Grand Junction, Colorado, 1975. 36 pp.

Hart, P. E. 1975. *A Computer Based Consultant for Mineral Exploration.* Menlo Park: SRI Int. 44 pp.

Hollister, V. F., Potter, R. R., Barker, A. L. 1974. Porphyry-type deposits of the Appalachian orogen. *Econ. Geol.* 69:627

Holser, A. F. 1976. *Manganese Nodule Resources and Mine Site Availability.* Ocean Mining Adm. U.S. Dept. Inter. 12 pp.

Horn, D. R., Delach, M. N., Horn, B. M. 1973. Metal content of ferromanganese deposits of the oceans. *Natl. Sci. Found. IDOE Tech. Rep. No. 3*

Ishihara, S., ed. 1974. Geology of Kuroko deposits. *Soc. Min. Geol. Jpn, Min. Geol. Spec. Issue No. 6.* 435 pp.

Kesler, S. E. 1976. *Our Finite Mineral Resources.* New York: McGraw Hill. 120 pp.

Kroft, D. J. 1976. The strategic position of the United States with respect to uranium supply and demand in the foreseeable future. PhD thesis. Stanford Univ., Calif.

Lowell, J. D. 1974. Regional characteristics of porphyry copper deposits of the southwest. *Econ. Geol.* 69:601–17

MacGregor, I. Nov. 1976. Mining and the continuing American revolution. *Min. Congr. J.* 62:102–5

Marsden, R. W., ed. 1975. *Politics, minerals, and survival.* Madison: Univ. Wisconsin Press. 86 pp.

Moores, E. M., Vine, F. J. 1971. The Troodos massif, Cyprus, and other ophiolites as oceanic crust: Evaluation and implications. *Philos. Trans. R. Soc. London Ser. A.* 268:443–66

National Committee on Materials Policy, Final Report 1973. Washington, D.C.: GPO

Park, C. F. Jr. 1975. *Earthbound.* San Francisco: Freeman, Cooper. 279 pp.

Patterson, J. A. 1975. *U.S. Uranium Situation.* Atomic Ind. Forum, Atlanta, Georgia. 23 pp.

Rensburg, W. C. J. van, Pretorius, D. A.

1977. *South Africa's strategic minerals—pieces on a continental chessboard.* Johannesburg: Colorpress. 156 pp.

Roberts, R. J. 1976. The genesis of disseminated and massive sulfide deposits in Saudi Arabia. *Saudi Arabian Proj. Rep. 207.* Prepared for Dir. Gen. Min. Res., Minist. Petrol. Min. Res., Jiddah, Saudi Arabia, by U.S. Geol. Surv.

Ruiz, F. C. 1965. *Geologia y yacimientos metaliferos de Chile.* Inst. Invest. Geol., Chile. 305 pp.

Sato, T. 1974. Distribution and geological setting of the Kuroko deposits. *Geology of the Kuroko deposits, Soc. Min. Geol. Jpn, Spec. Issue,* ed. S. Ishihara, No. 6, p. 1

Sawkins, F. J. 1972. Sulfide ore in relation to plate tectonics. *J. Geol.* 80:377–97

Sillitoe, R. H. 1972a. Relation of metal provinces in western America to subduction of oceanic lithosphere. *Geol. Soc. Amer. Bull.* 83:813–17

Sillitoe, R. H. 1972b. Formation of certain massive sulfide deposits at sites of sea-floor spreading. *Inst. Min. Metal. Trans. B,* No. 789. 81:B141–48

Sillitoe, R. H. 1975. Subduction and porphyry copper deposits in southwestern North America—A reply to recent objections. *Econ Geol.* 70:1474–77

Tatsumi, T., ed. 1970. *Volcanism and Ore Genesis* (Watanabe commem. vol.). Univ. Tokyo Press. 488 pp.

United States Bureau of Mines. 1976. *Minerals in the U.S. economy: Ten Year Supply-Demand Profiles for Mineral and Fuel Commodities* (1965–1974). 21 pp.

United States Bureau of Mines. *Minerals Yearbook,* published annually

United States Bureau of Mines. 1977. Commodity data summaries, 1977

White, D. E. 1974. Diverse origins of hydrothermal ore fluids. *Econ. Geol.* 69:954–73

Wyman, R. V. 1976. Ore genesis and mantle plumes. *Amer. Inst. Min. Eng. Preprint No. 76-I-51.* 9 pp.

Ann. Rev. Earth Planet. Sci. 1978. 6:325–51

ORGANIC MATTER IN THE EARTH'S CRUST ×10097

Philip H. Abelson
Carnegie Institution of Washington, 1530 P Street, N.W., Washington D.C. 20005

INTRODUCTION

There are about 830×10^{15} g of living matter on the land surface of the Earth and in the oceans. Annual productivity is estimated at 78×10^{15} g (Revelle & Munk 1977). We do not know how long this productivity has been maintained or how uniform the annual rate has been. A rough guess is that the total produced during the Earth's history has been between 5×10^{24} and 5×10^{25} g. Most of this has been consumed and returned to CO_2, but about 19×10^{21} g of organic or elemental carbon may be found in the crust of the Earth (Hunt 1972, 1977).

Most of this carbon is in sedimentary rocks. Some has been exposed for long periods of time to temperatures above 200°C and has been converted to graphite. Part has been deposited in relatively concentrated forms such as peat, lignite, or coal. Other portions of a different composition have given rise to petroleum. The processes involved are quite inefficient. Hunt (1972) estimated that only 2% of the carbon in sedimentary rocks becomes the carbon of petroleum. Of this, only 0.5% of the petroleum finds its way to a reservoir accumulation. Thus, the overall efficiency is only 0.01%. Nevertheless, even with so tiny an efficiency, reservoirs contain nearly 100 billion tons, with a value of trillions of dollars. Obviously, the efficiency of producing recoverable petroleum from organic matter varies with circumstances; in some instances nature is more efficient than in others. Understanding of such processes is intellectually and economically challenging.

About one part in a thousand of the living matter produced annually escapes being converted to CO_2. A small fraction of the chemicals persist relatively unchanged for long periods of time. They have been called *biochemical fossils*. Among the compounds that have sufficient stability to last are, for example, fatty acids. During the late 1950s and most of the 1960s a principal activity of organic geochemists was to discover biochemical fossils and to identify the chemical processes involved when compounds known to be present in living matter were converted to somewhat altered, but recognizable, forms. For example, in nature, chlorophyll is rapidly degraded. However, two of its components, phytol and the porphin structure, are preserved in only slightly modified form for hundreds of millions of years. Similarly slightly modified forms of sterols and terpenes can

325

be found. By 1970 a very fruitful period of study of geochemical fossils was drawing toward an end. At that time a number of review volumes and articles appeared summarizing the field as of that time. Two especially useful articles are by Kvenvolden (1975), on amino acids, and Maxwell, Pillinger & Eglinton (1971), who addressed the field broadly.

In pursuing biochemical fossils, the chemists were concentrating on an interesting, but extremely tiny, fraction of the world's organic matter. The overwhelming majority of such matter is made up of complex heteroatomic polymeric materials such as fulvic acid, humic acid, humus, and kerogen. These complex materials are formed by chemical reactions at whose nature we can only guess. Their chemical structure is largely a mystery. The mechanism by which they change with time and temperature in geologic settings is little known.

This review is devoted to a survey of research that has been directed toward providing a greater understanding of the formation, reactions, and behavior of these polymers. The first section is devoted to work related to thermal degradation of kerogen to form petroleum. Such reactions can occur in 180 m.y. at 60°C. These or related reactions can occur in the laboratory in a few minutes at 450°C. The work reported on formation of petroleum is at the stage where an intellectual framework has been built. The second portion of this review deals with processes involved in the formation of humic acids and kerogen. The involvement of carbohydrates and proteins has been fairly well demonstrated, but the roles of other participants have not been established.

Studies of the internal structure of the polymers and of their reactivities are only beginning, but progress has been made. An approach to the matter by using specific chemicals as probes seems promising. Such work, coupled with analysis of natural specimens, seems destined to bring new understanding of processes occurring in nature's great organic chemistry laboratory.

PETROLEUM FROM KEROGEN

The great economic importance of petroleum, natural gas, and coal has encouraged considerable scientific interest in these fuels. Coal has long been studied, and its evolution from plant materials was well understood many decades ago. Steps in the diagenesis of the materials were well known, including, for example, the loss of CO_2 and H_2O in the conversion of lignite to bituminous coal. In general, it was recognized that petroleum was formed in source beds that contained organic matter and that the oil migrated to reservoir rocks. Since the oil and, hence, the wealth was to be found in the reservoir rocks, these rocks and their contents drew major attention. But today the situation is that the easy-to-find reservoir rocks have already been discovered. If additional oil is to be found on land in such places as the contiguous 48 states of the United States, it will probably be discovered in stratigraphic traps and other places difficult to discover. One way of improving the success rate for drilling is to be certain that excellent source rocks are present in the stratigraphic sequence. This practical consideration has led recently to much increased attention to the organic matter in source rocks. Interest has been focused on changes in composition that occur as a function of time and temperature and on the relationship of these changes to the formation of petroleum.

Many organic geochemists pointed to kerogen, the principal organic constituent, as the primary source of petroleum (e.g. Dunton & Hunt 1962, Abelson 1963, Welte 1964). These workers demonstrated that heating kerogen gave rise to oil.

Philippi (1965) published results of detailed studies of a series of specimens taken at various depths in the Ventura and Los Angeles Basins. He found that the hydrocarbon content of the sediments at first increased only slowly with depth and age in Pliocene shales but more rapidly in deeper, warmer Miocene shales. At greater depths and at temperatures above 115°C the amount of petroleum per gram of noncarbonate carbon increased substantially; its composition gradually became very similar to waxy oils being pumped from the basins.

McIver (1967) published a seminal paper on the composition of kerogen and its role in the formation of petroleum. He pointed out how widely variable kerogens are in their carbon, hydrogen, and oxygen contents. This is true at the time of deposition, and, depending on burial history, the variations become even greater. McIver showed that the composition of kerogen in the Lower Cretaceous Manville Shale of Alberta changed progressively from carbon (68), hydrogen (5.1), and oxygen (26.9) at a burial depth of 200 m, to carbon (87), hydrogen (4.7), and oxygen (8.3) at 2150 m. He also noted that the maturation of kerogen and coal were closely related processes.

Petroleum in Toarcian Shales

Albrecht & Ourisson (1969) and Tissot et al (1971) made great advances toward a solid understanding of the genesis of petroleum from kerogen. The contributions of Tissot and his co-workers were particularly impressive and persuasive.

Tissot et al concentrated a large effort, involving a sizable team of geochemists and extending over about a decade, on a particularly favorable geologic setting, the Toarcian Shale (Jurassic) of the Paris Basin, and made detailed analyses of kerogen and associated petroleum as a function of depth of burial. Later, Tissot and his collaborators (1974) extended their observations to include measurements of infrared spectra, vitrinite reflectance, and electron spin resonance. In addition, they compared their results with those of others and set forth a general framework for the maturation of kerogens. This framework was made the more valuable because of the extensive and thorough manner in which they had conducted their work.

The Toarcian Shales were deposited 180 m.y. ago. The paleogeographic conditions were relatively uniform, and it is reasonable to assume that the nature of the original organic matter was homogeneous over what is now the Paris Basin. The composition of the clay minerals in the shales is fairly constant. Tissot et al (1974) fixed their attention on a particular member of the Toarcian sequence.

Originally the shales were deposited in shallow water. However, shortly after they were deposited, part of the basin subsided. In consequence, part of the shale became buried under 2500 m of sediments, while the same member elsewhere was less deeply buried. Thus it was possible for Tissot and his co-workers to hold constant most of the cogent variables bearing on formation of petroleum and to study the effect of variations in depth of burial, with attendant higher temperatures, on the production of petroleum from the shale. Precise values for temperature are not given, but the gradient is 35–40°C per km, with a present-day ground level average of 10°C. Near the surface, the rate of transformation of kerogen into petroleum was seen to be slow. At about 1200 m the rate of generation accelerated. This is especially true of the hydrocarbon fraction. The values for total petroleum and for hydrocarbons may be seen in Figure 1.

It can also be noted that the hydrocarbon fraction goes from about 10 mg g^{-1} at 500 m depth to nearly 100 mg at 2500 m. Correspondingly, the part due to resins plus asphaltenes increased only from 35 mg g^{-1} to about 75 mg g^{-1}. Thus, the generation of hydrocarbons predominated at the deeper and hotter place. With depth, saturated hydrocarbons increased with respect to aromatic hydrocarbons. Hydrocarbons lower than C_{15} were very scarce at shallow depth but at 2500 m they constituted a third of the hydrocarbons. Trends in the hydrocarbons with depth are shown in Figure 2. The increase in the amount of the C_1–C_5 fraction is especially dramatic—nearly a factor of 500 between a depth of 500 m and 2500 m.

In their extraction and analysis procedure, Tissot and co-workers prepared fractions of hydrocarbons, resins, asphaltenes, and a more polar fraction extracted from the kerogen by methanol (15), acetone (15), and benzene (70). At 500 m depth the MAB (methanol-acetone-benzene) fraction constituted about half the

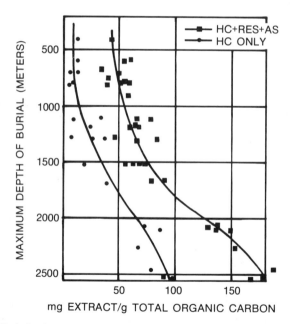

Figure 1   Variation in amount of hydrocarbons (HC only) and total chloroform extract— hydrocarbons + resins + asphaltenes (HC + RES + AS) as function of depth in lower Toarcian shales, Paris Basin (Tissot et al 1971).

total extract. With depth, it diminished in amount and importance. Trends in the abundance of the various fractions can be noted in Figure 3.

Another feature observed as an evolution of the naphtheno-aromatic molecules, i.e. those containing at least one aromatic ring together with saturated cycles, was

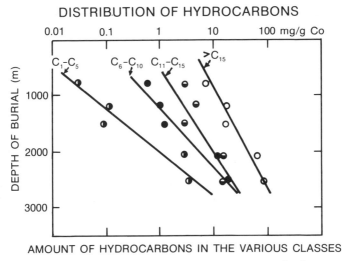

Figure 2 Distribution of hydrocarbons as a function of depth showing increase of light hydrocarbons (C_1-C_5, C_6-C_{10}) compared with medium and heavy ones (Tissot et al 1971).

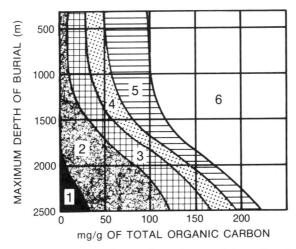

Figure 3 Evolution of various constituents of organic matter in lower Toarcian shales, as a function of depth: *1*, hydrocarbons $< C_{10}$; *2*, hydrocarbons $> C_{10}$; *3*, resins; *4*, asphaltenes; *5*, methanol (15), acetone (15), benzene (70) extract; *6*, insoluble organic matter (kerogen) (Tissot et al 1971).

a progressive increase in the ratio of aromatic to saturated cycles. In deep samples, the abundance of polycyclic structure lessened. The noncyclic structures were generally predominant, followed by monocyclic naphthenes and purely aromatic hydrocarbons. There was also a diminution of isoprenoids compared with n-alkanes.

In the studies in the Paris Basin, Tissot did not attempt a precise estimate of the temperature history. However, geologic events in the Basin were probably relatively simple, and there probably has not been much vertical motion or erosion since Jurassic time. In other areas the history may be more complex, and a present-day temperature or a present-day thermal gradient coupled with a present-day depth of burial may not provide helpful information about the time-temperature history experienced by the kerogen. This would be especially so if, for example, there were a hot intrusion. Thus petroleum geochemists have been led to seek indicators of maturation. A first, and rather practical, approach was to draw on the experience of coal petrologists.

Maturation of Kerogen in Toarcian Shales

With the establishment of a close relationship between the maturation of coal and kerogen, it was natural to draw comparisons of chemical constitutional changes during maturation. With coal, maturation proceeds with the escape of CO_2, H_2O,

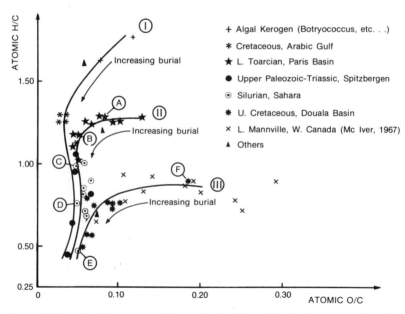

Figure 4 Examples of kerogen evolution paths. Path I includes algal kerogen and excellent source rocks from Middle East; path II includes good source rocks from North Africa and other basins; path III corresponds to less oil-productive organic matter, but may include gas source rocks. Evolution of kerogen composition with depth is marked by arrow along each particular path (Tissot et al 1974).

and CH_4. The residual solid ultimately approaches graphite. The beginning phases of maturation as well as the concluding phases in kerogen rather closely parallel those of coal.

Information about changes of kerogen with time and temperature was developed by a number of workers, at least some of them independently. Outstanding work was performed again by Tissot and his collaborators (1974), who simultaneously brought to bear on the problem a powerful battery of tools.

For the most part, kerogens are deposited in marine environments (kerogen of the Green River Shale is a notable exception). Typically, the organic content is of the order of 1–10%. Composition is variable, depending on the source and nature of the organic material. In some instances the sources may be carbohydrate-rich. In other instances the sources may be lipid-rich. The latter type of kerogen, which has a high H/C ratio and is rich in aliphatic structure, can give rise to oil and gas. The former, which is similar to coal in composition, generates mostly gas.

Tissot et al (1974) chose to emphasize three major evolutionary pathways in the maturation of kerogen, though stating that some kerogens would fall between the major trends. A diagram summarizing the maturation trends is shown in Figure 4. To emphasize the relation of kerogen and coal, a diagram from Van Krevelen (Tissot et al 1974) for coal macerals is also shown (Figure 5).

In Figure 4, path 1, with its high H/C ratio and low O/C ratio, can be most favorable for generation of oil. Some of the prolific source rocks of the Middle East fall along this trend line. In contrast, path 3, with its lower H/C ratio, produces gas and relatively modest amounts of oil.

During the changes in chemical composition shown in Figure 4, there are corresponding changes in physical properties. With increasing maturation the

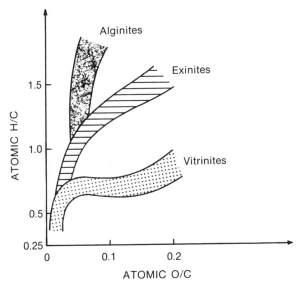

Figure 5 Evolution of coal macerals, after Van Krevelen 1961 (Tissot et al 1974).

infrared spectra evolve. During early maturation, when CO_2 is being eliminated, there is a marked decrease in contribution of the C=O band. In a following stage, in which hydrocarbons are formed and the H/C ratio drops, there is a decrease in aliphatic bands. These changes may be noted in Figure 6 (the letters $A-E$ refer to points shown in Figure 4 and are on evolution path 2). Tissot and co-workers (1974) also measured electromagnetic spin resonance (χ_p) on these same samples and vitrinite reflectances (R) for the various kerogens, which were: kerogen A, R, 0.5%, χ_p, 2.5×10^{-9}; kerogen B, R, 0.6%, χ_p, 3.8×10^{-9}; kerogen C, R, 0.8%, χ_p, 8×10^{-9}; kerogen D, R, 1.3%, χ_p, 7.1×10^{-9}; and kerogen E, R, 3.1%, χ_p, 15.5×10^{-9}.

Vitrinite reflectance has come to be widely used as a means of gauging the degree of maturation of both coal and kerogen. It is a relatively simple, quick way of obtaining information useful in petroleum exploration. Electron spin resonance (ESR) has not drawn as much attention so far. However, interest in this type of measurement has been increasing recently, and Pusey (1973) has found that it is a useful and highly practical tool in gauging paleotemperatures.

Laboratory Simulations of Maturation of Kerogen

Tissot et al (1974) also made a series of laboratory heating experiments in which aliquots of immature kerogen were subjected to a series of increasingly rigorous temperature regimes. They found that in a short time they could reproduce rather well changes apparently quite similar to those that required many millions of years in nature. The samples were heated in a nitrogen atmosphere with a constant temperature increase of $4°C$ min^{-1}; one sample was heated to 350°C, a second to 400°C, and so on, up to 600°C.

Weight loss of the samples was recorded, and the modified kerogens were subjected to optical examination, reflectance measurement, elementary, and infrared analysis. Weight loss for immature samples was as high as 70%. (In other experiments, a more mature kerogen lost only 10% of its weight.) Heating immature kerogen to 350°C at the relatively rapid rate employed resulted in a rather small weight loss, only modest darkening of the kerogen, and slightly increased reflectance. The major changes occurred in the 350–470°C temperature range. There the weight loss was maximal per unit time, and reflectance increased sharply. The elementary chemical composition changed in a way that was similar to the evolutionary trends for kerogens noted earlier and shown in Figure 4. That is, an initial oxygen loss was followed by a hydrogen loss. The succession of O/C and H/C ratios shown in Figure 4 was duplicated. A similar concordance was noted with respect to infrared measurements. The C=O band diminished first, followed by a decrease in both aliphatic and carbonyl groups in a way that paralleled quite closely the phenomena displayed in Figure 6 for natural maturation.

Tissot emphasized the significance of the close relationship between simulated laboratory degradations and natural maturation. The coincidence of the two processes permits an evaluation of the petroleum potential of a given formation in the buried parts of a sedimentary basin by use of shallow or outcrop samples of the same formation and simulation of their maturation in the laboratory.

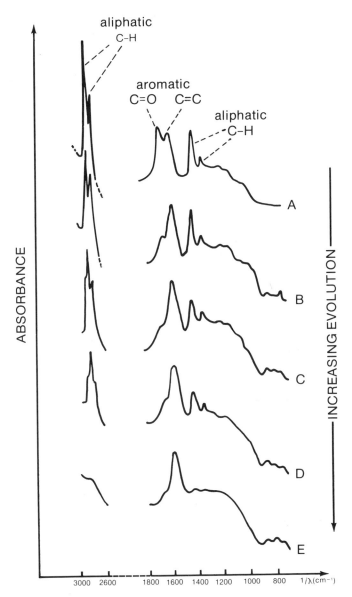

Figure 6 Evolution of kerogen structure during burial. Samples are taken along evolution path II marked *A–E* on Figure 4. Infrared spectra show progressive elimination of C=O, carbon dioxide and water formation; and aliphatic CH_2, CH_3 bands, hydrocarbons formation (Tissot et al 1974).

Petroleum in Logbaba Series of Douala Basin

Further detailed information about the diagenesis of kerogen as a function of depth of burial was obtained by others from studies of sediments of the Douala Basin (Cameroon). The age of the sediments and their organic content was different from those of the Toarcian Shales. The methods employed in studying the two kerogens were similar, however. This permitted a detailed comparison of the progression of

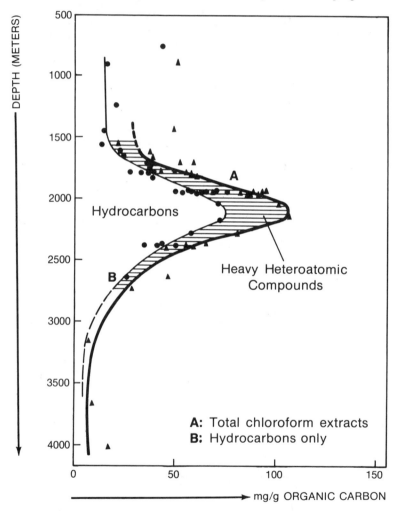

Figure 7 Evolution of total chloroform extract (*A*, ▲) and hydrocarbons [saturated + aromatic (*B*, ●)] with burial in the Logbaba series (Albrecht, Vandenbroucke & Mandengue 1976).

chemical events during maturation of the respective kerogens. In consequence, the utility of the Tissot approach was enhanced.

In the Douala Basin a column of sediments over 4000 m thick was deposited in late Cretaceous time. The sedimentation was detrital, apparently deltaic and very monotonous. The content of organic carbon has been examined in the cores all along the series and is also monotonous. The organic matter appears to be mainly of terrestrial origin and in composition is reminiscent of vitrinite. Albrecht & Ourisson (1969) examined the alkanes from this series. A more extensive study of the organic matter has been conducted by Albrecht, Vandenbroucke & Mandengue (1976), Durand & Espitalié (1976), and Vandenbroucke, Durand & Albrecht (1976). In a general way the results fit the scheme outlined by Tissot et al (1974); evolution of the organic matter roughly follows evolutionary path 3 of Figure 4.

From the surface to about 1500 m, the materials extracted by chloroform amount to about 50 mg g^{-1} of organic carbon. This amount is comparable to that often obtained from recent sediments. Extraction of a sample taken at 910 m with an aqueous solution of sodium hydroxide–sodium pyrophosphate yielded substantial amounts of humic and fulvic acids. Humic acids alone contained 56 mg of organic carbon. Between 1500 m and 2200 m the amount of chloroform extract increased rapidly. Beginning at 1200 m (temperature 65–70°C), evolution of hydrocarbons increased rapidly with depth. The total of hydrocarbons plus heavy heteroatomic molecules peaked at 2200 m (100–110°C). (No temperature is provided in the original papers, but a present-day gradient of 35–40 km^{-1} is mentioned). At greater depths both the total extractable and the hydrocarbon portion dropped off rapidly. The decrease was mainly due to thermal cracking, which resulted in light material that was not measured by the procedures employed. The results from these extractions are shown in Figure 7.

Albrecht, Vandenbroucke & Mandengue (1976) also determined the amounts of aliphatic and aromatic hydrocarbons as a function of depth. At depths less than 1000 m only tiny amounts of aliphatic hydrocarbons were present; the aromatics predominated. However, at depths below 1200 m the aliphatic component increased rapidly and was the dominant component at 2200 m. At greater depths, with cracking of the aliphatics, the aromatic hydrocarbons, which are more stable, dominated again. These trends are displayed in Figure 8.

Albrecht, Vandenbroucke & Mandengue (1976) conducted a laboratory heating experiment that buttressed their study. A sample taken at 1200 m had practically no alkanes. However, after being incubated for 12 days at 245°C the sample possessed a suite of alkanes that matched very well with one obtained from a depth of 1500 m.

Durand & Espitalié (1976) made physical and chemical measurements on the corresponding portions of Douala Basin rocks. Vitrinite reflectance of samples taken at progressively greater depths increased and came to resemble that of anthracite. The infrared spectra had C=O and aliphatic H peaks in the shallow cores; in the deeper samples these peaks had vanished. Kerogen composition was measured in samples at many depths. Values at three different depths illustrate the range of composition. At 910 m: carbon (78.84), hydrogen (5.46), oxygen (17.96);

at 2280 m: carbon (82.37), hydrogen (4.45), oxygen (8.27); at 4018 m: carbon (91.59), hydrogen (3.25), oxygen (2.86).

Vandenbroucke, Durand & Albrecht (1976) compared in detail the various components of the petroleum extracted from the Toarcian and Douala Shales. They found differences that, in large measure, could be explained by differences in the originally sedimented organic matter. For example, land-derived sediments of the Douala Basin contained larger amounts of chemical structures identifiable as originating from terpenes. In contrast, Toarcian Shales were relatively richer in structures related to the sterols, as might be expected from a situation where autochthonous marine organisms were principal contributors to the organic matter

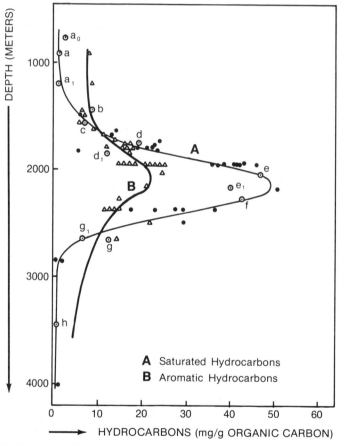

Figure 8 Evolution of saturated (*A*, ●) and aromatic (*B*, ▲) hydrocarbons with burial in the Logbaba series. The amounts of saturated hydrocarbons from chloroform extraction as well as benzene-methanol extraction have been plotted (Albrecht, Vandenbroucke & Mandengue 1976).

of the sediments. Despite such differences, the various types of physical and chemical measurements performed on Douala Shales and their associated petroleum again demonstrated a pattern of progressive, correlated changes that paralleled those found by Tissot et al (1974) in the Paris Basin.

Maturation of Kerogen in Tertiary Sediments

LaPlante (1974) determined the carbon, hydrogen, and oxygen content of many kerogens from Gulf Coast Tertiary sediments from different depths. Samples were taken at 75 m intervals. From the changes in composition, he calculated the amounts of gaseous products, e.g. CO_2, H_2O, and the amounts of hydrocarbons. In a Lower Miocene sediment, calculations showed that hydrocarbon generation occurred in the 3050–4575 m interval. The hydrocarbon-generating reactions were initiated at a carbonization level of carbon (76), hydrogen (5.5), and a temperature of 86°C. These particular sediments were taken from wells in the South Pecan Lake field. The field produces wet gas from intervals closely associated with the generation zone postulated by LaPlante. Another well which was studied penetrated the Pliocene–Upper Miocene section in the West Delta area of offshore Louisiana. The carbonization trends there were similar to those seen at South Pecan Lake. The 75% carbon level occurred at a burial depth of 4144 m and at a temperature of 96°C. The age of the sediments was 13.5 m.y. The chemical composition of the kerogen was carbon (75), hydrogen (6), and oxygen (17). Production from the well is principally gas.

In his studies and findings on kerogen, LaPlante took a different approach than did Tissot et al (1974). Yet there are basic similarities—an emphasis on the evolution of kerogen with time and temperature, an evolution that involves progress toward the stable chemicals CO_2, H_2O, and graphite.

Laboratory Maturation of a Recent Kerogen

A group led by I. R. Kaplan has made a considerable number of investigations of chemical and physical changes that occur when recent sediments are heated at various temperatures for various periods of time. Kerogen from the Tanner Basin off Los Angeles has been a particular object of study. Ishiwatari et al (1977) heated samples of the sediment under nitrogen in sealed tubes at single temperatures between 150 and 410°C for specified times in the range of 5–120 hr. Chemical analysis of the altered kerogens and of the products formed yielded results reminiscent of Tissot· et al (1974) and others in terms of the effect of long-term burial on kerogen. The unheated kerogen contained carbon (56.7), hydrogen (6.5), nitrogen (6), and oxygen (30.8). With increasing time and temperatures between 150 and 410°C, the carbon content of the kerogen increased to 82.1% and atomic H/C decreased from 1.37 to 0.49. Heating for 18.5 hr at 260°C was optimal for producing a liquid product but yielded little in the way of n-alkanes. In the process, the kerogen became more carbonaceous—carbon (68.8) and H/C (1.07). The liquid product was separated and then heated again to 330°C for 24 hr to produce paraffins and a highly aromatic kerogen. Results of the experiment supported the model hypothesis of Tissot and others for the production of n-alkanes. Related observations also gave concordant results. That is, in the heating of the kerogen the maximum amount of liquid products occurred when H/C ratio had dropped from 1.37 to 0.90. With further maturation to H/C = 0.80, the total liquid products dropped, while the normal alkane fraction rose sharply to a peak.

Another useful aspect of the investigations was determination of electron spin resonance on the treated kerogens. Changes in the ESR of the kerogens after heating follow the general pattern described by Pusey (1973) for kerogen and Van Krevelen (1961) for coals (see Figure 5). That is, the free radical concentration values are low in the unheated specimens and increase by almost a factor of 10

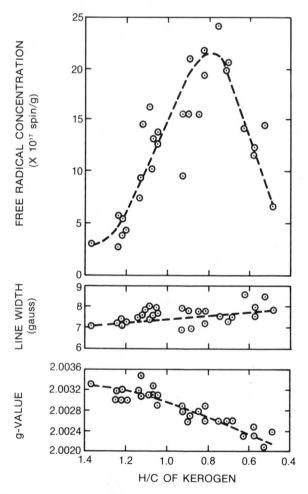

Figure 9   Relation between ESR properties of kerogen and its H/C ratios (Ishiwatari et al 1977).

when the H/C content drops to 0.8 (Figure 9). The g-value of kerogen decreased gradually from 2.0033 to 2.0023 with increasing thermal alterations. These observations are in accord with previous observations of ESR on kerogen and relate to studies of Pusey (1973), who observed systematic variations in ESR parameters of kerogen as a function of depth of burial along a geothermal gradient.

Numerous geochemists have commented on the high activation energies (58,000 cals) involved when pure hydrocarbons are cracked. However, activation energies as low as 15,000 cals have been calculated from observations made on the maturation of kerogen in nature. In the experiments of Ishiwatari et al (1977) an activation energy for production of n-alkane of 31,000 cals was calculated.

Additional observations by Peters et al (1977) on Tanner Basin kerogen were directed at studying the color of kerogen as an index of organic maturity. These studies on samples that had been heated for various periods at temperatures in the range of 150–410°C provided data on both the color of the kerogen and on its elemental constitution. They determined the length of time as a function of temperature necessary to bring about a change from, for example, the original light yellow color of the kerogen to a brown color. The results were analyzed by means of the Arrhenius equation. At the lower temperatures, where modest changes in color occurred, the activation energy was 15,000 cal mole^{-1}. To obtain progressively deeper darkening, higher temperatures were needed, and the activation energies rose from 15,000 cals mole^{-1} to 40,000–45,000 cals mole^{-1} to bring about a change to dark brown.

It is not surprising that at higher temperatures new mechanisms should predominate. But today we do not understand the mechanisms at either low or high temperatures.

Maturation of Green River Shale

A survey of research activities relating to maturation of kerogen should contain at least a passing reference to the Green River Shale. This formation, located in Colorado, Utah, and Wyoming, contains the world's largest potential source of oil, an amount overshadowing the ultimate production of the Middle East. The shale has been the subject of many geochemical searches for biochemical fossils, such as fatty acids, amino acids, porphyrins, and the like. A summary of many of the findings is contained in Burlingame et al (1969). The kerogen has also been extensively studied by workers at the Laramie Petroleum Research Center, formerly of the U.S. Bureau of Mines but in 1977 part of the Energy Research and Development Administration. Much of the work at Laramie has been directed toward practical methods of obtaining oil by aboveground retorting of the shale at temperatures of around 500°C and over. However, there has also been research devoted to seeking to understand the detailed constitution of the kerogen. These studies (Robinson 1969) have included partial oxidation, reduction, solvent treatment, hydrolysis, and functional group analysis. More recently it has become evident that underground (in situ) retorting of the shale poses less environmental and other problems than does the aboveground procedure. Moreover, petroleumlike fluids can be released at temperatures below 500°C, provided sufficient time is available. Thus, studies of the thermal degradation of the shale in the range 150–350°C have become of substantial practical interest. Such studies have been conducted by Cummins & Robinson (1972) and Cummins, Doolittle & Robinson (1974). The oil shale sample employed contained 30.2% organic carbon and assayed 65 gallons per ton of shale (210 liter per kiloton). A typical elemental analysis for such kerogen is carbon (80.5), hydrogen (10.3), nitrogen (2.4), sulfur (1.0), and oxygen (5.8). The H/C ratio is 1.6.

Table 1   Benzene soluble pyrolytic products from Green River Shale[a]

Days	Converted wt %	n-alkanes	Branched plus cyclic alkanes	Aromatic oil	Polar resins	Pentane insolubles
				Soluble products components (wt% of total)		
			150°C			
90	2.0	3.6	3.6	6.7	54.9	31.2
180	3.6	3.2	8.3	1.1	66.3	21.1
270	6.1	4.3	9.7	1.6	71.7	12.7
360	7.7	4.3	8.0	1.8	66.4	19.5
			200°C			
90	3.0	1.3	10.0	1.3	71.7	15.7
180	4.9	0.8	8.5	0.8	82.9	7.0
270	7.5	1.1	5.5	0.3	82.9	10.2
360	8.9	0.9	4.8	0.4	84.8	9.1
			250°C			
10	9.1	2.2	19.0	2.4	73.6	2.8
12	9.8	3.0	18.8	3.1	71.9	3.2
16	11.8	2.1	13.7	2.1	78.1	4.0
21	14.4	1.5	14.6	3.0	77.5	3.4
			300°C			
0.5	3.3	2.1	5.6	2.1	76.7	13.5
1.0	3.8	1.6	9.7	2.1	81.1	5.5
2.0	7.5	1.5	9.1	5.1	81.7	2.6
3.0	8.7	1.6	9.4	5.6	77.5	5.9
4.0	12.8	2.3	9.4	9.4	74.3	4.6
			350°C			
0.5	27.6	2.3	7.2	5.7	75.5	9.3
1.0	44.5	5.5	9.3	8.5	65.7	11.0
2.0	53.9	2.7	7.6	3.4	70.3	16.0
3.0	62.4	4.5	5.8	1.9	58.9	28.9
4.0	65.1	4.8	7.6	6.7	76.5	4.4

[a] Abstracted from data in Cummins, Doolittle & Robinson (1974).

Cummins and co-workers (1974) performed a series of heating experiments on the shale employing temperatures from 150–350°C and times from 0.5–360 days. They extracted the soluble pyrolytic products and fractionated them. The chemical composition of the various fractions was determined, and the individual major fractions were further split into subfractions that could be examined in more detail. The residual kerogens from the thermal degradation were also examined. Results of some of the studies are shown in Table 1. On heating at 350°C for 4 days, 65.1% of the organic carbon appears in a benzene-soluble product. This is a remarkable yield, one quite unlike that usually seen from typical kerogens by petroleum geochemists. In its composition, the kerogen is most closely related to path 1 on the Tissot diagram of Figure 4.

Heating of the kerogen for 4 days at 350°C released 80% of the carbon that can be released even on drastic heating. The residue from the 350°C heating had an H/C ratio of 0.64. With the usual kerogen, such an H/C ratio corresponds to the late stages of petroleum maturation in which polar materials have virtually disappeared and hydrocarbons are undergoing cracking. However, the composition of the pyrolytic products of the 4-day exposure at 350°C in percent was n-alkanes (4.8), branched plus cyclic alkanes (7.6), aromatic oil (6.7), resins (76.5), and pentane insoluble (4.4). In the heating experiments of Ishiwatari et al (1977) a 100-hr incubation at 350°C on a recent kerogen yielded a residue with H/C ratio of 0.72. Correspondingly, the amount of petroleum liquids had dropped to a third, and total n-alkanes strongly predominated.

In a sense, thermal degradation of the Green River Shale is similar to that of the kerogen of old and young marine sediments. When exposed to elevated temperatures for sufficient time, all kerogens trend toward graphite. For most of the kerogens studied, the pathway is remarkably similar. But such evidence as is at hand indicates that in detail the maturation of Green River Shale is different.

The general trends presented by Tissot and co-workers (1971, 1974) are surely basically correct and will have many applications. It is probable that minor modification of the findings may later prove desirable. That is, Tissot did not deal with the role of nitrogen or sulfur. In comparison with carbon, hydrogen, and oxygen, on an atomic basis, these are present at a level about an order of magnitude smaller.

The close correlation of results obtained from core samples and from thermal experiments has important practical applications in petroleum exploration. The new results also underline once again the potential usefulness of laboratory simulation experiments in approaching difficult problems posed by complex phenomena occurring in nature.

FORMATION OF KEROGEN AND HUMIC ACIDS

One of the central problems of organic geochemistry is the origin and formation of kerogens. These materials are variable in composition. They are insoluble in all the usual solvents and behave like polymers of very large molecular weight. Humic acids are related to the kerogens and are their predecessors. They are of variable composition but have intermediate molecular weights (e.g. 10,000). The humic acids are defined as being complex organic materials that are soluble in alkali and insoluble in acid. The humic acids and kerogens are relatively immune to biological attack, especially in anaerobic environments.

In the minds of soil chemists, humic acids are the residual degradation products of plant materials, which often contain lignin. In marine sediments, however, may be found organic matter that also meets the usual criteria of humic acids but does not contain lignin. Hedges & Parker (1976) showed that humic acids from recent marine sediments from the Gulf of Mexico did not include lignin derivatives. The humic materials had stable carbon isotope ratios consistent with an origin from marine plankton, which do not contain lignin.

Because of variabilities in composition, the problems of studying the natural materials and their synthesis are enormous. The geochemist collects a recent sediment and, after some processing, isolates a humic acid and a kerogen. What are the building blocks that went into these geopolymers? What were the reactions among the constituents?

If the geochemist makes a detailed study of a sample from a given locality, how can others verify the findings? One of the crucial factors in maintaining the integrity of science is that sufficient information usually is provided so that others can duplicate the work. Such an opportunity is not always readily available in studies of various kerogens. Another question concerning results in a specific sample is, how widely applicable are the findings?

The material that is now humic acid or kerogen once was part of living matter. But in a marine environment countless species of differing composition might have been involved. In the case of older occurrences some or most of the contributing organisms may be extinct. After death their chemicals were partially consumed by other organisms in a complex sequence of events. Thus one cannot hope to know the starting materials for most marine occurrences. One can obtain some limited information about humic acid and kerogen by heating and oxidative degradation. But such procedures furnish little information about mechanisms of formation. These are very difficult questions, which may be best approached indirectly. Thus, the most useful approach may be to produce, from known substances, materials having many or all the properties of geopolymers. The mechanisms of formation of such materials then can be studied. Various scientists can reproduce each other's work and observations. Ultimately it may be expected that general principles applicable to all such geopolymers would be discovered and their significance understood. In such a framework, cogent studies of natural materials would be made to determine ways in which the natural and the synthetic materials behaved alike. The following section describes observations on synthetic humic acids and kerogens both with respect to synthesis and to behavior of the products in comparison with natural geopolymers.

Synthetic Humic Acids and Kerogens

The principal constituents of living matter are carbohydrates, proteins, and lipids. In looking for the source materials for geopolymers, first priority ought to go to these major constituents. Chemical analysis of humic acids and young kerogens

shows that they contain C, H, N, and O in fractional amounts that correspond
to a mixture mainly of carbohydrates and amino acids. It is well known that these
two classes of compounds react to form complex materials. Maillard (1913)
recognized that the condensation of sugars and amino acids yielded products
similar to natural humic acids. Additional similarities were pointed out by Enders
(1943). An admirable review of reaction mechanisms involved in the combination
of carbohydrates and amino acids was published by Hodge (1953). Studies of
reactions between carbohydrates and amino acids, with emphasis on comparison
of products to humic acids, have been made by Abelson (1959), Abelson & Hare
(1971), Hoering (1973), and Hedges (1976).

Abelson & Hare (1971) broadly explored the products of reactions of carbohydrates and amino acids using d-glucose as the carbohydrate and most of the
conventional amino acids. They produced synthetic products that resembled in
many respects natural humic acids and kerogens. When solutions 1 M in glucose
and the respective amino acids are adjusted to pH 8.5 and brought to a boil, all the
solutions become brown almost immediately. (Related color changes also occur
slowly when glucose–amino acid mixtures are held at 4°C.) With some mixtures,
formation of humic-acid-like materials was rapid at 100°C. With glucose and
phenylalanine, 25% yields could be isolated after a day. Lysine and arginine reacted
with glucose very quickly.

Hydrophobic amino acids incubated with glucose tended to yield humic acids
faster than the hydrophilic ones. With longer incubations (a month or more)
insoluble solids were formed that fit the definition of kerogen. In this respect the
combination of lysine and glucose was different from other pairs. They formed
kerogen quickly. With some combinations, e. g. glucose and alanine, yields of only
20% humic acid were obtained, even after 6 months, and practically no kerogen was
formed.

Proportions other than equimolar were employed. Sets of components ranging
from 1 phenylalanine–5 glucose to 2 phenylalanine–1 glucose and including
intermediate compositions all yielded humic acids.

The C/N ratios of some of the synthetic materials were examined. Humic acids
that were produced from 1:1 proportions of amino acids and glucose, and were
refluxed at 100°C for 6 months, had the following C/N ratios: alanine, 6.5;
isoleucine, 9.3; phenylalanine, 13.5; tyrosine, 12.5. An equimolar glucose-lysine
kerogen, refluxed for 6 months, had a C/N ratio of 5.6. Humic acids, prepared in
10 days from various proportions of phenylalanine and glucose, had the following
C/N ratios: 2 phenylalanine–1 glucose, (12.5); 1 phenylalanine–3 glucose, (16), 1
phenylalanine–5 glucose, (25).

Hedges (1976) prepared a number of artificial humic acids by reacting three
amino acids respectively with glucose for a week at 100°C. The proportions
employed were 9:1, 1:1, and 1:9. Materials of molecular weight greater than 6000
were isolated by dialysis. A principal finding was that the weight of the product
was largely controlled by the amount of glucose present. These studies underline
a central role of carbohydrates in forming geopolymers.

The soluble artificial substances produced from glucose and amino acids fit within the rather loose operational limitations used to characterize natural humic acids. They are soluble in alkali and insoluble in acid. They have titratable acid groups much like the natural products. By choice of starting materials and length of incubation one can produce humic acids with a wide range of O/C ratios, C/N ratios, molecular weights, and titratable acid groups.

Given a natural humic acid, it would be feasible to construct a synthetic polymer that closely matches its major compositional factors.

Comparisons of Natural and Synthetic Humic Acids

Abelson & Hare (1971) and Hoering (1973, and personal communication) have shown that the similarity of the synthetic and natural materials goes much deeper than compositional factors. Hoering synthesized artificial humic acids by reacting d-glucose and amino acids at a 12 to 1 weight ratio for 4 days at just below boiling temperatures. He observed that freshly precipitated synthetic humic acid and natural humic acid form gels that have similar appearance, color, texture, and solubility. The gels bind great quantities of water and undergo a twenty-fold reduction in volume when dried at 110°C.

Hoering compared synthetic humic acids and humic acid by (a) elemental analysis, (b) functional group analysis, (c) infrared spectroscopy, (d) Mössbauer spectroscopy of iron complexes, and (e) degradation with chromic acid, chlorine, and alkaline copper oxide. In every case the behavior of the two materials was similar.

Comparable infrared absorption spectra were obtained from the synthetics and humic acid. The total acidity of the synthetics and humic acid were in close agreement. Hoering also observed the comparative behavior of ferric iron with humic acid and artificial humic acid at room temperature. In both instances the ferric iron was reduced to the ferrous state and bound to the polymer. Mössbauer spectra of both polymers were closely similar, indicating that iron was bound in the same type of linkages in both cases.

Hoering also investigated the behavior of humic acid and a synthetic when exposed to strong oxidizing and reducing agents. Again, the products were similar. The close relationship of the two polymers was made particularly evident by procedures involving treatment with chlorine and hypochlorite followed by ether extraction of the soluble acids. Comparable yields (30%) were obtained. The acids were converted to methyl esters and subjected to gas liquid chromatography. The emerging patterns were quite similar.

Amino Acids as Chemical Probes

The close relationship demonstrated by Hoering of fundamental properties of synthetic humic acids and of a natural humic acid was an extension of earlier observations by Abelson & Hare (1970, 1971), who used amino acids as chemical probes to compare the reactivities of artificial humic acids and kerogens with their natural counterparts. By incubating a small aliquot of a natural or a synthetic humic acid with a standard mixture containing most of the common amino acids, it was possible to conduct in one incubation many simultaneous tests of the behavior of the materials. All reacted with all amino acids, but at varying rates. Times of incubation were adjusted so that the differences among the individual reactivities were made apparent.

When the behavior of artificial and natural humic acids and kerogen was compared, it was found that in both instances lysine and arginine reacted rapidly, the more hydrophobic amino acids disappeared readily, and the dicarboxylic acids disappeared slowly.

In the course of these experiments, measurements of ESR of the geopolymers were made. Both natural and artificial types possessed ESR of comparable intensity. In both instances the intensity observed was sensitive to pretreatment. That is, after exposure to a reducing agent, $Na_2S_2O_4$, a smaller intensity was seen, while after exposure to oxygen the ESR was enhanced. After reducing and mild oxidizing treatments, the humic acids and kerogens were exposed to amino acid solutions. Again, a parallel behavior was observed. Both types of the oxidized geopolymers reacted more readily with the amino acids in a strictly comparable way, while the reduced ones showed markedly less reactivity.

An illustration of the kinds of effects that may be observed are shown in Table 2. This table was abstracted from part of the data provided by Abelson & Hare (1971). The effect of mild oxidation (bubbling oxygen through a suspension at room temperature) is clear-cut. There is qualitative similarity among all the geopolymers

Table 2 Percentage recovery of amino acids as a function of time and temperature from kerogen-amino acid mixtures[a]

Amino acid	25°C		52°C		80°C			110°C			
	1 hr	97 days	8 days	83 days	1 day	2 days	8 days	6 hr	2 days	4 days	8 days
Lysine	95	13	0	0	6	0	0	0	0	0	0
Histidine	90	10	tr	tr	11	0	0	5	0	0	0
Arginine	90	4	0	0	2	0	0	0	0	0	0
Threonine	98	64	77	27	65	65	36	60	24	13	6
Serine	98	71	79	37	68	69	42	66	31	18	9
Glutamic acid	98	97	100	95	87	90	90	91	76	61	60
Glycine	99	43	64	38	77	57	35	59	32	24	18
Alanine	99	80	81	61	61	60	52	65	35	22	15
Half cystine	29	0	0	0	0	0	0	0	0	0	0
Valine	83	80	62	57	61	54	26	50	19	9	3
Methionine	61	43	31	6	35	25	14	10	0	0	0
Isoleucine	82	66	47	38	46	36	11	33	11	0	0
Leucine	88	60	32	23	40	21	2	16	0	0	0
Tyrosine	36	19	9	3	29	6	0	10	0	0	0
Phenylalanine	30	12	2	1	16	2	0	7	0	0	0

[a] 0.2 μM of each amino acid originally. Data expressed as percentage recovery with aspartic acid normalized to 100%.

in their relative reactivity with lysine and arginine. There are quantitative differences among the geopolymers. There were also quantitative differences between the reactivities of the humic acids and kerogens derived from phenylalanine. The product formed from 5 glucose–1 phenylalanine was much the more active, and in its oxidized form its behavior closely paralleled that of humic acid from the Drummer Loam, which also was oxygenated in nature. In plants on land the carbohydrate content far overshadows that of amino acids. Thus the similar behavior of the natural polymer and the high glucose polymer is to be expected.

The data in Table 2 show that large variations in reactivity can be expected among both natural and synthetic polymers. Such variations depend on composition and on presence or lack of oxygen.

Thus it is apparent that the disappearance of amino acids is a function of the geopolymer and its previous history (e. g. exposure to oxygen) and of the particular amino acid involved. A high concentration of glucose in the humic acids is related to a greater rate of disappearance. Exposure to oxygen, which increases the level of free radicals, also fosters disappearance. The fast disappearance of cystine is probably related to its sulfur content. It is possible that conversion to cysteic acid (not measured) has occurred. The contrast in behavior of lysine and leucine points to a role of amino groups reacting with carbonyl groups. The hydrophilic monobasic amino acids are much less reactive than their hydrophobic counterparts. Thus, in the formation of kerogens there would be a tendency to discriminate against hydrophilic amino acids while favoring the incorporation of the hydrophobics. In view of the reactivity of the amino group, the ε-amino group of lysine probably has an important role in the incorporation of proteins and protein fragments into kerogen. Peptides, with their free amino group, are also reactive. An exploratory experiment with leucyl glycine and glycyl leucine resulted in uptake of the peptides (Abelson & Hare 1970).

Low Temperature Uptake of Amino Acids by Kerogen

In nature the reactivity of humic acids and kerogen must be an important mechanism in the disappearance of amino acids, peptides, and proteins. Most of the experiments with amino acids already cited were conducted at temperatures of around 100°C. If the results of studies are to be relevant to events occurring in natural environments, the gap between 100°C and ambient temperatures must be bridged. This was done in experiments that involved a crude kerogen that had been prepared from recent sediments collected in the San Pedro Basin off Los Angeles.

To determine the temperature dependence of interaction of kerogen with various amino acids, a series of experiments was performed on solutions of standard amino acids at temperatures from 25 to 100°C. Incubations were conducted for times between 0.5 and 97 days. The general pattern seen in Table 2 was also found in the low temperature incubations. Indeed, results of an incubation of 25°C for 97 days with the San Pedro kerogen were practically identical to those listed for the Drummer Loam in Table 2. Thus, it appears that major factors leading to disappearance of amino acids were essentially identical in the two cases. Temperature dependence of the disappearance indicates that the primary mechanism is reaction rather than adsorption. The overall results were consistent with a low activation energy for disappearance of the amino acids and with substantial rates of reaction, even at 0°C.

Products of Reactions of Kerogen with Amino Acids

To examine the extent to which the reactions of kerogen and amino acids are irreversible, samples of kerogen were hydrolyzed with 6 N HCl before and after treatment with amino acids. In every case, only a small fraction of the reacted amino acids could be recovered. In addition, the amount of ammonia found in the supernatant was significantly greater after reaction with the amino acid mixture, which shows that ammonia was produced from the degradation of the amino acids. Five amino acids—aspartic acid, isoleucine, phenylalanine, lysine, and arginine—were incubated separately at a concentration of 40 μM ml^{-1} with 50 mg kerogen at pH 8.5 for 5 days at 110°C. An aliquot of kerogen was similarly incubated with water, with the pH adjusted to 8.5. The amounts of amino acid were 10–14% of the weight of kerogen. In each instance, including the control with H_2O, some NH_3 appeared. The control value was 2.4 μM. In each incubation with amino acid some of the latter disappeared. Subtracting the amount of NH_3 contributed by the control from that observed, the adjusted results are as follows: Of an initial 40 μM, 11 μM aspartic acid disappeared and 4 μM NH_3 appeared.

Other corresponding values are 19 μM isoleucine, 10.5 μM NH_3; 25 μM phenylalanine, 10.5 μM NH_3; 29 μM lysine, 17 μM NH_3; and 32 μM arginine, 24 μM NH_3. In a control run with 40 μM NH_4SO_4, 12 μM NH_3 disappeared. Thus one would not expect to observe an exact correspondence of disappearance of amino acids with appearance of NH_3. From previous experiments the order of least reactive to most reactive is aspartic acid, isoleucine, phenylalanine, lysine, and arginine. This is exactly the order of ammonia production, with aspartic acid producing the least and arginine the most. In all instances there was an impressive incorporation of the amino acid, thus demonstrating that the kerogen had a substantial combining capacity.

The kerogen that had been incubated with arginine was later hydrolyzed for 22 hr with 6 N HCl. The supernatant contained 1.2 μM. Experience with long and repeated hydrolysis of kerogen is that prolonged hydrolysis might have brought off a total of 2.4 μM arginine out of the 32 μM bound. Clearly arginine had disappeared irrevocably. On the other hand, only trifling amounts of urea, citrulline, ornithine, or any other amino acid appeared during the original incubation. Thus, it appeared that arginine molecules were largely incorporated into the kerogen with a binding other than peptide linkage.

An analysis of the carbon and nitrogen values of the kerogen residue confirmed this observation. The C/N molar ratio of kerogen incubated with water was 14.4, whereas the C/N ratio of the kerogen incubated with arginine dropped to 8.2. To account for such a change, three of the nitrogen atoms of arginine must be incorporated in the kerogen. Most of the fourth nitrogen atom of arginine appeared in the supernatant as NH_3.

The behavior of the kerogen with respect to NH_3 indicates that NH_3 is released into the medium at some sites, while perhaps being taken up elsewhere. The arginine experiments indicate that most of that molecule has been incorporated.

Synthesis of Humic Acids in Nature

In the preceding pages a substantial amount of evidence was assembled, linking the formation of natural humic acids to reactions between carbohydrates and amino acids. It is possible to prepare from d-glucose and amino acids polymers that have compositions, physical properties, and reactivities that are closely similar to those of natural products. The properties and ease of formation of these synthetic materials vary somewhat, depending on the starting materials. One would expect that in nature a spread in composition would occur. From data provided by T. C. Hoering (personal communication) and Rashid & King (1970), it appears that on the average marine humic acids contain about twice as much nitrogen as do those in soils.

Other components of living matter are also potentially reactive, and they, too, may have a significant role. The composition of living matter in the marine environment differs from that on land. This is particularly true of sessile forms. On land the carbohydrates are by far the major constituents of living matter. In the sessile forms amounts of carbohydrates are roughly comparable, and lipids are present in larger proportions than they are on land. Thus, the mechanisms leading to humic acids on land and in the marine environment may differ qualitatively, and they surely differ quantitatively. Indeed, the mechanisms of formation and the yields of geopolymers must vary from organism to organism, depending on composition.

One important difference between land plants and sessile marine forms is in their content of nucleoproteins. Nucleoprotein content of land materials often amounts to less than 1% of dry weight. Content of sessile forms is typically much greater. The microorganisms of marine sediments and of terrestrial soils also contain as much as 35% of nucleoproteins. Thus, the role of this material in the formation of humic acids and kerogen is variable but could be of broad significance.

Nucleoproteins are made up of nucleic acids and proteins. The nucleic acids contain purines and pyrimidines, each of which is associated with a carbohydrate, either ribose or desoxyribose, together with a phosphate. The protein is highly basic in character and consists mainly of lysine and arginine. These two combine avidly with carbohydrates. After death of an organism and partial lysis of its contents, ribose and desoxyribose would be liberated in close proximity to the basic protein or partial hydrolysates of it. The abundant free amine groups present in either the protein or its breakdown products would be conveniently accessible reaction partners.

Lipids are likely to have a larger role in marine chemistry than that on land if only because of their greater relative abundance in sessile forms. Also relevant, at least to some degree, is the high abundance of chlorophyll in marine algae and diatoms relative to its content in land plants.

Further Experiments with Chemical Probes

The material in this section makes it evident that in the presence of recent humic acids and kerogens, amino acids disappear. The results indicate a major role for irreversible reactions binding the carbon chain to the matrix. However, other mechanisms may also have a role. There is the puzzling slow release of amino acids and kerogens even after five weeks of soxhlet extraction with 6 N HCl at 108°C. Is this to be explained by a tightly binding adsorption phenomenon with amino acids held in a very complex cross-linked structure? Or, alternatively, can small amounts of amino acids be synthesized and released from the polymers?

Many such questions could be addressed in further experimentation. It would be particularly useful to employ probing molecules containing C^{14}, C^{13}, N^{15}, or H^2 with all or part of the chains of the probing molecules tagged. Eglinton (1972) has made some preliminary experiments along these lines.

Nature has provided an example of the wealth of information that a chemical probe (chlorophyll) can elicit from kerogen. The chemical probe is more complex than one a chemist might use since chlorophyll may or may not undergo a number of reactions before it interacts with kerogen. These reactions include, for example, loss of magnesium and removal of the phytol chain. In chlorophyll, attached to the porphin nucleus, are side chains with differing functional groups. Porphyrins recovered from sediments and petroleums possess different modifications of the original side chains.

From a detailed study of porphyrin structures it was possible to demonstrate many of the reactions that occur in nature. These include decarboxylation, hydrogenation, and hydrolysis of esters.

The most interesting type of reaction is one in which alkyl chains from the kerogen are added to make higher molecular weight porphyrins

Thus, the use of the chlorophyll probe alerted geochemists to the fact that alkylation can occur at modest temperatures in nature. The topic has been reviewed by Blumer (1965) and Baker (1969).

Chlorophyll is an outstanding example of a natural chemical probe. Others available include sterols, terpenes, and phytol. However, living matter synthesizes more than one chlorophyll, sterol, and terpene. In the sediments various changes occur before incorporation into kerogen. Thus the chemical probe is a fuzzy one. The use of specific known entities would permit a much more definitive approach to understanding both the formation and reactions of kerogens.

CONCLUSION

Organic geochemistry is evolving into an increasingly important branch of science. During the past decade much attention has been devoted to study of petroleums and their relationships to kerogens. Both time and temperature are important factors in the maturation of kerogens and concomitant evolution of associated petroleums. Chemical and physical changes in kerogen that accompany the maturation process are now well-established, and they provide tools of great practical value in petroleum exploration.

Another trend in organic geochemistry is the increasing use of laboratory simulations. It has been established that events requiring 180 m.y. in nature can be paralleled very well in the laboratory in a few minutes at elevated temperatures. This insight is also beginning to have broad application in petroleum exploration. Thus, a suspected petroleum source rock can be heated and its products compared to a petroleum of unknown origin.

Other useful products of laboratory simulation are advances in the understanding of formation of humic acids and kerogens and their reactions. This work is still in its beginning phases, but with the marvelous tools of equipment and technique now available, future progress is assured.

ACKNOWLEDGMENTS

In preparing this article I enjoyed the friendly cooperation and advice of many geo-chemists. I am particularly indebted to Thomas Hoering, for many years a valued colleague.

I thank the *American Association of Petroleum Geologists Bulletin* for permission to reproduce Figures 1–6 from Tissot et al (1971, 1974), and *Geochimica Cosmochimica Acta* for permission to reproduce Figures 7–8 from Albrecht, Vandenbroucke & Mandengue (1976) and Figure 9 from Ishiwatari et al (1977).

Literature Cited

Abelson, P. H. 1959. Organic geochemistry: kerogen. *Carnegie Inst. Washington Yearb.* 58:181–85

Abelson, P. H. 1963. Organic geochemistry and the formation of petroleum. *Proc. World Pet. Congr., 6th,* Sect. 1, pp. 397–407

Abelson, P. H., Hare, P. E. 1970. Uptake of amino acids by kerogen. *Carnegie Inst. Washington Yearb.* 68:297–303

Abelson, P. H., Hare, P. E. 1971. Reactions of amino acids with natural and artificial humus and kerogens. *Carnegie Inst. Washington Yearb.* 69:327–34

Albrecht, P., Ourisson, G. 1969. Diagénèse des hydrocarbures saturés dans une série sédimentaire épaisse (Douala, Cameroon). *Geochim. Cosmochim. Acta* 33:138–42

Albrecht, P., Vandenbroucke, M., Mandengue, M. 1976. Geochemical studies on the organic matter from the Douala Basin (Cameroon)—I. Evolution of the extractable organic matter and the formation of petroleum. *Geochim. Cosmochim. Acta* 40:791–99

Baker, E. W. 1969. Porphyrins. In *Organic Geochemistry—Methods and Results,* ed. G. Eglinton, M. T. J. Murphy, Chap. 19, pp. 464–97. Berlin: Springer. 828 pp.

Blumer, M. 1965. Organic pigments: their long-term fate. *Science* 149:722–26

Burlingame, A. L., Haug, P. A., Schnoes, H. K., Simoneit, B. R. 1969. Fatty acids derived from the Green River Formation oil shale by extractions and oxidations—

a review. In *Advances in Organic Geochemistry 1968,* ed. P. A. Schenck, I. Havenaar, pp. 85–129. Oxford: Pergamon. 617 pp.

Cummins, J. J., Doolittle, F. G., Robinson, W. E. 1974. *Thermal degradation of Green River kerogen at 150° to 350°C: composition of products.* Wash. D.C.: U.S. Bureau of Mines, Rep. Invest. 7924. 18 pp.

Cummins, J. J., Robinson, W. E. 1972. *Thermal degradation of Green River kerogen at 150° to 350°C: rate of product formation.* Wash. D.C.: U.S. Bureau of Mines, Rep. Invest. 7620. 15 pp.

Dunton, M. L., Hunt, J. M. 1962. Distribution of low molecular-weight hydrocarbons in recent and ancient sediments. *Am. Assoc. Pet. Geol. Bull.* 46:2246–48

Durand, B., Espitalié, J. 1976. Geochemical studies on the organic matter from the Douala Basin (Cameroon)—II. Evolution of kerogen. *Geochim. Cosmochim. Acta* 40:801–8

Eglinton, G. 1972. Laboratory simulation of organic geochemical processes. In *Advances in Organic Geochemistry 1971,* ed. H. R. von Gaertner, H. Wehner, pp. 29–48. Braunschweig: Pergamon. 736 pp.

Enders, C. 1943. Wie entsteht der Humus in der Natur? *Die Chemie* 56:281–92

Hedges, J. 1976. The formation and clay mineral reactions of melanoidin. *Carnegie Inst. Washington Yearb.* 75:792–800

Hedges, J. I., Parker, P. L. 1976. Land

derived organic matter in surface sediments from the Gulf of Mexico. *Geochim.
Cosmochim. Acta* 40:1019–29

Hodge, J. E. 1953. Chemistry of browning
reactions in model systems. *Agric. Food
Chem.* 1:928–43

Hoering, T. C. 1973. A comparison of
melanoidin and humic acid. *Carnegie
Inst. Washington Yearb.* 72:682–90

Hunt, J. M. 1972. Distribution of carbon
in crust of the earth. *Am. Assoc. Pet.
Geol. Bull.* 56:2273–77

Hunt, J. M. 1977. Distribution of carbon
as hydrocarbons and asphaltic compounds in sedimentary rocks. *Am. Assoc.
Pet. Geol. Bull.* 61:100–4

Ishiwatari, R., Ishiwatari, M., Rohrback, B.
G., Kaplan, I. R. 1977. Thermal alteration
experiments on organic matter from
recent marine sediments in relation to
petroleum genesis. *Geochim. Cosmochim.
Acta* 41:815–28

Kvenvolden, K. A. 1975. Advances in the
geochemistry of amino acids. *Ann. Rev.
Earth Planet. Sci.* 3:183–212

LaPlante, R. E. 1974. Hydrocarbon generation in Gulf Coast tertiary sediments.
Am. Assoc. Pet. Geol. Bull. 58:1281–89

Maillard, L. C. 1913. Formation de matières
humiques par action de polypeptides sur
les sucres. *C. R. Acad. Sci.* 156:148–49

Maxwell, J. R., Pillinger, C. T., Eglinton, G.
1971. Organic geochemistry. *Q. Rev.
(London)* 25:571–628

McIver, R. D. 1967. Composition of kergoen;
clue to its role in the origin of petroleum.
Proc. World Pet. Congr., 7th, 2:25–36

Peters, K. E., Ishiwatari, R., Kaplan, I. R.
1977. Color of kerogen as index of organic
maturity. *Am. Assoc. Pet. Geol. Bull.* 61:
504–10

Philippi, G. T. 1965. On the depth, time and
mechanism of petroleum generation. *Geochim. Cosmochim. Acta* 29:1021–49

Pusey, W. D. 1973. Paleotemperatures in
the Gulf Coast using the ESR-kerogen
method. *Trans. Gulf Coast Assoc. Geol.
Soc.* 23:195–202

Rashid, M. A., King, L. H. 1970. Major
oxygen-containing functional groups
present in humic and fulvic acid fractions
isolated from contrasting marine environments. *Geochim. Cosmochim. Acta* 34:
193–201

Revelle, R., Munk, W. 1977. The carbon
dioxide cycle and the biosphere. In *Energy
and Climate,* pp. 243–81. Wash. D.C.:
Natl. Acad. Sci. 281 pp.

Robinson, W. E. 1969. Kerogen of the Green
River formation. See Baker 1969, Chap.
26, pp. 619–37

Tissot, B., Califet-Debyser, Y., Deroo, G.,
Oudin, J. L. 1971. Origin and evolution
of hydrocarbons in early Toarcian shales,
Paris Basin, France. *Am. Assoc. Pet. Geol.
Bull.* 55:2177–93

Tissot, B., Durand, B., Espitalié, J., Combaz,
A. 1974. Influence of nature and diagenesis of organic matter in formation
of petroleum. *Am. Assoc. Pet. Geol. Bull.*
58:499–506

Vandenbroucke, M., Durand, B., Albrecht,
P. 1976. Geochemical studies on the
organic matter from the Douala Basin
(Cameroon)—III. Comparison with the
early Toarcian shales, Paris Basin, France.
Geochim. Cosmochim. Acta 40:1241–49

Van Krevelen, D. W. 1961. *Coal—Topology,
Chemistry, Physics, Constitution.* New
York, Amsterdam: Elsevier. 514 pp.

Welte, D. H. 1964. Nichtflüchtige Kohlen-
wasserstoffe in Kernproben des Devons
und Karbons der Bohrung Münsterland
1. *Fortschr. Geol. Rheinl. Westfalen* 12:
559–68

Ann. Rev. Earth Planet. Sci. 1978. 6 : 353–75

QUANTIFYING BIOSTRATIGRAPHIC CORRELATION

✻10098

William W. Hay and John R. Southam

Division of Marine Geology and Geophysics, Rosenstiel School of Marine
and Atmospheric Science, University of Miami, 4600 Rickenbacker Causeway,
Miami, Florida 33149

INTRODUCTION

The historical development of concepts of biostratigraphic correlation has been reviewed recently by Hancock (1977).

Stratigraphic correlation had its origin in the very practical problems faced by the British civil engineer William Smith in the course of his work in Somerset. From 1793 he was busy classifying the strata in the region of Bath and relating them to the sequence in the north of England, using fossils as indicators to establish the original continuity of lithologic units. Smith's interest in fossils was wholly practical; they enabled him to discriminate between units having similar appearance, and to determine what the sequence might be in places where only small exposures were available to view. By 1799 he had prepared the first stratigraphic table of the English Mesozoic, and in 1815 he issued a geologic map of England in 15 sheets.

George Cuvier, a French contemporary of Smith's, provided the concept of catastrophism as the logical basis for stratigraphic correlation of France. Cuvier's idea of catastrophism was based chiefly on observations in the Paris Basin, where the successive marine invasions during the Tertiary each left homogeneous assemblages of organisms as a paleontologic record (Cuvier & Brongniart 1808). Within each of the transgressive-regressive sequences, there is no evidence for the gradual evolution of species, but between the cycles there are wholesale changes in fauna. It was a logical conclusion on Cuvier's part to suggest a series of global catastrophies that annihilated life, and a series of creations that produced new forms of life. Cuvier's theory of catastrophism would remain merely an interesting relict were it not for the fact that he exerted a considerable influence on the development of geologic thought in France and on that of d'Orbigny in particular, so that it became an integral part of the legacy of stratigraphy.

Alcide d'Orbigny created the term "stage" in 1851; he envisioned it as a particular landscape in the earth's history. By this he meant a particular distribution of land and

353

sea. His concept derived from that of Cuvier: the stage represents a particular creation with internal stability, and the boundaries between stages represent catastrophic episodes of change. Stratigraphy based on a succession of such stages is perfectly defined, and if all stages have been detected then all strata must be assigned to one stage or another.

D'Orbigny's concept of stage was followed closely by Albert Oppel's coining of the term "zone" (1856) by which he hoped to refine the stage concept and aid in correlation of strata over greater distances. Exactly what Oppel meant by the term "zone" was never quite clear from his own writings; at one point it seems that he intended something similar to what we term "assemblage zone"; at others he apparently intended something more closely akin to our modern "range zones."

Charles Darwin's *Origin of Species* was published in 1859, and soon the simple basis of stratigraphy founded in ideas derived from catastrophism began to disintegrate as paleontologic interest became focussed on establishing evolutionary trends and on the search for ancestors and descendants.

CLASSICAL BIOSTRATIGRAPHIC CORRELATION

Biostratigraphy has been built around the recognition of "zones" based on the occurrences of fossils; these are generally termed "biozones," although that term has been used by many authors with more restricted meaning. Once a sequence of zones has been determined, a given sample can then be assigned to one or another of them. The terminology of biostratigraphic zones has become complex and often confusing because of the differing definition of some terms. Two major categories of bio-stratigraphic zone based on taxa [American Commission on Stratigraphic Nomenclature (ACSN) 1961] or fossil groups (George et al 1969) have been recognized: the assemblage zone or cenozone [Van Hinte 1969, International Subcommission on Stratigraphic Classification (ISSC) 1972, 1976], defined as the body of strata characterized by the presence of an assemblage of fossils, and a family of "zones" based on the ranges of individual taxa. Among the zones based on the ranges of fossils are: 1. range zone (ACSN 1961), total range zone (George et al 1969), acrozone (Van Hinte 1969) or taxon range zone (ISSC 1972, 1976), defined as the body of strata comprising the total range of a taxon; 2. concurrent range zone (ACSN 1961, George et al 1969, ISSC 1972, 1976) defined as the body of strata characterized by the overlapping ranges of specified taxa; 3. partial range zone (George et al 1969), defined as the body of strata within the range of a taxon above the last appearance of the preceding taxon and below the first appearance of the next succeeding taxon; 4. consecutive range zone (George et al 1969), defined as the special case of 3 in which the zone is the body of strata within the range of a taxon such that it forms the first part of the range of that fossil group before the first appearance of its immediate evolutionary descendant; 5. phylozone (Van Hinte 1969) or lineage zone (ISSC 1972, 1976), based on the range of one of the member taxa in a phylogenetic lineage; and 6. interval zone or interbiohorizon zone (ISSC 1972, 1976), defined simply as the stratigraphic interval between two biostratigraphic markers (see Figure 1). The term local range zone (ACSN 1961, George et al 1969) or teilzone (Teichert 1958) or topozone (Van Hinte 1969), defined as the body of rock between the base and top of the occurrence of a particular taxon in a particular area, is an attempt to recognize in a qualitative way that the ranges of taxa in local sections may differ among themselves and may differ from the total range of the taxon compiled from all known sections.

There is a widely held belief that the finest biostratigraphic subdivisions can be obtained by using members of an evolutionary lineage (Van Hinte 1969), and this is certainly true for some groups in one geologic region. Kauffmann (1970) estimated the duration of the zones based on molluscs in the western interior of North America at 0.08 to 0.5 million years. Kauffmann (1977) has explored how the occurrence of rapidly evolving groups of fossils might be predicted and has investigated the controls on evolutionary rates. However, if the members of a single lineage are used to establish zonations, the problem becomes reduced to one of discriminating and recognizing patterns in a series. The number of patterns that can be distinguished in a continuous series is infinite, but in practice is limited by the number of samples and individuals available, as discussed by Drooger (1974). The effort involved in recognition increases markedly as the number of different patterns discriminated in a series increases, so that time and manpower constraints limit the number of forms that can be distinguished in a single lineage. Furthermore, interpretation of morphologic changes in terms of evolutionary lineages of taxa may be highly

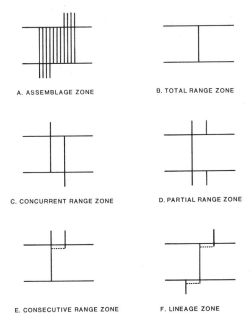

A. ASSEMBLAGE ZONE   B. TOTAL RANGE ZONE

C. CONCURRENT RANGE ZONE   D. PARTIAL RANGE ZONE

E. CONSECUTIVE RANGE ZONE   F. LINEAGE ZONE

Figure 1 Some biostratigraphic zones: stratigraphic ranges of species are indicated by vertical lines, zone boundaries by horizontal lines; *a* is a cenozone, *b–f* are acrozones, and *c–f* are interval zones.

subjective and depend upon the goals of the paleontologist (Shaw 1969). In one of the groups that has been most intensively studied, the planktonic Foraminifera, the time resolution attained by recognizing species along evolutionary lineages is in the order of one million years or longer. This is far from the theoretical resolution that might be achieved using all species of planktonic Foraminifera together.

Biostratigraphic methodology has worked well as a basis for stratigraphic correlation as long as biostratigraphic correlations were either limited to one geologic area or carried out by one specialist. Problems and disputes began to arise when many specialists began to study larger areas and investigations began to overlap. At first it seemed as though differences of opinion regarding the distribution of taxa could be resolved through restudy by a few select experts who could agree among themselves, but more recently it has become evident that there are real problems of a statistical nature involved in interpreting the distributions of fossils.

The situation in the 1950s and 1960s is summarized in papers by Jeletzky (1956) and Harrington (1965). Both point out the significance of biostratigraphy for stratigraphic correlation, but they do not suggest quantification of correlation techniques beyond calibration of biostratigraphic zonation schemes with absolute ages determined by radiometric or other techniques.

In 1960 the desire to develop quantitative expressions for stratigraphic correlation was eloquently expressed by Cheetham: "biostratigraphers need not face extinction if they can adapt themselves to the quantitative environment of modern science." The complexities of proper mathematical treatment of biostratigraphic correlation were recognized by Simpson (1960) in the course of a discussion of the properties of coefficients used to evaluate faunal similarity. He noted two particularly vexing problems: taxonomic homogeneity and sampling effects. Comparisons of faunas are not meaningful unless the taxonomy is homogeneous—that is, that one part of the data set has not been "split" or "lumped" more than the rest. The effects of sampling differences were noted to affect the coefficients of similarity and Simpson suggested (p. 308) that "for fully reliable comparisons, statistical confidence intervals should, of course, be established for any index used. ... this is an exceptionally difficult problem that has not yet been worked out."

QUANTITATIVE METHODS OF CORRELATION BY SIMILARITY OF FOSSIL ASSEMBLAGE

Biostratigraphic correlation based on comparison of assemblages of fossils was the first quantitative technique applied. Quantitative comparison by dividing the number of taxa common to both of two samples (C) by the total number of taxa in both samples (N_t) (the "percentage" method of paleontologists, later identified as the Jaccard coefficient of botany) had long been used, but Simpson (1947), recognizing that this measure is strongly influenced by differences in sample size, suggested replacing it with a coefficient of faunal resemblance by dividing C by the total number of taxa in the smaller sample (N_1). Sorgenfrei (1958) used a more elaborate correlation ratio involving the total number of taxa in the larger sample (N_2): $100C^2/N_1N_2$. He also proposed a technique for estimating the total number of species (n) in a population of which N_1 and N_2 are samples: $n = N_1 N_2/C$. Using this estimate, he formulated a second correlation ratio: $100\ N_1 N_2/n^2$, but this is algebraically the same as his first ratio. It was in a review of Sorgenfrei's paper that Cheetham (1960) made the comment on quantitative biostratigraphy quoted above.

Simpson (1960) reviewed and evaluated indices of faunal resemblance for both biostratigraphic and taxonomic comparisons, discussing the advantages and disadvantages of each but noting the problems of taxonomy and sampling and the need for development of methods for determining the confidence intervals associated with the different coefficients of similarity.

Cheetham & Hazel (1969) presented a review of some 22 binary similarity coefficients, describing their characteristics and properties. Evolution of the stratigraphic value of these coefficients ensued, and Hazel (1970) cited five that have come to be preferred in biostratigraphic work: the Jaccard, $C/(N_1 + N_2 - C)$; Dice, $2C/(N_1 + N_2)$; Simpson, C/N_1; Fager, $C/(N_1 N_2)^{1/2} - \frac{1}{2}(N_2)^{1/2}$, and Otsuka $C/(N_1 N_2)^{1/2}$. Noting that the Jaccard coefficient is the most widely used, he cites the Dice coefficient as an improvement suggested by Hall (1969) to reduce the effect of mismatches (species present in one sample but not the other). Hazel rejected the Simpson coefficient as a standard measure of biostratigraphy because it emphasizes similarity since the denominator is the smaller of two samples being compared. Because beds above a faunal discontinuity commonly contain relatively few species, they will have high similarity values with respect to more diverse younger samples. Hazel suggested that either the Fager or Otsuka coefficient is most appropriate for biostratigraphic work, and then used the latter in his examples.

A comparison of the Jaccard, Dice, Simpson, and Otsuka correlation ratios and that of Sorgenfrei for equal and unequal samples and varying proportions of taxa in common is presented in Table 1.

When a large number of samples are to be compared, the coefficients can be displayed as a matrix, but the array soon becomes bewilderingly large. To determine groupings within the coefficient matrix, Hazel (1970) proposed the use of cluster analysis with the product of the analysis displayed as dendrograms of the type shown in Figure 2. Hazel's dendrograms, which were based on real data from Deboo (1965), allowed the stratigraphic sections being analyzed to be subdivided into three major units, similar to those in Figures 2a and 2b. The dendrograms maintained proper order of the sample sequence in each section. A composite dendrogram for 62 samples from seven localities also showed three major clusters, but the samples assigned to each of these clusters were not always those assigned to the three major clusters in each of the stratigraphic section dendrograms, nor was the correct sequence of samples in single sections maintained; Figure 2c is a fictitious composite dendrogram indicating the nature of the problem. Hazel (1970) suggested the use of shaded trellis diagrams to aid interpretation of the cluster analysis. Hazel (1977) has explored the usefulness of other techniques for grouping the bioassociational coefficients — principal components analysis, principal coordinates analysis, and multidimensional scaling — and the reader is referred to this excellent review paper for a comparative evaluation of these methods.

One generic problem inherent in comparing samples by their fossil assemblage is

Table 1   Comparison of bioassociational coefficients

	Jaccard $C/(N_1+N_2-C)$	Dice $2C/(N_1+N_2)$	Simpson C/N_1	Otsuka $C/(N_1N_2)^{1/2}$	Sorgenfrei C^2/N_1N_2
For two equal assemblages, each containing 100 species					
$C = 0$	0	0	0	0	0
10	0.053	0.100	0.100	0.100	0.010
25	0.143	0.250	0.250	0.250	0.063
50	0.333	0.500	0.500	0.500	0.250
75	0.600	0.750	0.750	0.750	0.563
90	0.818	0.900	0.900	0.900	0.810
100	1.000	1.000	1.000	1.000	1.000
For two equal assemblages, each containing 10 species					
$C = 0$	0	0	0	0	0
1	0.053	0.100	0.100	0.100	0.100
5	0.333	0.500	0.500	0.500	0.250
9	0.818	0.900	0.900	0.900	0.810
10	1.000	1.000	1.000	1.000	1.000
For two equal assemblages, one containing 50, the other 100 species					
$C = 0$	0	0	0	0	0
10	0.071	0.133	0.200	0.141	0.020
25	0.200	0.333	0.500	0.354	0.125
40	0.364	0.533	0.800	0.566	0.320
50	0.500	0.667	1.000	0.707	0.500
For two unequal assemblages, one containing 25, the other 50 species					
$C = 0$	0	0	0	0	0
5	0.071	0.133	0.200	0.141	0.020
10	0.154	0.267	0.400	0.283	0.080
20	0.364	0.533	0.800	0.566	0.320
25	0.500	0.667	1.000	0.707	0.500
For two unequal assemblages, one containing 90, the other 100 species					
$C = 0$	0	0	0	0	0
10	0.056	0.105	0.111	0.105	0.011
25	0.152	0.263	0.278	0.264	0.069
50	0.357	0.526	0.556	0.527	0.278
75	0.652	0.789	0.833	0.791	0.625
85	0.810	0.895	0.944	0.896	0.803
90	0.900	0.947	1.000	0.949	0.900

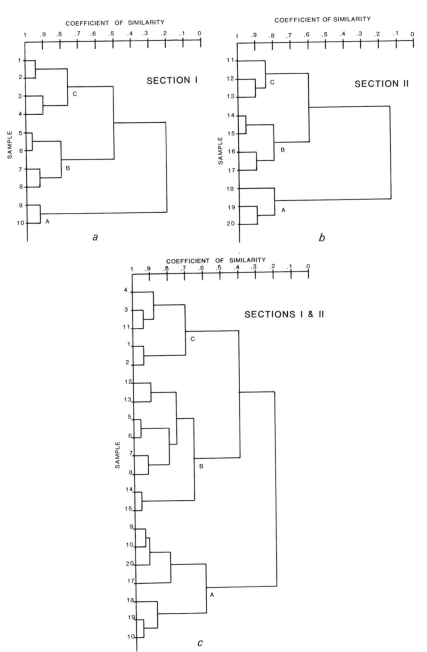

Figure 2 Schematic dendrograms, such as would be derived from cluster analysis of bioassociational coefficients; 2a and 2b are dendograms for two stratigraphic sections, and 2c is a composite dendogram combining the two sections.

that of the absence in a particular sample of a taxon known to occur both higher and lower in the section. Cheetham & Deboo (1963) proposed the "range through method" to obviate this problem; the presence of a species known to occur both higher and lower in a given section is assumed to be present even though not actually found in a given sample. Hazel (1970, 1977) has demonstrated that this is a useful and important adjustment of the data in the context of bioassociational coefficients used for biostratigraphic interpretation. The assumption reduces the effects of sampling error and facies changes. In the development of these quantitative techniques for establishing stratigraphic correlation based on assemblages of fossils, only presence-absence information has been used. Simpson (1960) had noted that coefficients that also take into account the abundances of individual taxa within assemblages provide a more accurate measure of faunal similarity, but recognized that assemblages with similar abundances are most likely to represent similar environmental conditions. Cheetham & Hazel (1969) and Hazel (1970, 1977) have purposely avoided considering coefficients that take into account the abundances of species within the assemblage. Curiously, although these attempts at quantification have offered a wholly new range of insight into the nature of biostratigraphic correlation based on entire assemblages of fossils, there has been little progress in refining stratigraphic zonation or in developing an unambiguous mathematical expression to quantify the likelihood of synchroneity. When Hazel (1970, 1977) applied the cluster analysis technique to data gathered by others, he was not able to recognize zonations finer than those established by the original workers on the basis of hand calculation and inspection of the data. Simpson's (1960) concern over sampling problems looms as a critical factor in the evaluation of correlation coefficients.

One of the problems with the commonly used correlation coefficients is immediately obvious from inspection of Table 1: they are wholly insensitive to sample size if the samples are equal in size. A second problem is evident in that it is not obvious which is the proper coefficient to use for biostratigraphic interpretation, suggesting that the problem has never been completely stated in formal mathematical terms. Clearly, correlation by the presence of a single taxon at each of two localities is less certain than correlation by the presence of a hundred species in each of two samples, all species being common to both samples. Confidence limits would show this difference and should be an integral part of a correlation coefficient.

QUANTIFICATION OF THE INDEX FOSSIL CONCEPT

Jeletzky (1965) criticized attempts to quantify biostratigraphic correlation by means of coefficients of similarity. He called attention to the increasing number of zonations based on evolutionary lineages as having a far greater potential for the development of biostratigraphy. He pointed out correctly that not all taxa in an assemblage are of equal value as biostratigraphic indicators but stated that it would be impossible to devise any quantitative expression for the biostratigraphic value of a taxon.

Cockbain (1966) was the first to take up Jeletzky's challenge. Citing Quastler (1958), he proposed using the Shannon-Weiner information function ("entropy function") on the basis that "a measure of information can be thought of as a measure of the amount of uncertainty eliminated by a particular event." The information function is

$$H(x) = -\sum_i p(i) \log_2 p(i),$$

where x is the event with categories i (in this case the categories are biostratigraphic zones), and each category has the probability $p(i)$. The units of the function are the number of binary digits (bits) needed to represent the event. As an example, Cockbain cited the case where an index fossil will identify a particular zone. Assuming that Phanerozoic time can be divided into about 250 biostratigraphic zones, and that the probability of finding each zone is equal, then, for a fossil that identifies a particular zone, the information function takes on the value $H(x) = \log_2 250 = 7.96$ bits; correspondingly, if a fossil places a sample into one of the ten systems, $H(x) = \log_2 10 = 3.32$ bits, etc. The problem arises in choosing the categories to be recognized and in assigning relative probabilities to them.

McCammon (1970) has presented a more rounded approach to the problem of expressing the value of a taxon for biostratigraphy. He proposed that the relative biostratigraphic value of a fossil species, RBV, be expressed by

$$RBV = \alpha(1 - \mu_v) + (1 - \alpha)\mu_h,$$

where α is a measure of the facies independence of the species, μ_v is the vertical range, and μ_h is the lateral persistence; each of these parameters is expressed on a scale from 0 to 1 so the RBV is also a number between 0 and 1. The methods by which α, μ_v, and μ_h are determined is left to the investigator. McCammon suggested that it would be more appropriate to establish α as the number of facies in which the species occurs divided by the number of facies being considered. The vertical range, μ_v, was established as the maximum of local vertical range values measured in stratigraphic thickness divided by the maximum stratigraphic thickness being considered. The lateral persistence, μ_h, was established as the number of localities where the species is present divided by the number of localities considered. Each of the factors is thus expressed as a proportion.

Analyzing the data of Deboo (1965), McCammon assigned a value of 0.5 for all cases because adequate data on facies distribution were not available, so that the analysis rests solely on comparisons of the vertical and lateral extent of the taxa. Eleven of the 194 species recognized by Deboo were found to have an $RBV > 0.8$, each of these being restricted to one of the four biostratigraphic units recognized by Deboo.

McCammon also proposed a classification function

$$Z_{ik} = \sum_{j=1}^{n_i} W_{ij} X_{jk},$$

which can be used to assign a particular sample (k) to a biostratigraphic unit after the characteristic species (those with high RBVs) have been identified. For the classification function, the subset of species characteristic for a biostratigraphic interval (i) is recognized (n_i) and each of their RBVs (w) normalized so that all total

1 ; i.e. if there are 5 characteristic species for the biostratigraphic interval, each having equal RBVs, the normalized relative biostratigraphic value of each is 0.2. The normalized relative biostratigraphic value of the jth species in biostratigraphic interval i is W_{ij}. The normalized relative biostratigraphic values of each characteristic species is summed for each characteristic species present. Presence or absence of species j in sample k is denoted by the term X_{jk}, the value of which is 1 if the species is present or 0 if the species is absent.

McCammon answered Jeletzky's arguments completely by providing an objective method of weighing the relative biostratigraphic value of each of the taxa in an assemblage. The subset of species "characteristic" of a biostratigraphic interval could be only those species restricted to the interval, could include all species occurring in the interval, or could be only those species with an RBV higher than a given level. Because each of the three factors, α, μ_v, and μ_h is represented by a proportion, confidence limits for each could be introduced to take sampling problems into account, casting the function in a probabilistic form. If this were carried through to the classification function, the term W_{ij} is then based on probabilities, and the term X_{jk} is easily converted into a probabilistic estimate of the probability of absence of the species due to small sample size, as discussed below.

Hazel (1970, 1977) has introduced two terms, biostratigraphic fidelity and constancy, to evaluate the biostratigraphic value of fossil taxa. Constancy (Pi) is defined simply as the percentage of samples from a particular stratigraphic unit that contain a particular taxon. The biostratigraphic fidelity (BF) of a species is calculated by dividing the percentage of occurrences of a species in a unit (constancy) by the sum of the percentages of occurrences of that species in all other biostratigraphic units within the limits of the problem. Because these terms are also based on proportions, confidence intervals could be introduced to cast the expressions into a probabilistic form. Hazel (1970) applied these terms to the Ostracoda in Deboo's (1965) data. Both Hazel and McCammon were successful in detecting the most restricted species, but because the two techniques are measures of different factors, strict quantitative comparison of their results is not possible.

QUANTITATIVE METHODS OF CORRELATION BASED ON THE RANGES OF FOSSILS IN STRATIGRAPHIC SECTIONS

Shaw (1964) proposed a new method for correlation based on graphical comparison of sections displayed at right angles or construction of a composite section and interpolation; the method has recently been further developed and discussed by Miller (1977). Stratigraphic events, such as the end points of the ranges of species, are plotted along an ordinate and abscissa, as shown in Figure 3a. If the stratigraphic sections in an area have been well enough studied to permit designation of a standard reference section, that section is usually plotted along the abscissa and others, to be compared with it, are plotted along the ordinate. The diagonal representing the line connecting the points to which the limits of ranges of taxa in the two sections are projected is called the "line of correlation."

Shaw incorrectly defined this line as a "rate of rock accumulation," as opposed to a "rate of sedimentation," thus, according to Miller (1977), implying no relationship to absolute time. However, the term "accumulation rate" has already been preempted in geology as the rate of accumulation of the solid phase of a sediment in a rock or sediment, expressed in terms of mass per unit time. Accumulation rate is independent of the degree of compaction or pore space in the sediment or rock and also independent of the nature of the fluid filling the pore space. In any case, the word "rate" involves a physical measure of mass or distance divided by time. As de Cserna (1972) stated, "rocks are measured in meters, time in days or years, and wine in liters. As long as a stratigraphic sequence is not described in liters, time in meters, and wine in days, there is no confusion whatsoever." Miller (1977) accepted Shaw's discussion of rate literally, concluding that, in this case, chronostratigraphy can be defined in terms of a single reference section or constructed composite reference section. This is true only if one is willing to accept a form of chronostratigraphy in which time is a nonlinear unknown function. Shaw's method is, however, important in using simple interpolation as a means of establishing the most likely correlations in the interval between correlatable biostratigraphic reference points in different sections.

Shaw (1964) used the "line of correlation" as a composite section, combining the first two sections compared. The composite section can then be used as an ordinate and a new third section plotted as the abscissa to produce a new composite section.

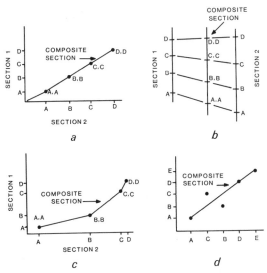

Figure 3 Graphical correlation of stratigraphic sections: 3*a* correlates the end points (*A, B, C, D*) of the ranges of taxa in two sections using Shaw's (1964) method; 3*b* correlates the same sections by an arithmetic average method; 3*c* illustrates the shape of the "line of correlation" or composite section when sedimentation rates in the two sections are unequal; 3*d* illustrates the situation when taxon range end points *B* and *C* are reversed in the two sections.

Using this method, any number of sections can be added together to produce a composite. The relative distance between stratigraphic events on the composite axis can be used as a basis for correlation by interpolation back to any single section. The expected position of events not found in a given section can be determined for that section; further, an infinite subdivision of the sections and an infinite number of correlation lines appear to be possible.

Figure 3a represents an ideal situation, where stratigraphic events occur in two sections, each of which has a constant sedimentation rate. The composite section plots as a straight line. However, it will be noted that because this line is the hypotenuse of a triangle formed with either section, distances along it are greater than in either section. This is why Shaw's composite section constructed by the graphic method must have an arbitrary scale. Figure 3b illustrates another method of constructing a composite section; in this case, the distance along the composite section axis has real meaning because it is in fact the arithmetic mean of the two sections that have been combined. An additional advantage of this method is that the standard error and standard deviation are readily calculated.

Figure 3c illustrates a different sort of case, in which one section has very different sedimentation rates. It is evident that the slope of the line is a measure of sedimentation rate between events. It will also be seen that by using the graphic method for such a case, the portions of sections having high sedimentation rate in effect determine the nature of the composite section. Low sedimentation rates do not affect the composite in a similar manner. This bias introduced by high sedimentation rates persists as more sections are added, so that ultimately the composite section becomes strongly biased by those sections or parts of sections that have high sedimentation rates. This is not necessarily a disadvantage, but it is a peculiarity of the method that should be noted.

If a composite section is constructed for three or more sections by using the graphic method consecutively, each new section has as much influence on the nature of the composite as all of the previous sections together (remembering, of course, that any high sedimentation rate in any section will introduce a special bias). If the sections are combined "all at once" using the arithmetic average method for the intervals between event pairs, the problem of undue influence of a single section can be avoided, but the biasing effect of high or low sedimentation rates is still significant, although less pronounced.

Figure 3d illustrates what happens if the sequence of events is different in the two sections. In this case, the events for which the sequence is reversed do not plot along a straight line, and may be far from the line defined by the majority of points. Two solutions to this problem are possible: 1. the investigator can decide which section contains spurious information and can delete it, or 2. a regression line can be drawn through the points, using the least squares method. Regression techniques are, in fact, a better means of combining information from multiple sections than is the arithmetic average method because it more closely approximates the central tendency and reduces the bias from sections having unusually high or low sedimentation rates; i.e. it will result in a more generally applicable solution. For a good account of linear regression applied to geologic problem, the reader is referred to Till (1974). In the classical regression technique, one axis is regressed against the other, so that for any set of data two lines are possible, that of x regressed on y or that of y regressed on x. Another possibility for handling the data, but one that can be used only when considering two sections, is the technique of reduced major axis line also discussed by Till (1974). The reduced major axis is in effect the line that splits the difference between x on y and y on x. It is a potentially useful technique, but it must be remembered that unless the standard errors are calculated along with the equation of the line it is possible for completely random data to produce an apparently meaningful line.

The standard error is not directly calculated using Shaw's graphic method; standard error and standard deviation are easily calculated using the arithmetic mean method; and standard error can be calculated using the regression methods. It is important to know the error, because if it should be a significant fraction of the interval being correlated from one stratigraphic section to the next, it indicates that the correlation is imprecise.

Shaw's method of correlation incorporates stratigraphic thickness as a unique piece of information not utilized in the other quantitative methods discussed here, with the exception of that of McCammon. That it works well as a tool in basin analysis is indicated by Miller's (1977) evaluation.

QUANTITATIVE METHODS INVOLVING SEQUENCE OF BIOSTRATIGRAPHIC OR OTHER EVENTS

These techniques are based on a statistical analysis that assigns to each ordering of interval end points, such as the limits of range of a species or boundaries of lithologic units, a probability and confidence interval. The generality of the methods makes them equally applicable to lithostratigraphic or biostratigraphic data or to any combination of the two types of information; the schemes for stratigraphic correlation remain objectively defined, but because they are an integrated description of knowledge of stratigraphic sequence, they are analogous to chronostratigraphic interpretation. The techniques have the additional feature of being applicable to cases where the number of stratigraphic sections available for study is small and the area under consideration is large.

Hay & Cepek (1969) suggested that if stratigraphic correlation is defined as an expression of the degree of probability that samples from two different sections occupy the same level in a known sequence of events, and if the assumption is made that no two events occurred simultaneously, then tests can be devised to determine the probability that the occurrence of events is nonrandom, and that the samples do occupy particular positions with respect to the sequence of events. Defined in this way, the probability of a correlation becomes the product of the probabilities of three factors: 1. the probability that the stratigraphic events defining a stratigraphic increment have been detected; 2. the probability that the true sequence of stratigraphic events is known; 3. the probability that the events that define stratigraphic increments have been correctly identified.

The first of these factors can be determined graphically from the tables of Shaw (1964) from the figure originally published in another context by Dennison & Hay (1967) and reprinted in modified form by Hay (1972), or from any of a variety of books of statistics with tables (e.g. Crow et al 1960). The second factor involves the construction of a matrix that permits the probability of sequence between pairs of events to be determined; this was the main subject of a paper by Hay (1972). The third factor is still under analysis, but a general discussion of some of the problems involved and potential means of solution have been presented by Hay (1971).

In developing the idea of probabilistic stratigraphy, a variety of terms needed to be rigorously defined. Hay (1972) defined a "stratigraphic event" as an occurrence of some importance in stratigraphy. The term "stratigraphic event" can be used with appropriate modifiers, such as "local stratigraphic event" or "regional stratigraphic event" or "global stratigraphic event," to indicate the areal extent of the occurrence. Special kinds of stratigraphic events, those pertaining to the ranges of fossils and those related to physical properties of the rock, can be recognized. From these considerations, a biostratigraphic event is defined as the lowest or highest stratigraphic occurrence of a taxon or fossil group.

The accuracy of the determination of lowest and highest stratigraphic occurrences depends on two factors: 1. the abundance of the fossil group in the total population, and 2. the size of the sample available for study. Although biostratigraphers are almost universally loath to admit it, biostratigraphy depends as much on the absence as it does on the presence of certain fossil groups. Determining the probability of a fossil group being absent from a population is very important in biostratigraphy; the Partial-Range Zone and Interval Zone rely on establishing the absence of the taxon or fossil group defining the next higher or next lower zone. The abundance of a fossil group in a population at the time of its origin or extinction may be assumed to be vanishingly small, and it may also be assumed that it is virtually impossible to determine the absolute level of its lowest or highest occurrence. Using the figures or tabulations cited above, however, it is possible to give a numerical value that indicates the probability that a fossil group is absent in a population, based on its known abundance in samples elsewhere and on the number of specimens in the sample investigated. To include all other possibilities, a physical stratigraphic event is defined as an occurrence of stratigraphic importance not based on the limits of ranges of fossils.

Examples of physical stratigraphic events are geomagnetic reversals and fluctuations of isotopic composition, such as the O^{16}/O^{18} ratio, which might be plotted; unusual layers, such as ash or coal beds; and the abundant occurrence of fossils. Physical stratigraphic events may be changes in lithology, chemical attributes, or physical characteristics such as sonic velocity or natural radioactivity; biological events included in this category may be the acmes of abundance, or change in ratios of abundance of species.

To be most useful, stratigraphic events must be changes; the events themselves do not extend over a stratigraphic interval, but are horizons that lie between two samples. If physical and/or biological events are used and interpreted to be isochronous throughout their extent, the resulting subdivision would be regarded as chronostratigraphic although the stratigraphic axis need not be linear with respect to time.

As noted above, physical stratigraphic events are by their nature rarely if ever unique. Biostratigraphic events are commonly unique; because of the irreversible nature of the evolutionary process, origins and extinctions of species are by definition unique. A sample having indications of reversed geomagnetic polarity may have been deposited during any of a large number of ages, but a sample containing *Globorotalia pseudomenardii* must represent material deposited after the origin of that species and before its extinction. Virtually all of the applications of Hay's ideas have been limited to considerations of biostratigraphic events.

Hay defined the stratigraphic distance between two stratigraphic events as a stratigraphic increment, a concept necessary to the definition of stratigraphic resolution. Worsley et al (1977) have defined the stratigraphic interval between the sample containing the highest or lowest known occurrence of a taxon in a section and the superjacent or subjacent sample from which the taxon is absent as a diastrat. The concept of the diastrat is important in relating biostratigraphic precision to stratigraphic thickness. "Stratigraphic resolution" was defined as the smallest stratigraphic increment that can be distinguished at a given level of probability. Considered in this way, stratigraphic resolution is analogous to resolution in optics. The theoretical limit of stratigraphic resolution, expressed in terms of the number of stratigraphic increments that can be recognized, is:

$$N_I = N_E + 1,$$

where N_I is the number of stratigraphic increments and N_E is the number of stratigraphic events.

The theoretical limit of biostratigraphic resolution, using only the levels of origin and extinction of species preserved as fossils, is:

$$N_i = 2N_{xs} + N_{ls} + 1,$$

where N_i is the number of stratigraphic increments, N_{xs} is the number of species that have originated and become extinct, and N_{ls} is the number of species that have originated and are still alive. Because the number of fossil species that can be recognized is in the order of tens of thousands, the potential time resolution for the approximately half billion years of Phanerozoic history is in the order of thousands of years.

Hay (1972), like Shaw (1964), found it convenient to assume that no two biostratigraphic events are simultaneous. If it is assumed that no two species ever originated or became extinct at exactly the same moment, and that each species spread instantaneously over the surface of the globe and occupied all areas and environments until the moment of its extinction, then the theoretical limits of biostratigraphic resolution should be reached everywhere. The first assumption is made only for the purpose of testing a body of data. The second assumption is obviously untrue because geographic and environmental restrictions limit the distribution of all species.

The occurrence of a fossil is the record of the intersection of a number of sets of geological parameters, including particularly: 1. geologic age, 2. paleobiogeographic situation, 3. environment occupied during life, 4. post-mortem changes, including transportation from the site where the organism lived to the site of burial, 5. environment of burial, 6. diagenetic changes, and 7. changes in the enclosing environment during exhumation. The number of geological parameters about which fossils may yield information is potentially very large, and there is no absolute meaning to the occurrence or nonoccurrence of any particular fossil, only an interpretation of the meaning of the occurrence or nonoccurrence.

From a logical context developed from these definitions, Hay (1972) demonstrated how to produce and order a matrix of sequence between pairs of biostratigraphic events. The ordered matrix reveals the most likely sequence for all the stratigraphic events considered. He used a modification of the unbiased coin test to calculate a value for the probability that each pair of biostratigraphic events is separable and known in true sequence, by assuming that the events are not sequential, but that the relations of all pairs are random. In this circumstance the probability of this hypothesis being untrue can be determined. If the sequence of a pair of biostratigraphic events is random, the probability of one biostratigraphic event preceding the other is $\frac{1}{2}$.

If the biostratigraphic events occur n times in the order predicted in a total of N sections in which the events occur sequentially, then the probability P that n would occur, assuming a random distribution, is

$$P = \frac{N!}{n!(N-n)!}(\tfrac{1}{2})^N.$$

This technique was used by Hay & Steinmetz (1973) to analyze the data on distribution of 144 species of calcareous nannofossil in 13 sections in California from Bramlette & Sullivan (1961) and Sullivan (1964, 1965). Although the standard zonation of Martini (1970) permits only seven zones to be recognized, Hay & Steinmetz (1973) noted that it was possible to recognize eleven zones based on lowest occurrences only and nine zones based on highest occurrences only. They

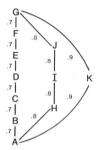

Figure 4 A schematic chain of events, of the sort developed by Hay & Steinmetz (1977) showing that alternative zonations may be developed at different levels of probability. The end points ($A-K$) of the ranges of taxa (biostratigraphic events) are in vertical sequence, with the probability of each pair link indicated. Alternative zonations are possible using events A, B, C, D, E, F, G at the 0.7 probability level; A, H, I, J, G at the 0.8 level; and A, K, G at the 0.9 level. The reliability indices $\hat{p}_{ij}/(1-p^i_{ij})$ of Southam et al can be used to replace the probability levels of Hay & Steinmetz; the geometric average of the Southam et al reliability indices for each chain is a very sensitive measure of the relative value of alternative zonation schemes.

demonstrated that many alternative zonation schemes were possible and presented these as "chains of events" in which the "links" in the chain are of different strength, as shown in Figure 4. The strength of a "link" was, of course, the probability of non-randomness of Hay (1972). They also devised a measure of the reliability and usefulness of a zonation; reliability being defined simply as the ratio of the number of times events occur in the predicted sequence to the number of times they occur in the reversed order and usefulness as the number of times events can be related to samples. The latter term is thus analogous to the index of lateral persistence of McCammon.

Worsley et al (1973) used the same technique to develop and correlate eight alternative zonations for Oligocene calcareous nannofossils in high, intermediate, and low latitudes, hemipelagic and open marine environments, and a general global zonation. Further development of this study of the Oligocene was presented by Worsley & Jorgens (1974), who used the correlated zonations to analyze paleobiogeographic distribution of the species.

Hay & Steinmetz (1977) related seven zonation schemes based on the California data to samples and stratigraphic thickness. One of the zonations was a composite based on highest and lowest occurrences, and divides the sequence into 17 units, the thickest being 337′ and the thinnest 9′. This compared very well with application of the standard zonation in which five zones of the sequence were not represented by any measurable stratigraphic sequence and the thickest was 1020′.

Rubel (1976) developed a similar technique for establishing and evaluating biostratigraphic sequence by adapting the concepts of Hay (1972) to the entire ranges of fossil taxa rather than the end points of the ranges. He introduced the notation and terminology of set theory. The lower limit of the range of a taxon A_i is denoted by $\underline{A}_i$, the upper limit by $\overline{A}_i$, and the range by $\overline{\underline{A}}_i$. In the local section, the taxon is a_i and its limits and range are $\underline{a}_i$, $\bar{a}_i$, and $\overline{\underline{a}}_i$. A point in the local section is designated by geographic coordinates x, y, and a stratigraphic coordinate t that increases downward. Thus

$$\underline{a}_i = \{a(t_{max}, x_p, y_g)\} \in A_i.$$

Ranges of pairs of taxa are grouped into three sets such that

$$\langle A_i, A_k \rangle \in P, \text{ if } \overline{\underline{A}}_i \cap \overline{\underline{A}}_k \neq 0,$$

$$\langle A_i, A_k \rangle \in Q, \text{ if } \overline{\underline{A}}_i \cap \overline{\underline{A}}_k = 0 \quad \text{and} \quad \overline{\underline{A}}_i > \overline{\underline{A}}_k,$$

$$\langle A_i, A_k \rangle \in R, \text{ if } \overline{\underline{A}}_i \cap \overline{\underline{A}}_k = 0 \quad \text{and} \quad \overline{\underline{A}}_i < \overline{\underline{A}}_k.$$

That is, if the range of A_i and A_k overlap (are not exclusive), the pair is defined as contemporaneous and belongs to the set P; if the range of A_i is below that of A_k, the pair is sequential and belongs to set Q; and if the range of A_i is above that of A_k, the pair is sequential and belongs to set R. A matrix of the range pair relations is prepared with 0 entered if the relation belongs to set P, $+$ if it belongs to R, and $-$ if it belongs to Q. The matrix can be ordered to aid in subdivision of a specific section, or data from multiple sections can be summed as the matrix elements.

Rubel defined the ranges of two taxa as "contacting" if they are sequential, and between their total ranges there are no total ranges of any other taxa that are sequential with regard to them. The interval between the range of two "contacting" sequential taxa is termed a "moment," designated in the form $\boxed{k}$, $\boxed{1}$, $\boxed{m}$. Rubel demonstrated how "moments" can be used in correlating stratigraphic sections and for determining whether a particular stratum in a particular section S_{ij} is above, below, or within a particular "moment."

Rubel's technique does not allow for the possibility of conflicting ranges in different sections, and thus avoids the problems of probability. He suggested that this sort of analysis should be carried out for each facies encountered in order to provide the greatest subdivision of the sections. In effect, Rubel's technique divides the section into a series of partial range zones.

A more sophisticated, and mathematically correct, technique was developed by Southam et al (1975). Initially, the formulation followed that of Hay (1972). Noting that a stratigraphic event can be represented as a point on a stratigraphic axis (i.e. a $-$ axis), stratigraphic data are represented by a series of points on the stratigraphic axis.

To properly order and interpret this sequence of events for multiple stratigraphic axes representing stratigraphic sections, it is necessary to consider the relationship between all possible pairs of events, i.e. for events i and j either $\tau_i < \tau_j$ or $\tau_j < \tau_i$. This pair-wise ordering of events is most conveniently represented by an $M \times M$ matrix where M is the total number of events being considered. The matrix elements are of the form $a_{ij} = n_{ij}/N_{ij}$, where n_{ij} is the number of stratigraphic sections in which events i and $j(i \neq j)$ satisfy $\tau_i < \tau_j$. N_{ij} is the number of stratigraphic sections in which either $\tau_i < \tau_j$ or $\tau_j < \tau_i$. The statistical model developed by Southam et al (1975) departs significantly from that of Hay (1972) and is based on three conditions: 1. the matrix elements are determined by independent repetitions of the experiment; 2. the outcome of the experiment is one of two mutually exclusive and exhaustive possibilities, i.e. either $\tau_i < \tau_j$ or $\tau_j < \tau_i$; 3. the probability p_{ij} of the outcome $\tau_j < \tau_i$ is the same on each repetition, hence the probability of the outcome $\tau_i < \tau_j$ is $p_{ji} = 1 - p_{ij}$. Conditions 1 and 2 are satisfied in all experimental situations, but for condition 3 to be satisfied requires spatial homogeneity.

The model predicts that the outcomes of repeated trials of the experiment are described by the binomial probability density function. Although p_{ij} is the parameter of interest in interpreting stratigraphic data and for use in prediction, the situation is such that only n_{ij} and N_{ij} are known and p_{ij} must be inferred. Exact knowledge of p_{ij} is prohibited by the limitation of finite sampling, and in order to estimate it and obtain a confidence interval, Southam et al introduced the maximum-likelihood estimator $\hat{p}_{ij} = x/N_{ij}$, where x can have the values 0, 1, 2, ..., N_{ij}; the value of the estimator $\hat{p}_{ij}$ obtained from N_{ij} repetitions of the experiments is $\hat{p}_{ij} = n_{ij}/N_{ij}$, a simple proportion and the elements of the matrix.

Determination of the confidence interval was more complex. Greenwood & Hartley (1962) included in their useful guide references to tables and graphs for confidence intervals in existence prior to 1962; unfortunately, those they cited are all for parameter values that are not well suited to the needs of stratigraphers. The usual methods for calculating the confidence intervals for the binomial distribution (Crow

1956) make use of the fact that the binomial probability density function can be approximated by the normal distribution for large numbers of repetitions (i.e. for large N_{ij}) when the distribution possesses a minimum amount of skewness (i.e. $p_{ij} = \frac{1}{2}$).

Stratigraphers cannot afford the luxury of large samples characteristic of many other areas of science. It is rare to have more than eight or ten sections for comparison when important decisions must be made affecting strategy for exploration for petroleum and other natural resources. Also, stratigraphers are interested in the special cases where the order of events is highly preferred, i.e. the statistical distribution is highly skewed. To overcome these problems, and to obtain exact values, Southam et al used a method for calculating the value of the lower end point p_{ij}^l and upper end point p_{ij}^u of the confidence interval such that $p(p_{ij}^l \leqq p_{ij} \leqq p_{ij}^u) = 0.90$, making use of the incomplete beta functions (Abramowitz & Stegun 1972). This permitted exact calculation of the values of p_{ij}^l and p_{ij}^u and the width of the confidence interval indicated in Figure 5. Southam et al presented values for p_{ij}^l and a for values of N from 1 to 16. Although their formulation of the conditions that must be satisfied by the upper and lower end points of the confidence interval was standard, the exact solution is not indicated in most textbooks. As Brownlee (1965) noted, "the solution of these equations from first principles would be tedious" (p. 130).

The results of the statistical analysis are summarized in Figure 4. The actual probability of the ordering, p_{ij}, and its observed estimator, $\hat{p}_{ij}$, lie within the confidence interval between p_{ij}^l and p_{ij}^u. The width of this interval, $\delta_{ij} = p_{ij}^u - p_{ij}^l$, is represented by the distance between the points p_{ij}^l and p_{ij}^u.

Southam et al then discussed quantification of biostratigraphic reliability, using the following argument:

The closer p_{ij} approaches one the more certain the relation $\tau_i < \tau_j$ between events i and j becomes; for quantification of the reliability of a stratigraphic zonation, it is desirable to have a parameter that expresses how closely p_{ij} approaches one and how strongly skewed the relation is. Two necessary conditions for p_{ij} to approach one are: 1. $\hat{p}_{ij}$ must approach one and 2. $1 - p_{ij}^l$ must approach zero. The single reliability parameter $\hat{p}_{ij}/(1 - p_{ij}^l)$ incorporates both of these conditions, so that the most reliable sequence of events is that set of events that maximizes $\hat{p}_{ij}/(1 - p_{ij}^l)$ between each pair of events. For sequences with the same number of events or stratigraphic levels, the most reliable is that which has the largest geometric average value of $\hat{p}_{ij}/(1 - p_{ij}^l)$. The largest values of the geometric average of $\hat{p}_{ij}/(1 - p_{ij}^l)$ will correspond to sequences with a small number of levels. As the number of levels

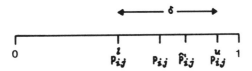

Figure 5   Relations between the probability of a pair of events occurring is in a particular order, p_{ij}, the maximum-likelihood estimator is $\hat{p}_{ij}$, the confidence interval, δ, the lower end point of the confidence interval, p_{ij}^l, and the upper end point of the confidence interval, p_{ij}^u.

increases, this reliability parameter tends to decrease, so that the stratigrapher must weigh the usefulness of a fine stratigraphic subdivision with a large number of levels against the greater reliability obtained by considering a smaller number of levels.

As an alternative, $\hat{p}'_{ij}/\delta_{ij}$ may be used as a reliability parameter. This would be most useful in exploring relationships where similarity rather than difference is sought.

In determining the end points of the ranges of taxa sampling noise becomes critical. Because one is searching for the highest or lowest occurrence of a fossil species when it is vanishingly rare, its occurrence becomes sporadic. Two methods have been used to overcome this problem. Hay & Steinmetz (1974) have experimented with simply deleting all rare occurrences from consideration in establishing the ranges of species. This is essentially recognition of the upper and lower limits at a low level of abundance rather than the true end points of the range. Worsley (personal communication) has applied the techniques of three-point rolling averages to smooth range charts in which several categories of abundance are recognized. Both of these techniques result in more consistent biostratigraphic zonations and higher values for sequential reliability.

CORRELATION BY POSITION IN THE MOST LIKELY SEQUENCE OF EVENTS

Worsley et al (1977) have devised a method of correlation using the most likely sequence of events, determined from the $M \times M$ matrix for a given data set. Each event of the most likely sequence is identified with a consecutive number and is displayed along an ordinate, the interval between each event being equal. The stratigraphic section is plotted along the abscissa. For a particular section (in this case, Deep Sea Drilling Project sites) the position of each biostratigraphic event is plotted; the result is a diagram of the sort shown in Figure 6a. A curve can then be fitted through these points and the scale of events stretched or shrunk as necessary to be calibrated to a radiometric or other geochronological scale. The end result is an

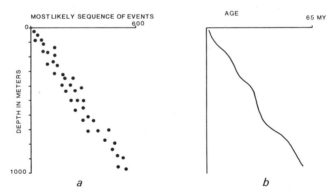

Figure 6 (*a*) Schematic event-depth. (*b*) Age-depth diagrams derived from probabilistic analysis of a large data array, as has been done for selected planktonic fossils from Pacific Ocean Deep Sea Drilling Project sites by Worsley et al (1977).

age depth diagram as shown in Figure 6b. Correlations from one section to another can be achieved either through the scale of events or through the geochronological scale. This technique is an analog of that of Shaw (1964) but, in this case, the sequence of events ("composite section" of Shaw) is generated from a matrix and stratigraphic thickness was not involved in its production.

CONCLUSIONS

Most quantitative methods for stratigraphic correlation are amenable to formulation in terms of the true probability, maximum likelihood estimators, and confidence intervals of Southam et al. The bioassociation coefficients are all derived as proportions and could be used as maximum likelihood estimators. Dividing by the confidence interval would then yield a probabilistic expression for similarity or difference between assemblages. McCammon's (1970) relative biostratigraphic value index, based on three proportions, is also amenable to expression in probabilistic terms. Bias inherent in Shaw's (1964) method has been removed by Worsley et al's (1977) matrix approach to construction of the composite most likely sequence, based in probability. Much progress has been made, but much remains to be done. Hay & Cepek (1969) suggested how a particular stratigraphic correlation might be expressed in terms of a numerical probability, but this goal still remains to be achieved in a universally accepted manner.

ACKNOWLEDGMENTS

This work was performed under Grant No. OCE-76-82198 from the National Science Foundation and ACS PRF9835-AC2 from the American Chemical Society, Petroleum Research Fund.

Literature Cited

Abramowitz, M., Stegun, I. A. 1972. *Handbook of Mathematical Functions*. Washington, D.C.: GPO 1046 pp.

American Commission on Stratigraphic Nomenclature. 1961. Code of stratigraphic nomenclature. *Bull. Am. Assoc. Pet. Geol.* 45(5): 645–55

Bramlette, M. N., Sullivan, F. R. 1961. Coccolithophorids and related nannoplankton of the Early Tertiary in California. *Micropaleontology* 7: 129–88

Brownlee, K. A. 1965. *Statistical Theory and Methodology in Science and Engineering*. New York: Wiley. 590 pp. 2nd ed.

Cheetham, A. H. 1960. Review of: Molluscan assemblages from the marine middle Miocene of South Jutland and their environments, by Theodor Sorgenfrei. *Bull. Am. Assoc. Pet. Geol.* 44: 1716–17

Cheetham, A. H., Deboo, P. B. 1963. A numerical index for biostratigraphic zonation in the mid-Tertiary of the eastern Gulf. *Trans. Gulf Coast Assoc. Geol. Soc.* 13: 139–47

Cheetham, A. H., Hazel, J. E. 1969. Binary (presence-absence) similarity coefficients. *J. Paleontol.* 43: 1130–36

Cockbain, A. E. 1966. An attempt to measure the relative biostratigraphic usefulness of fossils. *J. Paleontol.* 40: 206–7

Crow, E. L. 1956. Confidence intervals for a proportion. *Biometrika* 43: 423–35

Crow, E. L., Davis, F. A., Maxfield, M. W. 1960. *Statistics Manual*. New York: Dover. 279 pp.

Cserna, Z. de. 1972. Essay review of O. H. Schindewolf's Stratigraphie und Stratotypus. *Am. J. Sci.* 272: 189–94

Cuvier, G., Brongniart, A. 1808. Essai sur la géographie minéralogique des environs de Paris. *Ann. Mus. Hist. Nat. Paris* 11: 293–326

Darwin, C. 1859. *The Origin of Species*. London: Murray. 490 pp.

Deboo, P. B. 1965. Biostratigraphic correlation of the type Shubuta Member of the Yazoo Clay and Red Bluff Clay with their equivalents in southwestern Alabama. *Bull. Geol. Surv. Ala.* 80:1–84

Dennison, J. M., Hay, W. W. 1967. Estimating the needed sampling area for subaquatic ecologic studies. *J. Paleont.* 42(3):706–08

d'Orbigny, A. 1851. *Cours Elementaires de Paleontologie et de Geologie Stratigraphiques*, Vol. 2, fasc. 1. Paris: Masson. 382 pp.

Drooger, C. W. 1974. The boundaries and limits of stratigraphy. *Proc. K. Ned. Akad. Wet. Ser. B.* 77:159–76

George, T. N., Harlan, W. B., Ager, D. V., Ball, H. W., Blow, W. H., Casey, R., Holland, C. H., Hughes, N. F., Kellaway, G. A., Kent, P. E., Ramsbottom, W. H. C., Stubblefield, J., Woodland, A. W. 1969. Recommendations on stratigraphical usage. *Proc. Geol. Soc. London* 1656:139–66

Greenwood, J. A., Hartley, H. O. 1962. *Guide to Tables in Mathematical Statistics.* Princeton: Princeton Univ. Press. 1014 pp.

Hall, A. V. 1969. Avoiding informational distortion in automatic grouping programs. *Syst. Zool.* 18:318–29

Hancock, J. M. 1977. In *Concepts and Methods of Biostratigraphy*, ed. E. Kauffman, J. Hazel, pp. 3–22. Stroudsburg, Pa.: Dowden, Hutchinson & Ross

Harrington, H. J. 1965. Space, things, time and events—an essay on stratigraphy. *Bull. Am. Assoc. Pet. Geol.* 49:1601–46

Hay, W. W. 1971. In Heywood, V. H.: *Scanning Electron Microscopy: Systematic and Evolutionary Applications: Systematics Assoc. Spec. Publ. 4.* pp. 123–43. New York: Academic.

Hay, W. W. 1972. Probabilistic stratigraphy. *Eclogae Geol. Helv.* 65(2):255–66

Hay, W. W., Cepek, P. 1969. Nannofossils, probability, and biostratigraphic resolution. *Bull. Am. Assoc. Pet. Geol.* 53:721

Hay, W. W., Steinmetz, J. C. 1973. *Proc. Symp. Calcareous Nannofossils, Gulf Coast Sect.*, Soc. Econ. Paleontol. Mineral. pp. 58–70

Hay, W. W., Steinmetz, J. C. 1974. *Geol. Soc. Am.* 6:363 (Abstr.)

Hay, W. W., Steinmetz, J. C. 1977. Mem. Segundo Congr. Latinoamericano Geol., *Caracas, 1973.* 3:1529–40

Hazel, J. E. 1970. Binary coefficients and clustering in biostratigraphy. *Bull. Geol. Soc. Am.* 81:3237–52

Hazel, J. E. 1977. In *Concepts and Methods of Biostratigraphy*, ed. E. Kauffman, J.

Hazel, pp. 187–212. Stroudsburg, Pa.: Dowden, Hutchinson & Ross

International Subcommission on Stratigraphic Classification. 1972. Introduction to an international guide to stratigraphic classification, terminology, and usage. *Lethaia* 5(3):283–95

International Subcommission on Stratigraphic Classification. 1976. *International Stratigraphic Guide: A Guide to Stratigraphic Classification, Terminology and Procedure*, ed. H. Hedberg. New York: Wiley. 200 pp.

Jeletzky, J. A. 1956. Paleontology, basis of practical geochronology. *Bull. Am. Assoc. Pet. Geol.* 40:679–706

Jeletzky, J. A. 1965. Is it possible to quantify biochronological correlation? *J. Paleontol.* 39:135–40

Kauffman, E. G. 1970. *Proc. North Am. Paleontol. Conv.* Pt. F:612–66

Kauffman, E. G. 1977. In *Concepts and Methods of Biostratigraphy*, ed. E. Kauffman, J. Hazel, pp. 109–41. Stroudsburg, Pa.: Dowden, Hutchinson & Ross

Martini, E. 1970. Standard Paleogene calcareous nannoplankton zonation. *Nature* 226:560–61

McCammon, R. B. 1970. On estimating the relative biostratigraphic value of fossils. *Bull. Geol. Inst. Univ. Uppsala, N. S.* 2:49–57

Miller, F. X. 1977. In *Concepts and Methods of Biostratigraphy*, ed. E. Kauffman, J. Hazel, 165–86. Stroudsburg, Pa.: Dowden, Hutchinson & Ross

Oppel, A., 1856–58. *Die Juraformation Englands, Frankreichs und des Sudwestlichen Deutschlands.* Stuttgart, Germany: Verlag von Ebner und Seubert. 857 pp.

Quastler, H. 1958. In *Symposium on Information Theory in Biology*, ed. H. Yockey, pp. 3–49. London: Pergamon

Rubel, M. 1976. On biological construction of time in geology. *Eesti NSV Tead. Akad. Toim. Keem. Geol.* 25:136–44

Shaw, A. B. 1964. *Time in Stratigraphy.* New York: McGraw-Hill. 365 pp.

Shaw, A. B. 1969. Adam and Eve, paleontology, and the non-objective arts. *J. Paleontol.* 43:1085–98

Simpson, G. G. 1947. Holarctic mammalian faunas and continental relationships during the Cenozoic. *Bull. Geol. Soc. Am.* 68:613–88

Simpson, G. G. 1960. Notes on the measurement faunal resemblance. *Am. J. Sci.* 258A:300–11

Smith, W. 1799. Tabular view of the order of strata in the vicinity of Bath with their respective organic remains.

Smith, W. 1815. A map of the strata of England and Wales with a part of Scotland, exhibiting the collieries, mines and canals, the marshes and fen lands originally overflowed by the sea and the varieties of soil according to the variations in the strata. 15 sheets, 8'9" × 6'2", 400 copies. London.

Sorgenfrei, T. 1958. Molluscan assemblages from the marine middle Miocene of South Jutland and their environments. *Geol. Surv. Den.* II Ser(79) 503 pp.

Southam, J. R., Hay, W. W., Worsley, T. R. 1975. Quantitative formulation of reliability in stratigraphic correlation. *Science* 188:357–359

Sullivan, F. R. 1964. Lower tertiary nanno-plankton from the California coast ranges. *I. Paleocene. Univ. Calif. Publ. Geol. Sci.* 44:163–228

Sullivan, F. R. 1965. Lower tertiary nanno-plankton from the California coast ranges.

II. Eocene. Univ. Calif. Publ. Geol. Sci. 53:1–52

Teichert, C. 1958. Some biostratigraphic concepts. *Bull. Geol. Soc. Am.* 69:99–119

Till, R. 1974. *Statistical Methods for the Earth Scientists.* London: Macmillan. 154 pp.

Van Hinte, J. E. 1969. *Proc. 1st Int. Conf. Planktonic Microfossils, Geneva, 1967.* pp. 267–272. Leiden: Brill

Worsley, T. R., Blechschmidt, G., Ralston, S., Snow, B. 1973. *Proc. Symp. Calcareous Nannofossils, Gulf Coast Sect., Soc. Econ. Paleontol. Mineral.* pp. 71–79.

Worsley, T. R., Jorgens, M. L. 1974. Oligocene calcareous nannofossil provinces. *Soc. Econ. Paleontol. Mineral. Spec. Publ.* 21:85–108.

Worsley, T. R., Blank, R., Suchland, C. 1977. *Cenozoic Biostratigraphy and Age-Depth Relationships of Pacific Ocean Deep-Sea Drill Sites.* In press.

Ann. Rev. Earth Planet. Sci. 1978. 6 : 377–404
Copyright © 1978 by Annual Reviews Inc. All rights reserved

EVOLUTION OF OCEAN CRUST SEISMIC VELOCITIES

×10099

B. T. R. Lewis

Department of Oceanography and Geophysics Program, University of Washington, Seattle, Washington 98195

INTRODUCTION

In 1963 Vine & Matthews published a paper suggesting that the symmetrical nature of magnetic anomalies across mid-ocean ridges may be explained by alternately reversed and normally magnetized ocean crust, and that the reversals could be associated with the changes in polarity of the dipole component of the earth's field. About this time paleomagnetic studies on land were used to date the polarity changes in earth field (Cox et al 1963), and so was born a technique for dating the oceanic crust. The magnetic ages of the oceanic crust have since been verified by paleontological dating of the sediments just above basement that were obtained from Deep Sea Drilling Project holes (for example Maxwell et al 1970). Still unavailable, however, is a satisfactory model of what the relative contribution of parts of the oceanic crust are to the remanent magnetization causing the magnetic anomalies (Harrison 1976).

Since the recognition of mid-ocean ridges as accretionary plate boundaries, there has been a great deal of geophysical and geological research on "zero" age crust and its evolution. This research has revealed that the geological processes resulting in the formation of oceanic crust are more complicated than the simple conveyor belt model. For example, thermal models of an idealized cooling plate lead to conductive heat fluxes far greater than are observed. Detailed heat-flow studies show that near rise axes heat flow is generally very irregular, and this has led to the notion of convective cooling of the crust by penetration of sea water through cracks caused by cooling (Williams et al 1974, Lister 1972).

Evidence for the existence of an oceanic crust and information about its structure with depth comes predominantly from seismic refraction studies, although seismic reflection techniques are beginning to be applied to the study of sub-basement structure. Most of the early work in marine seismic refraction was done at Scripps Institution of Oceanography and Lamont Geological Observatory using near-surface hydrophones either attached to a buoy or a ship. These techniques are discussed by Shor (1963) and Hill (1963). Some early ocean-bottom seismograph techniques are described by Ewing & Ewing (1961). Many of these early refraction profiles were

377

0084-6597/78/0515-0377$01.00

interpreted by fitting straight-line segments to the first arrivals and computing velocities and depths using a standard formula for layered models. It was from these procedures that concepts of a layered oceanic crust consisting of layer 1 (sediments), layer 2 (basement), layer 3 (oceanic layer) and an upper mantle were derived. Raitt (1963) has summarized a number of experiments in this way, and lately Woollard (1975) has analyzed a large body of refraction data in terms of the change in crustal parameters with age and correlation with other geophysical parameters, for example heat flow.

With the advance of concepts of plate tectonics, it is now thought that certain sequences of rocks on land called ophiolites may represent obducted oceanic crust. Laboratory measurements of velocities of these rocks are now being made (for example, Christensen & Salisbury 1975) to see how they compare to oceanic refraction measurements. In order to make this comparison in a meaningful way, one must ask what the resolution of the refraction measurements are and how the techniques can be improved. The recent popularity of ocean bottom seismographs has also stimulated inquiries into the question of resolution and technique.

Because of the importance of refraction measurements in unraveling the detailed structure of the crust, this review concentrates on this aspect of marine geophysical research. First, resolution problems associated with refraction measurements are discussed. This is followed by a review of age-related velocity behavior in the upper crust, middle crust, lower crust, and upper mantle. Finally, geologic and geophysical models that might explain these data are discussed.

INTERPRETATION AND RESOLUTION OF SEISMIC REFRACTION DATA

For many years marine seismic refraction data have been interpreted in terms of a small number of constant-velocity layers separated by planar interfaces. The velocities and thicknesses of the layers are determined by fitting straight-line segments to the first arrival time-distance data. These interpretation techniques have led to the concept of a layered oceanic crust consisting of three or more layers and, consequently, it is now widely believed that the crust is layered. However, as is demonstrated, this technique produces a model that is by no means unique.

Kennett & Orcutt (1976) have clearly demonstrated the lack of uniqueness using extremal and linearized inversion methods. The starting point for these inversions is the time-distance data, and if we assume uniform plane layers and head-wave propagation the travel time T at a distance X corresponding to head-wave propagation along the top of the Nth layer with velocity α_N can be written as

$$T = \frac{x}{\alpha N} + \sum_{i=0}^{N-1} 2 h_i [1/\alpha_i^2 - 1/\alpha_N^2]^{1/2},$$

where h_i is the thickness of the ith layer and α_0 is the velocity at the top of the stack. The slope of this portion of the travel-time curve will be equal to $dt/dx = 1/\alpha N = P_N$, the ray parameter. The intercept time of this portion of the travel-time curve will be

$$\tau(P_N) = T - P_N X$$

$$= \sum_{i=0}^{N-1} 2\, h_i [1/\alpha_i^2 - 1/\alpha_N^2]^{1/2}.$$

As we let the layer thickness go to zero and approach a continuous velocity-depth function $\alpha(z)$, the intercept time

$$\tau(P) = 2 \int_0^{Z_p} [1/\alpha(z)^2 - p^2]^{1/2}\, dz.$$

The advantage of using $\tau(P)$ is that it is a single-valued monotonically decreasing function of P, as contrasted say with $X(P)$, which can be a multiple-valued function in the presence of triplications in the travel-time curve.

Gerver & Markusevitch (1966) have shown that if we knew $\tau(P)$ continuously and exactly, then we can recover the velocity-depth function using the relationship (from Kennett 1976)

$$z(P) = \frac{1}{\pi} \int_P^{u(0)} \frac{x(q)^{dq}}{[q^2 - p^2]^{1/2}} + \frac{1}{\pi} \sum_R \int_{Z_R^+}^{Z_R^-} \tan^{-1} \left[\frac{u^2(z) - P_R^2}{P_R^2 - p^2} \right]^{1/2} dz,$$

where the second term is the contribution from any low-velocity layers and $u(0)$ is the reciprocal of the surface velocity, Z_R^+ is the depth to the top of the Rth low-velocity zone, and Z_R^- is the depth to the bottom of the Rth low-velocity zone

$$x(q) = \frac{T(q) - \tau(q)}{q}.$$

In the case where we have discretely sampled T, X data with observational errors, Bessonova et al (1974) have shown how the limits on the function $\tau(P)$ can be mapped in limits of $\alpha(z)$, the velocity-depth function. Bessonova et al's method is known as an extremal inversion method since the limits on $\alpha(z)$ contain all possible velocity models. However, not all the possible models within these limits will satisfy the data (Kennett & Orcutt 1976). Various methods for establishing the limits of $\tau(p)$ from the data are given by Bessonova et al 1974, Kennett & Orcutt 1976, and Bates & Kanasewich 1976.

Another technique for inverting the function $\tau(P)$ to obtain the velocity-depth function is the linearized inversion using the formalism of Backus & Gilbert (1967). These methods have been applied to $\tau(P)$ inversions by Johnson & Gilbert (1972), Kennett (1976), and Kennett & Orcutt (1976). The linearized inversion starts with an assumed velocity model, and this model is then linearly and iteratively perturbed until the calculated $\tau(P)$ fit the observed $\tau(P)$ to within their estimated errors. A comparison of the extremal and linearized inversion methods is given by Kennett (1976).

An example of the extremal and linearized inversion applied to typical marine-refraction first-arrival data is shown in Figure 1 from Kennett & Orcutt (1976). These results clearly show just how little resolution is available from time-distance data with typical observational error. The model from the linearized inversion and the model from the layer solution fit the data equally well, and yet the models are entirely different. The linearized-inversion model shows no evidence for a Moho boundary or for a first-order discontinuity between layer 2 and layer 3. One must

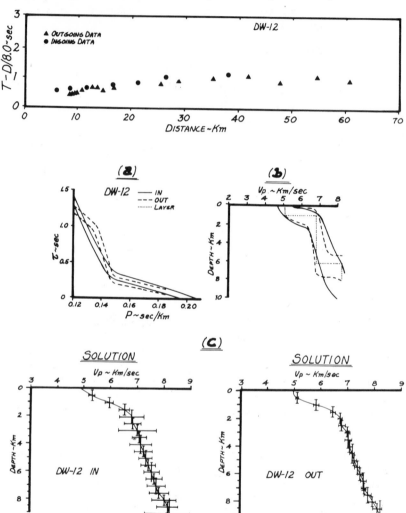

Figure 1 Comparison of extremal and linear inversions of typical refraction data (after Kennett & Orcutt 1976). Top graph shows the time-distance data; (*a*) shows the τ-*P* parameterization of the data; (*b*) shows the extremal inversion; and (*c*) shows the linearized inversion.

conclude from these results that the commonly assumed layered model of the oceanic crust is an artifact of the method of data interpretation, and one should be very wary of applying geologic significance to the layers.

This is not to say that significant information cannot be extracted from refraction data. Seismograms contain more information than just the first arrival times, and greater resolution can be obtained by using the additional information in the seismogram, such as second arrivals and amplitudes. One method of using this additional information that is gaining in popularity is the computation of synthetic seismograms and the comparison of the synthetic with the observed seismogram. Several algorithms for computing synthetics have been used. Helmberger & Morris (1969, 1970) have used a method based on generalized ray theory in which one specifies the rays and mode conversions one wishes to include. Head waves and reflections result from the generalized reflection coefficients. Another technique has been described by Fuchs & Müller (1971). This algorithm has been called the reflectivity method because one computes the complex reflectivity of a stack of layers for a given range of phase velocities and frequencies. This method includes all the mode conversions, reflections, and head waves occurring in the stack of layers.

To illustrate the usefulness of synthetic seismograms in distinguishing between models, Figure 2 (from Lewis & Snydsman 1977a) shows sections of the synthetic seismogram records for three different models. Given a standard deviation of about 0.05 secs as a measure of the error in determining the travel times of first arrivals, these models would be indistinguishable on the basis of the first arrivals. Model a is a typical layered model, and model b has a gradient in the upper crust, constant-velocity lower crust, and a sharp Moho. Model c is similar to model b but with a low-velocity zone at the base of the crust. One cannot distinguish between these models on the basis of first arrivals, but there are well-defined differences in the second arrivals. For example, the diagnostic feature of the model with a low-velocity zone is not the disappearance of the crustal arrivals (Pg) but the offset between Pg and the wide-angle mantle reflections PmP. The diagnostic of the 7.4 km sec^{-1} layer at the base of the crust in model a is the slope of the PmP branch. The PmP, Pn critical point is also diagnostic but more difficult to identify. It is shown later that these features are highly important if we are to meaningfully discuss the evolution of the oceanic crust.

In addition to the interpretive techniques outlined above, the experiment layout can be designed to improve the resolution. The most powerful innovation in the experiment design is the use of arrays of detectors to measure directly the ray parameter P or dt/dx. Arrays can be used not only for determining the apparent velocity of first arrivals but also for determining the P of later arrivals. In principle one can determine from a seismic array the $\tau(P)$ function directly since one can measure P, $X(P)$ and $T(P)$, and $\tau(P) = T(P) - P.X(P)$. In practice one is limited by the resolving power of the array, which is determined by the array length, number of sensors, and frequency. If the array is too long one will be averaging out the detail one is looking for, and if the array is too short one loses velocity resolution. For ocean-crust studies an array about 2 km long with 5 or more sensors using typical peak energy at 10 Hz provides a reasonable compromise between array-velocity resolution and velocity-depth resolution. Very few successful seismic array measurements have been made in the ocean. However, the few that have been made tend to indicate that the velocity in the upper crust increases in a gradient manner rather than discrete jumps as predicted by the layered model. Figure 3 shows an example from Lewis & Snydsman (1977a), where a buoy array was used in an experiment on the Cocos plate. This result is typical of measurements in this area.

Another experimental technique that can improve the resolution of travel-time data is the determination of topographic corrections. Whitmarsh (1975) and Kennett & Orcutt (1976) have shown that to make these corrections accurately one should ideally know the velocity-depth function before one can make the correction. However, because of the large velocity contrast between near-bottom basement and water (typically 4.5 km sec^{-1} to 1.5 km sec^{-1}), the correction is moderately insensitive to the basement velocity. A more important correction that is often not made is that due to sediment thickness. Because the sediment P velocity is close to water (1.5 to 1.6 km sec^{-1} for unconsolidated deep-sea sediments), this effect is as large as

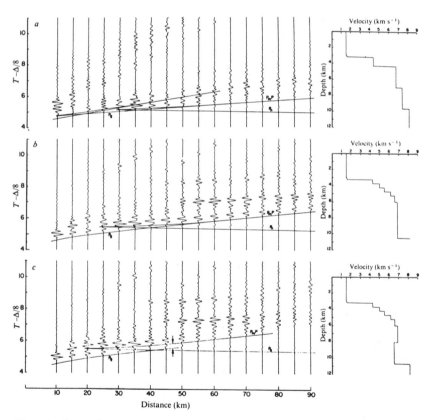

Figure 2 Comparison of synthetic seismogram record sections for three different models having similar first arrival travel times.

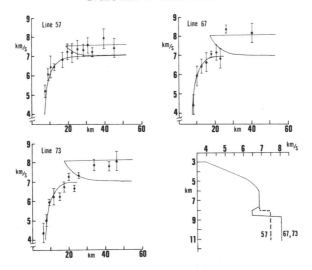

Figure 3 Apparent velocities across a five-element buoy array about 2 km long. For location of the lines see Figure 12.

water-depth variations. Because of the technical difficulty of towing a reflection-profiling system while shooting with explosives, some investigators do not routinely make sediment thickness estimates for each shot, and this must certainly show up as scatter in the data.

EVOLUTION OF THE UPPER CRUST

As was discussed in the previous section, the evidence for a layered crust, layered in the sense of having first-order discontinuities separating a small number of layers, is not convincing, except for the sediment-basement interface. In fact, the analysis of refraction data using the Tau inversion method (Kennett & Orcutt 1976), synthetic seismograms (Helmberger & Morris 1969, Orcutt et al 1975, Lewis & Snydsman 1977a), and array velocities (Lewis & Snydsman 1977a, b) suggests that velocity structure in the upper crust more nearly approximates a linear gradient. In order to retain continuity with previously published work, for example Raitt (1963), the terms layer 2 and layer 3 will be retained, but layer 2 will be taken to imply the upper 2 or 3 km of the igneous crust and the term layer 3a and 3b will be used to refer to the middle and lower sections of the crust respectively.

Because of the layer of ocean covering the oceanic crust, the upper 100–200 m of layer 2 are the most difficult to sense remotely using acoustic techniques and the easiest to sample using direct methods, dredging, and drilling. This is because in surface-to-surface refraction methods information pertaining to the top section often occurs behind the water-wave reflection arrivals and is masked by the energetic water waves. This situation can be partially relieved by placing the detector, say an ocean-bottom seismograph, on the bottom and completely relieved by placing the shots and receivers on the bottom.

Using sonobuoy receivers and air gun sources, Houtz & Ewing 1976 have analyzed a vast amount of surface-to-surface refraction data in terms of the apparent velocity of the first appearing refraction arrival as a function of age. Their results are shown in Figure 4 and indicate that the first observable refraction velocity increases systematically from about 3.5 km sec^{-1} at zero age to about 5.2 km sec^{-1} at 40 m.y. Beyond this age no appreciable change in velocity is observed. Systematic velocity measurements on sea-floor basalts by Christensen & Salisbury (1972, 1973) have shown that samples of hand specimen size show a slight decrease in velocity with age due to weathering.

Results from Deep Sea Drilling Project (DSDP) holes in young crust (Kirkpatrick 1977) indicate considerable differences between velocity measurements on core samples and velocities obtained from a downhole acoustic logging device. The basaltic core velocities are consistently above 5 km sec^{-1}, whereas the logged velocities are generally much lower and scattered, averaging about 4 km sec^{-1}. This difference can be explained by macroscopic cracks in the in-situ material, with the cracks being filled by sea water. In a hole in the older Atlantic Basin, near magnetic anomaly Mo, it was found that the logged velocities agreed with the core-sample velocities much more closely and that cracks in the rock were well cemented (M. H. Salisbury, personal communication).

Additional information about the shallow crust from DSDP holes is also reported by Hyndman et al (1976), who found that in holes in young crust (about 3.5 m.y.) the heat flow measured in the holes to depths of 300 m were an order of magnitude less than theoretically expected [0.6 HFU (heat flow units) compared to 6.4 HFU]. In another hole at 16 m.y. they found evidence that sea water was flowing down the hole into a permeable zone at a depth of 324 m.

A further datum relevant to the upper oceanic crust is the propagation of shear waves. Lewis & McClain (1977) have shown that in some parts of young Pacific crust the propagation of shear waves is correlated with the distribution of sediments. They find that converted refracted shear waves are clearly identifiable in sedimented areas and absent in unsedimented areas. A possible explanation for this phenomenon is that in unsedimented areas the shear waves are highly attenuated by cracks in

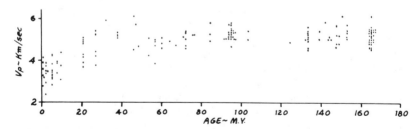

Figure 4 Velocities of first identifiable refraction arrivals as a function of age (from Houtz & Ewing 1976).

the rock, whereas in sedimented areas the cracks are sufficiently filled to allow efficient propagation of shear waves.

It seems reasonable to conclude on the basis of these data from the upper crust that the increase in the upper crustal velocity with age observed by Houtz & Ewing (1976) is indeed caused by a gradual sealing and cementing of the cracks and that this process is largely completed by 40 m.y. This mechanism would explain why Christensen & Salisbury (1975) find an apparent thinning of layer 2, which is caused by an increase in layer-2 velocity. It also suggests that the cracks are far more effective in controlling the bulk acoustic velocity than the weathering processes.

The depth to which cracking, and hence water penetration, extends is not yet resolved. Since the bulk seismic velocities are a function of the porosity, we might expect to see that the effect produced when cracks close with depth is comparable to the effect produced when cracks seal with age. The effects on rock velocity of water and cracks are also well known from laboratory measurements (for example, Schreiber & Fox 1977). Figure 5 shows the effect of pressure on a basalt and gabbro with air-filled cracks and water-filled cracks (Christensen & Salisbury 1975). An important conclusion from the results given in this figure is that the air-dried and water-saturated samples do not reach the same velocity until a pressure of about 2 kbars, which corresponds roughly to the pressure at the bottom of the crust (Woollard 1975). In this experiment the water was allowed to escape from the sample as the pressure was increased, in principle maintaining the pore water at atmospheric pressure. This implies that the cracks in the rock samples did not completely close until a pressure of about 2 kbars. In the oceanic environment one will expect the pore pressure to be somewhere between lithostatic and hydrostatic, depending on the permeability. With nonzero hydrostatic pressures one may therefore expect cracks

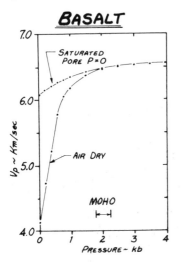
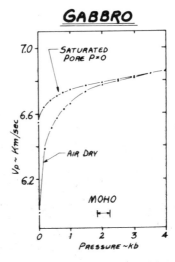

Figure 5 Compressional velocity versus pressure for a wet and dry basalt and gabbro (after Christensen & Salisbury 1975).

to remain open to a slightly greater pressure. These results suggest that if cracks exist in the crust they could remain open to depths corresponding to Moho. Lister (1974), in an extensive discussion of the theory of cracking of hot rock by sea water, concludes from an analysis of creep rates as a function of temperature and the elastic moduli of rocks that, in fact, cracking fronts could propagate to Moho depths. As we shall see, this argument is also relevant to the evolution of the lower crust and upper mantle.

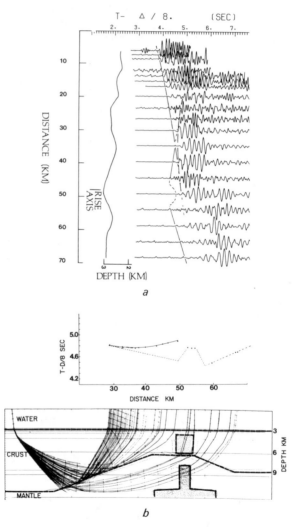

Figure 6 (*a*) Refraction data across the East Pacific Rise at 22°N. (*b*) Interpretation and model for this area. (After J. McClain et al, in preparation)

EVOLUTION OF THE LOWER OCEANIC CRUST AND UPPER MANTLE

Crustal Structure at Zero Age

A number of seismic refraction experiments have been undertaken on the East Pacific Rise and Mid-Atlantic Ridge to study the ocean crust structure at zero age. A few of these are reviewed to evaluate differences between different ridge segments.

On the East Pacific Rise the fastest spreading-rate segment to be examined in detail is at 9°N near the Siqueiros fracture zone where the spreading rate is 6 cm yr^{-1}. Here Orcutt et al (1976) find evidence for a narrow low-velocity zone in the lower crust. This evidence comes from an offset in the travel-time curve between the crustal arrivals and Moho reflection branch. Orcutt et al's data was from ocean bottom seismographs, and Rosendahl et al (1976) discuss results from sonobuoys in the same area. These lines were run parallel to the rise axis, and in similar trending lines at 2.9 and 5 m.y. no evidence for a low-velocity zone was found. However, in a more recent experiment at 1.6 m.y., Garmany, Dorman & Orcutt (1977) find offsets in the travel-time curves for lines parallel to the rise axis similar to those at the rise axis. In lines perpendicular to the axis no offsets were seen. Garmany et al interpret these data in terms of a low-velocity zone parallel to the rise but not perpendicular to the rise. Since no data have been published from this area for lines across the rise axis it is unclear whether this result implies that the lower oceanic crust is anisotropic and the low-velocity zone at the rise axis does not represent a magma chamber; or that a magma chamber exists at the rise axis and at 1.6 m.y.; or that the low-velocity zone at the axis and at 1.6 m.y. are due to different causes. These results do indicate that lower crust in this area is laterally heterogeneous.

In a University of Washington refraction experiment on the East Pacific Rise between 12°N and 15°N (spreading rate 4.5 cm yr^{-1}), one line that crossed the rise axis diagonally did not show any evidence of Pn time delays over the axis. In a Scripps experiment at 21°N, where the spreading rate is about 2.9 cm yr^{-1}, Reid, Orcutt & Prothero (1977) find evidence for a low-velocity zone at the rise axis from lines running parallel to the rise axis. In contrast to the area at 9°N, which is characterized topographically by an axial horst, the area at 21°N has subdued axial topography. At 22.5°N, south of the Tamayo fracture zone, the rise axis develops an axial valley, and J. McClain and I (in preparation) find evidence for a zone of low velocity beneath the axial valley. Our results were obtained primarily from lines running across the axis and looking at Pn and Pg delays over the axis. The seismic data and the interpreted axial structure for this area are shown in Figure 6 (after McClain & Lewis).

Further north on the East Pacific Rise at 49.5°N (the Explorer ridge off Vancouver Island) Clowes & Malecek (1976) find no anomalous effects from a reversed refraction line across the rise axis. From reversed lines parallel to the rise axis but about 30 km from the axis they find offsets in the travel-time curves at the same distance from each end of the line. They interpret these offsets as faults in Moho, but as they point out the data can be equally well explained by a low-velocity zone in the lower crust. The data in fact look quite similar to that found by Garmany et al (1977) at 9°N and 1.6 m.y.

These results along the East Pacific Rise suggest that low-velocity zones at the rise axis may occur for a variety of topographic situations, axial horsts, axial valleys, and axial zones of low relief. These low-velocity zones do not appear to be continuous along the rise axis and are also found off axis. Although Orcutt et al (1976) have suggested that these axial low-velocity zones may be due to magma chambers, the data are still sufficiently ambiguous to warrant further testing. Particularly, lines across the rise axis, if correctly planned, should allow one to differentiate between the presence or absence of narrow intrusive zones at the axis using mantle and crustal time delays.

Refraction experiments on the Mid-Atlantic Ridge are more difficult to undertake because of the rugged topography and the close spacing of the fracture zones. The rugged topography requires large topographic corrections, and the assumptions upon which these corrections are based become critical (for example, Fowler 1976). The close spacing of the fracture zones, often less than 30 km, has caused many investigators to restrict the length of their refraction lines to lie between fracture zones. This has resulted in many of the lines being too short for clear identification of refracted mantle arrivals. This appears to be especially true in the Famous area at 37°N. Several refraction experiments have been done in this area (Poehls 1974, Whitmarsh 1973, 1975, Fowler 1976), and in none of these experiments are the useful lines longer than about 40 km, which is only slightly greater than the crust-mantle crossover distance. As was seen in Figure 2, information critical to the structure of the lower crust is contained beyond 50 km. However, upper crustal interpretations have been made, and a discussion of the axial valley structure is given by Fowler (1976). Fowler concludes from the refraction data that there is probably no low-velocity zone under the axial valley, but the crust is thin, about 3 km, and thickens within 10 km of the axial valley.

Change in Middle and Lower Crustal Velocities with Age

The velocity in the middle of the crust (layer 3a) is probably the best known and shows the least amount of scatter. The reason probably is that the arrivals from rays bottoming in this part of the crust occur over a relatively large distance range, typically 15–30 km. In addition, most crustal models show the middle of the crust to have almost constant velocity because the travel-time curve from 15–30 km is usually fairly straight. This can be contrasted to the range from 5–15 km where the travel time curve is usually curved because of velocity gradients, and the velocity one measures depends on where one happens to have data. The velocity in the middle of the crust appears to be about 6.7–7.0 km sec^{-1} and shows little change with age or direction with respect to magnetic anomalies (Bibee & Shor 1976). The compiled data of Bibee & Shor for the middle crustal velocities are shown in Figure 7.

Determining the velocity structure for the lower oceanic crust is considerably more difficult than for the middle crust. As discussed in the seismic resolution section, diagnostic information on the lower crust is contained in information on second arrivals and amplitude. The velocity structure in the lower crust is also closely tied to the nature of the crust-mantle transition. A conclusion one can make at this time is that the velocity in the lower crust and the crust-mantle transition is highly variable. This conclusion is reached after comparing the results from Sutton et al (1971), who interpret their results with a high-speed layer at the base of the crust; Helmberger (1977), who has a 2-km transition layer between the middle crust and mantle; Lewis & Snydsman (1977a), who find a low-velocity zone at the base of the crust; and Kennett & Orcutt (1976), who can interpret the travel times with no distinct change between the crust and mantle.

Because of the difficulties of determining the velocity structure at the base of the crust, it is not possible at this time to look at the evolution of the lower crust except on a case-by-case basis. This is done in a following section using data from the Northern Cocos plate.

Change in Upper Mantle Velocities with Age

It was pointed out by Hess (1964) that seismic refraction observations in the North-eastern Pacific showed a correlation of mantle velocity with profile direction. Sub-sequently Raitt et al (1969), Morris et al (1969), Raitt et al (1971), Keen & Barrett (1971), and Shor et al (1973) have shown that the direction of maximum mantle velocity is approximately perpendicular to the observed magnetic anomalies.

In addition, Shor et al (1971) showed an apparent correlation of mantle velocity

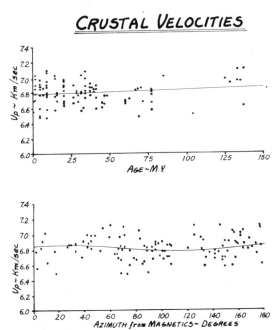

Figure 7 Compressional velocities for the middle crust as a function of age and azimuth with respect to magnetic lineations (after Bibee & Shor 1976).

with age. More recently, Bibee & Shor (1976) have summarized a large body of Pacific refraction data and show the trends in mantle anisotropy and mantle velocity with age. Figure 8 (after Bibee & Shor 1976) shows these trends. Bibee & Shor have interpreted the change in mantle velocity with age as reflecting the cooling of the upper mantle.

A University of Washington experiment on the Northern Cocos plate between 12° and 15°N also showed increases in mantle velocity with age (Snydsman et al 1975). Further analysis of these data also show that the anisotropy may change with age, as seen in Figure 9, which shows the change in velocity perpendicular and parallel to the ridge. The anisotropy decreases with age due to the greater increase in the velocity parallel to the ridge relative to that perpendicular to the ridge.

Interpreting the change in mantle velocity with age in terms of temperature changes requires that we know the velocity-temperature coefficient of possible mantle rocks. Not many experimental determinations of this coefficient have been made, but data from Anderson & Lieberman (1966) and Christensen (1975) indicate for dunite a value of about 0.1 km sec^{-1} per 100°C. This implies about a 400°C temperature change from the ridge crest areas ($V_p = 7.8$ km sec^{-1}) to old ocean basins ($V_p = 8.2$ km sec^{-1}). This should be regarded as a rough estimate until more experimental results on the temperature coefficients become available.

MANTLE VELOCITIES

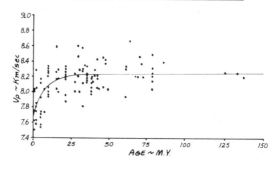

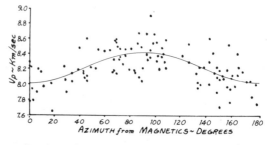

Figure 8 Compressional velocities for the upper mantle as a function of age and azimuth with respect to magnetic lineations (after Bibee & Shor 1976).

Change in Crustal Thickness with Age

Several investigators who have synthesized data from the Pacific have concluded that layer 3 thickens with age. Goslin et al (1972) and Christensen & Salisbury (1975) both find evidence for a thickening of the lower crust. However, this may not be too surprising since they are probably using very similar data sets. The analysis of Christensen & Salisbury is shown in Figure 10 and indicates a thickening of about 2 km in 40 m.y. However, the scatter in the data is considerable, and many earth scientists are skeptical of this result. The scatter in these data could be due to several sources: 1. Nonuniformity along the rise axis. 2. Misinterpretation of "layer 3". It has been shown in previous sections of this paper that one is probably dealing with gradients rather than discrete layers, and therefore it is difficult to define unambiguously a "layer 3". 3. Experimental error.

It is apparent that nonuniformity along the rise axis is a major contributor to the

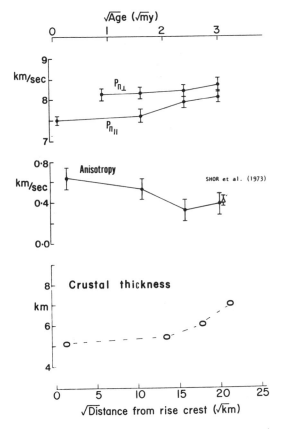

Figure 9   Change in upper mantle anisotropy and crustal thickness on the Cocos plate as a function of age. Data from Snydsman et al (1975). For location of the data see Figure 12.

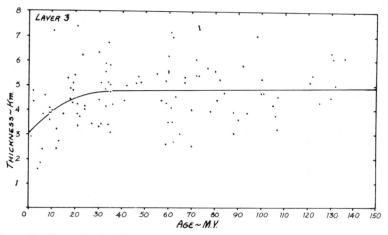

Figure 10   Change in the thickness of layer 3 with age (after Christensen & Salisbury 1975).

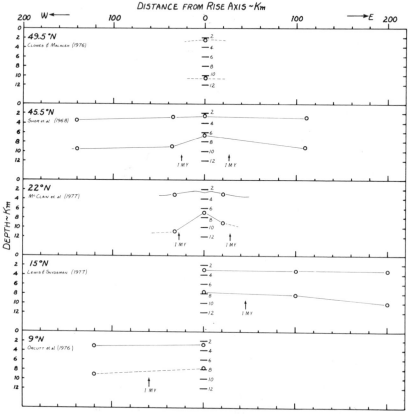

Figure 11   Change in depth to Moho with age from individual experiments on the East Pacific Rise.

scatter in Figure 10. To show this, results from individual experiments have been plotted in Figure 11. The depth to Moho and the depth to the sea floor have been plotted rather than the "layer 3" thickness because of the problem of defining "layer 3." It is clear from Figure 11 that the crust does thicken in some places with age, but not in all [for example the Explorer ridge (Clowes & Malecek 1976)]. It is also remarkable that although the Moho starts out having a variable depth at zero age, it rapidly approaches a uniform depth of about 10–11 km. The similarity in the scatter between Figures 10 and 11 suggests that the variability along the rise axis is the major cause of the scatter.

To investigate one of the thickening crust situations in greater detail, the University of Washington experiment between 12°N and 15°W has been chosen because it has the greatest density of data relevant to this specific problem. Figure 12 shows the location of refraction lines used in this experiment. Lewis & Snydsman (1977) showed data from lines 67, 47, and 8 of this experiment in support of a thesis that in this area the crust thickens by the development of a low-velocity zone at the base of the crust. Lewis & Snydsman (1977b) further developed this thesis, and Figures 13, 14, 15, 16, and 17 show data from lines 57, 66, 63, 19, and 22, together with the interpretations that are based on ray tracing and synthetic seismogram modeling. It is clear from these data that with increasing age the major changes are a reduction in the amplitude of Pg (crustal arrivals) and the development of an offset between Pg and PmP. Figure 2 shows that this is diagnostic of a low-velocity zone. It should also be pointed out that the part of line 57 that was at Pn ranges was also right over the rise axis. No time delays or peculiarities are seen, and we conclude that at this spot there is no magma chamber in the crust. Figure 18 (after Lewis & Snydsman 1977b) summarizes our conclusions from this area. As Bibee & Shor (1976) have pointed out, the increase in mantle velocity with age is probably associated with the decrease in temperature with age. If we knew the temperature at the base of the crust at say 10 m.y., we could

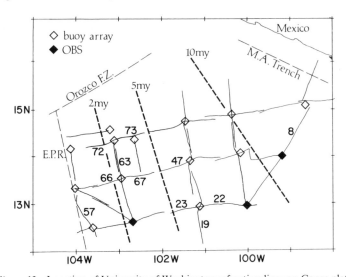

Figure 12   Location of University of Washington refraction lines on Cocos plate.

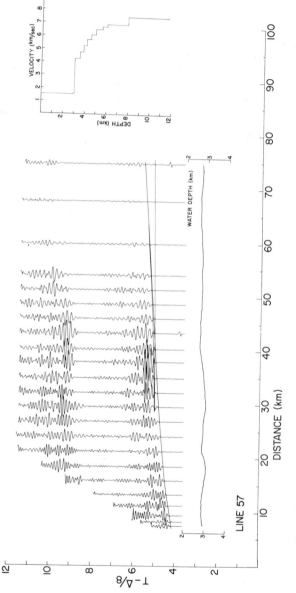

Figure 13   Record section from line 57 (Cocos plate). Age about 0.2 m.y. (after Lewis & Snydsman 1977b).

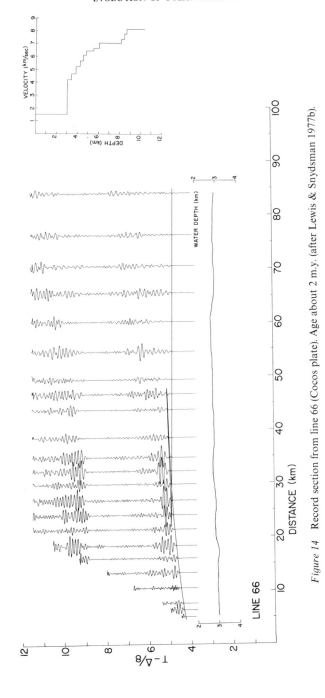

Figure 14   Record section from line 66 (Cocos plate). Age about 2 m.y. (after Lewis & Snydsman 1977b).

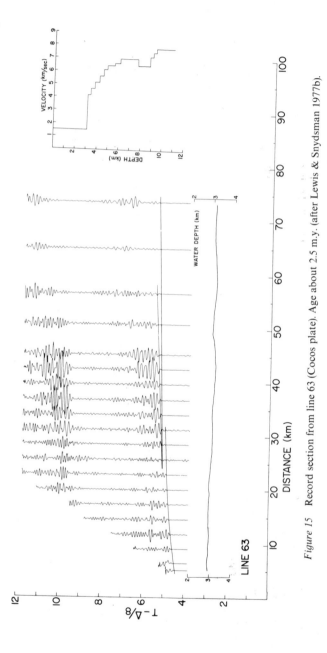

Figure 15   Record section from line 63 (Cocos plate). Age about 2.5 m.y. (after Lewis & Snydsman 1977b).

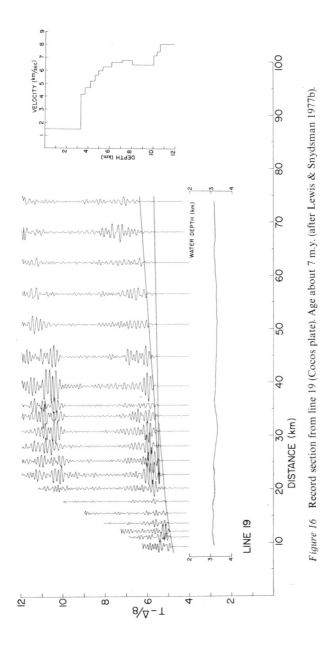

Figure 16 Record section from line 19 (Cocos plate). Age about 7 m.y. (after Lewis & Snydsman 1977b).

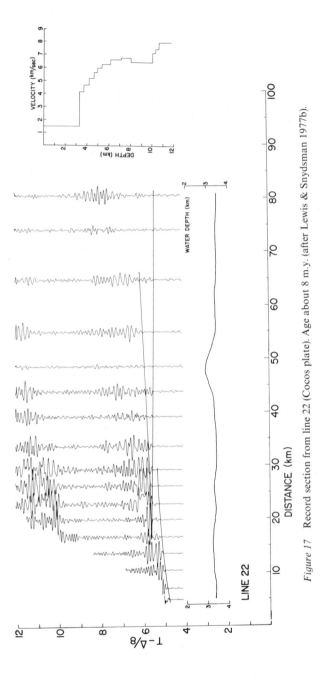

Figure 17 Record section from line 22 (Cocos plate). Age about 8 m.y. (after Lewis & Snydsman 1977b).

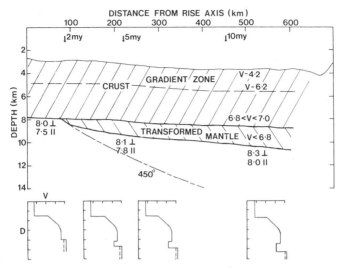

Figure 18 Cartoon summarizing results from the Cocos plate experiment (after Lewis & Snydsman 1977b).

use velocity temperature coefficients to estimate the temperature near the rise axis. However, estimating the temperature at the base of the crust is difficult because we do not know the relative contribution of conductive versus convective cooling and how deep the convective cooling extends. Anderson & Hobart (1976) have computed mean heat-flow values for the Pacific, and these data, especially from the Southeast Pacific, suggest that beyond about 12 m.y. there is an increase in heat flow that is associated with a transition from convective to conductive cooling regimes. This transition is also associated with the sediment thickness. High sedimentation rates promote the early sealing of cracks, reduction in permeability, and transition to conductive cooling. Anderson et al (1977) have estimated the temperature at the base of a 5-km-thick convecting layer and obtain a value of about 200°C. Since there is very little sediment in the area of the refraction study between 12° and 15°N this estimate may be appropriate for this area. If we use a value of 0.1 km sec^{-1} per 100°C for the velocity-temperature coefficient then we obtain a temperature difference of about 300°C going from velocities of 8.3 km sec^{-1} at 10 m.y. (perpendicular to the rise) to a velocity of 8.0 km sec^{-1} at the rise (perpendicular). This implies a temperature of about 500°C at the base of the crust near the rise axis in this area. This value and the Parker & Oldenburg (1973) cooling law are the basis for the 450°C isotherm given in Figure 18.

A MODEL OF THE EVOLVING OCEANIC CRUST

The principal results pertinent to a model of the evolution of the physical properties of the oceanic crust are:

1. Increase in upper crustal velocity with age

2. Lack of change of the middle crustal velocity with age
3. Variability of lower crustal velocities, particularly the possibility of low- and high-velocity layers at the base of the crust
4. Thickening of the oceanic crust with age and the consistency of the depth to Moho beyond a few million years
5. Increase in the mantle velocity with age
6. Possibility of increased anisotropy near zero age in the upper mantle associated with a decrease in velocity parallel to the rise axis
7. The increase in heat flow at some distance from the rise axis associated with a change from a predominantly convective to a predominantly conductive regime
8. Laboratory measurements on crack closure with pressure

The following model, based in part on Lister (1974) and Clague & Straley (1977), explains some of these results. Lister (1974) has shown that when hot rock at a spreading center comes into contact with sea water, the cooling of the rock causes cracking and consequent further cooling and cracking. A cracking front is developed that can propagate downwards at relatively high velocity, of the order of 30 m.y. Lister's analysis of creep data indicates that cracking should occur between 800 and 1000°K (about 600°C). The limitation on the cracking depth might be the static fatigue of polygonal columns isolated by the cracks. The laboratory measurements relevant to crack closing of Christensen & Salisbury (1975) (Figure 5) indicate that with low pore-water pressure cracks remain open until about 2 kb, which is equivalent to a depth of about 10 km below sea level. Higher pore pressures might cause this depth to be even greater.

This suggests that within a few kilometers of a spreading center and even at a spreading center the entire crust might be cracked and convectively cooled. The volume of the cracks will of course decrease with depth and depends on the strength of the rock. The cracks might have some preferential alignment if other stresses, for example plate driving forces, are acting to reduce the cracks in one direction. In this case the preferential alignment would be cracks perpendicular to the rise axis. This might explain the low values of the upper mantle velocities observed parallel to the rise axis since they could be the weighted average of the water velocity in the cracks and the rock velocity.

As this section of crust is transported away from the axis, two processes might occur.

1. Depending on the sedimentation rate and chemical reactions occurring in the crust, the cracks gradually seal and cause the velocities in the upper crust to increase.
2. Near the rise crest the acoustic Moho may represent a petrologic boundary between ultramafics and gabbro, both possibly having cracks and water to some degree. Away from the rise axis the temperature in the ultramafics decreases below about 450° allowing serpentinization of the ultramafics following a reaction of the sort

$$2\,Mg_2SiO_4 + 3\,H_2O \rightleftharpoons Mg_3Si_2O_5(OH)_4 + Mg(OH)_2$$
Forsterite   Water   Serpentinite   Brucite

This reaction is not sensitive to the partial pressure of water (Wyllie 1971).

If the brucite $[Mg(OH)_2]$ is not removed from the system this reaction results in a substantial volume increase of the solid material. Depending principally on the amount of olivine and water present, this reaction will cause a decrease in the velocity (Christensen & Salisbury 1975). Whether or not the velocity is reduced below about 6.8 km sec^{-1} (the velocity of the overlying gabbros and water) will depend on the details of the petrology and amount of serpentinization. It is conceivable that this process could result in low-velocity or high-velocity layers at the base of the crust. It should be pointed out that the gabbros in the middle of the crust will not undergo serpentinization because they contain no olivine.

The second of these processes could cause the apparent thickening of the crust by changing the acoustic Moho from a petrologic boundary at the rise axis to a hydration or serpentinization boundary with increasing age.

Clague & Straley (1977) have presented a very similar argument for the nature of the acoustic Moho based on evidence from ophiolites. A petrologic model (after Clague & Straley) is presented in Figure 19.

The principal arguments that have been made against the serpentinization model of the lower crust are:

1. The uniformity of layer 3 velocities and relatively low Poissons ratio for this layer compared to serpentinites (Christensen & Salisbury 1975). As has been shown, the uniform layer 3 velocities of about 6.8 km sec^{-1} probably represent the middle of the crust. The lower crustal velocities appear to be highly variable. The Poissons ratio of about 0.28 is based on refracted shear waves that correspond to the middle of the crust. For a P wave low-velocity zone at the base of the crust, a corresponding

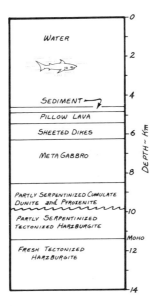

Figure 19 Possible petrologic model of the oceanic crust (after Clague & Straley 1977).

S wave low-velocity zone would be very hard to detect using refraction techniques. Surface-wave studies hold promise for resolving the shear velocity structure.

2. Isotopic composition of oxygen from some ophiolites seems to indicate a meteoritic origin for the water rather than sea water (Wenner & Taylor 1973). This problem is not yet resolved.

3. Another unresolved problem with this model relates to the velocity of rock + water and the degree of serpentinization. In this model the velocities of the middle crust, presumably a predominantly gabbro zone, would represent the velocity of water + gabbro, since the water would have to pass through this zone to reach the lower crust. Making a simple approximation that the velocity of water + rock equals the average of water velocity + rock velocity weighted by the relative proportion of each, we find that to reduce the velocity by 3% from 7.0 km sec^{-1} to 6.8 km sec^{-1} requires only about 1% of water. A more sophisticated analysis of the velocity of granite with fluid-filled cracks (Anderson et al 1974) also shows that for cracks of low aspect ratio (0.05) and a 1% porosity the effect of the cracks on the velocity is about 3%. This implies that probably less than 1% water is available at greater depths for serpentinization. However, this may not preclude greater amounts of serpentinization. The possibility exists of the convective cooling waters flushing out some products of the serpentinization reaction in solution, for example SiO_2 and MgO or Brucite [$Mg(OH)_2$], thereby allowing the reaction to proceed until the mechanical strength of the rock is decreased to the point where the rock flows and the cracks close. At this point this is speculative and further work on the physics and chemistry of this process is needed.

ACKNOWLEDGMENTS

This work was supported by an N.S.F. grant OCE-21503 and O.N.R contract N-00014-67-A-0103-0014. J. McClain kindly initially reviewed the manuscript.

Literature Cited

Anderson, D. L., Minster, B., Cole, D. 1974. The effect of oriented cracks on seismic velocities. *J. Geophys. Res.* 79:4011–16

Anderson, O. L., Lieberman, R. C. 1966. Sound velocities in rocks and minerals. *Versiac State of the Art Report.* Univ. Michigan

Anderson, R. N., Hobart, M. A. 1976. The relation between heat flow, sediment thickness, and age in the Eastern Pacific. *J. Geophys. Res.* 81:2968–89

Anderson, R. N., Langseth, M. G., Sclater, J. G. 1977. Mechanisms of heat transfer through the floor of the Indian Ocean. *J. Geophys. Res.* 82:3391–410

Backus, G. E., Gilbert, J. F. 1967. Numerical applications of a formalism for a geophysical inversion problem. *Geophys. J. R. Astron. Soc.* 13:247–76

Bates, A. C., Kanasewich, E. R. 1976. Inversion of seismic travel times using the tau method. *Geophys. J. R. Astron. Soc.* 47:59–72

Bessanova, E. N., Fishman, V. M., Ryaboyi, V. Z., Sitrikova, G. A. 1974. The tau method for the inversion of travel times. I. Deep seismic sounding data. *Geophys. J. R. Astron. Soc.* 36:639–42

Bibee, L. D., Shor, G. G. Jr. 1976. Compressional wave anisotropy in the crust and upper mantle. *Geophys. Res. Lett.* 3:639–42

Christensen, N. I. 1975. Ultrasonic velocities in minerals and rocks at high pressures and temperatures. *Geol. Soc. Am. Abstr. with Programs.* 7:1026

Christensen, N. I., Salisbury, M. H. 1972. Sea floor spreading, progressive alteration of layer 2 basalts, and associated changes in seismic velocities. *Earth Planet. Sci. Lett.* 15:367–75

Christensen, N. I., Salisbury, M. H. 1973.

Velocities, elastic moduli, and weathering-age relations for Pacific layer 2. *Earth Planet. Sci. Lett.* 19:416–70

Christensen, N. I., Salisbury, M. H. 1975. Structure and constitution of the lower oceanic crust. *Rev. Geophys. Space Phys.* 13:57–86

Clague, D. A., Straley, P. F. 1977. Petrologic nature of the oceanic Moho. *Geology.* 5:133–6

Clowes, R. M., Malecek, S. J. 1976. Preliminary interpretation of a marine deep seismic sounding survey in the region of Explorer Ridge. *Can. J. Earth Sci.* 13:1545–55

Cox, A., Doell, R. R., Dalrymple, G. B. 1963. Geomagnetic polarity epochs and pleistocene geochronometry. *Nature.* 198:1049–51

Ewing, M., Ewing, J. 1961. A telemetering ocean-bottom seismograph. *J. Geophys. Res.* 66:3863–78

Fowler, C. M. R. 1976. Crustal structure of the Mid-Atlantic Ridge Crest at 37°N. *Geophys. J. R. Astron. Soc.* 47:459–91

Fuchs, K., Müller, G. 1971. Computation of synthetic seismograms with the reflectivity method and comparison with observations. *Geophys. J. R. Astron. Soc.* 23:417–33

Garmany, J. D., Dorman, L. M., Orcutt, J. A. 1977. Analysis of survey data from a two dimensional array of ocean bottom seismometers. *Trans. Am. Geophys. Union.* Vol. 58, No. 6:509

Gerver, M., Markusevitch, V. 1966. Determination of seismic wave velocity from the travel-time curve. *Geophys. J. R. Astron. Soc.* 11:165–73

Goslin, J., Benzart, B., Francheteau, J., Le Pichon, X. 1972. Thickening of the oceanic layer in the Pacific Ocean. *Mar. Geophys. Res.* 1:418–27

Harrison, C. G. A. 1976. Magnetization of the oceanic crust. *Geophys. J. R. Astron. Soc.* 47:257–84

Helmberger, D. V. 1977. Fine structure of an Aleutian crustal section. *Geophys. J. R. Astron. Soc.* 48:81–90

Helmberger, D. V., Morris, G. B. 1969. A travel time and amplitude interpretation of a marine refraction profile: primary waves. *J. Geophys. Res.* 74:483–94

Helmberger, D. V., Morris, G. B. 1970. A travel time and amplitude interpretation of a marine refraction profile: transformed shear waves. *Bull. Seismol. Soc. Am.* 60:593–600

Hess, H. H. 1964. Seismic anisotropy of the uppermost mantle under oceans. *Nature* 203:629–31

Hill, M. N. 1963. In *The Sea,* ed. M. N. Hill,

3:39–45. New York: Wiley

Houtz, R., Ewing, J. 1976. Upper crustal structure as a function of plate age. *J. Geophys. Res.* 81:2490–98

Hyndman, R. D., Von Herzen, R. P. Erickson, A. J., Jolivet, J. 1976. Heat flow measurements in deep crustal holes on the Mid-Atlantic Ridge. *J. Geophys. Res.* 81:4053–60

Johnson, L. E., Gilbert, F. 1972. Inversion and inference for teleseismic ray data. *Methods Comput. Phys.* 11:321–66

Keen, C. E., Barrett, D. L. 1971. A measurement of seismic anisotropy in the Northeast Pacific. *Can. J. Earth Sci.* 8:1056–64

Kennett, B. L. N. 1976. A comparison of travel-time inversions. *Geophys. J. R. Astron. Soc.* 44:517–36

Kennett, B. L. N., Orcutt, J. A. 1976. A comparison of travel time inversions for marine refraction profiles. *J. Geophys. Res.* 81:4061–70

Kirkpatrick, J. 1977. *Initial rep. of the Deep Sea Drilling Proj. for Leg 46,* ed. J. R. Hiertzler, L. V. Dmitriev

Lewis, B. T. R., McClain, J. S. 1977. Converted shear waves as seen by ocean bottom seismometers and surface buoys. *Bull Seismol. Soc. Am.* 67:1291–302

Lewis, B. T. R., Snydsman, W. E. 1977a. Evidence for a low velocity layer at the base of the oceanic crust. *Nature* 266:340–44

Lewis, B. T. R., Snydsman, W. E. 1977b. Fine structure of the lower oceanic crust. Submitted to *Tectonophysics*

Lister, C. R. B. 1972. On the thermal balance of a mid-ocean ridge. *Geophys. J. R. Astron. Soc.* 26:515–35

Lister, C. R. B. 1974. On penetration of water into hot rock. *Geophys. J. R. Astron. Soc.* 39:465–509

Maxwell, A. E., Von Herzen, R. P., Hsu, K. J., Andrews, J. E., Saito, T., Percival, S. F. Jr., Milow, E. D., Boyce, R. E. 1970. Deep sea drilling in the South Atlantic. *Science* 168:1047–59

Morris, G., Raitt, R. W., Shor, G. G. Jr. 1969. Velocity anisotropy and delay time maps of the mantle near Hawaii. *J. Geophys. Res.* 74:4300–16

Orcutt, J. A., Kennett, B. L. N., Dorman, L. M. 1976. Structure of the East Pacific Rise from an ocean bottom seismometer survey. *Geophys. J. R. Astron. Soc.* 45:305–20

Orcutt, J. A., Kennett, B. L. N., Dorman, L. M., Prothero, W. 1975. Evidence for a low velocity zone underlying a fast-spreading rise crest. *Nature* 265:475

Parker, R. L. Oldenburg, D. W. 1973. Thermal model of ocean ridges. *Nature* 242:137–39

Poehls, K. A. 1974. Seismic refraction on the Mid-Atlantic Ridge at 37°N. *J. Geophys. Res.* 79:3370–73

Raitt, R. W. 1963. See Hill 1963, pp. 85–102 Wiley

Raitt, R. W., Shor, G. G. Jr., Francis, T. J. G., Morris, G. B. 1969. Anisotropy of the Pacific upper mantle. *J. Geophys. Res.* 74:3095–109

Raitt, R. W., Shor, G. G. Jr., Morris, G. B., Kirk, H. K. 1971. Mantle anisotropy in the Pacific Ocean. *Tectonophysics* 12:173–86

Reid, I., Orcutt, J. A., Prothero, W. A. 1977. Seismic evidence for a narrow intrusive zone of partial melting underlying the East Pacific Rise at 21°N. *Geol. Soc. Am. Bull.* 88:678–82

Rosendahl, B. R., Raitt, R. W., Dorman, L. M., Bibee, L. D. 1976. Evolution of ocean crust. I. A physical model of the East Pacific Rise Crest derived from seismic refraction data. *J. Geophys. Res.* 81:5294–304

Schreiber, E., Fox, P. J. 1977. Density and p-wave velocity of rocks from the FAMOUS region and their implication to the structure of the oceanic crust. *Geol. Soc. Am. Bull.* 88:600–8

Shor, G. G. Jr. 1963. See Hill 1963, pp. 20–38

Shor, G. G. Jr., Merard, H. W., Raitt, R. W. 1971. Structure of the Pacific Basin. In *The Sea*, ed. M. N. Hill, Vol. 4. New York: Wiley

Shor, G. G. Jr., Raitt, R. W., Henry, M., Bently, L. R., Sutton, G. H. 1973. Anisotropy and crustal structure of the Cocos plate. *Geofis. Int.* 13:337–62

Snydsman, W. E., Lewis, B. T. R., McClain, J. 1975. Upper mantle velocities on the northern Cocos plate. *Earth Planet. Sci. Lett.* 28:46–50

Sutton, G. H., Maynard, G. L., Hussong, D. M. 1971. Widespread occurrence of a high velocity basal layer in the Pacific crust found with repetitive sources and sonobuoys. *Geophys. Monoar. Am. Geophy. Union* 14:193–207

Vine, F. J., Matthews, D. H. 1963. Magnetic anomalies over oceanic ridges. *Nature* 199:947–49

Wenner, D. B., Taylor, H. P. Jr., 1973. Oxygen and hydrogen isotopes studies of the serpentinization of ultramafic rocks in oceanic environments and continental ophiolite complexes. *Am. J. Sci.* 273:207–39

Whitmarsh, R. B. 1973. Median valley refraction line, Mid-Atlantic Ridge at 37°N. *Nature* 246:297–99

Whitmarsh, R. B. 1975. Axial intrusion zone beneath the Median Valley of the Mid-Atlantic Ridge at 37°N detected by explosion seismology. *Geophys. J. R. Astron. Soc.* 42:189–215

Williams, D. L., Von Herzen, R. P., Sclater, J. G., Anderson, R. N. 1974. The Galapagos spreading center: lithosphere cooling and hydrothermal circulation. *Geophys. J. R. Astron. Soc.* 38:587–608

Woollard, G. P. 1975. The interrelationships of crustal and upper mantle parameter values in the Pacific. *Rev. Geophys. Space Phys.* 13:87–137

Wyllie, P. J. 1971. *The Dynamic Earth: Textbook in Geosciences.* New York: Wiley. 416 pp.

Ann. Rev. Earth Planet. Sci. 1978. 6: 405–36

STATE OF STRESS IN THE EARTH'S CRUST

×10100

A. McGarr[1]

U.S. Geological Survey, Office of Earthquake Studies, Menlo Park, California 94025

N. C. Gay

Bernard Price Institute of Geophysical Research, University of the Witwatersrand, Johannesburg, South Africa

INTRODUCTION

Measurements of the stress field within the crust can provide perhaps the most useful information concerning the forces responsible for various tectonic processes, such as earthquakes. Advances in knowledge of the state of stress at mid-crustal depths are essential if further progress is to be made toward solving a broad class of problems in geodynamics.

Most stress measurements have been, and will continue to be, motivated by engineering needs rather than the needs of geologists engaged in fundamental research. Knowledge of the state of stress is critical to the design of underground excavations for mining and for nuclear waste disposal (e.g. Jaeger & Cook 1969, pp. 435–64). The massive hydraulic fracturing of formations in oil and gas fields to stimulate production is another application for which knowledge of the stress field at depth is very important and, in fact, many of the deeper stress determinations have been by-products of these "hydrofrac" operations (e.g. Howard & Fast 1970). A recent and exciting application of hydraulic fracturing is the Hot-Dry-Rock Geothermal Energy Program (Aamodt 1977). Heat is extracted from the rock by circulating fluid down a pipe into hot rock and then up through a second pipe. A large fracture connecting the two pipes serves as the heat exchanger. Knowing the state of stress is critical in the solution to the problem of creating and maintaining such a crack. There is little argument about the applicability of information on the state of stress to these and many other engineering problems.

The application of stress measurements to the solution of problems in tectonics is not so straightforward as in engineering design. Whereas the engineer is concerned with the stress field affecting the rock, the geologist attempts to deduce the processes that might have caused the stresses. Before the measured stress field can be related

---

[1] On leave from the Bernard Price Institute of Geophysical Research, University of the Witwatersrand, Johannesburg, South Africa

405

0084-6597/78/0515-0405$01.00

to tectonic processes it must be correctly interpreted and analyzed, because the total field is influenced by many events in the geologic history of the rock as well as the contemporary tectonic and gravitational forces. Residual stresses are imposed on a rock according to its history of processes such as burial, lithification, denudation, heating, cooling, and past tectonic events. These *residual* stresses persist to some extent after the rock is freed of boundary loads (e.g. Friedman 1972), and their existence complicates the analysis of stress observations considerably.

In spite of these difficulties stress measurements are assuming a steadily increasing role in the solution of geodynamics problems such as the prediction and/or control of earthquakes (e.g. Raleigh, Healy & Bredehoeft 1976), the origin of forces driving plate tectonics (e.g. Sykes & Sbar 1973), and the origin of the forces responsible for regional deformation (e.g. the Rhinegraben in Germany, Greiner & Illies 1977). Almost all of the geodynamics studies to date have involved the analysis of stress directions; stress magnitudes have only been used in a few studies (e.g. Raleigh, Healy & Bredehoeft 1972). This review summarizes all of the data known to us about the *contemporary* state of stress in North America, Southern Africa, central Europe, Australia, and Iceland. We have elected not to discuss indicators of "paleo-stress" such as observations of deformation lamellae, stylolites, grain size, etc, (e.g. Carter & Raleigh 1969, Friedman & Heard 1974).

Because stress measurements are becoming increasingly important in tectonic analyses it is important to identify the significant gaps in our knowledge of the stress field and then to decide on the types of measurements that will contribute the most to our understanding of geodynamic processes. One notable gap in our understanding involves the magnitude of the stresses driving the movement on tectonic faults such as the San Andreas fault in California. Are the shear stresses of the order of 1 or 100 MPa (megapascals)? At present, no one is in a position to answer this question convincingly.

Several review articles have been published recently that summarize many of the stress observations and draw broad conclusions from the data. Ranalli & Chandler (1975) reviewed measurements made all over the world and emphasize observations made using strain relief techniques: Haimson (1977) summarized stresses measured using the hydraulic fracturing technique within the United States. Hast (1969, 1973) reviewed stress measurements throughout much of the world using his stressmeter (Hast 1958). As we shall show, the strain relief methods and the hydrofrac technique have different advantages and drawbacks, so the two sets of observations complement each other. Our primary intent here is to present each type of observation to its maximum advantage and to avoid presenting the least certain components of each data set. We decided not to present comprehensive tables of observations because a monograph is currently in preparation that is intended to be a compilation of these data (K. Hadley, in preparation; Riecker 1977).

Most of the stress magnitudes quoted in this paper are in units of megapascals (MPa). For comparison with other stress units commonly used in the literature $1\,MPa = 10\,bars = 145\,psi = 10.2\,kgf\,cm^{-2}$. The Système International (SI) unit of stress is $N\,m^{-2}$ (Pascal), but this unit is so small that it must be multiplied by 10^6 to be of the order of stresses commonly observed in the crust.

The stress field at a point can be represented as three principal stresses (e.g. Jaeger & Cook 1969, p. 20), which we denote as S_1, S_2, and S_3 for the maximum, intermediate, and minimum stresses, respectively. Because stresses measured in the crust generally are compressive rather than tensile, compressive stresses are taken as positive here. Often one of the principal stresses is oriented in the vertical, or at least a near-vertical, direction. We represent this stress as S_v and the two horizontal principal stresses as S_{Hmin} and S_{Hmax}.

STRESS MEASURING TECHNIQUES

There exist a number of references (e.g. Leeman 1964, Fairhurst 1968, Jaeger & Cook 1969, pp. 363–73) that review the available techniques for measuring in situ stress in rocks. Here we summarize some of these techniques and compare their advantages and disadvantages.

Stress Relief Methods

These methods involve measuring the change in strain that occurs after relieving the ambient stress acting on the rock. The stress relief is achieved by an overcoring or trepanning process. Normally a device capable of monitoring deformation is either inserted into a borehole or attached to a prepared surface. The deformation associated with the stress relief is measured and then can be related to the ambient stress field.

BOREHOLE DEFORMATION CELLS   A borehole deformation cell measures changes in the dimensions of one or more diameters of a borehole. Ideally it should be able to measure changes in diameter as small as 0.025 mm. The United States Bureau of Mines (U.S.B.M.) deformation cell (Hooker, Aggson & Bickel 1974) contains three beryllium-copper cantilevers that push with a negligible load against the interior of the borehole. Deflections of the cantilevers during borehole deformation, as a result of overcoring, can be used to calculate the principal stresses acting in a plane perpendicular to the axis of the borehole. Measurements have to be made in three nonparallel boreholes to determine the complete stress tensor.

BOREHOLE STRAIN GAUGE CELLS   In these cells, one or more strain rosettes are bonded directly to the surface of the borehole. The most commonly used cell of this type is the C.S.I.R. (Council for Scientific and Industrial Research) "doorstopper" cell (Leeman 1969) consisting of a single-strain rosette that is stuck onto the flat-end surface of a borehole and then overcored. The four gauges making up the rosette record the resultant changes in diameter of the borehole end in various directions. With this cell, it is necessary to make several measurements along three nonparallel boreholes to obtain the complete stress-relief field, which is, in turn, related to the ambient stress field by empirically determined stress-concentration factors (e.g. Crouch 1969, Coates & Yu 1970).

The C.S.I.R. triaxial strain cell (Leeman & Hayes 1966) consists of three four-gauge strain rosettes that are glued to the wall of the borehole at known orientations and positions. On overcoring, the rosettes record enough changes in the components of strain for the entire stress relief to be calculated at the point of measurement.

DIRECT STRAIN-GAUGE TECHNIQUE   This method, used to measure the horizontal components of the stress field at a free surface, consists of simply overcoring foil resistance strain gauges bonded directly to a prepared rock-surface (Swolfs, Handin & Pratt 1974, Engelder & Sbar 1976).

BOREHOLE INCLUSION STRESS METERS   Inclusion stressmeters are usually stiff devices with elastic moduli significantly greater than those of rock so that they can measure stresses directly rather than deformations (Abel & Lee 1973, Hast 1958). The instrument used by Hast (1958) consists of a spool of a magnetostrictive nickel alloy around which a solenoid of permalloy is wound. The spool is placed diametrically across the borehole and prestressed against the walls. During overcoring the stress is relieved, resulting in a drop of potential across the solenoid that can be correlated with the change in stress. Four measurements with different orientations are required to calculate the stresses normal and parallel to the borehole, and measurements in two nonparallel, nonperpendicular holes are needed for the determination of the complete state of stress.

The photoelastic stressmeter (Roberts et al 1964) or strain gauge (Hawkes & Moxon 1965) is an inclusion stressmeter that is not rigid but can be used to measure stress directly. The stressmeter consists of a solid cylinder of glass that is preloaded diametrically in a borehole. On overcoring, the change in load in the cylinder is monitored by measuring changes in the photoelastic fringe pattern. Measurements must be made in three nonparallel boreholes for a complete determination of the stress tensor.

DISCUSSION   The most serious drawback of strain-relief measurements is that they are operationally limited to distances of 30 to 50 m from a free surface. Furthermore, to obtain reliable results that are not unduly influenced by small-scale inhomogeneities in the rock properties or the stress field, it is necessary to make a series of measurements along each borehole, a time-consuming and costly process.

Probably the most satisfactory tool in terms of cost is the C.S.I.R. triaxial cell (Leeman & Hayes 1966), which allows the complete stress field to be determined at each measuring point. However, this instrument has to be overcored with a relatively large-diameter drilling crown, so it can only be used in very good ground; in broken or highly stressed rock it is impossible to obtain an annulus of rock suitable for the strain-relief measurements. Perhaps the easiest to use is the U.S.B.M. deformation cell, but the necessity to obtain a large diameter core also limits its versatility.

Accurate determinations of the elastic constants of the rock are also required to correlate the strain reliefs with the stresses. Ideally these should be determined under confining pressure and temperature conditions similar to those prevailing at the points of measurement; for this purpose devices such as the C.S.I.R. borehole simulator (Leeman 1969) have been developed. The correct determination of Poisson's ratio is particularly important as this modulus affects the calculation of the stress concentration factors markedly. Stressmeters have an advantage over strain gauge cells in that they do not depend on accurate determinations of the elastic moduli. However, photoelastic cells are temperature dependent, and difficulty may be experienced in reading the fringe pattern at small stresses (see Kotze 1970).

A potential problem with strain-relief measurements at depth is the influence of the excavation-induced stresses on the observations. Most underground measurements are made from tunnels. If the stress determination is made more than a tunnel diameter into the rock, it should not be influenced significantly by the stresses induced by the tunnel (Herget, Pahl & Oliver 1975). In this review we have tried to present only observations that were not affected by nearby excavations. We note that it is possible to account for the excavation-induced stresses (e.g. Salamon, Ryder & Ortlepp 1964) to recover the ambient stress field but this adds another level of interpretation to the data.

Estimates of the error in measurements have been reported for the various devices as follows: for the doorstopper, 5% (Van Heerden 1971) to 20% (Pallister 1969); for the triaxial strain cell, 5% (Herget 1973b), and for Hast's stressmeter, 2–4% (Hast 1969). We feel that some of these reported estimates of uncertainty are overly optimistic, but we are not in a position to present more realistic figures. Van Heerden & Grant (1967) concluded that both the U.S.B.M. deformation cell and the doorstopper cell were equally reliable. This contrasts with the results of de la Cruz & Raleigh (1972), who measured stresses at surface sites near Rangely, Colorado, using five different methods and found that the U.S.B.M. cell was the most convenient and reliable gauge for determining near-surface, in-situ stresses.

Hydrofrac Technique

Hydraulic fracturing is the only method currently in use that enables measurements to be made at large distances from a free surface. The experimental procedure, discussed in detail by Fairhurst (1968), Haimson & Fairhurst (1970), Haimson 1974, and Zoback et al (1977), consists of isolating a section of a borehole over a known depth interval by means of inflatable packers and then pressurizing this section by pumping in fluid while recording the pressure-time history of the hydraulic fluid.

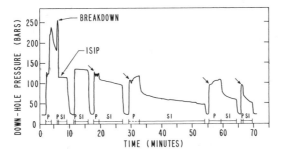

Figure 1 Hydrofrac pressure time history recorded at a site near San Ardo, California at a depth of 240 m (after Zoback, Healy & Roller 1977). *P* and *SI* indicate pumping and shut in, respectively. The inferred stress field was $S_1 = 22.5$ MPa (N 15 E), $S_2 = 11.4$ MPa (N 75 W), and $S_3 = 5.1$ MPa (vertical). (Courtesy of M. D. Zoback).

When the pressure increases to the level at which a tensile fracture occurs, a break-down pressure P_b (Figure 1) is recorded. If the pump is shut off immediately and the hydraulic circuit is kept closed, an "instantaneous shut-in" pressure (ISIP) is recorded (Figure 1); this is the pressure that is just sufficient to hold the fracture open. The orientation of the tensile fracture is established by inflating an impression sleeve against the borehole wall and determining the orientation of the impression with a downhole compass or by means of a more sophisticated device such as an ultrasonic borehole televiewer (e.g. Bredehoeft et al 1976).

The theory developed by Hubbert & Willis (1957) and Kehle (1964) for fracture around a pressurized borehole is used to relate P_b and the ISIP to the ambient stress field. For a vertical borehole, the tensile fracture should be oriented in a direction perpendicular to S_{Hmin}, and the magnitude of S_{Hmin} is equivalent to the ISIP. S_{Hmax} is then determined from

$$P_b = T + 3S_{Hmin} - S_{Hmax} - P, \tag{1}$$

where T is the tensile strength of the rock and P is the static pore pressure in the rock surrounding the borehole. As T and P can be determined independently, Equation (1) allows S_{Hmax} to be determined. Assuming that one of the principal stresses is oriented vertically, the third principal stress can be estimated from $S_v = \rho g H$, where ρ is the average density, g is gravity, and H is the depth to the interval that is isolated by packers.

If the vertical stress is the minimum principal stress, a horizontal tensile fracture might be expected to form (Kehle 1964). However, Zoback et al (1977) argue that the inflatable packers inhibit this and that horizontal fractures form only if fluid penetrates along pre-existing planes of weakness; this contention is supported by laboratory experiments reported by Haimson & Fairhurst (1970). Haimson (1976b) suggested that if $S_v = S_3$ then the induced fractures initiate in a vertical plane and then become horizontal as they propagate.

Quite a number of problems and complications are associated with the interpretation of hydrofrac data, and some of these are illustrated in Figure 1, which shows six pumping cycles of pressure at a depth of 240 m in a borehole near San Ardo, California adjacent to the San Andreas fault (Zoback, Healy & Roller 1977). The pressure drop preceding breakdown was caused by momentarily shutting off the pump. Breakdown, P_b, occurred at 25.2 MPa and the ISIP was 11.4 MPa. On cycle 2 the well was shut in before any breakdown actually occurred. Cycles 3 to 6 show that the shut-in pressure is characterized by a decay that is initially fast and then much slower. The slowly decaying shut-in pressure seems to approach an asymptotic value in the latter cycles that happens to be close to the expected overburden pressure. Zoback, Healy & Roller (1977) suggested that the fracture, initially vertical (oriented N 15° E from an impression sleeve record), was possibly turning into the horizontal plane as it propagated away from the borehole. Thus, the ISIP recorded on the first cycle is S_{Hmin} and the asymptotic shut-in pressure (ASIP) is the minimum principal stress of 5.1 MPa, previously established as being oriented vertically.

Another interesting feature of Figure 1 is that the secondary breakdown pressure (with $T = 0$ in Equation 1) has approached 9.3 MPa by the sixth cycle. The high "zero-strength" breakdown pressures seen in cycles 3 and 4 are the result of using a fracturing fluid of high viscosity (Zoback & Pollard 1978). The use of high-viscosity fluids was advocated by Zoback et al (1977) as a means of avoiding fluid penetration into pre-existing fractures or into permeable rock before the generation of a tensile fracture; these effects lead to rate-dependent, and thus spurious, estimates of P_b. Zoback & Pollard (1978), however, noted that a drawback of using high-viscosity fluids is that the fracture may initiate significantly in advance of the observed break-down in the pressure-time history because the fracture propagation is stable until the time at which the fracture attains some critical size. This effect leads to over-estimates of P_b, as seen in cycles 3 and 4 of Figure 1.

Another potential ambiguity in the interpretation of hydrofrac data is the possibility that a shear, rather than tensile, fracture may be generated from the bore-hole if the fluid injection rate is too slow (Lockner & Byerlee 1977); this would lead to a misinterpretation of P_b. This effect might be important in areas where rocks with high permeability are in a state of high tectonic shear stress.

In general there is, at present, considerable controversy over the correct interpret-ation of P_b, and recent theoretical developments (e.g. Clifton et al 1976, Zoback & Pollard 1978) suggest that a number of existing pressure-time histories from various sites should be re-analyzed.

More positively, there seems to be little argument about the significance of the ISIP among the various practitioners of the hydrofrac technique, and so it appears that the estimates of S_{Hmin} are probably on a firm basis. With the amount of attention currently being devoted to the hydrofrac method, it seems reasonable to hope that the other aspects of the data will also be amenable to an unambiguous interpretation in the near future. At this time it seems particularly important to publish the actual pressure-time histories in view of the possibility of re-interpretation at a later date.

Earthquake Source Studies

Analysis of the seismic waves radiated from earthquakes can indicate the *orientations* of the three principal stresses as well as the stress *change* associated with an earthquake.

FAULT PLANE SOLUTIONS   The radiation pattern of P (compressional) waves from an earthquake is quadrantal about the source. In two of the quadrants the initial motion of the P waves, as plotted on a stereographic or equal-area projection of the "focal sphere", is toward the source, and in the other two the first motion is away from the source. These quadrants are termed dilational and compressional, respectively, and are separated by nulls in the radiation pattern, termed nodal planes. According to Scheidegger (1964), the directions of S_1 and S_3 are centered in the dilational and compressional quadrants, respectively, and these directions are called the P and T axes. S_2 lies along the intersection of the nodal planes and this direction is the B axis.

It is not clear that the directions of S_1 and S_3 should necessarily coincide with the P and T axes. For example, Sbar & Sykes (1973) chose S_1 to be in a direction 30° from the inferred direction of motion on one of the nodal planes, toward the P axis, on the basis of laboratory experiments of fracture. McKenzie (1969) noted that the possible presence of pre-existing faults allows the direction of S_1 to be anywhere within the dilational quadrant. So far, however, studies comparing directions of S_1 and S_3 determined from in situ measurements with the directions of the P and T axes have shown good agreement between the results of the two techniques (e.g. Raleigh, Healy & Bredehoeft 1972, de la Cruz & Raleigh 1972, Ahorner 1975).

Perhaps the principal advantage of estimating stress orientations using seismic radiation patterns is that the inferred directions are representative of stresses at substantial depths within the crust over regions comparable in size to the earthquakes. The principal drawback is that only stress directions and not magnitudes are estimated.

STRESS DROPS   The source parameters of earthquakes that are commonly determined are the seismic moment, M_0 (Aki 1966), and the source dimension, r_0. If an earthquake is due to an average slip, D, across a fault of area A, then its seismic moment is given by $M_0 = GAD$, where G is the modulus of rigidity. M_0 is the most straightforward measure of the total deformation of an earthquake and its measured value, proportional to the long-period level of the spectrum of the seismic radiation, is quite independent of the choice of earthquake source models.

The source dimension r_0, as estimated from analysis of the seismic radiation, is highly model-dependent. The source model most widely used in recent years has been that of Brune (1970, 1971), who assumed a circular fault of radius r_0. With this model the stress drop is given by

$$\Delta\tau = (7/16)M_0/r_0^3. \tag{2}$$

Generally Brune's (1970, 1971) model seems to yield source dimensions that are in good agreement with those measured by other means, such as from the distribution of aftershocks (e.g. Hanks & Wyss 1972). It therefore seems unlikely that stress drops estimated from Equation (2) are systematically in error.

Hanks (1977) summarized most of the studies to date of seismic source parameters, and these results are illustrated in Figure 2. We see that over a very broad range in earthquake size nearly all of the points are between the two lines indicating constant stress drops of 1 bar (0.1 MPa) and 100 bars (10.0 MPa), with no apparent dependence of stress drop on M_0 or r_0.

The most controversial aspect of seismic stress drops is the question of how they are related to the absolute state of stress in the seismogenic regions of the crust. If earthquakes release a significant fraction of the ambient stress, then we expect $\Delta\tau$ to be indicative of the strength of the crust (e.g. Chinnery 1964). If this is so, Figure 2 would indicate that the average shear strength of the crust is of the order of several tens of bars (several MPa's). Tucker & Brune (1977) suggested that the peak stress drops of about 300 bars for the San Fernando aftershocks (Figure 2) might be indicative of the ambient tectonic stress. At the other extreme, Hanks (1977), on the basis of laboratory experiments of rock friction (Brace 1972, Byerlee 1977), has argued that earthquake stress drops are not indicative of ambient tectonic shear stresses which are of the order of kilobars, but rather may be related to the shear stresses applied to the base of the lithosphere.

Measurement of Residual Stress

Most of the measurements of residual stress, the stress present in a rock after removal of the boundary loads, have been by means of strain-relief techniques (e.g. Nichols 1975, Swolfs, Handin & Pratt 1974, Engelder, Sbar & Kranz 1977). Normally, a rock, which has been freed of its boundary loads, is progressively dissected and the

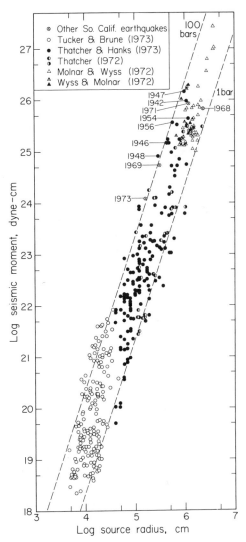

Figure 2   Earthquake moment as a function of source radius. The dashed lines are lines of constant stress drop. (Courtesy of T. C. Hanks).

corresponding strain changes recorded. If the deformation during the strain relief is elastic, then the recorded strain changes can be related to the residual stresses. Tullis (1977a) indicated some problems, based on St. Venant's principle (Love 1934, p. 132), with the interpetation of strain changes observed during the overcoring of isolated blocks. He suggested that some of these strain changes are due to inelastic processes and so it seems that the relationship of these "residual strains" to stresses is, in some cases, questionable.

Friedman's (1967) X-ray diffraction technique involves the measurement of inter-atomic d-spacings in quartz grains near the polished surface of an oriented sample. Comparison of the measured d spacing to that of strain-free quartz permits the calculation of residual strains and stresses.

Friedman (1972) reviewed the techniques and results of residual stress studies in considerable depth and also suggested some models to explain the existence of residual stresses.

THEORETICAL STRESS FIELDS

At any point in the earth's crust the observed state of stress is generally influenced by many factors such as present topography, present tectonics, man-induced conditions, paleotectonics, paleotopography, and thermal history. To analyze the stress measurements it is desirable to be able to account properly for all of these effects. Here we briefly discuss some of the simpler effects that can be calculated, mostly from the theory of elasticity.

Gravitational Loading

Nearly all theories of the state of stress assume that one of the principal stresses is oriented vertically and of magnitude $S_v = \rho g H$, where ρ is the *average* density of the overburden, g is the acceleration of gravity, about 9.8 m sec^{-1}, and H is the depth. As will be seen (Figure 3), the observations indicate that this is generally a valid assumption.

HEIM'S RULE   This theory is based on the assumption that stresses at depth are lithostatic. That is, $S_1 = S_2 = S_3 = \rho g H$ (Jaeger & Cook, 1969, p. 355). Heim's justification for this rule was that rocks tend to creep over long periods of time due to differences in the principal stresses. This state of stress is rarely, if ever, observed in rock, even limestone. In fact, principal stress differences generally increase with depth (e.g. Figure 8), at least in the upper 3–5 km of the crust. The lithostatic state does serve as a convenient reference, however, because departures from this state indicate the stresses available to drive geologic processes, such as folding and faulting.

LATERAL CONSTRAINT   If a region of the crust is subject only to the vertical force of gravity and the horizontal displacements are constrained to zero, then (Jaeger & Cook 1969, p. 356) the horizontal stresses are

$$S_H = [v/(1-v)]S_v = [v/(1-v)]\rho g H, \tag{3}$$

where v is Poisson's ratio, typically about 0.25. Thus S_H is only about a third of S_v.

This state of stress has rarely been observed, probably because the assumptions are not realistic. Even in relatively undisturbed sedimentary basins the horizontal stresses tend to be much higher than predicted by Equation (3).

Changes in the Depth of Overburden

A number of workers have appealed to Voight's (1966) denudation effect to explain anomalously high values of the horizontal stresses measured at shallow depths. This hypothesis considers the mechanical effect of removing a layer of overburden of thickness ΔH. The vertical stress is reduced by $\rho g \Delta H$ and from Equation (3) the horizontal stress is reduced by $[v/(1-v)]\rho g \Delta H$. Thus, after erosion, near-surface values of S_H might be quite high relative to S_v, depending on the state of stress before erosion.

Voight & St. Pierre (1974) considered the thermal, as well as mechanical, effects due to the removal of overburden and found that for normal thermal gradients within the crust, the thermal effect predominates, resulting in a *reduction* in S_H relative to S_v.

Haxby & Turcotte (1976) analyzed stress changes caused by the addition or removal of overburden, taking the effect of isostatic uplift or subsidence into account as well as thermal and mechanical effects. As before, the total change in the vertical stress is $\Delta S_v = \rho g \Delta H$. The effect of uplift or subsidence on S_H is $\Delta S_H^a = [Y/(1-v)](\rho_s/\rho_m)\Delta H/a$, where Y is Young's modulus, ρ_s is the density of the overburden, ρ_m is the density of the mantle at the depth of compensation, and a is the radius of the earth. Negative ΔH corresponds to a decrease in the depth of overburden (erosion) and a decrease in S_H.

The effect of temperature change, ΔT, on S_H is $S_H^{TH} = \alpha Y \Delta T/(1-v)$, where α is the thermal expansivity. ΔT is estimated from the geothermal gradient and a typical value for continents is $\partial T/\partial H = 25°C\,km^{-1}$. Thus, removal of 1 km of overburden reduces the temperature by 25°C and reduces S_H because ΔS_H^{TH} is tensional.

Haxby & Turcotte (1976) demonstrated that for a variety of rock types the overall effect of 1 km of erosion on the state of stress is a considerable reduction in S_H relative to S_v due to the effects of uplift and temperature decrease.

Theoretical Models

Price (1966), Seagar (1964), and Price (1974) have presented models, some involving complicated scenarios of geologic events, intended to predict changes in the state of stress in sedimentary basins. These various analyses indicate a broad range of possibilities for the predicted stress field depending on the specific geologic history of the basin.

THE STATE OF STRESS

The following discussion of various aspects of the stress field at depth (Figures 3 to 8) is based largely on the stress measurements listed in Table 1. Because of space limitations we have had to be selective with regard to our regional coverage of the state of stress. We first consider the common assumption that the vertical stress is due to the weight of overburden.

Table 1 Stress determinations in southern Africa, North America and Australia

Locality	Depth m	S_1 [a] MPa	Azimuth Degrees	Plunge Degrees	S_2 [a] MPa	Azimuth Degrees	Plunge Degrees	S_3 [a] MPa	Azimuth Degrees	Plunge Degrees	Reference [b]
Roodepoort, Transvaal, South Africa	2500	88.0	332	18	58.0	112	67	34.0	238	15	1
Boksburg, Transvaal	2400	40.3	024	67	31.5	136	9	19.5	230	21	1
Carletonville, Transvaal	2320	62.5	285	70	40.5	030	5	19.5	120	15	1
Roodepoort, Transvaal	2300	70.0	112	72	52.0	292	18	39.0	203	1	1
Carletonville, Transvaal	1770	55.2	280	70	30.6	126	26	13.0	028	11	1
Evander, Transvaal	1577	49.5	270	88	37.2	081	2	26.4	171	1	1
Virginia, Orange Free State	1500	33.5	176	81	19.3	024	8	13.5	294	4	1
Carletonville, Transvaal	1320	46.0	310	60	19.5	100	25	11.5	200	15	1
Evander, Transvaal	1226	38.6	100	79	31.2	257	10	31.0	345	5	1
Evander, Transvaal	508	16.5	164	2	13.9	284	85	11.0	074	5	1
Copperton, Cape Province	410	13.0	330	6	9.6	098	78	6.4	239	6	29
Copperton, Cape Province	279	22.4	004	22	8.8	123	48	2.5	260	33	29
Drakensberg, Natal	150	12.4	297	13	10.2	206	8	5.9	086	75	2
Drakensberg, Natal	111	8.7	060	3	6.8	150	2	3.0	090	87	2
Ruacana, South West Africa	115	8.8	192	3	6.9	111	7	3.9	308	83	2
Shabani, Rhodesia	350	17.3	279	13	16.1	013	33	8.4	170	57	1
Kafue Gorge, Zambia	160	17.3	291	10	13.7	197	26	7.1	039	62	1
Kafue Gorge, Zambia	400	27.5	275	10	19.4	177	32	12.2	021	55	1
Elliot Lake, Canada	350	21.0	East		18.0	North		11.0	Vertical		3
Elliot Lake	300	37.0	NE		20.0	NW		11.0	Vertical		3
Elliot Lake	700	37.0	East		23.0	North		17.0	Vertical		3
Timmins, Canada	853	61.5	078	13	44.6	170	8	25.7	287	71	4
Timmins	488	33.1	094	6	26.8	186	23	10.7	350	66	5
Timmins	732	72.6	258	19	64.7	358	25	34.4	135	58	5
Timmins	853	53.3	250	10	51.9	342	8.5	19.1	112	77	5
Sudbury Basin, Canada	1219	80.7	243	6	38.6	358	76	36.6	150	22	6
Sudbury Basin	1707	128.8	249	10	100.8	350	52	62.3	152	37	6

Sudbury Basin	2134	79.5	270	20	61.2	013	32	37.4	152	51	6
Sudbury Basin	1219	60.3	250	13	45.7	348	35	34.3	144	52	6
Wawa, Canada	366	21.4	118	12	20.1	027	12	16.1	230	78	7
Wawa	366	42.5	133	33	34.3	229	9	15.1	332	56	7
Wawa	479	30.0	251	11	27.7	343	8	18.7	110	76	7
Wawa	573	47.2	222	17	34.1	315	9	26.7	070	70	7
Wawa	573	31.6	162	11	27.9	070	12	21.5	295	74	7
Wawa	573	19.9	224	4	16.6	315	6	14.6	100	83	7
Wawa	573	38.3	356	22	29.5	090	11	21.4	206	66	7
San Ardo, California, U.S.A.	240.2	22.5	N15E		11.4	N75W		5.1	Vertical		8
Alma, New York	512	22.3	N77E		14.7	N13W		(13.3)	Vertical		9
Nevada Test Site	380	8.8	N35E		7.0	Vertical		3.5	N35W		10ᶜ
Nevada Test Site	380	8.0	N44E		6.0	Vertical		2.4	N46W		10ᵈ
Henderson Project, Colorado	624	18.2	Vertical		12.2	308	0	8.1	218	0	11
Henderson Project	785	33.8	338	15	27.7	240	25	22.5	096	60	11
Henderson Project	1131	40.7	321	38	25.0	213	21	22.0	101	44	11
Rangely, Colorado	1914	59.0	N70E		(43.4)	Vertical		31.4	N20W		12,13
Barberton, Ohio	701	44.8	N90W		24.1	Vertical		23.4	North		14
Falls Township, Ohio	815	28.0	N64E		(21.2)	Vertical		15.0	N26W		15
Fenton Hill Site, GT1, New Mexico	765	(18.0)	Vertical					14.7			16
GT2	1990	(50.4)	Vertical					33.3			16
EE1	2930	(75.3)	Vertical					36.7			16
Michigan Basin	5110	135.0			(127.8)	Vertical		95.0			17
Michigan Basin	3660	(91.5)			90.0			67.0			17
Michigan Basin	2806	(70.2)			56.0			42.0			17
Michigan Basin	1230	48.0			(30.8)	Vertical		29.5			17
Sierra Nevada Mtns., Calif.	300	9.5	N25E		(8.2)	Vertical		5.4	N65W		18
Oconee County, South Carolina	230	23.0	N60E		16.0	N30W		(6.0)	Vertical		18
Montello, Wisconsin	135	16.0	N63E		7.0	N27W		(3.5)	Vertical		18
Near Charleston, South Carolina	194	4.8	N51E		(4.4)	Vertical		3.0	N39W		19
Silver Summit Mine, Idaho	1670	105.1	N25E		56.8	Vertical		37.5	N65W		20
South of Vernal, Utah	2750	(65.0)	Vertical		56.0	N65W		51.5	N25E		21

Table 1 *continued*

Locality	Depth m	S_1[a] MPa	Azimuth Degrees	Plunge Degrees	S_2[a] MPa	Azimuth Degrees	Plunge Degrees	S_3[a] MPa	Azimuth Degrees	Plunge Degrees	Reference[b]
Southeast of Farmington, N.Mexico	2150	(58.0)	Vertical		(83.0)			32.0	N35W		22
RMA Well, Denver, CO	3671							36.2	ENE		23
Marble Falls, Texas	346	28.3	N67W		(8.5)			7.6	N23E		24
Lead, South Dakota	1890	(55.2)	Vertical		35.9	N50E		18.0	N40W		25
New Mexico	934				(21.1)	Vertical		14.5			26
Wyoming	2769				(62.6)	Vertical		52.0	N65W		26
Wyoming	4484				(115.1)	Vertical		81.3			26
North of Denver, CO	2303				(52.1)	Vertical		35.4	N70E		26
North of Denver	2322				(52.5)	Vertical		40.4	N75E		26
Piceance Basin, Colorado	453	(10.2)	Vertical		9.5	N87E		7.1	N03W		28[e]
Warrego Mine, Australia	241	12.0	248	3	8.9	339	15	3.5	146	75	27
Warrego Mine	319	24.5	063	17	11.2	315	46	8.3	167	40	27
Mt. Isa Mine	664	21.6	090	45	16.4	000	0	12.4	270	45	27
Mt. Isa Mine	1089	24.8	East		17.7	North		16.3	Vertical		27
Mt. Isa Mine	1000	40.0	095	27	30.0	284	62	20.0	007	4	27
Cobar Mine	366	14.8	086	37	11.2	176	0	4.6	267	55	27
Cobar Mine	588	31.2	108	28	24.6	014	7	10.3	273	61	27
North Broken Hill Mine	1098	42.7	088	25	28.3	180	5	16.5	278	63	27

[a] Parentheses indicate that the stress was calculated from the weight of the overburden.

[b] 1. Gay (1975), 2. Van Heerden (1976), 3. Eisbacher & Bielenstein (1971), 4. Herget (1976), 5. Miles & Herget (1976), 6. Herget, Pahl & Oliver (1975), 7. Herget (1973a), 8. Herget, Healy & Roller (1977), 9. Haimson (1974), 10. Haimson et al (1974), 11. Hooker, Bickel & Aggson (1972), 12. Raleigh, Healy & Bredehoeft (1972), 13. Haimson (1973a), 14. Obert (1962), 15. Haimson & Stahl (1970). 16. Aamodt et al (1977), 17. Haimson (1976c), 18. Haimson (1976b), 19. Zoback & Healy (1977), 20. Chan & Crocker (1972), 21. Brechtel, Abou-Sayed & Jones (1977), 22. Swolfs (1975), 23. Healy et al (1968), 24. Roegiers & Fairhurst (1973), 25. E. Hoskins (unpublished data), 26. H. S. Swolfs (private communication), 27. Denham, Alexander & Worotnicki (1976), 28. Bredehoeft et al (1976), 29. Gay (1977).

[c] Measured using the hydrofrac technique.

[d] Measured using a U.S.B.M. borehole deformation gauge.

[e] This is one of the deepest of 34 measurements reported by these authors.

Vertical Stress at Depth

Figure 3 shows the observed variation of S_v with depth based on the available data from strain-relief measurements. All but one of the observations scatter about the straight line representing the stress gradient due to the overburden with an average density of 2.7 gm cm^{-3}. The high value at 1.7 km was located next to an extensive "sheared zone" (G. Herget, private communication).

The data of Figure 3 are generally consistent with the assumption that the vertical stress corresponds to the weight of the overburden, but localized departures from this assumption are also indicated.

Principal Stress Orientations

We now assess the validity of another common assumption that one of the principal stresses is oriented vertically. To this end we have plotted principal stress directions for all of the measurements made in Southern Africa on an equal area projection of the lower hemisphere (Figure 4). If one of the principal stresses were always oriented vertically we would expect to see a cluster of points about the center of the projection with the remaining points plotting close to the circumference. As seen in Figure 4 there is a loose cluster of points about the center of the projection and most of the rest of the directions tend to plot near the circumference, but it is clear from these data

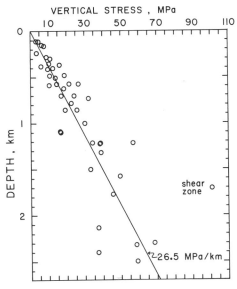

Figure 3   Vertical component of stress for depths greater than 100 m. The line corresponding to an average density of overburden of 2.7 gm cm^{-3} is shown for comparison. The sources of data were Gay (1975), Hooker, Bickel & Aggson (1972), Chan & Crocker (1972), Denham, Alexander & Worotnicki (1976), Herget (1973a, 1976), Herget, Pahl & Oliver (1975), Eisbacher & Bielenstein (1971), and Miles & Herget (1976).

that departures from the assumption are common (Table 1). Gay (1972, 1975) showed that over much of the Witwatersrand basin S_1 tends to be oriented closer to the vertical direction than the horizontal, but that over fairly broad areas the direction of S_1 shows a consistent and significant departure from verticality. Most of the maximum principal stresses fall within a circle of radius 30° about the vertical axis, however. Stress measurements made in deep mines in Canada, Australia, and the United States support the conclusion illustrated in Figure 4 that departures from the assumption that one of the principal stress directions is vertical are significant. Most of these data, however, were obtained in mines, often in regions of complex geology, and so it is perhaps not surprising that the observed principal stress directions

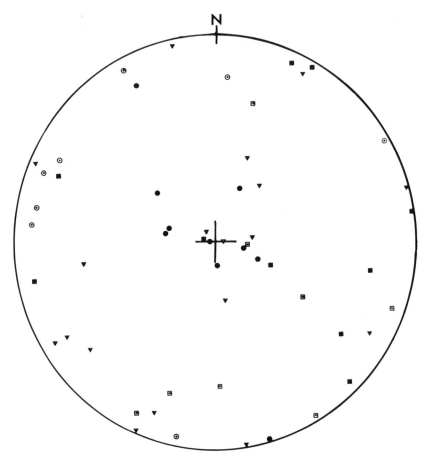

Figure 4 Orientation of principal stresses measured in southern Africa after Gay (1975, 1977) and Van Heerden (1976). Filled symbols refer to sites within the Witwatersrand system and open symbols to sites elsewhere. Circles denote S_1, squares, S_2, triangles, S_3. This is an equal area projection of the lower hemisphere.

show so much scatter. Orientations of stresses measured at depth in sedimentary basins might be expected to conform more closely to the assumption that one of the principal stresses is oriented vertically.

Horizontal Stress Magnitudes

As the vertical stress seems to be fairly predictable from the weight of overburden, measurements of the horizontal components of the stress field are of prime interest because these components could depart substantially from S_v. The extent of departure of S_{Hmin} and S_{Hmax} is limited only by the strength of the rock and, as mentioned before, very little is known about the strength of the crust. Unlike S_v, the magnitudes S_{Hmin} and S_{Hmax} are not constrained to 0 at the surface and, in fact, could show very high values at shallow depths.

The question of the relationship between surface measurements of the horizontal stresses and measurements at depth is important because considerable effort continues to be spent on obtaining near-surface data. To address this question we have plotted horizontal stress components as a function of depth for three regions where stress measurements have been made over a range of depths extending to at least 2 km. In the following sections principal stresses oriented within 30° of horizontal are considered to be "horizontal principal stresses," S_{Hmin} and S_{Hmax}. In two of the regions, Southern Africa and Canada, all of the measurements were made using strain-relief techniques, and so the complete state of stress was determined. In the third region, consisting of some sedimentary basins in the United States, most of the measurements were made using the hydrofrac technique.

SOUTHERN AFRICA   Figure 5 shows S_{Hmin} and S_{Hmax} as a function of depth for sites in Southern Africa (Table 1). All of the measurements below 500 m were made at sites in deep gold mines in the Precambrian quartzites of the Witwatersrand basin. These quartzites tend to be strong and brittle, and have a low value of Poisson's ratio, typically about 0.15.

We see that both components of the horizontal stress field generally increase with depth from near surface values of the order of 10 MPa to values centered about 30 MPa at depths between 2 and 2.5 km. Although the data show considerable scatter, all of the points, except one, below 500 m are to the left of the line indicating the stress due to the weight of the overburden. Above 500 m at least one and sometimes both of the horizontal stresses exceed the overburden stress. Between 500 m and 1200 m the stress field changes orientation with S_1 oriented horizontally above 500 m and vertically below 1200 m. Thus, in southern Africa stress measurements in the upper 500 m or so are not indicative of the state of stress at greater depths.

The influence of residual tectonic stresses may explain some of the apparent scatter in the data of Figure 5. For example, Gay (1975) argued that the high horizontal stresses measured by Cahnbley (1970) at the Durban Roodepoort Deep Mine, at depths of 2300 and 2500 m (Table 1), reflect a large component of residual tectonic stress associated with the folding and flattening of the strata.

CANADA   The stress determinations in Canada all indicate horizontal components of the stress field in excess of the overburden stress (Table 1). As seen in Figure 6, the

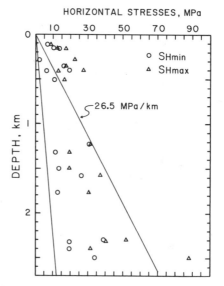

Figure 5   Horizontal stresses measured in southern Africa. For comparison, the line of the expected vertical stress corresponding to a stress gradient of 26.5 MPa km^{-1} is shown. The other line, down the left of the plot, is the horizontal stress predicted from Equation 3. The sources of data were Gay (1975, 1977) and Van Heerden (1976).

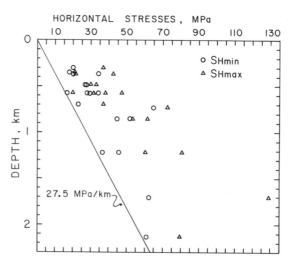

Figure 6   Horizontal stresses measured in Canada. The sources of data were Herget (1973a, 1976), Herget, Pahl & Oliver (1975), Eisbacher & Bielenstein (1971), and Miles & Herget (1976). The very high value of S_{Hmax}, near the lower right-hand corner of the figure, seems to represent a localized region of high stress on the basis of other measurements in the Creighton mine (Sudbury basin).

horizontal stresses show a general, though irregular, increase with depth to 2.1 km. As with the measurements in southern Africa, part of the spread in the distribution of S_{Hmin} and S_{Hmax} with depths is probably due to residual tectonic stresses. Eisbacher & Bielenstein (1971) suggested, for instance, that in the Elliot Lake area the high eastward component of the stress field contains a significant remanent component of the tectonic stress field that existed at the time of intense deformation in the area.

Another factor contributing to the apparent scatter in the stress data appears to be variations in rock properties. Herget (1973a) determined the state of stress in tuff, chert, metadiorite and siderite at the G. W. MacLeod Mine, Wawa, Ontario, and showed that the magnitudes of the stresses depend to some extent on the elastic moduli of the rocks; higher elastic moduli correspond generally to greater stresses.

The measurements in the deep Canadian mines, all near the margin of the Canadian shield, indicate a state of stress substantially different from that of Southern Africa (Figure 5). At any given depth the horizontal stresses in the Canadian mines are typically a factor of two or more greater than those within the Witwatersrand mines. This contrast is qualitatively consistent with the tectonics of the two regions in that normal faulting and subsidence were the predominant mode of deformation in the Witwatersrand basin whereas thrust faulting and folding accounted for most of the deformation along the edge of the Canadian shield.

U.S. BASINS   A large number of stress measurements using the hydrofrac technique have been made in the oil and gas fields in Colorado, New Mexico, Utah, and Wyoming and in the Michigan basin at depths extending to 5.1 km. This data set is of particular interest because nearly all of the measurements have been made in sandstones and shales in conditions of reasonably homogenous tectonics. The surface measurements, reported by de la Cruz & Raleigh (1972) and H. S. Swolfs, C. E. Brechtel, and H. R. Pratt (in preparation), were made at sites near Rangely, Colorado using the U.S.B.M. borehole deformation gauge and the direct strain gauge technique in the Mesa Verde sandstone. The intense distribution of measurements at depths between 37 and 475 m were made in seven oil-shale test holes in the Piceance Basin of northwest Colorado by Bredehoeft et al (1976). The deeper measurements in shales and sandstones were reported by Haimson (1973), Raleigh, Healy & Bredehoeft (1972), H. S. Swolfs (private communication), Haimson & Stahl (1970), and Haimson (1976c) (Table 1).

Only the measurements of S_{Hmin} have been plotted in Figure 7 because at many of the sites S_{Hmax} was either not determined or, somewhat uncertain, as discussed previously. We see that in the upper 2.3 km the observations fall remarkably close to the line corresponding to a gradient of 15 MPa km^{-1}, empirically determined from the results of many hydraulic fracturing operations (Howard & Fast 1970). From below 2.3 km to 5.1 km the measurements of S_{Hmin} appear to follow a gradient intermediate to the "oilfield" gradient and one corresponding to the average weight of overburden.

As seen in Figure 7, the "soft rock" measurements show much less scatter than those in hard rock (Figures 5 and 6). Part of this reduction in scatter may be attributed to the hydrofrac technique, but the surface measurements made using strain-relief techniques also show very little spread in the magnitudes. This suggests that the state of stress is inherently more homogeneous in soft rocks, such as shales and sandstones, than in hard rocks, such as granites and quartzites.

Another feature, which deserves more comment (Figure 7), is the departure of the data from the "oilfield" gradient at depths below about 2.3 km. The departure of S_{Hmin} from the average gradient of $15\,MPa\,km^{-1}$ may be due to the inability of the

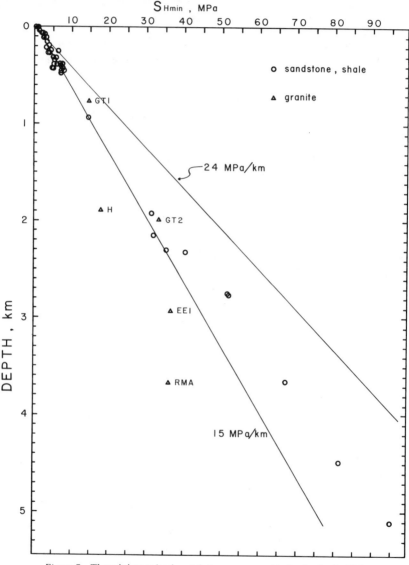

Figure 7   The minimum horizontal stress measured in basins in the U.S.

sandstones and shales to withstand stress differences much in excess of 20–40 MPa. Admittedly, more data below 2.5 km are necessary to confirm this suggestion.

The data in Figure 7 probably provide the best indication of the relationship of magnitudes of stresses measured at the surface to those at depth. The stress magnitudes at the surface fall on the same straight line as the rest of the measurements in the upper 2.3 km. Thus, in this case we can say that the magnitude of the surface stress is consistent with those at depth, although it is close to zero.

For comparison with the "soft rock" stresses we have plotted some estimates of S_{Hmin} in precambrian granite from hydrofrac measurements. GT1 (Geothermal Test 1), GT2, and EE1 (Energy Extraction) (Figure 7) refer to holes drilled in connection with the Hot-Dry-Rock Geothermal Energy Program at the Fenton Hill site, New Mexico (Aamodt et al 1977). RMA represents the Rocky Mountain Arsenal well near Denver, Colorado (Healy et al 1968), and the remaining measurement (H) was made in South Dakota by E. Hoskins (unpublished data).

The measurements in granite do not show such a regular increase of stress with depth as the measurements in sandstones and shales. The low value of S_{Hmin} for the point labeled EE1 has immediate engineering significance for the Hot-Dry-Rock Project because a gigantic crack is most easily propagated in conditions of low S_{Hmin}.

Shear Stress

The observed variation of the maximum component of shear stress, $(S_1 - S_3)/2$, is shown in Figure 8, which includes most of the available data from sites below 100 m depth (Table 1). The data have been divided into soft rock measurements—shales, sandstones, limestones, etc.—and hard rock measurements—granites, quartzites, norites, etc. Many of the points for sites in soft rock were determined from hydrofrac measurements and so S_1, as discussed previously, may be uncertain, which makes the maximum shear stress correspondingly uncertain.

The high value of shear stress at 1670 m (Figure 8) was measured in quartzite in the Coeur d'Alene mining district, Idaho (Chan & Crocker 1972) using a U.S.B.M. borehole deformation gauge, and the high value at 1707 m was determined from measurements in the Sudbury basin, Canada (Herget, Pahl & Oliver 1975) using the "doorstopper" method. The high Canadian value of shear stress almost certainly reflects some localized stress concentration because shear stresses measured above and below this site in the same mine are substantially lower, as seen from the points labeled with S in Figure 8.

The data of Figure 8 show tremendous scatter but, even so, some conclusions can be drawn. First, the shear stress shows a general increase with depth. The increase appears to be much more rapid in the upper one or two kilometers than at greater depths. The data for the "soft" rocks in Figure 8 indicate that the gradient of the shear stress is substantially less at depths below a kilometer than at shallower levels.

The shear stresses measured in the "hard" rocks (Figure 8) are, for the most part, significantly higher at a given depth than those in "soft" rock. The gradient in shear stress also appears to be diminishing with depth for the hard rock, but this tendency is not very well established because of a lack of data below 3 km.

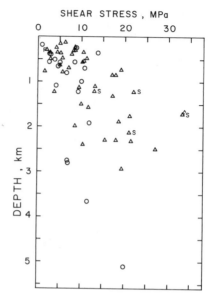

Figure 8 Maximum shear stress, $(S_1 - S_3)/2$. All of these values were derived from observations listed in Table 1. Measurements in soft rock such as shale and sandstone are indicated as circles, and those in granites, quartzites, etc, are shown as triangles. The P near the upper-left of the figure denotes the general level of shear stress for the measurements in the Piceance basin (Bredehoeft et al 1976). Symbols marked S indicate the same mine in the Sudbury basin, Ontario.

The data of Figure 8 suggest some lower limits for the magnitudes of regional shear stresses at mid-crustal depths (say 20 to 40 MPa), but much more data at depths below 3 km are necessary to check this suggestion. In any case, it is clear that near-surface estimates of the shear stress are not indicative of shear stresses throughout the crustal section.

Horizontal Stress Orientations

Stress orientations are intrinsically much more amenable to analysis than magnitudes because measurements of stress directions at all depths can be meaningfully compared; in addition, measured stress orientations can be compared to directions from earthquake fault-plane solutions and directions inferred from geologic indicators of stress.

Over certain broad regions the horizontal stress orientations appear to be quite homogeneous, although localized anomalies do occur. Over other regions the horizontal stress directions seem to be completely incoherent from site to site. Both of these situations are described in the following discussion of horizontal stress orientations by region.

NORTH AMERICA   Sbar & Sykes (1973) presented the results of many in situ stress measurements, earthquake fault-plane studies, and geological observations made in eastern North America and concluded that the maximum compressive stress trends east to northeast from west of the Appalachian Mountain system to the middle of the continent and from southern Illinois to southern Ontario (Figure 9). More recent data, especially from the hydraulic fracturing experiments, tend to support this generalization (e.g. Haimson 1977). Most recently, Sbar & Sykes (1977) specified more exactly the eastern boundary of the "stress domain" for which S_{Hmax} trends ENE on the basis of some recent fault-plane solutions, additional in-situ stress measurements, and observations of the orientations of glacial "pop-ups."

Raleigh (1974) commented on possible relationships between the driving forces of plate tectonics and stress orientations observed in the U.S. by means of in-situ measurements, earthquake fault-plane solutions, and some observations of dyke orientations. The stress orientations in the western U.S. are consistent with a state of right-lateral shear along the boundary between the Pacific and North American plates (Figure 9). Essentially all of these orientations were on the basis of fault-plane solutions and indicate that S_{Hmax} is generally oriented NNE to NE in California and Nevada becoming NNW to N in orientation from northern California up into Washington. More recent hydrofrac measurements by Haimson (1976a) in the Sierra Nevada Mountains and by Zoback, Healy & Roller (1977) near the San Andreas Fault in central California also indicate a NNE orientation for S_{Hmax} in California.

Some Surface Measurements

Here we briefly describe results from selected studies of surface stresses and their bearing on generalizations based on deeper measurements, earthquake fault-plane solutions, and tectonic considerations. Some of the surface measurements have yielded results in good agreement with results from other types of data.

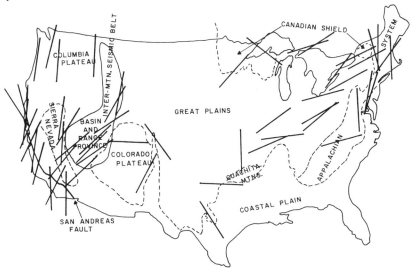

Figure 9   Orientations of S_{Hmax} in the U.S. as measured from in situ stress measurements and earthquake fault-plane solutions. [Adapted from Raleigh (1974)].

De la Cruz & Raleigh (1972) found that near-surface stress directions at sites near Rangely, Colorado were consistent with those inferred from earthquake fault-plane solutions (Raleigh, Healy & Bredehoeft 1976) and from existing joint patterns.

Engelder & Sbar (1976) measured stress orientations using the direct strain gauge technique at five sites in the Potsdam sandstone in northern New York and found that the average direction of S_{Hmax} at four of the sites is N 78° W and at the fifth the average direction is N 18° E. Over broad areas of the formation, but not the entire formation, the stress seems to be uniformly oriented. They were not able to conclude whether any components of their measured stresses were caused by the broad-scale applied stress trending E to NE hypothesized by Sbar & Sykes (1973) for the eastern U.S.

Newman & Clark (1977) reported near-surface measurements made at three quarries in western Ohio that showed consistent directions for measurements within a particular quarry but inconsistent directions between quarries only a few kilometers apart; at the three quarries S_{Hmax} was oriented N 40° E, N 14° W, and N 81° W. They noted that these results do not support the hypothesis of a regional orientation for S_{Hmax} in the eastern U.S.

Hooker & Johnson (1969) summarized the results of many near-surface measurements at sites in New England, the Appalachian Piedmont, Missouri, Oklahoma, and Texas. At all but two of these sites they found that S_{Hmax} seems to be aligned parallel to the fold axis of the major tectonic structure.

Most recently Tullis (1977b) and Sbar et al (1977) have measured stresses at surface sites adjacent to the San Andreas fault near Palmdale in southern California. Tullis used a U.S.B.M. deformation gauge in shallow holes and Sbar et al used the C.S.I.R. "doorstopper" technique. The two groups made measurements at a number of common sites for comparison of the two methods and found that the measured stress directions were consistent. Interestingly, the direction of S_{Hmax} in one area shows considerable, although systematic, changes in orientation through about 90° over distances of the order of 1 km.

CENTRAL EUROPE   In-situ stress measurements and earthquake fault-plane solutions indicate that within Switzerland and Germany to the north of the Alps the direction of S_{Hmax} is NW. This is a region where shallow stress measurements at many sites have yielded stress directions consistent with those deduced from analyzing a large number of crustal earthquakes. Both the fault-plane solutions (e.g. Ahorner 1975) and the in-situ measurements (e.g. Greiner 1975, Greiner & Illies 1977) show that S_1 and S_3 are horizontal (strike-slip faulting).

Most of the stress measurements were at depths less than 500 m using the "door-stopper" strain-relief technique, but at one site in southwest Germany Rummel & Jung (1975) determined the state of stress at a depth of 25 m using the hydrofrac technique for comparison with the stresses previously determined by Greiner (1975). Both the magnitudes and directions of the stresses measured using the two methods were in good agreement.

Geological evidence taken in conjunction with observations of the present-day stress field has been interpreted by Illies (1975) as indicating a counterclockwise rotation of the direction of S_{Hmax} through an angle of 60° beginning in early Miocene times. The Rhinegraben rift system was formed in the pre-existing stress field and the reorientation of the stresses has converted the Rhinegraben into a left-lateral shear zone according to Illies & Greiner (1976), who attribute this change in the stress direction to the development of the Alpine collision front.

AUSTRALIA   Earthquake fault-plane solutions (e.g. Fitch, Worthington & Everingham 1973, Mills & Fitch 1977) and in-situ stress measurements (e.g. Stephenson & Murray 1970, Endersbee & Hofto 1963, Denham, Alexander & Worotnicki 1976) show that S_{Hmax} is oriented close to E–W throughout much of Australia with the exception of one region extending from the central portion of the continent southward within which S_{Hmax} is oriented approximately N–S (e.g. Stewart & Denham 1974).

The stresses were measured using strain relief techniques of various types, most commonly using overcoring methods (Hoskins 1967). Most of the measurements, which were made at depths ranging from near-surface to slightly more than 1 km, and all of the earthquake fault-plane solutions indicate a state of stress in which the minimum principal stress is S_v (thrust faulting).

SOUTHERN AFRICA   The directions of S_{Hmin} and S_{Hmax} in this region (Gay 1977) appear to be broadly distributed as seen in Figure 4 although there may be a slight tendency for these directions to be aligned either N–S or E–W. Even if only the measurements made within the Witwatersrand basin are considered, the horizontal stresses do not show any preferred directions.

ICELAND   In-situ stress measurements in Iceland by Hast (1969, 1973) and by Haimson & Voight (1977) have yielded results that are somewhat anomalous in view of the earthquake data and plate-tectonics concepts. From the results of near-surface measurements Hast concluded that the horizontal stresses are high and compressive and not indicative of a spreading ridge. Haimson & Voight (1977) measured stresses using the hydrofrac technique in two boreholes and found that the direction of S_1 changes from horizontal to vertical at a depth of about 250 m in one of the boreholes but not in the other. At all depths the orientation of S_{Hmax} is roughly perpendicular to the axis of rifting. This measured state of stress is consistent with neither the earthquake fault-plane solutions nor the geology of the area (Ward, 1971, Klein, Einarsson & Wyss 1977). The re-orientation of S_1 observed in one of the boreholes may indicate that stresses measured in the upper half kilometer or so in Iceland are simply not indicative of the principal stress directions well within the crust. Haimson & Voight (1977) suggested that their results could be explained in terms of a combination of thermoelastic mechanisms associated with the accretion and cooling of spreading lithosphere.

Stress Gradients

Von Schonfeldt, Kehle & Gray (1973) produced maps showing the rate of increase of stress with depth for the oil and gas fields of the United States. They considered data from hydraulic-fracturing treatments of nearly 3000 wells to estimate the regional distribution of stress gradients. They found that the gradient of the minimum horizontal stress varies from 11.3 to 33.9 MPa km^{-1} and for the maximum horizontal stress from 15.8 to 29.4 MPa km^{-1}.

One of the regions where the gradient, γ, of S_{Hmin} is as low as 11.3 MPa km^{-1} is along the Gulf Coast in southern Louisiana and Mississippi where the sediments of the Mississippi embayment are known to be essentially in a state of failure. Normal faulting occurs such that the sediments tend to slump into the Gulf of Mexico. Hubbert & Willis (1957) noted that the observed failure of these sediments can be explained on the basis of the Coulomb (1773) failure criterion if the effect of pore pressure is taken into account.

According to Coulomb (1773), failure occurs at a level of shear stress across the plane of failure given by $|\tau| = \tau_0 + \mu S_n$, where τ_0 is the cohesive strength, μ is the coefficient of friction, and S_n is the compressive stress acting in a direction normal to the plane of failure.

If pore fluid is present at a pressure P, then the failure criterion becomes

$$|\tau| = \tau_0 + (S_n - P)\mu. \tag{4}$$

In general, $\tau = (S_1 - S_3)(\sin 2\theta)/2$ and $S_n = (S_1 + S_3)/2 + (S_1 - S_3)(\cos 2\theta)/2$, where θ is the angle between the normal to the failure plane and the direction of S_1. Along the Gulf coast S_1 is oriented vertically and presumably has magnitude $\rho g H$ at depth H. If we assume that the faults associated with the sediment failure dip 60° then $\theta = 60°$. We also assume that the pore pressure, P, is given by $P = \rho_w g H$, where ρ_w is the fluid density, taken as 1 gm cm^{-3} here. This estimate of P is correct if the water table is near the surface. Finally, if we can neglect the cohesive strength, τ_0, of the poorly consolidated sediments, then we can rewrite Equation (4) in terms of stress gradients since all terms are proportional to H. The critical gradient for failure, γ_c, for S_{Hmin} ($S_{Hmin} = \gamma H$) is

$$\gamma_c = g[(\rho - 1)(0.43 - 0.25\mu)/(0.43 + 0.75\mu) + 1].$$

Table 2 lists values of γ_c corresponding to various values of ρ and μ. We see that even if μ is as high as 1.0 we still expect the sediments to be in a state of failure along that part of the Gulf Coast where $\gamma \leq 11.3$ MPa km^{-1}. Probably the most reason-

Table 2 Critical gradients of S_{Hmin}

γ_c (MPa km^{-1})	ρ (gm cm^{-3})	μ
12.9	2.0	0.6
12.0	2.0	0.8
11.3	2.0	1.0
13.6	2.2	0.6
12.5	2.2	0.8
11.6	2.2	1.0
14.2	2.4	0.6
12.9	2.4	0.8
11.9	2.4	1.0

able estimates of ρ and μ, for the Gulf Coast sediments, are $2.2\,\text{gm cm}^{-3}$ and 0.6, respectively. In this case we expect normal faulting for $\gamma \leq 13.6\,\text{MPa km}^{-1}$ (Table 2), and this criterion is met over a fairly broad region of the Gulf Coast (von Schonfeldt, Kehle & Gray 1973).

M. D. Zoback, J. H. Healy, and G. S. Gohn (in preparation) have measured stress at various depths up to 491 m in two wells near Charleston, South Carolina using the hydrofrac technique. They found that in the Coastal Plain sediments the gradient of S_3 is very low, and from an analysis similar to that for the Gulf Coast, these sediments also appear to be in a state of incipient failure.

DISCUSSION AND CONCLUSIONS

High Horizontal Stresses

One of the most common observations emphasized by those who make in-situ stress measurements is the ubiquity of high, horizontal stresses relative to the vertical stress at shallow depths (e.g. Hast 1973). As discussed earlier, none of the usual theories involving gravity, uplift, erosion, or temperature changes can explain the relatively high values of S_{Hmin}. In particular we wish to emphasize that although the mechanical effect of erosion can lead to relatively high values of the horizontal stresses (Voight 1966), this effect necessarily occurs in conjunction with a reduction in temperature as well as probably some isostatic uplift. The net result is a reduction in the horizontal stresses relative to S_v (Voight & St. Pierre 1974, Haxby & Turcotte 1976).

The data suggest that, at any given depth, if S_{Hmin} falls below a certain value, some sort of inelastic process occurs to increase its magnitude. For example, Price (1974) suggested that in the presence of pore pressure, natural hydraulic fracturing would occur in sedimentary basins undergoing erosion and uplift if S_{Hmin} fell below the hydrostatic stress minus the tensile strength of the rock. Voight & St. Pierre (1974) suggested that the relaxation of residual stresses during erosion could impose substantial compressive stresses on the rock mass.

Reference State of Stress

Although stress orientations have proved amenable to geologic analysis, magnitudes of stresses have, for the most part, been quite enigmatic. This is because no one knows what magnitudes to expect from the horizontal stress field. A number of workers have considered the horizontal stresses calculated on the basis of lateral constraint (Equation 3) as a reference state, but this seems like a very poor choice because few of the observed horizontal stresses are even close to this predicted state. For example, some of the lowest values of S_{Hmin} at depths greater than 1 km have been observed in southern Africa, but we see in Figure 5 that all of the data fall well to the right of the line calculated from Equation 3 for $v = 0.15$.

Although the vertical stress at depth is reasonably predictable (Figure 3), it is not possible to generalize about the magnitudes of the horizontal stresses within the crust except to say that they usually increase with depth. In some regions the horizontal components of stress tend to be less than the vertical component (e.g.

Figure 5) and in others the horizontal stresses exceed S_v (e.g. Figure 6). Thus, no reference state of stress exists that actually resembles the stress field in any particular region and yet has world-wide applicability. The lithostatic state of stress (Heim's Rule) serves as a convenient point of departure in discussing stresses at depth, even if it does not actually represent the stress field found in any particular part of the crust.

In this regard it is interesting to note that the stress field as measured in the upper 5 km of the crust does not show any tendency to approach a lithostatic state with increasing depth. In fact, as seen in Figure 8 the shear stress generally increases with depth throughout this range.

Strength of the Crust

Earlier, in the discussion about seismic stress drops (Figure 2), we mentioned some of the conjecture by different workers regarding the shear strength of the crust. From the observed shear stresses (Figure 8) we feel that we can begin to narrow down the limits of the possible range of crustal strengths. Because granite is a very important constituent in continental crustal sections the data for sites in hard rock are probably more indicative of the shear stresses to be found throughout the continental crust than the observations in softer sedimentary rock. These data (Figure 8) suggest a shear strength of at least 20 MPa, and probably much greater, for the continental crust. If so, then the seismic stress drops (Figure 2) represent, in general, only a small fraction of the ambient shear stress.

ACKNOWLEDGMENTS

Many of the authors cited in this review generously supplied reprints, preprints and many useful suggestions, for which we are grateful. H. Swolfs was especially helpful in supplying the results of a considerable number of stress measurements, many of which were unpublished. T. Fitch kindly provided much of the Australian data, and J. B. Walsh supplied background material. Extensive discussions with M. Zoback, J. Healy, C. B. Raleigh, H. Pratt, T. Hanks, D. Pollard, and T. Tullis were useful in clarifying some of the issues discussed here. D. Bailey and W. Seiders provided considerable editorial assistance. M. Zoback, D. Pollard, and J. H. Healy reviewed the manuscript. N. C. Gay was supported in part by the Chamber of Mines of South Africa while this review was being prepared.

Literature Cited

Aamodt, R. L. 1977. Hydraulic fracture experiments in GT-1 and GT-2. *Tech. Rep. LA-6712*, Los Alamos Sci. Lab., Los Alamos, N.M.

Aamodt, R. L., Brown, D. W., Lawton, R. G., Murphy, H. D., Potter, R. M., Tester, J. W. 1977. *Ann. Rep. LA-6525-PR*, pp. 60–63, Los Alamos Sci. Lab., Los Alamos, N.M.

Abel, J. F., Lee, F. T. 1973. Stress changes ahead of an advancing tunnel. *Int. J. Rock Mech. Min. Sci. Geomech. Abstr.* 10: 673–97

Ahorner, L. 1975. Present-day stress field and seismotectonic block movements along major fault zones in Central Europe. *Tectonophysics* 29: 233–49

Aki, K. 1966. Generation and propagation of G waves from the earthquake of 16th June 1964. *Bull. Earthquake Res. Inst. Tokyo Univ.* 44: 73–88

Brace, W. F. 1972. Laboratory studies of stick-slip and their application to earthquakes. *Tectonophysics* 14: 189–200

Brechtel, C. E., Abou-Sayed, A. S., Jones,

A. 1977. *Terra Tek Rep.*, Terra Tek Inc., Salt Lake City, Utah

Bredehoeft, J. D., Wolff, R. G., Keys, W. S., Shuter, E. 1976. Hydraulic fracturing to determine the regional in situ stress field, Piceance Basin, Colorado. *Geol. Soc. Am. Bull.* 87:250–58

Brune, J. N. 1970. Tectonic stress and the spectra of seismic shear waves from earthquakes. *J. Geophys. Res.* 75:4997–5009

Brune, J. N. 1971. Tectonic stress and the spectra of seismic shear waves from earthquakes; correction. *J. Geophys. Res.* 76:5002

Byerlee, J. D. 1977. Friction of rocks. In *Proc. Conf. II, Experimental Studies of Rock Friction with Application to Earthquake Prediction*, ed. J. F. Evernden, pp. 55–77. Menlo Park, Calif: U.S. Geol. Surv.

Cahnbley, H. 1970. *Grundlagenuntersuchungen über das Entspannungsbohrverfahren wahrend des praktischen Einsatzes in grosser Tiefe.* Dissertation. Tech. Univ. Clausthal, Germany

Carter, N. L., Raleigh, C. B. 1969. Principal stress directions from plastic flow in crystals. *Geol. Soc. Am. Bull.* 80:1231–64

Chan, S. S. M., Crocker, T. J. 1972. A case study of in situ rock deformation behavior in the Silver Summit Mine, Coeur D'Alene Mining district, *Proc. 7th Can. Rock Mech. Symp., Edmonton, 1971*, pp. 135–60, Mines Branch, Dept. of Energy, Mines and Res., Ottowa

Chinnery, M. A. 1964. The strength of the Earth's crust under horizontal shear stress. *J. Geophys. Res.* 69:2085–89

Clifton, R. J., Simonson, E. R., Jones, A. H., Green, S. J. 1976. Determination of the critical stress-intensity factor K_{1c} from internally-pressured thick-walled vessels. *Exp. Mech.* 16:233–38

Coates, D. F., Yu, Y. S. 1970. A note on the stress concentrations at the end of a cylindrical hole. *Int. J. Rock Mech. Min. Sci.* 7:583–88

Coulomb, C. A. 1773. Sur une application des règles de maximus et minimus à quelques problèmes de statique relatifs à l'architecture. *Acad. R. Sci. Mem. Math. Physique Divers Savans* 7:343–82

Crouch, S. L. 1969. A note on the stress concentrations at the bottom of a flat-ended borehole. *J. S. Afr. Inst. Min. Metall.* 70:100–2, 386k

De la Cruz, R. V., Raleigh, C. B. 1972. Absolute stress measurements at the Rangely anticline, northwestern Colorado. *Int. J. Rock Mech. Min. Sci.* 9:625–34

Denham, D., Alexander, L. G., Worotnicki, G. 1976. *Stress measurement proposals for*

Western Australia, Rec. 1976/1, Dept. Min. and Energy, Bur. Miner. Res., Geol. and Geophys., Australia

Eisbacher, G. H., Bielenstein, H. U. 1971. Elastic strain recovery in Proterozoic rocks near Elliot Lake, Ontario. *J. Geophys. Res.* 76:2012–21

Endersbee, L. A., Hofto, E. D. 1963. Civil engineering design and studies in rock mechanics for Poatina underground power station, Tasmania. *J. Inst. Eng. Aust.* 35:187–206

Engelder, J. T., Sbar, M. L. 1976. Evidence for uniform strain orientation in the Potsdam sandstone, northern New York, from in situ measurements. *J. Geophys. Res.* 81:3013–17

Engelder, J. T., Sbar, M. L., Kranz, R. 1977. A mechanism for strain relaxation of Barre granite: opening of microfractures. *Pure Appl. Geophys.* 115:27–40

Fairhurst, C. 1968. *Methods of determining in-situ rock stress at great depths, Tech. Rep. No. 1–68*, U.S. Army Corps Eng., Missouri River Div., Omaha, Neb.

Fitch, T. J., Worthington, M. H., Everingham, I. B. 1973. Mechanisms of Australian earthquakes and contemporary stress in the Indian Ocean plate. *Earth Planet. Sci. Lett.* 18:345–56

Friedman, M. 1967. Measurement of the state of residual elastic strain in quartzose rocks by X-ray diffractometry. *Norelco Rep.* 14:7–9

Friedman, M. 1972. Residual elastic strain in rocks. *Tectonophysics* 15:297–330

Friedman, M., Heard, H. C. 1974. Principal stress ratios in Cretaceous limestones from Texas Gulf coast. *Bull. Am. Assoc. Pet. Geol.* 58:71–78

Gay, N. C. 1972. Virgin rock stresses at Doornfontein Gold Mine, Carletonville, South Africa. *J. Geol.* 80:61–80

Gay, N. C. 1975. In-situ stress measurements in Southern Africa. *Tectonophysics* 29:447–59

Gay, N. C. 1977. Principal horizontal stress in Southern Africa. *Pure Appl. Geophys.* 115:3–10

Greiner, G. 1975. In-situ stress measurements in southwest Germany. *Tectonophysics* 29:265–74

Greiner, G., Illies, J. H. 1977. Central Europe: active or residual tectonic stresses. *Pure Appl. Geophys.* 115:11–26

Haimson, B. C. 1973. Earthquake related stresses at Rangely, Colorado. In *New Horizons in Rock Mechanics, Proc. 14th Symp. Rock Mech.*, ed. H. Hardy, R. Stefanko, pp. 689–708. New York: ASCE

Haimson, B. C. 1974. A simple method for estimating in situ stresses at great depths.

In *Field Testing and Instrumentation of Rock, ASTM Spec. Tech. Publ. 554*, pp. 156–82

Haimson, B. C. 1976a. Preexcavation deep-hole stress measurements for design of underground chambers—case histories. In *Proc. 1976 Rapid Excavation and Tunneling Conf.*, ed. R. D. Coulon, R. D. Robbins, pp. 699–714. New York: Soc. Min. Eng. AIME

Haimson, B. C. 1976b. The hydrofracturing stress measuring technique-method and recent field results in the U.S. *Int. Soc. Rock Mech. Symp. Invest. Stress in Rock, Sydney, Australia*

Haimson, B. C. 1976c. Crustal stress measurements through an ultra deep well in the Michigan Basin. *EOS Trans. Am. Geophys. Union* 57: 326

Haimson, B. C. 1977. Crustal stress in the continental United States as derived from hydrofracturing tests. *The Earth's Crust, Geophys. Monogr. Am. Geophys. Union*, ed. J. C. Heacock 20: 576–92

Haimson, B., Stahl, E. 1970. Hydraulic fracturing and the extraction of minerals through wells. In *Proc. 3rd Symp. on Salt*, pp. 421–32, Northern Ohio Geol. Soc., Cleveland, Ohio

Haimson, B. C., Fairhurst, C. 1970. In situ stress determination at great depth by means of hydraulic fracturing. In *Rock mechanics—Theory and Practice, Proc. 11th Symp. Rock Mech.*, ed. W. Somerton, pp. 559–84. New York: AIME

Haimson, B. C., Lacomb, J., Jones, A. H., Green, S. J. 1974. Deep stress measurements in tuff at the Nevada test site. In *Adv. Rock Mech., Proc. 3rd Congr. Int. Soc. Rock Mech.* II-A: 557–62

Haimson, B. C., Voight, B. 1977. Crustal stress in Iceland. *Pure Appl. Geophys.* 115: 153–90

Hanks, T. C., Wyss, M. 1972. The use of body wave spectra in the determination of seismic source parameters. *Bull. Seismol. Soc. Am.* 62: 561–90

Hanks, T. C. 1977. Earthquake stress drops, ambient tectonic stresses and stresses that drive plate motions. *Pure Appl. Geophys.* 115: 441–58

Hast, N. 1958. The measurement of rock pressure in mines. *Sver. Geol. Under. Ser. C.* 52: 1–183

Hast, N. 1969. The state of stress in the upper parts of the earth's crust. *Tectonophysics* 8: 169–211

Hast, N. 1973. Global measurements of absolute stress. *Phil. Trans. R. Soc. London Ser. A* 274: 409–19

Hawkes, I., Moxon, S. 1965. The measurement of in-situ rock stress using the photo-elastic biaxial gauge with the core-relief method. *Int. J. Rock Mech. Min. Sci.* 2: 405–19

Haxby, W. F., Turcotte, D. L. 1976. Stresses induced by the addition or removal of overburden and associated thermal effects. *Geology* 4: 181–84

Healy, J. H., Rubey, W. W., Griggs, D. T., Raleigh, C. B. 1968. The Denver earthquakes. *Science* 161: 1301–10

Herget, G. 1973a. Variation of rock stresses with depth at a Canadian iron ore mine. *Int. J. Rock Mech. Min. Sci.* 10: 37–51

Herget, G. 1973b. First experiences with the C.S.I.R. triaxial strain cell for stress determinations. *Int. J. Rock Mech. Min. Sci.* 10: 509–22

Herget, G. 1976. Field testing of modified triaxial strain cell equipment at Timmins, Ontario. *Rep. MRP/MRL77–2 (IR)*, Elliot Lake Lab., Canada Cent. Miner. Energy Technol.

Herget, G., Pahl, A., Oliver, P. 1975. Ground stresses below 3000 feet. *Proc. 10th Can. Rock Mech. Symp., Queens Univ., Kingston* 1: 281–307

Hooker. V. E., Johnson, C. F. 1969. Near-surface horizontal stresses including the effects of rock anisotropy. *U.S. Bur. Mines Rep. Invest.* 7224 29 pp.

Hooker, V. E., Bickel, D. L., Aggson, J. R. 1972. In situ determination of stresses in mountainous topography. *U.S. Bur. Mines Rep. Invest.* 7654 19 pp.

Hooker, V. E. Aggson, J. R., Bickel, D. L. 1974. Improvements in the three-component borehole deformation gage and overcoring techniques. *U.S. Bur. Mines Rep. Invest.* 7894 29 pp.

Hoskins, E. 1967. *Field and laboratory experiments in rock mechanics.* PhD thesis. Australian National Univ., Canberra.

Howard, G. C., Fast, C. R. 1970. *Hydraulic Fracturing, Monogr. Ser. Soc. Petrol. Eng. of AIME* 2: 1–23

Hubbert, M. K., Willis, D. G. 1957. Mechanics of hydraulic fracturing. *AIME Trans.* 210: 153–68

Illies, J. H. 1975. Recent and paleo-intraplate tectonics in stable Europe and the Rhinegraben rift system. *Tectonophysics* 29: 251–64

Illies, J. H., Greiner, G. 1976. Regionales stress-feld und neotektonik in Mitteleuropa. *Oberrheinische Geol. Abh.* 25: 1–40

Jaeger, J. C., Cook, N. G. W. 1969. *Fundamentals of Rock Mechanics.* London: Methuen. 513 pp.

Kehle, R. O. 1964. Determination of tectonic stresses through analysis of hydraulic well fracturing. *J. Geophys. Res.* 69: 259–73

Klein, F. W., Einarsson, P., Wyss, M. 1977. The Reykjanes Peninsula, Iceland, earthquake swarm of September 1972 and its tectonic significance. *J. Geophys. Res.* 82:865–88

Kotze, T. J. 1970. Virgin rock stress measurements in the Evander gold field. *C.O.M. Res. Rep. 30/70*, Chamber of Mines of South Africa, Johannesburg. 33 pp.

Leeman, E. R. 1964. The measurement of stress in rock, II. *J. S. Afr. Inst. Min. Metall.* 65:82–114

Leeman, E. R. 1969. The "doorstopper" and triaxial rock stress measuring instruments developed by the C.S.I.R. *J. S. Afr. Inst. Min. Metall.* 69:305–39

Leeman, E. R., Hayes, D. J. 1966. A technique for determining the complete state of stress in rock using a single borehole. *Proc. 1st Congr. Int. Soc. Rock Mech., Lisbon, 1966* 2:17–24

Lockner, D., Byerlee, J. D. 1977. Hydrofracture in Weber sandstone at high confining pressure and differential stress. *J. Geophys. Res.* 82:2018–26

Love, A. E. H. 1934. *A Treatise on the Mathematical Theory of Elasticity.* Cambridge Univ. Press. 643 pp.

McKenzie, D. P. 1969. The relation between fault plane solutions for earthquakes and the directions of the principal stresses. *Bull. Seismol. Soc. Am.* 59:591–601

Miles, P., Herget, G. 1976. Underground stress determinations using the doorstopper method at Timmins, Ontario. *Rep. MRP/MRL76–148 (TR)*, Elliot Lake Lab., Can. Cent. Min. Energy Technol.

Mills, J. M., Fitch, T. J. 1977. Thrust faulting and crust-upper mantle structure in east Australia. *Geophys. J.* 48:351–84

Molnar, P., Wyss, M. 1972. Moments, source dimensions and stress drops of shallow focus earthquakes in the Tonga-Kermadec arc. *Phys. Earth Planet. Inter.* 6:263–78

Newman, D. B., Clark, B. R. 1977. Near-surface in situ stress measurements, Anna, Ohio earthquake zone. *EOS Trans. Am. Geophys. Union.* 58:493

Nichols, T. C. 1975. Deformations associated with relaxation of residual stresses in a sample of Barre granite from Vermont. *U.S. Geol. Surv. Prof. Pap. 875*, 32 pp.

Obert, L. 1962. In situ determination of stress in rock. *Min. Eng.* 14:51–58

Pallister, G. F. 1969. *The measurement of virgin rock stress*, MSc thesis. Univ. Witwatersrand, Johannesburg, South Africa

Price, N. J. 1966. *Fault and Joint Development in Brittle and Semi-brittle Rock.* London: Pergamon. 176 pp.

Price, N. J. 1974. The development of stress systems and fracture patterns in undeformed sediments. In *Advances in Rock Mechanics, Proc. 3rd Congr. Int. Soc. Rock Mech.* IA:487–96

Raleigh, C. B. 1974. Crustal stress and global tectonics. See Price 1974, pp. 593–97

Raleigh, C. B., Healy, J. H., Bredehoeft, J. D. 1972. Faulting and crustal stress at Rangely, Colorado. In *Flow and Fracture of Rocks, Am. Geophys. Union Monogr.* 16:275–84

Raleigh, C. B., Healy, J. H., Bredehoeft, J. D. 1976. An experiment in earthquake control at Rangely, Colorado. *Science* 191:1230–37

Ranalli, G., Chandler, T. E. 1975. The stress field in the upper crust as determined from in situ measurements. *Geol. Rundsch.* 64:653–74

Riecker, R. E. 1977. State of stress in the lithosphere. *EOS Trans. Am. Geophys. Union* 58:597–99

Roberts, A., Hawkes, I., Williams, F. T., Dhir, R. K. 1964. A laboratory study of the photoelastic stressmeter. *Int. J. Rock Mech. Min. Sci.* 1:441–58

Roegiers, J. C., Fairhurst, C. 1973. The deep stress probe—a tool for stress determination. In *New Horizons in Rock Mechanics, Proc. 14th Symp. Rock Mech.*, ed. H. Hardy, R. Stefanko, pp. 755–60. New York: ASCE

Rummel, F., Jung, R. 1975. Hydraulic fracturing stress measurements near the Hohenzollern Graben—structure, S.W. Germany. *Pure Appl. Geophys.* 113:321–30

Salamon, M. D. G., Ryder, J. A., Ortlepp, W. D. 1964. An analogue solution for determining the elastic response of strata surrounding tabular mining excavations. *J. S. Afr. Inst. Min. Metall.* 65:115–37

Sbar, M. L., Marshak, S., Engelder, T., Plumb, R. 1977. Near surface *in situ* stress measurements near the San Andreas fault, Palmdale, California. *EOS Trans. Am. Geophys. Union.* 58:1123

Sbar, M. L., Sykes, L. R. 1973. Contemporary compressive stress and seismicity in eastern North America: an example of intraplate tectonics. *Geol. Soc. Am. Bull.* 84:1861–82

Sbar, M. L., Sykes, L. R. 1977. Seismicity and lithospheric stress in New York and adjacent areas. *J. Geophys. Res.* In press.

Scheidegger, A. E. 1964. The tectonic stress and tectonic motion direction in Europe and western Asia as calculated from earthquake fault-plane solutions. *Bull. Seism. Soc. Am.* 54:1519–28

Seagar, J. S. 1964. Pre-mining lateral

pressures. *Int. J. Rock Mech. Min. Sci.* 1:413–19

Stephenson, B. R., Murray, K. J. 1970. Application of the strain rosette relief method to measure principal stresses throughout a mine. *Int. J. Rock Mech. Min. Sci.* 7:1–22

Stewart, I. C. F., Denham, D. 1974. Simpson Desert earthquake, Central Australia, August, 1972. *Geophys. J.* 39:335–41

Swolfs, H. S. 1975. Determination of *in situ* stress orientation in a deep gas well by strain relief techniques. *Terra Tek Rep. TR 75–43*

Swolfs, H. S., Handin, J., Pratt, H. R. 1974. Field measurements of residual strain in granitic rock masses. See Price 1974, IIA: 563–68

Sykes, L. R., Sbar, M. L. 1973. Intraplate earthquakes, lithospheric stresses and the driving mechanism of plate tectonics. *Nature* 245:298–302

Thatcher, W. 1972. Regional variations of seismic source parameters in the northern Baja California area. *J. Geophys. Res.* 77:1549–65

Thatcher, W., Hanks, T. C. 1973. Source parameters of southern California earthquakes. *J. Geophys. Res.* 78:8547–76

Tucker, B. E., Brune, J. N. 1973. Seismograms, S-wave spectra, and source parameters for aftershocks of San Fernando earthquake. In *San Fernando, California Earthquake of February 9, 1972, Geological and Geophysical Studies*, 3:69–122. U.S. Dept. Commer.

Tucker, B. E., Brune, J. N. 1977. Source mechanism and M_b–M_s analysis of aftershocks of the San Fernando earthquake. *Geophys. J.* 49:371–426

Tullis, T. E. 1977a. Reflections on measurement of residual stress in rock. *Pure Appl. Geophys.* 115:57–68

Tullis, T. E. 1977b. Stress measurements by shallow overcoring on the Palmdale uplift. *EOS Trans. Am. Geophys. Union.* 58:1122

Van Heerden, W. L. 1971. Stress measurements in coal pillars. *Proc. 2nd Congr. Int. Soc. Rock Mech., Belgrade, 1970.* II:4–16

Van Heerden, W. L. 1976. Practical application of the C.S.I.R. triaxial strain cell for rock stress measurements. In *Exploration for Rock Engineering*, ed. Z. T. Bieniawski. 1:189–94

Van Heerden, W. L., Grant, F. 1967. A comparison of two methods for measuring stress in rock. *Int. J. Rock Mech. Min. Sci.* 4:367–82

Voight, B. 1966. Beziehung zwischen grossen horizontalen Spannungen in Gebirge und der Tektonik und der Abtragung. *Proc. 1st Congr. Int. Soc. Rock Mech., Lisbon, 1966* II:51–56

Voight, B., St. Pierre, B. H. P. 1974. Stress history and rock stress. See Price 1974, IIA: 580–82

Von Schonfeldt, H. A., Kehle, R. O., Gray, K. E. 1973. Mapping of stress field in the upper earth's crust of the U.S. *Final Tech. Rep. USGS (14-08-0001-122278)* 40 pp.

Ward, P. L. 1971. New interpretation of the geology of Iceland. *Geol. Soc. Am. Bull.* 82:2991–3012

Wyss, M., Molnar, P. 1972. Source parameters of intermediate and deep focus earthquakes in the Tonga arc. *Phys. Earth Planet. Int.* 6:279–92

Zoback, M. D., Healy, J. H. 1977. In-situ stress measurements near Charleston, South Carolina. *EOS Trans. Am. Geophys. Union* 58:493

Zoback, M. D., Healy, J. H., Roller, J. C. 1977. Preliminary stress measurements in Central California using the hydraulic fracturing technique. *Pure Appl. Geophys.* 115:135–52

Zoback, M. D., Rummel, F., Jung, R., Raleigh, C. B. 1977. Laboratory hydraulic fracturing experiments in intact and pre-fractured rock. *Int. J. Rock Mech. Min. Sci. Geomech. Abstr.* 14:49–58

Zoback, M. D, Pollard, D. D. 1978. Hydraulic fracture propagation and the interpretation of pressure-time records for in-situ stress determinations. Submitted to *19th Symp. Rock Mech., Lake Tahoe, Calif., May 1978*

Ann. Rev. Earth Planet. Sci. 1978. 6 : 437–56

VOLCANIC EVOLUTION OF   ✖10101
THE CASCADE RANGE

Alexander R. McBirney

Center for Volcanology, University of Oregon, Eugene, Oregon 97403

INTRODUCTION

Much of the recent interest in andesites has been stimulated by the hope that these rocks, which are so characteristic of regions of plate convergence, may provide a much needed test of the basic concept of subduction. If oceanic lithosphere is being returned to the mantle and continents at rates that balance its formation by seafloor spreading and sedimentation, and if, as many believe, the igneous processes associated with subduction provide the mechanism by which crustal components are returned to the continent, it is obviously essential to evaluate the dynamic aspects of orogenic volcanism in the light of the earth's total geochemical balance.

The Cascade Range of the northwestern United States is an almost ideal place to do this. Although the system lacks some of the features considered typical of convergent plate boundaries, it has a well-preserved record of older rocks and provides a rare perspective of the development of such a system through time. In addition, there are marked variations in the crustal structure along the length of the system that make it possible to examine the effects of differing crustal features on the nature and composition of igneous activity, both in space and time. Numerous detailed geologic and geochemical studies have provided much new information on this system and have opened fresh insights into the magmatic evolution of the continental margin during much of Cenozoic time.

EVOLUTION OF THE CASCADE SYSTEM

The modern volcanic chain that extends from British Columbia to northern California is only the most recent of several igneous belts or zones that have followed the Pacific margin of North America since late Paleozoic time. There is no visible record of Precambrian igneous activity and little evidence for volcanism prior to the last part of the Paleozoic era, but it is clear that there was an important volcanic episode that began during the Permian period and continued well into Early Triassic time. Submarine lavas of this age are exposed in northern California, northeastern Oregon, and northern Washington, and Gilluly (1963) may well be

437

0084-6597/78/0515-0437$01.00

correct in stating that this outpouring of basalt exceeded that of any other period before or after. The nature of the activity was largely oceanic, however, and it is uncertain whether it has much in common with modern orogenic volcanism.

The earliest igneous activity that was clearly orogenic in nature dates from the last half of the Mesozoic era. It was long thought that the plutonic rocks that were emplaced in such large volumes during this episode had little associated volcanism, but detailed studies of contemporaneous sedimentary deposits, notably by Dickinson and his students (Dickinson 1962, 1970) have shown quite clearly that the batholithic rocks that are so conspicuous today are in fact the roots of a deeply eroded volcanic belt that may not have been very different from the modern High Cascades.

Segments of uplifted plutonic and weakly metamorphosed sedimentary and volcanic units are exposed from British Columbia diagonally across the state of Washington into northeastern Oregon and in isolated windows that extend across central Oregon to connect with the major axis of the Sierra Nevada system in northwestern California. This large sigmoidal belt and the embayment it forms near what is now the central part of the High Cascade Range is one of the major structural features of the crust in the Pacific Northwest. Hamilton (1969) has proposed that during the period of strongest activity the western Cordilleran axis may have resembled the modern Andes, especially if one reverses post-Cretaceous deformation that is postulated to have accentuated the kink and segmented it into discontinuous blocks. Unfortunately, little is known about the geochemical or geologic relations of the volcanic rocks, because most of them have been removed by erosion or are buried beneath a Cenozoic cover.

Eocene Episodes

The Cenozoic igneous history of western North America has recently been reviewed and summarized by Armstrong (1978), who shows that it followed a complex pattern of episodic activity as it migrated across broad regions. Intense and widespread volcanism began with what Armstrong refers to as the Challis episode. Starting in early Eocene time, it reached its peak between 54 and 44 million years ago before declining and finally coming to an end near the beginning of the Oligocene epoch.

The Eocene record of the Pacific Northwest can be conveniently divided into Early, Middle, and Late Eocene stages. Although Early Eocene igneous rocks are primarily basaltic and have compositions similar to those of Hawaiian rocks (Snavely et al 1968), they do not seem to have been laid down in a deep oceanic environment. Lavas and shallow intrusions occur within a thick eugeosynclinal series that accumulated in a shallow subsiding basin, the eastern margin of which was near the present Cascade Range. If there was a trench at this time it must have been far to the west, and although Snavely & Wagner (1963) inferred that andesitic volcanoes were active east of the southern part of the geosyncline, andesitic material has yet to be found in Lower Eocene horizons. Swamp deposits, including coal and shallow estuarine beds, are common, but there are no coarse sediments to indicate a nearby region of high relief. The nature of weathering and vegetation in lacustrine sediments of central Oregon show that the climate of that region was more humid

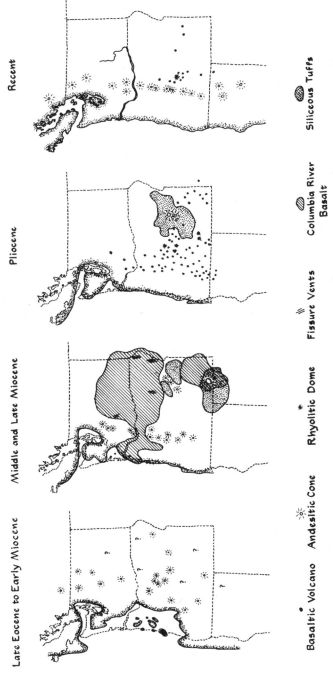

Figure 1   Generalized reconstruction of the distribution of volcanic centers during four Cenozoic episodes. Compiled from various sources including Armstrong (1978), Snavely & Wagner (1963), and Walker (1970).

than it has been since the elevation of the Cascade Range brought about more arid conditions in the rain shadow of its leeward side.

A marked unconformity between Lower and Middle Eocene rocks reflects a strong tectonic event that caused extensive folding and possibly thrusting in south-western Oregon (Baldwin 1965) and a change of sedimentation in the northern part of the province (Rau 1966). The Middle Eocene episode was marked by uplift and erosions of the Klamath Mountain region, together with volcanism, mainly of basaltic character, in localized centers along an axis extending from the partly emergent Coast Range of Oregon along the shallow basin of the Willamette–Puget Sound Depression. At least two volcanic complexes north and south of the present Columbia River erupted tholeiitic rocks that reached moderately high levels of differentiation (Snavely et al 1965, 1968).

There is sparse evidence that differentiated calc-alkaline rocks began to appear about this time in a belt that was about 100 miles wide and extended nearly 1000 miles from British Columbia through Washington and into central Oregon. Certainly by Late Eocene time there must have been a swarm of small volcanoes along this axis. The eruptive centers are marked today by andesitic lavas and subvolcanic intrusions of the Clarno formation in central Oregon and by stocks and batholiths where uplift and erosion have exposed deeper levels in Washington and British Columbia.

Much of the Oregon Coast Range was emergent by Late Eocene time, and the Willamette–Puget Sound Depression had become localized near its present axis. Although there were still scattered central-vent volcanoes in the area of the rising Coast Range, their importance had greatly diminished.

Oligocene to Early Miocene Volcanism

The calc-alkaline volcanism that began in Late Eocene time in Washington and central Oregon increased in intensity through the Oligocene epoch and spread westward and southward until it covered a broad zone across most of western Oregon and Washington. It is still uncertain when activity began along the Cascade axis. The oldest andesitic rocks from this region that have been dated by radiometric methods are less than 40 million years old (Table 1). Andesitic lavas of the Colestin, Fisher, and equivalent formations in southern Oregon were once thought to have come from Late Eocene volcanoes, but dating has shown that they are somewhat younger. Instead, there is a very marked angular unconformity separating the Eocene formations from Oligocene calc-alkaline rocks, and the deformation that occurred at that time seems to have coincided with the onset of orogenic igneous activity in the Cascade region.

Certainly by Oligocene time there were many eruptions of andesite and more siliceous rocks from centers in the area of the Western Cascades. Andesitic flows and siliceous tuffs interfinger with Oligocene sediments along the east side and southern end of the Willamette Valley and are abundant in the Oligocene sections of the Clarno and John Day formations east of the Cascades (Hay 1963, Fisher & Rensberger 1972). Within the Western Cascade Range, rocks of this age range include the thick Mehama and Little Butte formations and numerous intrusive bodies that probably represent the eroded remnants of volcanic centers.

Table 1 Potassium-argon age determinations of basal-members of the Tertiary volcanic series of the Western Cascades, determined by J. F. Sutter (work in progress) and J. Dymond.

Sample No.	Rock type	Location	Apparent age (m.y.)
DMS-43	Breitenbush Tuff, (plagioclase)	44°46′30″ N, 121°59′35″ W	19.74 ± 0.24
CP-173	Gabbro (whole rock) Little Butte Formation	43°55′30″ N, 122°48′05″ W	21.34 ± 0.55
BX-99	Andesite (whole rock) Little Butte Formation	44°46′0″ N, 122°1′33″ W	24.22 ± 0.27
DMS-133	Basalt, Colestin Formation (whole rock)	42°59′00″ N, 122°51′50″ W	24.83 ± 0.50
DMS-144	Andesite, Little Butte Formation (whole rock)	42°28′10″ N, 122°48′05″ W	28.75 ± 0.34
MS-270	Andesite, Colestin Formation (whole rock)	42°02′25″ N, 122°36′15″ W	29.11 ± 0.34
M-343	Andesite, Fisher Formation (whole rock)	43°38′ N, 123°05′ W	34.3 ± 0.46

The distribution of these centers has not been well defined, but it appears that by the end of Oligocene time small volcanoes were scattered over a broad zone along much of the Pacific continental margin. In Washington, the locus of activity has been identified in shallow subvolcanic intrusions and associated pyroclastic units near the present Cascade Range, but the full width of the zone cannot be determined there owing to the extensive cover of younger rocks east of the Cascades. In Oregon, Oligocene and early Miocene centers extend from the Coast Range well into central Oregon. In the Coast Range they are marked by subvolcanic intrusions of gabbro and nepheline syenite, while in the Western Cascades and central Oregon, flows, tuffs, and volcanic sediments are associated with vent complexes and shallow stocks, dikes, and sills.

The spatial distribution of the rocks is unlike that of modern volcanic belts in that the most alkaline compositions are closest to the ocean. Most of the Oligocene rocks of the Coast Range are alkali dolerites, gabbros, and nepheline synites, whereas the main volcanic series of the Western Cascades and central Oregon are strongly subalkaline, and although a few alkaline rocks have been reported from the John Day formation in central Oregon, they are very subordinate in volume.

As volcanism spread during the Oligocene epoch it tended to become less basaltic and more differentiated with time. The andesites and basalts that dominate the lower parts of the sections in the Western Cascades and make up most of the Clarno formation in the east give way upward to siliceous pyroclastic rocks that by the end of the episode reached enormous volumes.

Despite the large volumes of eruptive material, there seem to have been few large volcanoes during this period but rather a scattering of small basaltic and andesitic cones, and a few low-rimmed calderas or broad volcano-tectonic depressions (Walker 1970). Some of the most voluminous eruptions seem to have come from fissures or from unroofing of shallow intrusions that stoped their way toward the surface. The Coast Range at this time must have formed a peninsula or low-island chain, and behind it the Willamette-Puget Sound Depression formed a shallow arm of the sea. There seems to have been no pronounced topographic barrier near the present Cascade axis. Instead, great volumes of tuffaceous debris were deposited in a system of estuaries and shallow basins to form the Eugene formation and equivalent units of the Willamette Valley and the lacustrine beds of the John Day formation. Coarse detritus and deep erosional channels are notably rare, and the relief on the volcanic landscape could not have been great.

Mid-Miocene (Columbian) Volcanism

An episode of faulting, uplift, and erosion occurred during the later part of Early Miocene time throughout much of central Oregon. There is a marked erosional unconformity at the top of the John Day formation, and in places erosion cut down well into the underlying Clarno rocks before the Mid-Miocene basalts of the Columbia River Group were laid down. The intensity of this disturbance seems to have diminished westward, because it is not conspicuous in rocks immediately east of the High Cascades (Peck 1964) or in the Western Cascades (Peck et al 1964). A general decline in the intensity of volcanism about this same time is reflected in a relative scarcity of igneous rocks with ages of between 20 and 16 million years. Following this interval, however, volcanism increased greatly. In fact, the Mid-Miocene, or Columbian episode, as it is sometimes called, was by far the most important igneous event to occur in the Pacific Northwest during the Cenozoic era.

Rocks of this age in the Cascades have been assigned to the Sardine formation of Thayer (1937) and constitute what is probably the thickest and most voluminous assemblage of andesites in the region. If one adds to the Sardine formation the flood lavas of the Columbia River and Steens Mountain Groups and the calc-alkaline rocks of the Strawberry Mountains, all of which were erupted in central and eastern Oregon and Washington during the same time interval, the total volume of Mid-Miocene rock becomes enormous. This large volume is even more remarkable when one considers the brief time span in which it was laid down.

Volcanic centers of Mid-Miocene age have been relatively well delineated along two belts, one trending slightly east of north along the Western Cascade axis and possibly connecting with a similar belt that curves toward the southeast through northern California and southwestern Nevada (Noble 1972), and a second shorter belt that is marked by three Middle to Late Miocene volcanic centers trending northeast through the Strawberry Mountain complex and other igneous centers in east-central Oregon (Robyn 1977). Of these two chains, that of the Western Cascades was the most extensive and produced the largest volume of calc-alkaline rocks. Its axis is marked by stocks, mainly of quartz diorite, and by broad aureoles of hydrothermal alteration and mineralization. The forms of the cones can be inferred from the lithologic zones that reflect the conditions under which lavas and pyroclastic rocks accumulated on the slopes and lower flanks of large composite volcanoes. These centers seem to have been the first to have the form and alignment that is conventionally associated with andesitic belts. They probably resembled volcanoes of the modern High Cascades, except that the presence of shallow-water sediments between them indicates that the belt was a chain of volcanic islands or a broad shelf area, at least during its early stages of development.

There is surprisingly little correlation of Mid-Miocene units on opposite sides of the High Cascades. Despite their great thickness in the Western Cascades, andesitic rocks are scarce, if not altogether missing along the eastern base of the modern range. A few thin flows of basalt within the Sardine formation have been correlated with the Columbia River Group to the east (Peck et al 1964, White & McBirney 1978) and indicate that tongues of flood lavas flowed between the andesitic cones, but it is difficult to visualize the topographic configuration that could account for the limited interfingering of two adjacent units of such great thicknesses. The problem is made more difficult by the shallow level of erosion and the extensive cover of younger rocks that limit exposures on the east side to small areas and shallow depths.

Most of the products of Mid-Miocene volcanism in the Western Cascades were andesitic. Siliceous pyroclastic rocks are much less important than they were in the preceding episode, and basalts were still quite subordinate (Table 2). The relationship, if any, between this andesitic volcanism and the great outpouring of flood lavas of

Table 2 Weighted average compositions of cascade volcanic rocks

	Northern California		Central Oregon			
					Mid-Late Miocene	Oligocene-Miocene
	Quaternary		Pliocene			
SiO_2	52.2		52.7	52.7	57.3	62.4
TiO_2	1.2		1.4	1.3	1.1	0.9
Al_2O_3	18.0		17.4	17.3	16.6	15.7
ΣFeO	9.6		9.1	8.6	7.4	6.1
MnO	0.1		0.1	0.1	0.1	0.1
MgO	5.5		5.3	5.6	3.7	2.3
CaO	9.1		8.4	8.7	6.7	5.3
Na_2O	3.2		3.8	3.5	3.5	3.4
K_2O	0.9		0.9	0.9	1.3	1.7
P_2O_5	0.2		0.3	0.2	0.3	0.2
Total	100.0		99.4	98.9	98.0	98.1
	No. anal.	Vol. %	No. anal. / Vol. %	No. anal. / Vol. %	No. anal. / Vol. %	No. anal. / Vol. %
Basalt	7	69	33 / 85	17 / 90	20 / 39	6 / 10
Andesite	22	29	112 / 13	76 / 9	99 / 41	25 / 45
Dacite-Rhyolite	35	2	31 / 2	6 / 1	17 / 20	10 / 45
Total vol., km^3	3095		4600	2150	24,850	> 10,000

the Columbia River Group has never been explained. The same brief episode during which these large volumes of volcanic rocks were erupted was also marked by strong activity in Central America, the southwestern Pacific, and other parts of the Circum-Pacific system (McBirney et al 1974, Kennett et al 1977).

Late Miocene (Andean) Volcanism

The Columbian episode declined sharply about 13 to 14 million years ago and was followed by moderately strong deformation throughout much of the Pacific Northwest. Strong faulting and tilting occurred east of the Cascade Range, where an important angular unconformity separates Pliocene rocks from the underlying older units. In the Cascade region, broad folds developed along axes that closely parallel the trend of the earlier Mid-Miocene volcanoes.

A brief Late Miocene pulse of activity, dated around 9 to 10 million years ago, has recently been recognized in the Western Cascades (McBirney et al 1974) and appears to have been synchronous with strong volcanism elsewhere in the Circum-Pacific region, especially the Andes. The rocks have only been separated from the older Sardine formation in a few areas where detailed studies have been carried out. They appear to have been erupted from small cones around the lower flanks of the large eroded remnants of Mid-Miocene volcanoes and from scattered vents east of the Cascades. Known centers in the Cascades are too few to permit an interpretation of the distribution of vents or their relationship to regional conditions. In central Oregon, however, there appears to be a systematic relation between the location of rhyolitic eruptions that began about this time and migrated with time toward the Cascades (G. W. Walker, N. S. MacLeod, personal communication). In addition, rhyolitic and dacitic ignimbrites and scattered basaltic lavas were erupted throughout much of central and eastern Oregon (Walker 1970). Taken as a whole, the episode was characterized by andesitic and basaltic lavas in the Cascades and dominantly silicic ignimbrites and domes toward the east.

Pliocene (Fijian) Volcanism

Kennett and his co-workers (1977) have applied the name Fijian to the important volcanic episode that occurred between about 3 and 6 million years ago. The name was taken from the islands in the southwestern Pacific where the episode was first established by systematic dating, but volcanism was widespread at this time throughout much of the Circum-Pacific region. Rocks of this episode are common throughout the Cascade Range, central Oregon, northern California, and the Basin and Range Province, but it is often difficult to distinguish them from Pleistocene and Holocene units, because they are only moderately affected by weathering and erosion.

Activity in Oregon during this period produced mainly basaltic lavas and rhyolitic domes and ignimbrites. Small monogenetic cones and thin but extensive flows of basalt broke out over a broad region extending from the western side of the Cascades across central Oregon almost to the Idaho border. Rhyolitic ignimbrites were also discharged from centers east of the Cascades, and the chain of rhyolitic domes that began to develop in Late Miocene time continued its westward migrations toward the Cascade Range. Another region of basaltic and rhyolitic activity developed in southern Idaho and migrated eastward along the Snake River Plain toward Yellowstone.

Andesites seem to have been subordinate to basalt and rhyolite, and there were no conspicuously large composite cones near the axis of the Cascades. By this time most of the large Miocene volcanoes had probably been leveled by erosion. Coarse andesitic debris is widespread in alluvial and lacustrine deposits along the eastern side of the Cascades where it was deposited in pediments and a number of subsiding basins.

Quaternary (Cascadian) Volcanism

The recent episode of volcanism that has been responsible for the familiar volcanoes of the modern High Cascades has been studied in greater detail than any other Cenozoic period, but even today much remains to be learned about the volcanoes and their relations to the system as a whole. This last period of activity, from which the Cascadian episode takes its name, followed closely and, in places, merged with the preceding Pliocene episode. Its main feature was a marked narrowing of the focus of volcanism to form a well-defined chain of large composite cones extending from British Columbia to northern California.

The earliest Pleistocene activity resembled that of the preceding Pliocene episode in that it was characterized by basaltic cones, flows, and low overlapping shields. With time, activity became more localized in persistent centers from which progressively more differentiated magmas were discharged. Most of the large andesitic cones that form the crest of the High Cascades began to rise during Pleistocene time about one million years ago and reached their present elevations by rapid growth during a brief period of intense activity. The fact that few of the lavas have reversed magnetic polarities, even in the lowest levels of deeply glaciated cones, indicates that by far the greatest volumes must have been discharged since the present period of normal magnetic polarity began about 670,000 years ago.

Block-faulting occurred concurrently with volcanism in the central Cascade Range and resulted in uplift and westward tilting of the Western Cascades. At the same time, the basement below the active volcanoes of the High Cascades began to

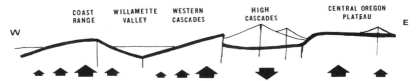

Figure 2 Generalized structural section across the Coast Range, Willamette Valley, and Cascade Range of central Oregon. The differing vertical displacements have been attributed to hydration and melting of the mantle above a descending slab of oceanic lithosphere (Fyfe & McBirney 1975), but similar structures are characteristic of most of the continental margins and island arcs of the Circum-Pacific system, including regions such as California, where active subduction has not been postulated during Quaternary time.

subside to form a shallow graben, much of which has been filled by the products of Quaternary volcanoes (Figure 2). Depression of the Cascade graben has been most pronounced in the central Cascades where volcanism has been strongest; it dies out toward the north and south where individual volcanoes are large but widely spaced, and the total volume of Quaternary volcanic rocks is small.

The topographically imposing volcanoes of the High Cascades give the impression that andesite is the dominant rock type in the modern range. In places this is probably true, but if one considers the total volume of rocks produced in the system as a whole, andesitic cones are seen to account for a very subordinate amount of the erupted volumes. The proportion of andesite is high only in those parts of the chain where the total volume of Quaternary rocks is small, namely in Washington and northern California. In one part of the central Oregon Cascades where absolute volumes have been estimated (McBirney et al 1974), it has been found that basaltic lavas beneath and between large andesitic cones account for about 85% of the total volume of Quaternary rocks. Glaciated shield lavas were found to total about

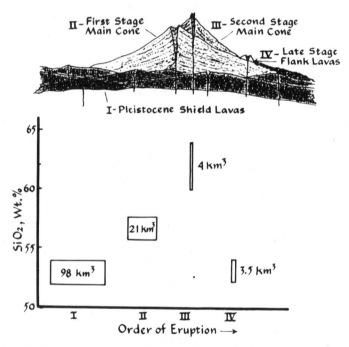

Figure 3a The Quaternary evolution of Mount Jefferson includes four main stages, each of which was characterized by distinctive rocks. Volumes of rocks in each stage (measured by Sutton 1974) are shown by the relative areas of rectangles in the lower diagram. The midpoint on the vertical dimension of the rectangles is placed at the mean value of silica for the rocks of that stage, and the vertical length of the edge indicates one standard deviation from the mean silica value. Total ranges of measured silica values for the four stages are: I, 51.6–54.3, II, 54.5–58.2, III, 60.1–64.3, IV, 51.5–54.3 (After White & McBirney 1978).

1282 km³, whereas large composite cones in the same area account for about 189 km³, and the very recent cinder cones and lava flows amount to about 55 km³. Unfortunately, most geological and petrological studies have been concentrated on the high cones, and the great volume of underlying rocks and smaller volcanoes have been largely ignored.

In recent years, several High Cascade volcanoes have been examined in considerable detail, but much of this work is still incomplete, and there are few places where the entire volcanic and petrologic development of a large cone can be traced. The most complete data are probably those for Mount Jefferson (Thayer 1937, Walker et al 1966, Greene 1968, Condie & Swenson 1973, Sutton 1974, White & McBirney 1978). Four stages of activity have been recognized (Figure 3a). The earliest Pleistocene eruptions formed a broad base of basaltic shield lavas on which the main cone was then built in two separate stages of andesitic activity. Finally, in very recent time, small flows of basalt were discharged from satellite vents on the lower flanks of the main cone. There was a general increase in the silica content through the three main stages of growth and a steady decline in the volumes of erupted rocks, but the flank eruptions of the last stage reverted to a more basic composition similar to that of the Pleistocene shield lavas. The magma of each of the four stages had its own distinctive geochemical character and does not seem to have had a direct genetic relation to the others (Figure 3b).

Similar patterns of development have been followed by most of the other large Quaternary cones, but in many places late-stage eruptions have produced not only basaltic rocks but rhyolitic or dacitic domes and pumice as well. This pattern in which compositions evolve through andesite and then diverge into more or less contemporaneous basalt and rhyolite in the latest stages of activity is most pronounced in the southern and northern parts of the chain.

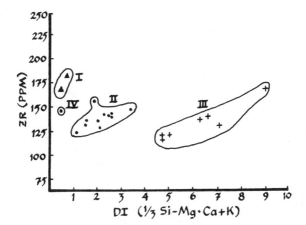

Figure 3b Concentration in ppm of Zr in Quaternary volcanic rocks of Mount Jefferson plotted against a modified Larsen index ($\frac{1}{3}$ Si $-$ Mg $-$ Ca $+$ K). Stages of activity are the same as in Figure 3a. Zr values were determined by Sutton (1974); figure from White & McBirney (1978).

Summary of Geologic Development of the Cascade System

The foregoing descriptions have been very brief and are far from complete, but it is apparent even from this short summary that the Cascade system has evolved to its present form through a varied succession of igneous and tectonic events and that the setting of volcanism today is by no means characteristic of that in earlier periods. Calc-alkaline volcanism first appeared around the close of the Eocene epoch east of the present Cascade axis and gradually spread across a broad region during Oligocene and Early Miocene time. A well-defined line of composite volcanoes did not develop until the Mid-Miocene Columbian event, and following that episode there was no new chain of large andesitic volcanoes until the modern High Cascades began to rise about a million years ago.

Tectonic disturbances occurred at several times in different regions. The strongest deformation seems to have occurred in the southern Coast Ranges around the end of Middle Eocene time, but conspicuous faulting and uplift also took place east of the Cascades shortly before the Mid-Miocene volcanic episode and again shortly before the Late Miocene and Pliocene episodes. Basin and Range faulting extended into central Oregon toward the end of Pliocene time and has continued down to the recent past.

TRENDS OF MAGMATIC EVOLUTION

Some of the broad geologic aspects of the igneous rocks produced during the various Cenozoic events have already been mentioned in discussing individual episodes. Certain general trends can be seen when these features are viewed in the perspective of the Cenozoic sequence as a whole (White & McBirney 1978).

Although the volumes of volcanic and intrusive rocks produced during the early episodes are difficult to estimate, they were certainly much greater than those of more recent times. There has been a somewhat irregular decline of volcanism with each successive episode since the mid-Tertiary pulses, which were by far the most intense and widespread to occur anywhere near the present Cascade axis. The time span in which most of the Oligocene and Early Miocene rocks were erupted seems to have been of the order of 10 million years, but the Mid-Miocene episode was much shorter, possibly only two or three million years. Hiatuses in which there was little or no volcanism have separated each of the subsequent volcanic episodes down to the present.

The proportion of basaltic rocks has increased steadily with time (Table 2). The Oligocene-Early Miocene episode produced the largest volume of siliceous rocks, mainly rhyolite and dacite; andesite was the dominant rock type produced in the Cascades during the Mid-Miocene (Columbian) episode, and since that time, basalt has outweighed all other rock types combined. In the modern High Cascade chain, andesite is important only in the southern and northern parts of the chain where the intensity of Cenozoic volcanism has been relatively mild. Elsewhere, there are large composite cones composed largely of andesite, but they constitute a relatively small part of the total erupted volume.

Knowing the relative proportions of the different rock types in each age group and the average compositions of the individual members of each series, it is a simple matter to calculate the average compositions of volcanic rocks produced during successive igneous episodes. This has been done for the central Oregon Cascades by White & McBirney (1978), who obtained the results shown in Table 2. For comparison purposes, data are also shown for Quaternary rocks of northern California.

The most notable feature of these averages is the decline of silica and potash with time. The sodium content is remarkably uniform, mainly because the concentration of that element in basalts has increased by an amount that balances the increased proportions of mafic rocks. Similarly, the average of Quaternary rocks in the central Oregon Cascades does not differ markedly from that of northern California, even though dacites and rhyolites are much more abundant in the southern part of the chain. The reason for this apparent inconsistency lies in the fact that the basalts of northern California tend to be more basic than those of central Oregon.

In addition to the change in the proportions of rock types with time, there have been systematic changes in the nature of the rock suites (Figure 4). If basalts of each of the major eruptive episodes are compared, they are seen to become progressively more sodic with time. The younger rocks have lower iron-magnesium ratios and decreased abundances of certain trace elements, notably rubidium. There is a strong correlation between the rate of decline of Rb contents of each successive suite and the amount of volcanism that has occurred in a given part of the Cascade chain (Figure 5). In those regions where there has been a large amount of Tertiary igneous activity, the modern basalts are more depleted in this element than are those

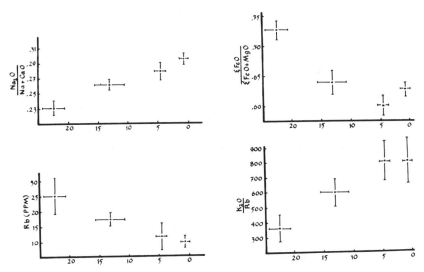

Figure 4 Variations of Fe, Na, Rb, and K/Rb with time in basaltic rocks of the central Oregon Cascades. Error bars show one standard deviation in chemical data and range of radiometric ages. (After White & McBirney 1978).

450   McBIRNEY

erupted in regions where the magnitude of earlier volcanism was less. At the same time, however, the isotopic composition of strontium has remained essentially constant at a low level of about 0.7028 through the entire Cenozoic section of the central Oregon Cascades, regardless of the age or composition of the rocks.

Relations such as these indicate that the source of Cascade magmas has been somewhere in the mantle, possibly between the postulated subduction zone and the overlying continental lithosphere, but a multistage process is required to explain all of the temporal and composition variations in the system as a whole. The decline of iron content and the increase of sodium indicate that the depth at which the basalts last equilibrated with the mantle has increased with time. Experimental studies of the compositions of melts in equilibrium with crystalline phases at various pressures (Kushiro 1973, Mysen 1973, Osborn & Watson 1977) have shown that mantle liquids probably become richer in sodium and poorer in iron with increasing depth and pressure.

One hypothesis that is consistent with these relations is illustrated schematically in Figure 6. A partially molten mass that is mobilized in the low-velocity zone of the mantle could rise to the base of the lithosphere where it spreads laterally and under-plates the thickening layer of relatively brittle rocks. Basaltic liquids segregated at this level may rise toward the surface and differentiate to differing degrees at shallow crustal levels, possibly within the volcanic substructure.

During this process there must be progressive depletion of the source region in the mantle, re-equilibration of each large batch of magma at a greater depth than the preceding one, and subsequent differentiation at high levels with little contribution from older crustal material. Progressive melting of a single source in the mantle is consistent with the decline of volumes, iron-magnesium ratios, silica, potassium, and rubidium with time, but these same variations are also consistent with successive batches of magma passing through the same rocks overlying the source

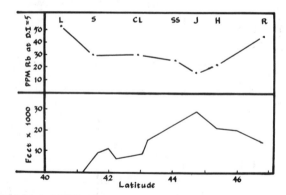

Figure 5 Relations of rubidium contents of Quaternary volcanic rocks of the High Cascades to measured stratigraphic thickness of Tertiary volcanic sections in the same areas. Rb values are taken at a uniform differentiation index (DI) of 5. Volcanoes are shown in order from south to north starting with Lassen Peak then proceeding to Shasta, Crater Lake, South Sister, Jefferson Hood and ending at Rainier. (After White & McBirney 1978).

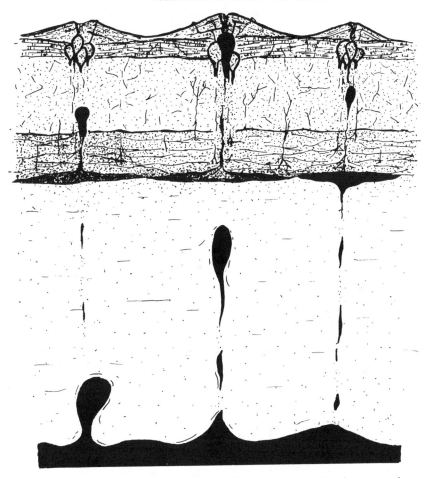

Figure 6 A schematic representation of a multistage process for the rise, segregation, and shallow differentiation of magma beneath the Cascades. The process is consistent with presently available data on rocks of the central Oregon Cascades.

and depleting them in the components that would be fractionated into the liquid. Although there is no sure way of making a distinction between these two possibilities, the consistently low strontium isotopic ratios rule out major contamination by old continental crust, at least in the central Oregon Cascades.

RELATIONS TO SUBDUCTION AND CRUSTAL CONTAMINATION

Remarkably little evidence has been found to relate the Cascade magmas to oceanic lithosphere or sediments. Nearly all workers who have so far examined the

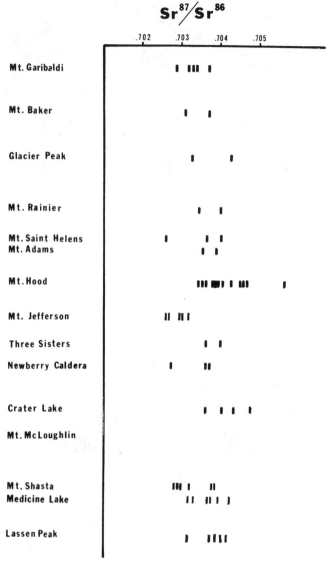

Figure 7 Summary of strontium isotopic ratios for rocks of the Cascade Range as a function of their relative positions starting with Mt. Garibaldi in the north and extending to Lassen Peak in the south. Note that the range of values for an individual volcano is in some cases greater than the differences of average values for the chain as a whole. There are no apparent differences between lavas of volcanoes standing on contrasting types and thicknesses of crust. [Compiled from data by R. L. Armstrong (personal communication), Church & Tilton 1973, Peterman et al 1970, Wise 1969, and White & McBirney 1978.]

petrologic relations, trace elements, or isotopic compositions of the rocks have concluded that they reflect a primary mantle origin and little if any contribution from subducted crustal rocks (e.g. Smith & Carmichael 1968, Peterman et al 1970, Church & Tilton 1973, Condie & Swenson 1973, Church 1976, White & McBirney 1978). It has been suggested, however, that the nature of the crust through which the magmas have risen may have influenced the volumes and compositions of the erupted rocks. Assimilation of sialic crust would be likely to vary widely in different parts of the chain; volcanoes at the southern and northern ends of the chain stand on thick continental crust (Dehlinger et al 1965) and have relatively small total volumes but large proportions of andesite. Moreover, andesites from these volcanoes tend to be more potassic than those of the central Cascades. The average K_2O content (normalized to 60 % SiO_2) for representative volcanoes in the northern, central, and southern regions are:

Rainier (10 analyses)	1.66% K_2O
Hood and Jefferson (41 analyses)	1.43% K_2O
Lassen and Shasta (10 analyses)	1.59% K_2O

Because the volumes also differ, the K_2O contents tend to vary directly with the proportion of andesite in each part of the chain.

There is an even greater variation in rubidium contents, which, as already noted, appear to be inversely correlated with the amount of Tertiary volcanism in the same region. It is more likely that the differences result from progressive depletion of source rocks than from greater or lesser amounts of contamination with continental crust, because there is no systematic variation in the strontium isotopic ratios from different volcanoes along the length of the chain (Figure 7). In fact, the strontium in Cascade lavas is far less radiogenic (average = 0.7036, SD = 0.0005) than that of the Columbia River basalts, which have ratios of about 0.705 to 0.706.

No correlation has been found between the rates of production of igneous rocks in the Cascades and subduction along the adjacent plate boundary. The wide variation in the tempo of Cenozoic volcanism, not only in the Cascades but in the Circum-Pacific as a whole, is in marked contrast to the nearly constant rates of seafloor spreading deduced from the spacing of magnetic anomalies on the seafloor (Kennett et al 1977). The same appears to be true of the intensity of recent volcanic activity, which varies widely from place to place with no detectable relationship to calculated subduction rates or the amount of crustal material that could be consumed in trenches (McBirney 1971). The lack of such relations seems to argue against generation of calc-alkaline magmas by flux-melting when water is released by dehydration of subducted hydrous phases and rises into the overlying mantle. Fyfe & McBirney (1975) have pointed out that there must be a direct relation between the amount of water introduced into the mantle and that of phlogopite reaching depths of 100 to 150 km in the down-going slab. Because the stability of phlogopite is directly dependent on the amount of potassium in the rocks, there should be a good correlation between the amount of magma generated and the rate at which the system recycles potassium (as well as other components that have high solubilities in hydrous fluids at elevated temperatures and pressures).

One can compare the amount of potassium entering the oceans from the continents with that being returned to the continents in calc-alkaline magmas and demonstrate a crude balance, at least in orders of magnitude over a period of 20 million years (McBirney 1976), but any such calculation is no better than the assumptions on which it rests, and in this case it is little more than an attempt to fit inadequate data to an unproven hypothesis.

CONCLUSIONS

It is still too early to offer a comprehensive synthesis of the complex igneous history of the Cascade region, and it would be even more premature to propose that the causes of orogenic volcanism can be discerned in the dim record of events as they are now seen. At best, one can only note a few salient features that have emerged from recent studies.

Igneous activity has been strongly episodic with distinct pulses occurring between periods in which there was little volcanism. Some of the episodes seem to have occurred in unison in widely separated parts of the Circum-Pacific region. The volumes and compositions of rocks produced during successive periods have varied widely, but there appears to have been an overall decline in the magnitude of volcanism and a gradual change toward more basic compositions. Most petrologic evidence points toward a mantle source that is being depleted with time.

The relations between volcanism and tectonic events are still obscure, and probably constitute the most serious gap in present knowledge. There have been few attempts to relate the changing mode and focus of volcanism to patterns of deformation and crustal structure.

Perhaps the most important conclusion that can be drawn from the evidence now available is that there is little direct relation, other than a spatial one, between volcanism in the Pacific Northwest and the subduction that is commonly thought to be associated with it. Even the spatial relation is somewhat ambiguous, because, as the Tertiary record shows, andesitic volcanism has seldom been concentrated in long linear belts near the continental margin as it is today. Even during Quaternary time, andesitic volcanoes are by no means confined to island arcs and continental margins; many are found well within the continental interior.

The record of Cenozoic events in the Cascades offers no obvious solution to this dilemma, but it may provide a better perspective of the nature of volcanism near convergent plate boundaries, and if further studies are undertaken with the knowledge that current concepts of orogenic volcanism are oversimplified and strongly biased toward modern conditions, there may be a better chance of finding rational relationships in what now seems to be a confusing record of tectonic and igneous events.

Literature Cited

Armstrong, R. L. 1978. Cenozoic igneous history of the U.S. Cordillera from 42° to 49° N latitude. *Geol. Soc. Amer. Mem.* In press.

Baldwin, E. M. 1965. Geology of the south end of the Oregon Coast Range Tertiary basin. *Northwest Sci.* 39:93–103

Church, S. E. 1976. The Cascade Mountains revisited: a re-evaluation in light of new lead isotopic data. *Earth Planet. Sci. Lett.* 29:175–8

Church, S. E., Tilton, G. R. 1973. Lead and strontium isotopic studies in the Cascade Mountains: bearing on andesite genesis. *Geol. Soc. Amer. Bull.* 84:431–54

Condie, K. C., Swenson, D. H. 1973. Compositional variation in three Cascade stratovolcanoes: Jefferson, Rainier, and Shasta. *Bull. Volcanol.* 37:205–30

Dehlinger, P., Chiburis, E. F., Collver, M. M. 1965. Local travel-time curves and their geologic implications for the Pacific Northwest states. *Seism. Soc. Amer. Bull.* 55:587–608

Dickinson, W. R. 1962. Petrogenetic significance of geosynclinal andesitic volcanism along the Pacific margin of North America. *Bull. Geol. Soc. Amer.* 73:1241–56

Dickinson, W. R. 1970. Relations of andesites, granites, and derivative sandstones to arc-trench tectonics. *Rev. Geophys. Space Phys.* 8:813–60

Fisher, R. V., Rensberger, J. M. 1972. Physical stratigraphy of the John Day formation, central Oregon. *Univ. Calif. Publ. Geol. Sci.*, Vol. 1. 33 pp.

Fyfe, W. S., McBirney, A. R. 1975. Subduction and the structure of andesitic volcanic belts. *Amer. J. Sci.* 275-A:285–97

Gilluly, J. 1963. The tectonic evolution of the western United States. *Quart. J. Geol. Soc. London* 119:133–74

Greene, R. C. 1968. Petrography and petrology of volcanic rocks in the Mount Jefferson area, High Cascade Range, Oregon. *U.S. Geol. Surv. Bull.* 1251-G. 48 pp.

Hamilton, W. 1969. The volcanic central Andes—a modern model for the Cretaceous batholiths and tectonics of western North America. *Proc. Andesite Conf., Oregon Dept. Geol. Miner. Ind. Bull.* 65:175–84

Hay, R. L. 1963. Stratigraphy and zeolitic diagenesis of the John Day Formation of Oregon. *Univ. Calif. Publ. Geol. Sci.* 42:119–262

Kennett, J. P., McBirney, A. R., Thunell, R. C. 1977. Episodes of Cenozoic volcanism in the Circum-Pacific region. *J. Volcanol. Geotherm. Res.* 2:145–63

Kushiro, I. 1973. Partial melting of garnet lherzolites from kimberlite at high pressures. In *Lesotho Kimberlites*, ed. P. H. Nixon, pp. 294–9. Maseru, Lesotho: Lesotho Nat. Dev. Corp.

McBirney, A. R. 1971. Thoughts on some current concepts of orogeny and volcanism. Comments on Earth Sciences. *Geophysics* 2:69–76

McBirney, A. R. 1976. Some geologic constraints on models for magma generation in orogenic environments. *Can. Miner.* 14:245–54

McBirney, A. R., Sutter, J. F., Naslund, H. R., Sutton, K. G., White, C. M. 1974. Epidosic volcanism in the central Oregon Cascade Range. *Geology* 2:585–89

Mysen, B. O. 1973. Melting in a hydrous mantle: Phase relations of mantle peridotite with controlled water and oxygen fugacities. *Carnegie Inst. Washington Yearb.* 72:467–78

Noble, D. C. 1972. Some observations on the Cenozoic volcano-tectonic evolution of the Great Basin, Western United States. *Earth Planet. Sci. Lett.* 17:142–50

Osborn, E. F., Watson, E. B. 1977. Studies of phase relations in subalkaline volcanic rock series. *Carnegie Inst. Washington Yearb.* 76. In press.

Peck, D. L. 1964. Geologic reconnaissance of the Antelope-Ashwood area, north-central Oregon. *U.S. Geol. Surv. Bull. 1161-D.* 26 pp.

Peck, D. L., Griggs, A. B., Schlicker, H. G., Wells, F. G., Dole, H. M. 1964. Geology of the central and northern parts of the Western Cascade Range in Oregon. *U.S. Geol. Surv. Prof. Pap. 449.* 56 pp.

Peterman, Z. E., Carmichael, I. S. E., Smith, A. L. 1970. Sr^{87}/Sr^{86} ratios of Quaternary lavas of the Cascade Range, northern California. *Geol. Soc. Amer. Bull.* 81:311–18

Rau, W. W. 1966. Stratigraphy and foraminifera of the Satsop River area, southern Olympic Peninsula, Washington. *Washington Dep. Conserv., Div. Mines Geol. Bull.* 53, 66 pp.

Robyn, T. L. 1977. *Geology and petrology of the Strawberry Mountain volcanic series, Central Oregon.* PhD thesis. Univ. Oregon, Eugene.

Smith, A. L., Carmichael, I. S. E. 1968. Quaternary lavas from the southern Cascades. *Contrib. Petrol.* 19:212–38.

Snavely, P. D., Wagner, H. C. 1963. Tertiary history of western Oregon and Washing-

456 McBIRNEY

ton. *Washington Dep. Conserv., Div. Mines Geol., Rep. Invest.* 22. 25 pp.

Snavely, P. D., Wagner, H. C., MacLeod, N. S. 1965. Preliminary data on compositional data variations of Tertiary rocks in the central part of the Oregon Coast Range. *The Ore Bin* 27: 101–17

Snavely, P. D., Wagner, H. C., MacLeod, N. S. 1968. Tholeiitic and alkalic basalts of the Eocene Siletz River volcanics, Oregon Coast Range. *Amer. J. Sci.* 266: 454–81

Sutton, K. G. 1974. *Geology of Mt. Jefferson.* MS thesis. Univ. Oregon, Eugene. 119 pp.

Thayer, T. P. 1937. Petrology of later Tertiary and Quaternary rocks of the north-central Cascade Mountains in Oregon. *Geol. Soc. Amer. Bull.* 48: 1611–52

Walker, G. W. 1970. Cenozoic ash-flow tuffs of Oregon. *The Ore Bin* 32: 97–115

Walker, G. W., Greene, R. C., Pattee, E. C. 1966. Mineral resources of the Mt. Jefferson Primitive Area, Oregon. *U.S. Geol. Surv. Bull. 1230-D.* 32 pp.

White, C. M., McBirney, A. R. 1978. Some quantitative aspects of orogenic volcanism in the Oregon Cascades. *Geol. Soc. Amer. Mem.* In press.

Wise, W. S. 1969. Geology and petrology of the Mt. Hood area: a study of High Cascade volcanism. *Geol. Soc. Amer. Bull.* 80: 969–1006

Ann. Rev. Earth Planet. Sci. 1978. 6:457–94

TEMPORAL FLUCTUATIONS OF ATMOSPHERIC ¹⁴C: CAUSAL FACTORS AND IMPLICATIONS

Paul E. Damon, Juan Carlos Lerman, and Austin Long

Laboratory of Isotope Geochemistry, Department of Geosciences, University of Arizona, Tucson, Arizona 85721

INTRODUCTION

In this review we consider the time variations of the atmospheric concentration of ¹⁴C, a radioisotope induced by cosmic rays and also known as radiocarbon. Radiocarbon dating is well known as the method that revolutionized the study of prehistory and late Quaternary geology. One of the basic assumptions of radiocarbon dating is that the concentration of radiocarbon in atmospheric carbon dioxide has been constant in the past. Willard Libby, in his classic book, *Radiocarbon Dating*, was quite cautious about this assumption: "We have seen how uncertain the experimental information on the present rate of production is and are therefore forced to conclude that agreement between this rate and the radioactivity of modern material, which of course reflects the production rate as of some 8000 years ago—since the carbon atoms now found in modern wood, for example, are 8000 years old on an average—is not very firm proof of the constancy and intensity of the cosmic radiation. It does, however, agree with this postulate. The rather satisfactory agreement between the predicted and observed radiocarbon contents of organic materials of historically known age (Figure 1) is somewhat more reassuring." (W. F. Libby 1955, p. 33). Libby's concern was not unfounded. Suess (1955) soon demonstrated a decrease in atmospheric radiocarbon concentration resulting from the combustion of fossil fuels, and by 1957 it was apparent that the atmospheric ¹⁴C was increasing as a result of atomic bomb tests (Rafter & Fergusson 1957). Moreover, de Vries (1958, 1959) demonstrated that atmospheric radiocarbon concentrations had fluctuated due to natural causes by about 2% during the Little Ice Age from the 16th through the 19th centuries. It is these fluctuations, their causes and consequent implications, with which we are concerned in this review. An investigation of this subject leads one into many fields of study, including geology and anthropology, astronomy, solar physics, nuclear and cosmic-ray physics, meteorology, climatology, oceanography, biology, chemistry, pedology, and history. Because of the diversity of disciplines interacting with radiocarbon research,

457

0084-6597/78/0515-0457$01.00

we have found it to be an excellent example of the unity of the sciences (Damon 1970a, 1977a).

Other sources to which the reader may refer for various aspects of this subject are the proceedings of the Twelfth Nobel Symposium, entitled *Radiocarbon Variations and Absolute Chronology* (Olsson 1970a), and the *Proceedings of the Eighth International Conference on Radiocarbon Dating* (Rafter & Grant-Taylor 1972). The reader might also wish to refer to the excellent reviews by Olsson (1968, 1974) and Aitken (1974).

The Egyptian Chronology

For his first checks on the reliability of the radiocarbon dating method, Libby chose to "date" Egyptian samples of known age. Using detectors or counters he and his associates developed, which had carbon black walls formed from the sample carbon itself, Libby (1955, Figure 1) demonstrated that the measured ^{14}C activities agreed within experimental error with those calculated from the known ages. The

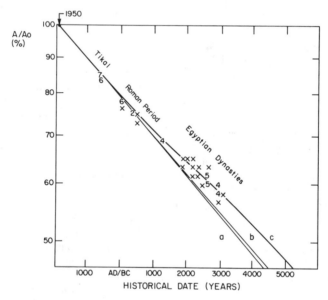

Figure 1 Radiocarbon activity of historical samples. ^{14}C activity (percent of "modern") vs historical date (AD and BC). This figure is an update of the classical comparison of historically well-dated samples and their remaining radiocarbon activity when measured at the present time (W. F. Libby 1955). The ordinate scale is logarithmic. The time scale is "reversed" to simulate the plot one obtains when studying the decay of a radioactive isotope. The lines indicate (*a*) a sample decaying with the "old" or "Libby half-life" of 5568 years; (*b*) a sample decaying with the presently accepted value of 5730 years (Godwin 1962, Olsson 1970a), and (*c*) the smoothed tree-ring curve plotted from the calibration table published by Damon et al (1974). The figures indicate the number of coincident samples.

^{14}C dating method had passed the crucial test. Later technological improvements in radiocarbon dating, consisting largely of the use of gas proportional counters rather than solid carbon-source counters, allowed significantly more precise measurements. In 1960, precise dating of Egyptian samples revealed radiocarbon dates 400 to 700 years younger than expected for the early Egyptian dynasties (3rd through 12th dynasties) (Damon & Long 1962). Since then, a large body of radiocarbon dates on Egyptian material has been obtained and compared with dendrochronologically dated wood, confirming the discrepancy.

Figure 1 presents a new version of Libby's illustration, including all historically well-dated Egyptian samples for which published radiocarbon dates are available. Non-Egyptian samples of well-known archaeological or historical ages are also included in this figure.

From Figure 1, it is apparent that historical and ^{14}C dates are in general agreement for the last 2500 years, but from 500 to 3000 B.C., neither the "Libby" half-life (a) nor the "new" half-life (b) satisfy the data points. The Egyptian dynasty points cluster about the tree-ring calibration curve (c).

Both half-lives are actively used in radiocarbon dating, but not because two camps fervently refuse to budge as sometimes happens in science. The "Libby" value of 5568 years was the best available value up to about 1960, when three laboratories made a concerted effort to make more precise and accurate determinations. They essentially agreed on a new value of 5730 ± 40 years (Godwin 1962). But since ^{14}C dates calculated on the new half-life value still do not agree with historical or tree-ring dates (Figure 1), and calibration corrections must be made regardless of the half-life used, it seemed pointless to recalculate all existing ^{14}C dates. Therefore, at present, all published radiocarbon dates, unless otherwise noted, are calculated using the Libby half-life, and can be compared with each other on the "conventional radiocarbon scale." Most ^{14}C dates acquired for geophysical investigations are now calculated using the new half-life.

The de Vries Effect

Preliminary but inconclusive measurements by Münnich (1957) suggested that even on a time scale of a few decades, the atmospheric radiocarbon activity may not have been constant during the past. Hl. de Vries and colleagues in The Netherlands initiated a careful, precise study of atmospheric ^{14}C activity by using radiocarbon to date tree rings that had been dendrochronologically dated. They began by remeasuring wood from the oak tree grown in Germany that had been measured by Münnich. They also analyzed a second oak from Germany, two Douglas firs from the United States, and two samples of historically dated wheat (de Vries 1958, 1959, Lerman, Mook & Vogel 1970). De Vries attempted then to correct the radiocarbon analyses for the isotope fractionation effect (cf p. 471), thus further increasing the precision of the radiocarbon assays. He examined trees dated (dendrochronologically) from AD 1525 to AD 1935 and concluded that the abundance of atmospheric radiocarbon as deduced from ^{14}C in the trees deviated by up to 2% above the average level, and that maxima occurred for the years AD 1500 and AD 1700, and minima for AD 1600 and AD 1800. De Vries then hypothesized that the radiocarbon variations are caused by two effects (de Vries 1958, 1959) (see Figure 2):

1. A secular effect, which has been called "de Vries effect" (Damon & Long 1962). The variations produced by this effect (also known as "wiggles," "wriggles," and "secular variations") are variations of the abundance of radiocarbon in atmospheric carbon dioxide on a time scale of one decade to a few centuries within an amplitude of a few percent of the average radiocarbon activity. De Vries hypothesized that these variations correlated with climate changes.

2. A geographical effect that de Vries thought was apparent in the comparison of the measurements of trees from Europe and North America. This effect was later

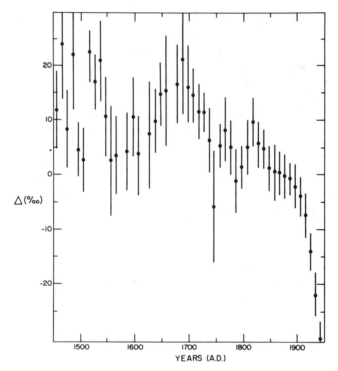

Figure 2 De Vries effect. Initial radiocarbon activity ($\Delta \pm 2\sigma$) as measured in tree rings dated from the last five centuries, expressed in $^{\circ}/_{\circ\circ}$ (per mil) with reference to the NBS standard activity (0.949 of the activity of the NBS oxalic acid sample) and age corrected with the ^{14}C half-life $T_{1/2} = 5730$ years. From analyses by five laboratories (cf data set of Figure 5). The measurements have been averaged by decades. Between AD 1500 and 1850, two maxima, two minima and other minor variations known as the de Vries effect are observed. The fast decrease seen after AD 1850 is partially due to the industrial effect (Suess effect) and partially to the de Vries effect. The time span covers the Little Ice Age, and the ^{14}C maxima correspond to the Maunder and Spörer minima of solar activity.

found to be an artifact resulting from the lack of precise stable isotope analyses (Lerman, Mook & Vogel 1970).

Publication of the de Vries results initiated an extensive survey of radiocarbon in tree rings, both in time and space. Although many of the subsequent publications in the 1960s have been superseded by more precise measurements, they are important for signaling the extent of the radiocarbon variations and for suggesting the different factors that may cause these variations. Especially significant were the analyses by Broecker, Olson & Bird (1959), Willis, Tauber & Münnich (1960), and Ralph & Stuckenrath (1960). Those publications were the precursors of later more extensive analyses now spanning seven millennia of dendrochronologically dated tree rings. These trees are mainly sequoia from the U.S. Pacific Northwest and bristlecone pine from the White Mountains in California, U.S.A. (Suess 1961, 1965, 1967, 1968, 1970a, Berger, Fergusson & Libby 1965, Damon & Long 1962, Damon, Long & Sigalove 1963, Damon, Long & Grey 1966, 1970, Damon, Long & Wallick 1972, Stuiver 1965, 1969, Mielke & Long 1969, Olsson, El-Gammal & Goksu 1969, Ralph & Michael 1970, Houtermans 1971, Michael & Ralph 1972, Ralph, Michael & Han 1973).

Other trees have been analyzed in different parts of the world to investigate other radiocarbon problems (Kigoshi & Hasegawa 1966, Kigoshi & Kobayashi 1966, Cowan, Atluri & Libby 1965, Lerman, Mook & Vogel 1967, Jansen 1962, 1970, Baxter & Walton 1971, Farmer & Baxter 1972, Damon, Long & Wallick 1973a, Cain & Suess 1976, Ferguson, Gimbutas & Suess 1976). Studies on the uniformity of the world-wide distribution of radiocarbon have also been carried out by comparison of tree rings from different continents and latitudes. These studies are discussed later.

THE GEOCHEMICAL CYCLE OF CARBON

The Reservoirs and Transfer between Reservoirs

Figure 3, which has been adapted and extensively modified from Bolin (1970), shows the carbon cycle in nature as it was prior to the advent of extensive detonation of nuclear weapons (about AD 1950). Natural radiocarbon is produced by an n, p reaction and decays back to ^{14}N by beta emission with a half-life of 5730 ± 40 years: ^{14}N(n, p) ^{14}C $\xrightarrow{\beta^-}$ ^{14}N. The rate of production during the three solar cycles from AD 1937 to AD 1967 was 2.2 ± 0.4 ^{14}C cm$_e^{-2}$ sec (132 ^{14}C cm$_e^{-2}$ min) (Lingenfelter & Ramaty 1970). When it is first produced, radiocarbon is a hot atom that rapidly forms CO_2 and equilibrates with atmospheric CO_2. Measurement of bomb-produced radiocarbon reveals that radioactive CO_2 then rapidly distributes itself throughout the atmosphere. According to Nydal (1968), the residence time (τ) in the stratosphere before transfer to the troposphere is 2.0 ± 0.5 years. To define τ, Nydal assumes that the number of atoms, dN, removed in time interval, dt, is proportional to the total number of atoms, N, present:

$$dN/dt = -kN, \tag{1}$$

where k is the transfer coefficient. The relationship between k and τ is

$$\tau = 1/k. \tag{2}$$

Thus,

$$\tau = \frac{N}{dN/dt}. \tag{3}$$

According to Nydal (1968), the transfer between the northern and southern tropospheres requires a residence time of only 1.0 ± 0.2 years, whereas direct strato-sphere-stratosphere transfer between the two hemispheres involves a much longer residence time, i.e. 5.0 ± 1.5 years. However, for the purposes of discussion of the geochemical cycle in Figure 3, we have considered the atmosphere as a single

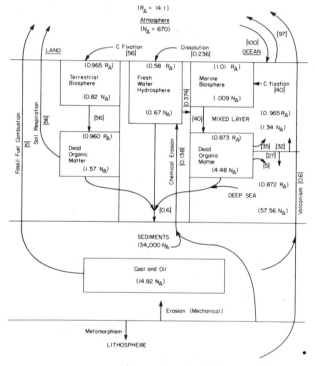

Figure 3 The carbon cycle in nature prior to large-scale detonation of nuclear weapons, i.e. AD 1950. See Table 1 for mass of carbon in reservoirs and Table 2 for radiocarbon activity of reservoir. C fluxes (10^{15} g yr^{-1}) are derived from the following sources: C fixation in terrestrial biosphere (Keeling 1973); dissolution and chemical erosion (Garrels & Mackenzie 1971); deep sea to mixed layer estimated from ^{14}C activity (Oeschger et al 1975); volcanic and sedimentary flux estimated from sea-floor spreading cycle (Damon & Wallick 1972); all other flux estimates from Bolin (1970). All masses are in units of 10^{15} g. Activities (R) are in units of dpm gC^{-1}.

reservoir. For other modeling purposes, it must be subdivided into four reservoirs (boxes).

Considering the atmosphere as a single reservoir, the residence time of ^{14}C in the atmosphere is only about four years before it is transferred by the carbon-oxygen cycle to the terrestrial biosphere, exchanged with the oceanic mixed layer, or converted to the HCO_3^- anion in the hydrosphere by chemical erosion of terrestrial rocks (Table 1). Essentially, the transfer within the terrestrial biosphere constitutes a separate cycle distinct from the oceanic cycle. The residence time for carbon in the terrestrial biosphere is ten years. This model includes all the carbon in living terrestrial organisms as well as wood in living trees, even though only the outer growth ring and rays contain living cells. An alternative would be to include non-living cells in trees within the dead organic matter box. In such a model, the residence time in the terrestrial biosphere would be much shorter. The residence time for decaying organic matter is about 19 years. Most of this carbon is oxidized and returned to the atmosphere except for a small portion that is permanently stored in sediments.

During the weathering and erosion process, CO_2 as carbonic acid is converted to the bicarbonate anion HCO_3^-, transported to the sea, and deposited as calcareous sediment. Typical weathering reactions are, for feldspar:

$$2KAlSi_3O_8 \quad + \quad CaSiO_3 \quad + \quad 3H_2CO_3 \quad \leftrightarrows$$

(K-feldspar) + (Wollastonite) + (Carbonic acid) $\leftrightarrows$

$$2K^+ \qquad + \qquad 2HCO_3^- \qquad + \qquad CaCO_3 \qquad +$$

(Aqueous K cation) + (Bicarbonate anion) + (Calcium carbonate) +

$$Al_2Si_2O_5(OH)_4 \quad + 5SiO_2, \tag{4}$$

(Kaolinite)   + (Silica), $\Delta G_R^\circ = -14.7$ Kcal mole^{-1};

and for calcite:

$$H_2CO_3 + CaCO_3 \leftrightarrows Ca^{++} + 2HCO_3^-,$$
$$\Delta_R^\circ = -7.1 \text{ Kcal mole}^{-1}, \tag{5}$$

where Δ_R° is the Gibb's standard free energy of the reaction. Under the commonly prevailing slightly acid weathering conditions, both reactions proceed spontaneously to the right. Under alkaline marine conditions, the reaction in Equation 5 proceeds to the left with the formation of calcareous sediments along with the clays and silica produced by the reaction in Equation 4. Also under marine conditions, reaction 4 may allow some "reverse weathering," but slow reaction rates seem to preclude directly observable effects except under diagenetic conditions. The residence time of carbon in the fresh-water reservoir, which consists of rivers, lakes, ice and ground-water, i.e. primarily the pore water in sediments, is relatively long (about 1200 years) before the carbon is transferred to the sea.

During the marine cycle, a steady state is approached between CO_2 in the

Table 1 Residence times of carbon in global reservoirs

Reservoir[a]	C in reservoir (N) Units = 10^{15} g	Flux (ρ) Units = 10^{15} g yr^{-1}	Residence time (τ)
Atmosphere	670	156	4 (yr)
Terrestrial biosphere	550	56	10 (yr)
Humus (dead terrestrial organic matter)	1,050	56	19 (yr)
Hydrosphere (fresh water)	449	0.37	1,213 (yr)
Hydrosphere (mixed layer of ocean)	900	124	7 (yr)
Hydrosphere (deep sea)	37,670	32	1,180 (yr)
Biosphere (marine)	6	40	2 (months)
Oceans (dead organic matter)	3,000	40	75 (yr)
Sediments	90,000,000	0.6	150 (10^6 yr)

[a] Source of reservoir data: Atmosphere (Verniani 1966), terrestrial biosphere and humus (Keeling 1973), fresh water hydrosphere (Garrels & Mackenzie 1971), mixed layer of ocean (Keeling 1973), deep sea (Suess 1965), biosphere (Bolin 1970), dead organic matter in the oceans (Skopintsev 1950, Hamilton 1965, Bolin 1970), sediments (Ronov & Yaroshevsky 1969).

atmosphere and dissolved carbon species (H_2CO_3, HCO_3^-, and some $CO_3^=$). The steady state is dominated by the tendency towards approach to equilibrium in the partitioning of CO_2 between the atmosphere and the oceans. This partitioning is a function of the temperature and volume of seawater and, to a lesser extent, salinity. For example, the solubility of CO_2 decreases with temperature by 5.8% per degree Celsius (Eriksson 1963). Carbon is rapidly cycled through marine organisms, primarily phytoplankton, which have a short lifetime. Upon death, the organic compounds of living organisms are broken down to dehydroascorbic acid, carbohydrates, citric acid, malic acid, fatty acids, amino acids, and plant hormones, most of which are recycled in the mixed layer (Provasoli 1963); however, a portion of the organic carbon is lost to the deep sea as dissolved organic matter and a lesser amount as suspended organic matter. Shells and other organic debris rain down from the mixed layer, falling slowly through the deep sea to accumulate in ocean sediments. Because of disturbances in the upper 10 cm of ocean sediments, this thin layer of sediments must also be considered part of the active carbon reservoir (Baes et al 1977). Dissolved carbon in the mixed layer also diffuses into the deep sea or is carried to the depths by cold sinking sea water, later to be released by back-diffusion or upwelling primarily in coastal areas. Upwelling creates very fertile ocean areas rich in nutrients. The residence time in the mixed layer of the oceans, before transfer back to the atmosphere, to the deep sea, or to sediments, is short (about 7 years). The residence time in the deep sea, as would be expected intuitively, is much longer (about 1200 years). To close the circle, after a residence time of about 150 million years in the sedimentary reservoir, volcanism returns CO_2 to the atmosphere. This slowest of the revolutions in the carbon cycle is controlled largely by the sea-floor spreading cycle (Damon & Wallick 1972).

The Radiocarbon Inventory

If the carbon content and average specific activity of radiocarbon are known for each reservoir, the radiocarbon content expressed in units of dpm cm_e^{-2} (^{14}C disintegrations per minute per cm^2 of earth surface) can be calculated and summed as in Table 2. The inventory obtained in this manner is 119 dpm cm_e^{-2}, which is about 11% higher than calculated by Suess (1965). Our greater inventory is primarily due to our assumption of a three-fold greater sedimentation rate and secondarily due to the assumption of a higher carbon content in the humus reservoir. Thirdly, Suess did not consider the fresh-water reservoir, which, however, contributes only 0.6% to the increase.

According to Lingenfelter & Ramaty (1970), the production rate during three solar cycles (1937–1967) was 132 ^{14}C cm_e^{-2} min^{-1}, which is 10.7% higher than our calculated inventory and 21.7% higher than calculated by Suess (1965). Actually, considering the uncertainties involved, our calculated inventory is not significantly different from the production rate calculated by Lingenfelter & Ramaty or from the inventory calculated by Suess.

Residence Times and Radiocarbon Models

The degree of complexity of the reservoir system required to model radiocarbon fluctuations is a direct function of the rapidity of the fluctuation being modeled. Slow, sinusoidal changes in production rate occurring with a period > 100 years may be modeled by the simple two-exchange-box model with a sedimentary sink in Figure 4 (Houtermans, Suess & Oeschger 1973). In this model, the ambient reservoir consists of all these reservoirs that have an exchange time of ten years or less: the

Table 2 Radiocarbon inventory

Reservoir	C in reservoir (N) g cm_e^{-2}	^{14}C activity[a] in reservoir (R) dpm gC^{-1}	Decay rate in reservoir dpm cm_e^{-2}
Atmosphere	0.131	14.1	1.85
Terrestrial biosphere	0.108	13.6	1.47
Humus (dead terrestrial organic matter)	0.206	13.5	2.78
Hydrosphere (fresh water)	0.088	8.2	0.72
Hydrosphere (mixed layer of oceans)	0.180	13.6	2.44
Hydrosphere (deep sea)	7.386	12.3	90.85
Biosphere (marine)	0.001	14.2	0.01
Oceans (dead organic matter)	0.588	12.3	7.23
Sediments	0.972	12.3	11.97
	$\Sigma = 9.660$ g cm_e^{-2}	$\bar{R} = 12.3$ (dpm g^{-1})	$\Sigma = 119.3$ dpm cm_e^{-2} = (1.99 dps gm_e^{-1})

[a] Carbon-14 content of recent wood (Karlén et al 1964) with corrections for isotopic fractionation (Schwarcz 1969) and residence times (Table 1).

atmosphere, biosphere, and mixed layer of the ocean. All other reservoirs are included with the deep sea.

The residence time of the ambient reservoir can be derived from Equation 3:

$$\tau_A = \frac{N_A}{dN_A/dt}. \tag{6}$$

Using values for N_A and dN_A/dt from Figure 4 and Table 1, the residence time, τ_A, is 65 years. If the terrestrial biosphere carbon that returns to the atmosphere through decay in the humus layer is removed from the ambient reservoir, the residence time, τ_A, becomes 48 years. Clearly, the residence time is a function of the model used. For the model in Figure 4, carbon transfers back and forth between the

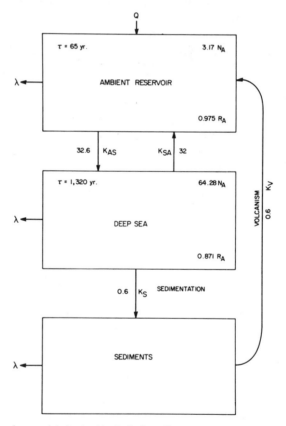

Figure 4   Three-box model obtained by including all reservoirs that have an exchange time of ten years or less (atmosphere, biosphere, and mixed layer of ocean) in an ambient reservoir. All other reservoirs are included with the deep sea. Fossil fuel combustion has not been included. Fluxes (K) are in 10^{15} g yr^{-1}; activities (R) in dpm gC^{-1}, $\lambda =$ radiocarbon decay constant, $Q =$ rate of cosmic-ray production of ^{14}C.

mixed layer and deep sea on the average six or seven times before being transferred to the deep sea. Unfortunately, considerable confusion has accumulated in the literature concerning this simple concept of residence time since it was clearly defined for radiocarbon by Craig (1957a) over two decades ago. Quite obviously, residence times are defined by the model used and controversies concerning residence time discrepancies are mostly apparent and not real problems.

Perturbation by Human Activity: Nuclear Technology and Combustion of Fossil Fuels

The products of industrial and domestic activity have made their impact on almost every natural reservoir. Carbon, in particular fossil carbon, which is devoid of ^{14}C, has been the main contribution of industrial activity to the atmosphere, biosphere, and hydrosphere. Because this industrial CO_2 injected into the atmosphere has essentially no ^{14}C, the effect has been to decrease the atmospheric radiocarbon activity. This effect is noticeable in tree rings grown at the beginning of this century, and caused a dip in the ^{14}C activity of about 2% before being over-whelmed at mid-century by another effect—thermonuclear devices exploded in the atmosphere produced radiocarbon very efficiently. These explosions doubled the natural radiocarbon activity of the atmosphere in the 1960s. Both types of human impact left their isotopic scars in our natural environment.

The effects of this uncontrolled experiment on our global ecology can only be surmised at this moment, but may range from genetic alterations from ^{14}C produced by bombs and reactors (Pauling 1958, Sakharov 1958) to climate modification from the CO_2 "greenhouse effect" (e.g. see Sellers 1974). But nothing is all bad. The "bomb effect" has provided us with an artificial tracer experiment on a global scale, allowing the evaluation of residence times and transfer rates of ^{14}C and CO_2 within and between the carbon reservoirs, data essentially unobtainable by other means. These results will be of crucial value in predicting the effects of continued burning of fossil fuels.

These admixtures of artificial carbon were first observed 20 years ago when Suess (1955) reported that wood dating from the last 100 years was depleted in radiocarbon when compared with pre-industrial (i.e. pre-AD 1890) wood, and ascribed this effect to the admixture of fossil or "dead" combustion products to the atmospheric reservoir. This depletion of the atmosphere was called the "industrial" or "Suess" effect. As the amount of released "fossil CO_2" can be very well estimated from the annual consumption of coal and oil, the measurements of the radiocarbon activity of tree rings spanning the last century have given an insight into the partitioning of CO_2 between the atmosphere, the biosphere, and the hydrosphere. This information is needed in order to estimate the consequences of the increase of atmospheric CO_2 contents for the earth's future climate (Baes et al 1977). The atmosphere has received an artificial input of about 30% of its original CO_2 content since AD 1850 and has retained more than 15%, the rest going mostly to the oceans with some part to land vegetation. The slope of the observed decrease in ^{14}C was about $0.35°/_{oo}$ yr^{-1} at AD 1900 (Lerman, Mook & Vogel 1967) and the largest depletion observed was ca. $-25°/_{oo}$ at AD 1950.

Unfortunately, two types of problems had prevented this approach from yielding much information about the carbon dioxide cycle up to now. First, not all of the decrease of ^{14}C has been caused by the industrial effect (Lerman, Mook & Vogel 1967, Figure 1). As the Wolf numbers (cf p. 484) have been increasing since AD 1900, a substantial part of the decrease in ^{14}C must be due to the solar modulation effect. Some estimations of the relative magnitudes of both contributions can be seen in the models by Grey (1969) and Oeschger et al (1975). Second, all the existing models fail to represent and predict simultaneously the behavior of all three isotopes ^{14}C, ^{13}C, and ^{12}C. Again, we encounter the problem that the residence time is model-dependent.

Nydal (1968) made the most exhaustive measurements of atmospheric ^{14}C in the post-nuclear-bomb era. After the United States and the U.S.S.R. tests of 1961, the northern hemisphere doubled its radiocarbon concentration in a couple of years. The time and space variations of ^{14}C in the atmosphere and oceans during the last 25 years, as is discussed in detail later, are prime data for designing and shaping models for atmosphere and ocean circulation and carbon cycle behavior. The better we understand the carbon cycle, the more valid will be our predictions of man's impact on our environment. Nydal (1968), Rafter & O'Brien (1970, 1972), and particularly Oeschger et al (1975) have effectively utilized the field measurements to derive useful model parameters.

DENDROCHRONOLOGY AND THE CALIBRATION OF THE RADIOCARBON TIME SCALE

Dendrochronology or tree-ring dating emerged about half a century ago as a consequence of the search for evidence concerning solar cycles begun by A. E. Douglass in 1901 (Douglass 1919, Bannister 1969). It has now evolved into a science that not only employs ring-thickness variations for very accurate age determinations, but also studies climatic and edaphic causes for these variations. Thus tree rings of accurately known age, since they contain carbon reflecting the isotope composition of the atmosphere in which they grew, are ideal samples for investigating temporal radiocarbon variations.

Dendrochronology

Dendrochronology and ^{14}C research have been companions since the early stages of the development of radiocarbon dating. W. F. Libby and coworkers checked their first radiocarbon dates with dated wood as well as Egyptian samples. Since then, most of the known age samples used for calibration of the ^{14}C time scale have been dendrochronologically dated wood.

Dendrochronologists study the annual growth layers or tree rings of woody plants. These layers of xylem are more or less distinct rings in transverse sections of tree trunks. Rings of many species are datable because their thicknesses vary from year to year, depending mostly on climatic factors. Ideally, a distinct pattern of widths develops for each climatically distinct geographic area. An outstanding example is the bristlecone pine (Schulman 1956), the oldest known living tree, which has been extensively used in radiocarbon calibration work. The longest bristlecone pine chronologies extend to 8400 years BP (Ferguson 1970a,b). Many other local chronologies have been constructed mainly in the U.S. Southwest by the Laboratory of Tree-Ring Research at the University of Arizona. Other regions of the earth have produced shorter chronologies, usually only a few centuries long, many of which have also been used for the calibration of the radiocarbon time scale. Huber and co-workers (see review by Huber 1970) discuss Central European "floating" chronologies extending into the Pleistocene, which when absolutely dated will provide valuable samples for radiocarbon calibration.

Retrospective Monitoring of Atmosphere Radiocarbon by the Tree-Ring Record: The Method and Its Reliability

As described above, the method most widely used for calibration of the radiocarbon scale is the comparison of the radiocarbon age of well-dated tree rings with their dendrochronological age. The reliability of this method is based on two assumptions: 1. that the radiocarbon activity of the wood accurately represents the activity of the atmosphere when the wood was formed; and 2. that the age of the tree ring(s) is accurately known.

These assumptions imply that we are able to determine the original radiocarbon activity of the sample, and with this as a basis we can then retrieve the value of the atmospheric radiocarbon activity at the time the wood formed. This is in itself based on two other assumptions:

1. That we are able to correct for the isotope fractionation effect that takes place during the photosynthetic process. Photosynthesis discriminates against the heavy isotopes of carbon (^{14}C and ^{13}C). Even within a species or an individual tree, the discrimination effect may vary by two or three per mil in ^{13}C, and twice that in ^{14}C (cf e.g. Lerman 1972, Olsson & Osadebe 1974, Lerman & Long, in preparation). The measured ^{14}C activity has to be normalized to a common standard by using measurements of the stable isotopes of carbon (^{12}C and ^{13}C) (Craig 1954, Lerman 1974b, Olsson & Osadebe 1974, Stuiver & Robinson 1974).

2. That all the carbon analyzed from the wood is of the age represented by the tree-ring material. Cellulose or cellulose plus lignin seem to be the best wood components for ^{14}C analysis (Cain & Suess 1976, Long et al 1978). Non-cellulose extractives left in the sample might have a higher (or lower) radiocarbon activity than that of the cellulose in the tree ring.

Another potential source of spurious carbon that deserves more exploration in light of the new developments in plant physiology is the possibility of fixation via the roots of carbon derived from soil limestone by the C_4 carbon assimilation pathway as discussed by Lerman (1974a,b). This carbon in most cases would be depleted of ^{14}C. Preliminary experiments have been performed by Olsson, Klasson & Abd-el-Mageed (1972). They analyzed trees growing in soils with different limestone content and began a series of laboratory experiments with tomato plants. They did not observe any significant difference within the standard deviation of their measurements ($10°/_{oo}$). Also, the possibility has been suggested of "in situ" production of radiocarbon in tree rings by the larger neutron flux present at high elevations (Baxter & Farmer 1973) and by neutrons generated by lightning (L. M. Libby & Lukens 1973). However, subsequent investigations have refuted these mechanisms as significant sources of error in radiocarbon dating (Damon, Long

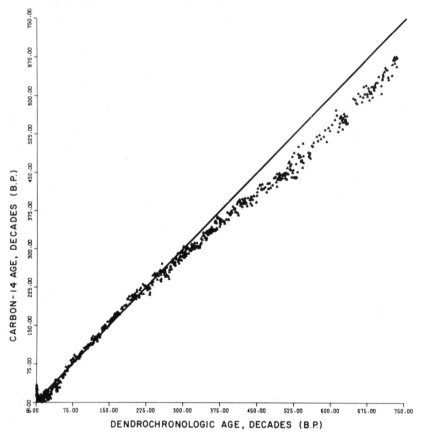

Figure 5   Radiocarbon dates vs. tree-ring dates. Radiocarbon age (BP decades) vs. tree-ring date (in calendar years BP). This figure shows the relationship between the radiocarbon dates and the historical or calendar dates as represented by dendrochronological dates. The ^{14}C dates in radiocarbon decades (BP) have been calculated on the basis of the 5730 year half-life. The standard activity corresponding to 0 year BP ($=$ AD 1950) is the NBS standard activity, which, as described in the text, is corrected for the industrial and nuclear-bomb effects. Most of the points represent bristlecone pine samples supplied by the Laboratory of Tree-Ring Research, Arizona, and most of them consist of ten tree rings. The results were obtained by the radiocarbon laboratories of Arizona, Groningen, La Jolla, Pennsylvania, and Yale (Damon, Long & Grey 1970, Damon, Long & Wallick 1972, Lerman, Mook & Vogel 1970, Ralph & Michael 1970, Stuiver 1969, Suess as published by Houtermans 1971).

& Wallick 1973b, Fleischer, Plumer & Crouch 1974, Fleischer 1974, Harkness & Burleigh 1974, Cain & Suess 1976).

The second assumption requires independent dating of the analyzed wood. This is precisely what dendrochronology provides with assurance. Mere tree-ring counting, by contrast, does not consider the possibility of presence of false (or extra) rings or of missing rings. In the case of suitable tree species in favorable localities, such as the bristlecone samples, the dendrochronologically determined ages for the last 5700 years are accurate to the year (LaMarche & Harlan 1973).

When attempting to cross-date tree-ring specimens in a particularly long chronology, such as the bristlecone pine chronology in the Inyo National forest, it is often expedient to obtain a ^{14}C date to give the time range in which the tree-ring pattern should fit. The symmetry of the dendrochronology-^{14}C relationship continues as we shall see: wood specimen of unknown age, then uncorrected ^{14}C date, then exact tree-ring date, then finally ^{14}C date calibration leading to more accurate ^{14}C dates.

The Results of Retrospective Monitoring of Radiocarbon

Bristlecone pine samples dendrochronologically dated at the University of Arizona Laboratory of Tree-Ring Research and analyzed for radiocarbon at five laboratories comprise most of the 549 data points in Figure 5. They span nearly 7500 years. Of the 589 analyses originally available, 7% were rejected according to the criteria of Damon et al (1974), most of these because the samples contained more than 25 growth rings.

The discrepancies between tree-ring or "true" age and ^{14}C age, and the concept of calibration are easily visualized in Figure 5. In this figure, the 45° line represents agreement between calendar age and conventional ^{14}C age. A best-fit curve, if drawn through the points, would be a calibration curve. Its exact shape may change, however, as more and better data become available. For geophysical purposes, the preferred data format is as shown in Figure 6. This has the same data and abcissa as Figure 5, but shows the deviation from the agreement line in per mil, normalized to AD 1950 or "Δ" values. For discussion of Δ, see Broecker & Olson (1959, 1961) and Damon (1970b). Symbolically,

$$\Delta = \delta^{14}C - 2(\delta^{13}C + 25)(1 + 10^{-3}\delta^{14}C), \tag{7}$$

where

$$\delta^{14}C = \left(\frac{A_x - A_0}{A_0}\right) \times 1000 \tag{8}$$

and

$$\delta^{13}C = \left[\frac{(^{13}C/^{12}C)_x - (^{13}C/^{12}C)_{PDB}}{(^{13}C/^{12}C)_{PDB}}\right] \times 1000. \tag{9}$$

A_x is the age-corrected, but not fractionation-normalized, ^{14}C activity of the sample. Equation 7 is a close approximation. See Olsson & Osadebe (1974), Stuiver &

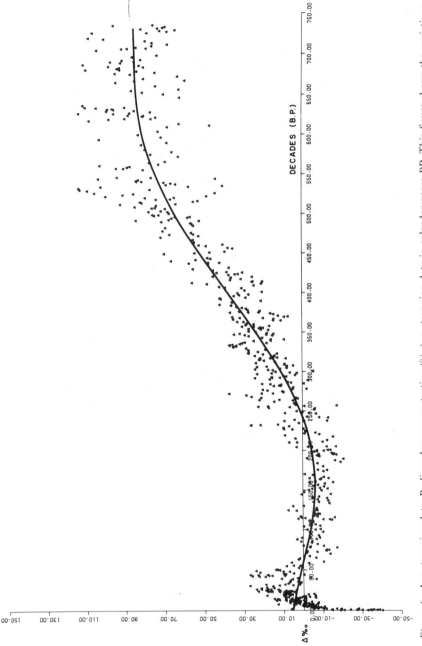

Figure 6 Δ vs. tree-ring date. Radiocarbon concentration (°/₀₀) vs. tree-ring date in calendar years BP. This figure shows the variations with time of the atmospheric concentration of radiocarbon as measured in tree rings. The data consist of the same data set of Figure 5. The plotted line was obtained by a regression using orthogonal polynomials. Deviations from the line for the recent centuries are known as de Vries and Suess effects as described in Figure 2.

Robinson (1974), and Lerman (1974b) for more precise formulae. A_0 is the ^{14}C activity of an hypothetical tree ring grown in AD 1950, but not affected by the industrial or bomb effect. $(^{13}C/^{12}C)_x$ and $(^{13}C/^{12}C)_{PDB}$ are stable carbon isotope ratios in, respectively, the sample and the PDB (Pee Dee belemnite) reference standard (Craig 1957b). An age-corrected ^{14}C activity for a sample is the activity the tree ring, or average tree ring in case of multi-ring samples, had the year it grew. The primary standard is wood grown in 1890 (before industrial and bomb perturbations). Its activity is normalized to $\delta^{13}C = -25.0°/_{oo}$ and to the year AD 1950, the "zeroth" year for radiocarbon dating. All laboratories use for a "working standard" a sample from the same batch of NBS (National Bureau of Standards) oxalic acid radiocarbon dating standard, the ^{14}C activity of which is related to the zeroth year by the factor 0.949. (The factor 0.950, which is used by most authors, corresponds to a zeroth year of 1958.)

Radiocarbon activities of all standards and samples are normalized to a standard $^{13}C/^{12}C$, or $\delta^{13}C$ value on the PDB scale: $-25.0°/_{oo}$ for organic matter and $-19.0°/_{oo}$ for the oxalic acid standard. This normalization compensates for environmental, biological, and laboratory isotope fractionation.

Sample pretreatments are customary, but not standardized. Most laboratories have followed modifications of a procedure suggested by de Vries, Barendsen & Waterbolk (1958), which consists of rather harsh acid and base treatments. Recently, ^{14}C analyses of untreated wood compared with de Vries-treated wood as well as with chemically separated fractions have shown that in some pre-bomb wood ^{14}C activities of wood extractives are significantly higher than cellulose, implying mobility of certain components into older tree rings (Olsson, Klasson & Abd-el-Mageed 1972; Cain & Suess 1976; Long et al 1978). Although the de Vries pretreatment seems effective in removing nearly all the noncontemporaneous, carbon-containing material in tree rings, we recommend more exacting chemical separations for future work.

Typical precision within a single laboratory is $\pm 5°/_{oo}$, or about 40 years, 1σ. That is to say, that if the laboratory were to repeat complete counting analyses on the same gas volume, two-thirds of the final values on each replicate would lie within $5°/_{oo}$ of the mean value. The most precise results emerge from researchers who count samples for long periods of time using the largest counters with the lowest stable backgrounds (see p. 485). Precisions as low as $1.5°/_{oo}$ 1σ have been attained (Lerman, Mook & Vogel 1970). The precision of radiocarbon counting has been thoroughly discussed by Currie (1972).

Ideally, the stated precision within a laboratory should also be the precision on replicate analyses performed by several laboratories. Indeed, this proved to be the case for a group of laboratories participating in a standard calibration test (H. Polach, written communication). Damon (1970b) has systematically compared ^{14}C analyses of same-year specimens from two laboratories, La Jolla and Arizona. He found that the actual precisions of analyses are closer to $\pm 10°/_{oo}$ between laboratories, and concluded that any overall systematic difference between laboratories must be less than $5°/_{oo}$.

Calibration of the Radiocarbon Time Scale

The temporal variations of the atmospheric radiocarbon activity make it necessary to have calibration curves for the purpose of expressing dating results in terms of historical or calendar dates.

Several radiocarbon calibration schemes have been published: Suess (1970a), Stuiver & Suess (1966), Lerman, Mook & Vogel (1970), Olsson (1970b), Wendland & Donley (1971), Michael & Ralph (1972), Damon, Long & Wallick (1972), Clark & Renfrew (1972), Ralph, Michael & Han (1973), Switsur (1973), Damon et al (1974), Clark (1973, 1975), McKerrell (1975).

All these calibration curves follow the same overall trend, but they are sufficiently different to have delayed the formulation of the "official calibration curve" and induced some hesitation among archaeologists in accepting the radiocarbon dating method. Most of the severe critiques of the radiocarbon method came (and still come) from European archaeologists who had several decades ago formed opinions about the relative chronology of the different European cultures. This well-established "model" of Old World inhabitants and their connections, relationships, and

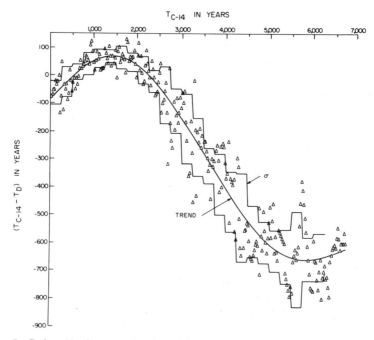

Figure 7   Carbon-14 minus tree-ring age ($T_{C-14} - T_D$) vs. radiocarbon age (T_{C-14}) for the last seven radiocarbon year millennia. Triangles are the 25-year interval averages; trend curve is determined by curvilinear regression analysis using orthogonal polynomials; σ is the square root of the variance of the 25-year interval averages about the trend curve calculated in 250-year intervals (modified from Damon, Long & Wallick 1972).

exchanges made difficult the acceptance of a method that shuffled (or rearranged) the prehistory and attempted to reverse some presumed sequences of cultural events. Currently under re-evaluation is the diffusionist theory of how culture travelled from East to West. By accepting calibrated ^{14}C dates, one arrives at the novel concept that Stonehenge preceded the Egyptian culture.

The fact that the initial radiocarbon activity has varied with time means that dates calculated in the usual manner do not correspond to historical dates. To make radiocarbon dating more useful to the archaeologists working in time periods that can be easily related to historical or calendar times, several graphs, tables, and equations have been published. One such radiocarbon calibration curve is shown in Figure 7. This particular curve was constructed by fitting a Tchebyschev orthogonal polynomial through the composite data set shown in Figure 5, and was published in tabular form by Damon et al (1974). This has proven to be a good smoothed curve describing the major radiocarbon variations up to the last 400 years. Shorter-term natural and artificially induced ^{14}C variations are clearly defined only for the last four centuries.

High-resolution ^{14}C dating in this time period is, in theory, possible if ^{14}C is measured in a sequence of closely spaced samples of known time range, but unknown absolute age; rings within a wood beam would be ideal. The resulting ^{14}C activity vs time curve may be uniquely matched to the standard curve, in a manner analogous to tree-ring dating. Post-bomb samples have been distinguished from pre-bomb cultural samples in this manner (see Tamers 1969).

Archaeological and Geological Implications

Radiocarbon has played a very important role in archaeology and geology. Together with the other isotopic dating methods, radiocarbon dating contributed to the establishment of an absolute time framework within which all the biologic and geologic processes could be studied and quantified. Quantitative knowledge of the time elapsed since events such as the emergence of agriculture and the last continental glaciation forced scientists to develop theories and models to explain the timing of these events, frequently revising prevailing ideas.

According to Renfrew (1973) there have been two radiocarbon revolutions. The first occurred when radiocarbon proved the greater antiquity of the European and Near East Neolithic period, setting its beginning some 3000 years earlier than previously hypothesized within the traditional "short chronology." Radiocarbon was instrumental in introducing a long chronology. The second revolution took place, as previously mentioned, when the calibration of dates obtained on archaeological materials indicated that the European cultures did not postdate the Middle Eastern cultures as the diffusionists expected. This second revolution has been harder to accept than the first. All these controversies are described and discussed in articles and books published by several authors (see, among others: Neustupný 1970a,b,c, Clark 1965, Renfrew 1973, Watkins 1975).

We should mention here that it was not only what Watkins (1975) calls the "dream that the ^{14}C dates were absolute" that caused the misunderstandings between radiocarbon producers and radiocarbon users, but also the fact that radiocarbon dates represent the submitted objects, which may or may not correspond to the cultures or events the anthropologist wishes to date. Obviously, the materials may be of older age, e.g. as described by Tauber (1958); younger ages, e.g. due to intrusion, are also possible. Another difficulty in the interpretation of the radiocarbon dates is the inevitable fact that a radiocarbon date has a statistical uncertainty, so we only know the probability that the true ^{14}C age lies within a certain range.

Thus, for European archaeology it seems that, as Fleming (1975) says when closing the volume edited by Watkins (1975), what seemed to be a technicality such as a difference of a few hundred years is ultimately forcing anthropologists to re-examine their theories and confront for the first time problems that have existed since pre-radiocarbon times.

Unprejudiced by the complexity of conflicting historical records and theories on the development of cultures and their interrelationships, New World anthropologists were more accepting of radiocarbon results. Radiocarbon work in the West established for the first time a time scale to describe the local cultures and attempted to establish the earliest appearance of man on the American continent.

Table 3   Possible causes of radiocarbon fluctuations.

I. Variations in the rate of radiocarbon production in the atmosphere.
  1. Variations in the cosmic-ray flux throughout the solar system.
    a. Cosmic-ray bursts from supernovae and other stellar phenomena.
    b. Interstellar modulation of the cosmic-ray flux.
  2. Modulation of the cosmic-ray flux by solar activity.
  3. Modulation of the cosmic-ray flux by changes in the geomagnetic field.
  4. Production by antimatter meteorite collisions with the earth.
  5. Production by nuclear weapons testing and nuclear technology.
II. Variations in the rate of exchange of radiocarbon between various geochemical reservoirs and changes in the relative carbon dioxide content of the reservoirs.
  1. Control of CO_2 solubility and dissolution as well as residence times by temperature variations.
  2. Effect of sea-level variations on ocean circulation and capacity.
  3. Assimilation of CO_2 by the terrestrial biosphere in proportion to biomass and CO_2 concentration, and dependence of CO_2 on temperature, humidity and human activity.
  4. Dependence of CO_2 assimilation by the marine biosphere upon ocean temperature and salinity, availability of nutrients, up-welling of CO_2-rich deep water, and turbidity of the mixed layer of the ocean.
III. Variations in the total amount of carbon dioxide in the atmosphere, biosphere and hydrosphere.
  1. Changes in the rate of introduction of CO_2 into the atmosphere by volcanism and other processes that result in CO_2 degassing of the lithosphere.
  2. The various sedimentary reservoirs serving as a sink of CO_2 and ^{14}C. Tendency for changes in the rate of sedimentation to cause changes in the total CO_2 content of the atmosphere.
  3. Combustion of fossil fuels by human industrial and domestic activity.

Thus, it seems that the broad issues introduced by the calibration curve can be resolved satisfactorily. However, McKerrell (1975) claims that a correction by means of the dendrochronological-calibration curve yields good agreement only for the Older Kingdom of Egypt. According to him, pre-Christian Egyptian samples younger than 2000 BC yield calibrated ages that are older by up to three centuries. Thus, only when the fine structure of the radiocarbon variations are agreed upon will the resulting calibration curve become useful for more precise dating such as by the "wriggle matching method" suggested by Ferguson, Huber & Suess (1966).

Before the entrance of radiocarbon dating, absolute time values on late Pleistocene and Holocene geologic processes were little more than guesswork. Only varve counting was quantitative, but it was also tedious, tricky, and very limited in application. Varve chronologies did not cross the Atlantic very successfully. Radiocarbon was quickly applied to such problems as the ages of continental glacial events and the rates of ocean sedimentation. For the first time, geologists could quantitatively gauge the rates of processes molding our landscape. Palynologists could tell when and how rapidly plant communities responded to climatic and environmental changes. The beginning of the Holocene was found to be more or less synchronous at about 11,000 BP throughout the northern hemisphere land and sea.

Typically, scientific breakthroughs open doors for many other studies, often in completely unforeseen directions. Quantitative dates revived interest in theories of climate change, especially the major Pleistocene glaciation/deglaciation cycles. At last astronomical cause theories could be tested. The presently favored Milankovitch Theory (Hays, Imbrie & Shackleton 1976) owes part of its time test to radiocarbon. As we see in the following section, radiocarbon and climate may be even more directly related. (See also L. M. Libby & Pandolfi 1976.)

^{14}C FLUCTUATION AND ITS CAUSES

The Nature of the Fluctuation

We have seen (Figure 6) that the radiocarbon fluctuations prior to perturbation by industrial activity consist of a quasi-sinusoidal variation with an apparent period of about 10^4 years, which is modulated by de Vries-type fluctuations such as occurred during the Little Ice Age. The de Vries-type fluctuations occur over centuries, whereas the long-term fluctuation occurs over millennia. It is not difficult to conceive of various phenomena that could affect the radiocarbon concentration of the atmosphere and, hence, the radiocarbon concentration in wood samples that derive their carbon from the atmosphere. Table 3, modified from Grey & Damon (1970), outlines a priori possibilities for causing fluctuations in the radiocarbon concentration of the atmosphere.

It is conceivable that the galactic cosmic-ray flux has varied significantly during the past, although data on the abundance of the radioactive and stable nuclides produced by cosmic rays in meteorites and lunar samples indicate that the galactic cosmic-ray flux has been constant within a factor of two during a period of several million years (Arnold, Honda & Lal 1961, Lal 1965, 1974). However, these data do not rule out the possibility of smaller variations or short-term variations that could have been produced by gamma-ray bursts associated with supernovae events (Konstantinov & Kocharov 1965, 1967). Supernovae explosions in the vicinity of the solar system would also cause a much longer-term increase and subsequent decrease resulting from the enhancement of the galactic cosmic-ray flux by the arrival of cosmic rays accelerated in the explosion. In their detailed discussion of possible increases in the radiocarbon production within the atmosphere from supernovae and gamma-ray bursts, Konstantinov & Kocharov (1967) have suggested that measurable quantities of radiocarbon could be produced by that effect. For example, they have estimated a 0.5 to 40°/₀₀ increase due to the gamma-ray bursts from the supernova, Tycho Brahe, which was seen in 1572 AD.

Lingenfelter & Ramaty (1970) have calculated the increment to the terrestrial cosmic-ray flux due to the arrival of the diffuse wave of relativistic cosmic rays from a hypothetical supernova explosion. For an energy of approximately 10^{50} ergs, the flux from a nearby supernova source would rise rapidly over a period of several thousand years, increasing the terrestrial cosmic-ray flux by an amount that could equal or somewhat exceed the current cosmic-ray flux and then decay away over a period of tens of thousands of years. The supernovae observed in historical time are all so distant from the Earth that the cosmic rays from them have not had time to reach the Earth. The presently available data (see Figure 6) show no evidence for the

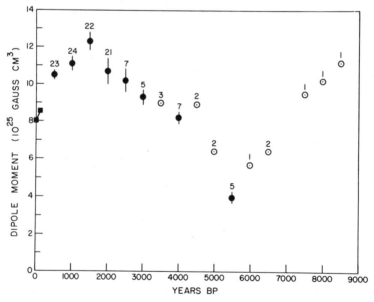

Figure 8   Dipole moment during last nine millennia BP (after Cox 1968). Variation during the last 130 years is known from observatory data. Earlier data are paleomagnetic (P. J. Smith 1967a,b,c). The dipole moments are averaged over 500-year intervals. The number of data averaged per interval is shown above the point. Vertical lines are standard deviations for intervals with sufficient data for calculation of σ.

arrival of cosmic rays from a nearby supernova. However, the possibility exists of a small decaying increment of cosmic rays resulting from a supernova that might have exploded several tens of thousands of years ago. Investigation of this possibility would require a longer time series of tree-ring data than is available at the present time.

It is now well known that the intensity of the Earth's dipole moment has varied during the past nine millennia (see Figure 8). The maximum that occurred near the beginning of the Christian Era is three times more intense than the minimum that occurred in the sixth millennium BP. At the Twelfth Nobel Symposium, which was held in Uppsala, Sweden in August, 1969 (Olsson 1970a), Bucha and Suess pointed out the close correlation between the variation of the intensity of the Earth's dipole moment and the long-term change in the radiocarbon concentration of the atmosphere recorded in tree rings. In the same symposium, Lingenfelter & Ramaty (1970) and also Damon (1970b) presented the results of calculations of the geomagnetic field effect based on two-exchange-reservoir models such as shown in Figure 4. The results of these two calculations were in good agreement with each other and with the hypothesis that the long-term change in radiocarbon activity is the result of changes in the terrestrial geomagnetic field intensity. However, both calculations indicated a best-fit for a geomagnetic field intensity peaking a few centuries earlier than suggested by the data on geomagnetic field intensity.

More recently, Sternberg & Damon (1976) determined the sensitivity of these two exchange–reservoir models to different values for seven independent geomagnetic and reservoir parameters. Their model was also used to calculate theoretical ^{14}C fluctuation curves and theoretical values for the ^{14}C inventory due to the geomagnetic effect. Several of their calculated curves agreed closely with the trend curve in Figure 6, provided that the maximum value of the geomagnetic dipole moment occurred at 2500 BP. There was a tendency for the model fluctuation curve and the data to diverge prior to 6000 years BP. This emphasizes the need for more paleomagnetic and radiocarbon data to establish tighter boundary conditions for theoretical models. The inventories calculated from the model (greater than 122 dpm cm_e^{-2} and less than 133 dpm cm_e^{-2}) were somewhat higher than the estimate from the radiocarbon inventory given in Table 2 (119 dpm cm_e^{-2}). However, the difference between the theoretical and calculated inventories is within the precision of the estimates. Thus, no great discrepancy between production rate and calculated inventory exists.

Solar magnetic activity, as first reported by Forbush (1954), also measurably changes the intensity of galactic cosmic rays incident upon the Earth's atmosphere. This effect modulates the galactic cosmic-ray intensity in antiphase with solar activity and results in the well-known 11-year cycle of neutron production due to modulation of the cosmic-ray flux (Lanzerotti 1977). The exact mechanism for modulation of the cosmic-ray flux is now in dispute (see, for example, Feldman et al 1977). Stuiver (1961, 1965) was the first to show a convincing relationship between the de Vries effect fluctuations and solar activity. Damon (1977b) has recently reviewed the effect of solar activity, as measured by the Wolf Sunspot Number R (Waldmeier 1961), on the radiocarbon concentration of the atmosphere. The reader may refer to that paper for a more detailed discussion. There seems to be general agreement that the de Vries effect during the Maunder and Spörer minima (Eddy 1976a) is the result of the very low solar activity during these minima (see, for example, discussions by Damon, Lingenfelter and Ramaty, and Suess in the Twelfth Nobel Symposium volume, Olsson 1970a). However, the smaller variations in the radiocarbon concentration of the atmosphere, which occurred during the 18th and 19th centuries AD following the Maunder minimum, are more controversial (Damon 1977b). Ekdahl & Keeling (1973) point out that more complicated radio-carbon reservoir models than used by the University of Arizona group, such as their five- and six-reservoir models, do not predict these fluctuations following the Maunder minimum. Oeschger et al (1975) have pointed out that box models that assume first-order exchange between well-mixed reservoirs produce exchange coefficients derived from the natural radiocarbon distribution that do not agree with those derived from the response of the different reservoirs to the bomb radiocarbon input and the fossil fuel CO_2 input. For example, the residence time (first-order exchange) of radiocarbon in the mixed layer of the ocean before transfer into the deep sea is 24 years, as determined from the data in Figure 3. On the other hand, Nydal (1968) has shown that the rapid transfer of radiocarbon produced by nuclear weapon testing from the mixed layer into the deep sea requires a residence time an order of magnitude lower. The problem, according to Oeschger et al, is the assumption of first-order exchange between the mixed layer and the deep sea. The deep sea is not a well-mixed box and cannot be described satisfactorily by a first-order exchange model. Using a box-diffusion model, these authors achieved satisfactory agreement for the dynamic transfer of radiocarbon from nuclear weapons tests and the increase of CO_2 from combustion of fossil fuels. Their model consists of a well-mixed atmospheric reservoir coupled to a biosphere that contains long-term storage of radiocarbon and a well-mixed surface ocean reservoir that transfers radiocarbon by diffusion into the deep sea. G. Lazear and P. E. Damon (in prepara-tion), using a box-diffusion model similar to that of Oeschger et al (1975), successfully modeled the small, short-term fluctuations produced by solar activity during the 18th and 19th centuries (Figure 9).

During recent solar cycles, there has been, approximately, a 22% variation in neutron flux and consequent ^{14}C production. Because the atmosphere-ocean system acts as a low-pass filter (de Vries 1958, 1959, Houtermans 1966), one expects measured radiocarbon in tree-ring sequences to show a 100-fold attenuation of the production amplitude in the 11-year cycle. Measured values seemed to be compatible with such an attenuation until the work of the University of Glasgow group was reported in 1971 (Baxter & Walton 1971). These workers analyzed wine, spirits, and plant seeds comprising annual growth of organic matter during years from AD 1897 to AD 1953 and reported an 11-year radiocarbon cycle with an overall ^{14}C variation of about 30°/oo peak to trough. In an attempt to corroborate this interesting result, the University of Arizona ^{14}C research group made measurements on annual tree rings for the 15-year period from AD 1940 through 1954. Their results were compatible with a much greater attentuation, yielding about a 3°/oo peak to trough amplitude. The reader may refer to papers published simultaneously in *Earth and Planetary*

Science Letters for the status of that controversy as of the summer of 1973 (Baxter & Farmer 1973, Baxter, Farmer & Walton 1973, Damon et al 1973a,b). Since then, results of Stuiver (1974) and the La Jolla research group (Cain & Suess 1976) have failed to confirm the large variation of atmospheric radiocarbon during the 11-year cycle reported by Baxter & Walton (1971). These more recent results are also compatible with the large attenuation predicted by theory.

Following speculation that the Tunguska meteor had introduced antimatter into the terrestrial environment, Cowan, Atluri & Libby (1965) and Vinogradov, Devirts & Dobkina (1966) made measurements of the radiocarbon concentration of tree rings that formed immediately before, during and after the Tunguska event. They predicted that, if the Tunguska meteor had contained antimatter, an increase in the radiocarbon concentration of the atmosphere would be detectable in tree rings. Their results were negative when compared with their own theoretical calculations and those of Marshall (1966). However, a different calculation by Gentry (1966) suggested that a much lower ^{14}C increase would be caused by that hypothesized antimatter explosion. Lerman, Mook & Vogel (1967) remeasured individual tree rings dated

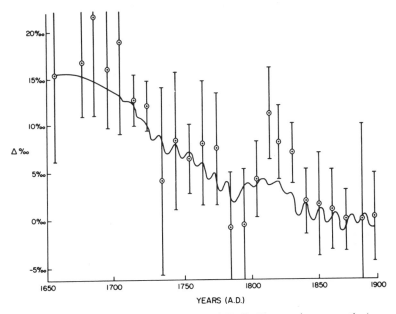

Figure 9 Box-diffusion model (G. Lazear and P. E. Damon, in preparation) curve compared with 10-year averages of data (with $\pm 2\sigma$ error bars) from same data base as in Figure 5. The box diffusion includes atmosphere (*A*), biosphere (*B*), mixed layer (*M*), sedimentary sink (*S*), and deep sea divided into seven boxes. Eddy diffusion transfer is assumed from mixed layer to upper deep sea and within the deep sea. The atmosphere, biosphere, and mixed layer are assumed to be well-mixed with first-order transfer between them. The following parameters are assumed: $\tau_{AM} = 5.7$ yr, K (eddy diffusion constant) = 4480 M^2 yr^{-1}, $R_M/R_A = 0.96$, $N_B/N_A = 1.8$, $\tau_{BA} = 60$, $K_S = 1.04 \times 10^{15}$ g yr^{-1}.

from the same period of time and found no deviations larger than $3°/_{oo}$ around the time of the Tunguska event, but possibly some correlation of Δ with the sunspot cycle (Lerman, Mook & Vogel 1967, Lerman 1970).

Lingenfelter & Ramaty (1970) have predicted atmospheric ^{14}C production from solar flares. Their calculations indicate that the solar flare of February 23, 1956, may have produced an increase in the radiocarbon concentration of the atmosphere by as much as $7.5°/_{oo}$. According to these authors, the production of radiocarbon from all solar flares during the previous solar cycle may have increased the radiocarbon activity of the atmosphere by as much as $11°/_{oo}$. The increment of radiocarbon in the atmosphere due to a single large solar flare has not yet been demonstrated by measurements of the radiocarbon concentration of tree rings.

Following the suggestion by de Vries (1958), various authors have supposed that climate changes may have affected the transfer rates and distribution of radiocarbon between reservoirs, thus producing a change in the radiocarbon activity of the atmosphere (e.g. Damon, Long & Grey 1966, Damon 1968, 1970b, Labeyrie, Delibrias & Duplessy 1970, Suess 1970b). The potential magnitude of this effect can be seen when it is realized that an increase of 1°C in the temperature of the mixed layer of the ocean could result in a 5.8% increase in the CO_2 content of the atmosphere (Eriksson 1963). However, Damon (1970b) modeled the effect of climate change on the radiocarbon concentration of the atmosphere and found that an increase in temperature causes the CO_2 concentration in the atmosphere to rise, but the atmospheric residence time of radiocarbon also lengthens, counter-balancing the larger atmospheric reservoir effect. Thus, the direct effect of climate on the radiocarbon concentration of the atmosphere is probably much less than the geomagnetic effect. Indirect climate effects such as changes in wind velocity and consequent changes in atmosphere-ocean gas exchange rates and ocean circulation rates were not explicitly modeled. Lerman and his colleagues have shown that the de Vries effect is synchronous in both northern and southern hemispheres during the last 500 years. However, the absolute ^{14}C content appears to be depleted by $4.5 \pm 1°/_{oo}$ in the southern hemisphere (Lerman et al 1969, Lerman, Mook & Vogel 1970). They have accounted for this phenomenon by a model that invokes differential atmosphere-sea exchange between northern and southern hemispheres due to the different extent of the ocean areas and to a larger wind agitation in the southern hemisphere "roaring forties," which would largely increase the CO_2 exchanges (Kanwisher 1963).

It is not yet known whether or not the larger climatic fluctuations during the last glaciation had a measurable effect on the radiocarbon concentration of the atmosphere. Again, this problem awaits the construction of a longer dendrochronological time series. To the authors' knowledge, there has not been a definitive study of the effect of a climatically induced change in the biomass on the radiocarbon concentration of the atmosphere.

RADIOCARBON, CLIMATE AND SOLAR ACTIVITY

As pointed out by a number of authors (de Vries 1958, 1959, Damon 1968, Damon 1970b, Suess 1970b), there is a distinct relationship between the radiocarbon concentration of the atmosphere and temperature variations in the northern hemisphere during both the Medieval Warm Epoch (12th and 13th centuries) and the Little Ice Age (15th through 18th centuries). During those epochs when northern hemisphere temperatures were high, the radiocarbon concentration of the atmosphere was low and vice versa (Figure 10). As previously pointed out, during the Little Ice Age, two episodes occurred when the sunspot activity was extremely low and for many years completely absent (Eddy 1976a). The earliest episode occurred between about AD 1450 to 1550 and the second episode of low sunspot activity occurred between AD 1640 and 1715. The reader will observe in Figure 10 that radiocarbon concentrations were unusually high during both the Spörer and Maunder minima. According to Schove (1955), solar activity was unusually high during the Medieval Warm Epoch, and Figure 10 shows a ^{14}C minimum during that period of time. The variations in radiocarbon concentration can be predicted from the relationship between radiocarbon production and solar activity (Lingenfelter & Ramaty 1970). When solar activity is high, the radiocarbon production is low, and vice versa. The inverse relationship between northern hemisphere temperatures and atmospheric radiocarbon concentrations suggests that the sun is responsible for the climatic change as well as for the change in the radiocarbon production. This relationship is in accord with the relationship between solar energy and solar activity,

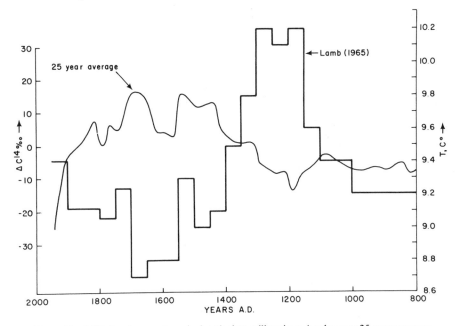

Figure 10   Δ (°/oo) vs. temperature during the last millennium. Δ values are 25-year averages from data base in Figure 5. Temperatures are Lamb's (1965) estimates for average annual temperature prevailing in central England. Note that during the Little Ice Age, when solar activity was low, cold temperatures prevailed and the radiocarbon activity (Δ) was high. Conversely, during the Medieval Warm Epoch, when solar activity was high, the radiocarbon activity was low, and high temperatures prevailed (modified from Damon 1968).

which is suggested by the studies of Abbott (1966) and more recently by Kondratyev et al (1971). These studies suggest that the "solar constant" increases by 2–2.5% as the Wolf Sunspot Number increases, reaching a maximum at a sunspot number (R) of approximately 80–100, after which it decreases to an intermediate value at the highest values of R ($R = 10g + f$, where f is the total number of sunspots regardless of size, and g is the number of sunspot groups). From various personal communications, we are aware that this relationship is not accepted by most solar physicists because of the severe difficulties in accurately measuring the solar constant (for example, see E.v.P. Smith & Gottlieb 1975). This points out the importance of precisely defining fluctuations due to the de Vries effect during the entire time for which we have a precise tree-ring chronology. A close correlation between the de Vries effect and climate throughout that tree-ring record would suggest that, in fact, changes in solar activity are responsible for at least short-term climatic change as well as ^{14}C production. Recently, Eddy (1976b) has proposed a mechanism whereby "a slow and subtle variation of the solar constant modulates the amplitude of the sunspot cycle." Eddy suggests that this variation would be felt directly on Earth in climate changes of roughly 100-year durations, and registered in radiocarbon history through modulation of the long-term envelope of solar activity. He further suggests that this hypothesis would not require a relationship between the 11-year solar cycle and terrestrial climate.

THE HISTORY OF SOLAR ACTIVITY DURING THE LAST EIGHT MILLENNIA

De Vries–effect fluctuations of $\pm 15°/oo$, which occurred during the Medieval Warm Epoch through the Little Ice Age (12–19 centuries AD), have been amply confirmed by a number of laboratories (see Figure 2). As stated previously, these fluctuations can be accounted for by variations in solar activity that have been historically recorded. However, because of the difficulties encountered in sustaining high precision with currently available techniques and the uncoordinated nature of the research to date, no single de Vries–type fluctuation (prior to the Medieval Warm Epoch) has been confirmed by two or more different laboratories. "Wiggles" (de Vries–effect fluctuations) reported by Suess (1970a) at the Twelfth Nobel Symposium have not been confirmed. Pearson et al (1977) have measured precisely the radiocarbon content of tree rings representing 20-year intervals on a floating tree-ring chronology from Northern Ireland. These authors attained a precision for radiocarbon measurements of $\pm 3°/oo$, including stable isotope composition corrections. We have shown that this floating chronology, which is 1140 years in length, corresponds very closely to the interval from 4100 BC to 5240 BC. The de Vries–type fluctuation permitted by the data of the Queens University group (Pearson et al 1977) are almost an order of magnitude less than those reported by Suess (1970a). More recent work of Suess (1978) has reduced the amplitude of the wiggles during a portion of the extant tree-ring record.

Records of daily sunspot observations date back only to the year AD 1700 when the Zurich Observatory first began keeping its records (Waldmeier 1961). Qualitative results, based upon historical reports of sunspots and aurorae, are available for the period from 649 BC to the beginning of the recording of the daily record in AD 1700 (Schove 1955). We now know that the radiocarbon concentration of tree rings contains a record of solar activity that can be retrieved with careful work using present techniques. However, we need more rapid, precise, and accurate techniques to retrieve this highly significant record of solar activity. The correlation observed during the Little Ice Age and Medieval Warm Epoch between radiocarbon, solar activity, and climate and the potential for climate prediction emphasizes the value of and need for high-precision analyses.

RADIOCARBON FLUCTUATIONS DURING THE LAST ICE AGE

Two hand-in-hand reasons for continuing the study of ^{14}C variations into the Pleistocene are (a) to extend the calibration curve and (b) to get more information about the history of the geophysical parameters that modulate the radiocarbon production. The first purpose becomes less important for ages greater than about 10,000 to 20,000 years, beyond which there are no more ties between the events to be dated and the historical time scale. But the second purpose becomes even more important. Radiocarbon analyses are one of the few possibilities contributing to reconstruction of past climate, geomagnetic field, solar activity, etc. Precise dating of these events is very important when one desires, for example, to check astronomic theories of climate change. One of the burning questions that has been asked and even wagered upon (Damon 1970b) is: what were the ^{14}C values doing before 8000 BP?

Can the present data base be extended into the past? We have seen that the bristlecone pine data reach back to about 8000 BP. There is hope of finding enough older wood to allow for cross-dating and perhaps constructing a 10,000-year tree-ring chronology. For the earlier period, other types of samples have been proposed and tried, without any conclusive results. The problem lies mainly in precise independent dating of the samples (cf. discussion of Vogel's paper in the Nobel Symposium). Varve and varve-like lake sediments have been ^{14}C dated, but their results did not produce a concordant picture (see Stuiver 1971, for review, and Yang & Fairhall 1972).

NEED FOR MORE PRECISE AND ACCURATE METHODS OF RADIOCARBON COUNTING

The "factor of merit," which is the net count rate for a modern ^{14}C standard divided by the square root of the background, is the commonly accepted measure of the precision and maximum age obtainable in radiocarbon determinations in a given period of counting time. Five different methods have been proposed for obtaining high factors of merit (Table 4): 1. Use of large proportional counters, operated at high pressures in an elaborate shielding and anticoincidence system. 2. Placing the counter (with shielding and anticoincidence as above) in an underground location. 3. Liquid scintillation counting of benzene prepared from the carbon in large wood samples. 4. Counting the ^{14}C atoms directly as negative ions in an electrostatic

Table 4   Factor of merit $(S/B^{1/2})^a$ obtained by different ^{14}C counting methods

Method	$S/B^{1/2}$	Laboratory
Gas proportional counting with anticoincidence and good shielding in underground laboratory.	70	Quaternary Research Laboratory, University of Washington, Seattle.
Liquid scintillation counting of radiocarbon in benzene.	40	Palaeoecology Laboratory, Queen's University, Belfast.
Negative ion mass spectrometry	0.1	Mass Spectrometry Research Center, Stanford Research Institute, Palo Alto.
Negative ion counting with a tandem electrostatic accelerator and dE/dx detector.	420	University of Rochester Nuclear Structure Laboratory, Rochester, New York.

[a] S = net standard count per unit time, B = background count per unit time. For a given counting time, the maximum measureable age depends only on $S/B^{1/2}$; consequently, it is referred to as a *factor of merit* (Moljk, Drever & Curran 1957). Its square, S^2/B, is a measure of precision referred to as the figure of merit (Oeschger & Wahlen 1975).

accelerator, or 5. Counting the ^{14}C atoms directly by negative ion mass spectrometry of ^{14}CN molecules.

The most recent attempt to achieve a high factor of merit by elaborate shielding-anticoincidence techniques and placing the counter in an underground location was reported by Gulliksen & Nydal at the Ninth International Radiocarbon Conference in 1976 (Gulliksen & Nydal 1978). For their experiments they used a proportional counter that had an effective volume of 1.2 liters filled to 2 atm with CO_2. Their best ground-level arrangement, which included three guard shells and 6.5 cm of old lead, yielded a background of 0.59 ± 0.01 cpm. This corresponds to a factor of merit of 19.5. By moving the counter to a location 380 m below ground, they reduced this background by 0.15 cpm, thus increasing the factor of merit to 22.6. Stuiver (see Oeschger & Wahlen 1975), using a proportional counter containing 13.2 liters of CO_2 in a location about 12 m underground, achieved a background of 1.4 cpm. This corresponds to a factor of merit of about 70.

By using a scintillation mixture consisting of 1 cc of toluene added to 15 cc (13.1 g) of benzene produced from standard oxalic acid, Pearson et al (1977) obtained a net standard count of 122.8 cpm with a background of about 9.3 cpm. This corresponded to a factor of merit of 40.

In a paper delivered at the Twelfth Nobel Symposium, Oeschger et al (1970) pointed out that, if all of the ^{14}C atoms in a sample could be counted rather than only those that disintegrate during the measurement, the figure of merit for ^{14}C counting could be greatly improved. They suggested that attempts should be made to develop a more sensitive ^{14}C determination technique based upon isotope enrichment and mass spectrometric detection. Direct measurement of ^{14}C by mass spectrometry was first attempted by researchers at the Mass Spectrometry Research Center at Stanford Research Institute (Schnitzer et al 1974, Aberth,

Schnitzer & Anbar 1975, Schnitzer & Anbar 1976). They used negative ion mass spectrometry to take advantage of its inherently low background and a two-stage mass spectrometer to suppress problems from scattering. The negative ions were produced by introducing a mixture of CO_2 and N_2 into a duoplasmatron negative ion source. CN^- ions were produced as the predominant ion. These workers achieved adequate sensitivity for ^{14}C dating. However, contamination proved to be the major obstacle. Elimination of background due to hydrocarbons enabled them to obtain a background that was within two orders of magnitude of the level required, but silicon isotopes evaporated from the metals within the duoplasmatron negative ion source proved to be the ultimate limitation. These workers proposed to build a duoplasmatron source out of very pure zone-refined nickel. Lack of funding forced them to discontinue the work before they had pursued it to its ultimate conclusion. Recently, Muller (1977) suggested the direct determination of ^{14}C atoms by means of a cyclotron. The major source of background in a positive ion source is ^{14}N from residual nitrogen in the sample and in the ion source of the cyclotron. The residual nitrogen beam in the 88-inch cyclotron at Berkeley would have to be reduced by a factor of at least 10^6 in order to use dE/dx counters. Muller suggested various ways of reducing the residual nitrogen beam. These included the use of stripping foils, the use of molecular ions rather than ionized atoms, and elimination of nitrogen from the ion source of the cyclotron. No results have yet been reported.

However, Bennett et al (1977) and Nelson, Korteling & Stott (1977) solved the problem by accelerating negative carbon ions from an ion sputter source in a tandem electrostatic accelerator. In their first experiment, Bennett et al achieved a figure of merit of 420 from a charcoal contemporary sample compared to a petroleum-based graphite sample. With the graphite sample, they achieved a 10 μA beam of $^{12}C^-$ ions. However, the compressed charcoal was not as suitable a source of negative ions as the graphite yielding only 1.2 μA of $^{12}C^-$ ions from the ion source. By using graphite instead of charcoal, these researchers at the University of Rochester Nuclear Structure Research Laboratory should be able to achieve a figure of merit of 5000. With such a figure of merit, they will be able to date radiocarbon samples back to 70,000 years without isotope enrichment. In order to use such an instrument for the precise measurements needed to define solar activity–induced variations of radiocarbon in tree rings, it will be necessary to achieve high precision as well as sensitivity. It seems possible to obtain precision comparable to or better than that obtainable by gas proportional counting, e.g. by simultaneously counting ^{12}C and ^{14}C. This may represent the most significant breakthrough in ^{14}C dating within the last two decades.

SUMMARY AND PERSPECTIVE

What at first appeared as erroneous ages resulting from a flaw in the basic assumptions of radiocarbon dating has become a study of the variations of past atmospheric ^{14}C concentrations, with ramifications at least as far-reaching as the radiocarbon dating method itself. Historical records of "blemishes" on the Sun, vacillation of the Earth's magnetic field, instruments of war, and industrial expansion have all contributed to changes in this atom's abundance in the atmosphere, and in return these and other phenomena have yielded valuable information about themselves.

We have learned that atmospheric ^{14}C concentrations as well as climate respond to solar activity. The radiocarbon content of tree rings provides a record of prehistoric solar activity, and radiocarbon may eventually tell us about the history of not only solar output, but climate as well. In earth science, prediction of the future depends heavily on knowledge of the past. The Earth's magnetic field intensity appears to play a moderating role on the solar-induced variations, as well as affecting long-term production rates of ^{14}C.

But on Earth, it seems that everything is interrelated. To understand why ^{14}C in the atmosphere behaves as it does is to understand every sphere of the Earth that carbon participates in and how each sphere dynamically relates to each other one. The more quantitative, dynamic, and detailed the carbon cycle model is, the better predictive qualities it has. Radiocarbon from atomic bombs has provided a tracer to follow carbon in its journey throughout its cycle and to time its rate of transfer.

So we find ourselves at a threshold in the geophysics of natural radiocarbon studies. Present state of the art just barely allows us to measure the most recent and obvious ^{14}C fluctuations induced by solar activity. Sample requirements are large and counter precision is at best one order of magnitude less than the amplitude of the largest of the short-term fluctuations. The tantalizing possibility is that important and revealing but more subtle short-term fluctuations occur and have occurred for longer than sunspots have been historically recorded. Our confidence in predicting solar activity will be directly related to the length and quality of the sunspot record. At present, radiocarbon in tree rings seems to be our only hope of lengthening the record.

On the other side of the radiocarbon coin are accurate dates for anthropological and geological studies. Calibrated dates have overturned several notions concerning the history of mankind; rates of sedimentation based on corrected ^{14}C dates are more exact. Extension of the calibration into the late Pleistocene is eagerly awaited by all. Improved precision of analyses with smaller sample size would allow dating of valuable or rare specimens not sacrificable or simply unobtainable heretofore.

But to improve the quality of the radiocarbon data, a measurement breakthrough is needed. At this writing, it appears that direct ^{14}C atom measurement with high-energy accelerators may open the door for us.

ACKNOWLEDGMENTS

We are grateful to Ms. Sandra Harralson for her excellent work in typing and proofing this manuscript. Our research was supported by National Science Foundation Grant DES76-22629.

Literature Cited

Abbott, C. G. 1966. Account of the astrophysical observatory of the Smithsonian Institution, 1904–1953. *Smithson. Misc. Collect.* 148:1–16

Aberth, W. H., Schnitzer, R., Anbar, M. 1975. Carbon dating mass spectrometry—background reduction techniques. *Proc. 23rd Ann. Conf. on M.S. and Allied Topics*, p. 279

Aitken, M. J., ed. 1974. Radiocarbon dating. In *Physics and Archaeology*, Chap. 2, pp. 26–84. Oxford: Clarendon. 2nd ed.

Arnold, J. R., Honda, M., Lal, D. 1961. Record of cosmic-ray intensity in the meteorites. *J. Geophys. Res.* 66:3519–32

Baes, C. F., Goeller, H. E., Olson, J. S., Rotty, R. M. 1977. Carbon dioxide and climate: the uncontrolled experiment. *Am. Sci.* 65: 310–20

Bannister, B. 1969. Dendrochronology. In *Science in Archaeology*, eds. D. Brothwell, E. Higgs, pp. 191–205. London: Thames & Hudson. 2nd ed.

Baxter, M. S., Farmer, J. G. 1973. Radiocarbon: short-term variations. *Earth Planet. Sci. Lett.* 20:295–99

Baxter, M. S., Farmer, J. G., Walton, A. 1973. Comments on "On the Magnitude of the 11-Year Radiocarbon Cycle" by P. E. Damon, Austin Long, and E. I. Wallick. *Earth Planet. Sci. Lett.* 20:307–10

Baxter, M. S., Walton, A. 1971. Fluctuations of atmospheric carbon-14 concentrations during the past century. *Proc. R. Soc. London Ser. A.* 321:105–27

Bennett, C. L., Beukens, R. P., Clover, M. R., Gove, H. E., Liebert, R. B., Litherland, A. E., Purser, K. H., Sondheim, W. E. 1977. Radiocarbon dating using electrostatic accelerators: negative ions provide the key. *Science* 198:508–10

Berger, R., Fergusson, G. J., Libby, W. F. 1965. UCLA radiocarbon dates IV. *Radiocarbon* 7:336–71

Bolin, B. 1970. The carbon cycle. *Sci. Am.* 223(3):125–32

Broecker, W. S., Olson, E. A. 1959. Lamont radiocarbon measurements VI. *Radiocarbon* 1:111–32

Broecker, W. S., Olson, E. A. 1961. Lamont radiocarbon measurements VIII. *Radiocarbon* 3:176–204

Broecker, W. S., Olson, E. A., Bird, J. 1959. Radiocarbon measurements on samples of known age. *Nature* 183:1582–84

Cain, W. F., Suess, H. E. 1976. Carbon 14 in tree rings. *J. Geophys. Res.* 81:3688–94

Clark, J. G. D. 1965. Radiocarbon dating and the expansion of farming over Europe. *Proc. Prehist. Soc.* 31:58–73

Clark, R. M. 1973. Tree-ring calibration of radiocarbon dates and the chronology of ancient Egypt. *Nature* 243:266–70

Clark, R. M. 1975. A calibration curve for radiocarbon dates. *Antiquity* 49:251–66

Clark, R. M., Renfrew, C. 1972. A statistical approach to the calibration of floating tree-ring chronologies using radiocarbon dates. *Archaeometry* 14:5–19

Cowan, C., Atluri, C. R., Libby, W. F. 1965. Possible anti-matter content of the Tunguska meteor of 1908. *Nature* 206: 861–5

Cox, A. 1968. Lengths of geomagnetic polarity reversals. *J. Geophys. Res.* 73: 3247–60

Craig, H. 1954. Carbon-13 in plants and the relationships between carbon-13 and carbon-14 variations in nature. *J. Geology* 62:115–49

Craig, H. 1957a. The natural distribution of radiocarbon and the exchange time of carbon dioxide between atmosphere and sea. *Tellus* 9:1–17

Craig, H. 1957b. Isotopic standards for carbon and oxygen and correction factors for mass-spectrometric analysis of carbon dioxide. *Geochim. Cosmochim. Acta* 12: 133–49

Currie, L. A. 1972. The evaluation of radiocarbon measurements and inherent statistical limitations in age resolution. See Rafter & Grant-Taylor 1972, pp. H1–H15

Damon, P. E. 1968. The relationship between terrestrial factors and climate. *Meteorol. Monographs* 8:106–11

Damon, P. E. 1970a. Radiocarbon as an example of the unity of science. See Olsson 1970a, pp. 641–44

Damon, P. E. 1970b. Climatic versus magnetic perturbation of the atmospheric C 14 reservoir. See Olsson 1970a, pp. 571–93

Damon, P. E. 1977a. El carbono 14 y la unidad de las ciencias. Presented at *25th Ann. Meet., Asociacion Venezolana para el Avance de la Ciencia* (Aso VAC), 1975. *Acta Cient. Venez.* 28:249–56

Damon, P. E. 1977b. Variations in energetic particle flux at earth due to solar activity. In *The Solar Output and Its Variations*, ed. O. R. White, pp. 429–48. Boulder: Colorado Assoc. Univ. Press.

Damon, P. E., Ferguson, C. W., Long, A., Wallick, E. I. 1974. Dendrochronologic calibration of the radiocarbon time scale. *Am. Antiq.* 39:350–66

Damon, P. E., Long, A. 1962. Arizona radiocarbon dates III. *Radiocarbon* 4:239–49

Damon, P. E., Long, A., Sigalove, J. J. 1963. Arizona radiocarbon dates IV. *Radiocarbon* 5:283–301

Damon, P. E., Long, A., Grey, D. C. 1966. Fluctuation of atmospheric C^{14} during the last six millennia. *J. Geophys. Res.* 71: 1055–63

Damon, P. E., Long, A., Grey, D.C. 1970. Arizona radiocarbon dates for dendrochronologically dated samples. See Olsson 1970a, pp. 615–18

Damon, P. E., Long, A., Wallick, E. I. 1972. Dendrochronologic calibration of the carbon-14 time scale. See Rafter & Grant-Taylor 1972, pp. A28–A43

Damon, P. E., Long, A., Wallick, E. I. 1973a. On the magnitude of the 11-year radiocarbon cycle. *Earth Planet. Sci. Lett.* 20:300–6

Damon, P. E., Long, A., Wallick, E. I. 1973b. Comments on "Radiocarbon : Short-Term Variations" by M. S. Baxter and J. G. Farmer. *Earth Planet. Sci. Lett.* 20:311–14

Damon, P. E., Wallick, E. I. 1972. Changes in atmospheric radiocarbon concentration during the last eight millennia. In *Contributions to Recent Geochemistry and Analytical Chemistry*. Moscow : Nauka Publ. Off., pp. 441–52 (In Russian; preprints in English)

de Vries, Hl. 1958. Variation in concentration of radiocarbon with time and location on earth. *K. Ned. Akad. Wet., Proc. Ser. B.* 61:94–102

de Vries, Hl. 1959. Measurement and use of natural radiocarbon. In *Researches in Geochemistry*, ed. P. H. Abelson, pp. 169–89. New York : Wiley

de Vries, Hl., Barendsen, G. W., Waterbolk, H. T. 1958. Groningen radiocarbon dates II. *Science* 127:129–38

Douglass, A. E. 1919. Climatic cycles and tree growth, Vol. 1. *Carnegie Inst. Washington Publ. 289*

Eddy, J. A. 1976a. The Maunder minimum. *Science* 192:1189–1202

Eddy, J. A. 1976b. The sun since the Bronze Age. In *Physics of Solar Planetary Environments*, ed. D. J. Williams, 2: 958–72. Washington, D.C.: Amer. Geophys. Union

Ekdahl, C. A., Keeling, C. D. 1973. Atmospheric CO_2 in the natural carbon cycle: I. Quantitative deductions from records at Mauna Loa Observatory at the South Pole. In *Carbon and the Biosphere*, 24th Brookhaven Symp. Biol., pp. 51–85. Springfield, Va : Natl. Tech. Inf. Serv.

Eriksson, E. 1963. Possible fluctuations in atmospheric carbon dioxide due to changes in the properties of the sea. *J. Geophys. Res.* 68:3871–76

Farmer, J. G., Baxter, M. S. 1972. Short-term trends in natural radiocarbon. See Rafter & Grant-Taylor 1972, pp. A58–A71

Feldman, W. C., Asbridge, J. R., Bame, S. J., Gosling, J. T. 1977. Plasma and magnetic fields from the sun. See Damon 1977b.

Ferguson, C. W. 1970a. Dendrochronology of bristlecone pine, *Pinus aristata*. Establishment of a 7484-year chronology in the White Mountains of eastern-central California, U.S.A. See Olsson 1970a, pp. 237–59

Ferguson, C. W., 1970b. Concepts and techniques of dendrochronology. In *Scientific Methods in Medieval Archaeology*, ed. R. Berger, pp. 183–200. Berkeley : Univ. Calif. Press

Ferguson, C. W., Gimbutas, M., Suess, H. E. 1976. Historical dates for Neolithic sites for southeastern Europe. *Science* 191: 1170–72

Ferguson, C. W., Huber, B., Suess, H. E. 1966. Determination of the age of Swiss lake dwellings as an example of dendrochronologically-calibrated radiocarbon dating. *Z. Naturforsch.* 21a:1173–77

Fleischer, R. L. 1974. Neutrons from lightning? *Gen. Electr. Tech. Inf. Ser. Rep. No. 74CRD260.* 9 pp.

Fleischer, R. L., Plumer. J. A., Crouch, K. 1974. Are neutrons generated by lightning? *J. Geophys. Res.* 79:5013

Fleming, A. 1975. The implications of calibration. In *Radiocarbon: Calibration and Prehistory*, ed. T. Watkins, Edinburgh : Edinburgh Univ. Press, pp. 101–8

Forbush, S. E. 1954. Worldwide cosmic-ray variations 1937–1952. *J. Geophys. Res.* 59:525–45

Garrels, R. M., Mackenzie, F. T. 1971. *Evolution of Sedimentary Rocks*, New York : Norton. 397 pp.

Gentry, R. V. 1966. Anti-matter content of the Tunguska meteor. *Nature* 211:1071–72

Godwin, H. 1962. Radiocarbon dating. *Nature* 195:984

Grey, D. C. 1969. Geophysical mechanisms for ^{14}C variations. *J. Geophys. Res.* 74: 6333–40

Grey, D. C., Damon, P. E. 1970. Sunspots and radiocarbon dating in the Middle Ages. See Ferguson 1970b, pp. 167–82

Gulliksen, S., Nydal, R. 1978. Further improvement of counter background and shielding. *Proc. 9th Int'l. Radiocarbon Conference, Los Angeles, San Diego, 1976.* In press

Hamilton, E. I. 1965. *Applied Geochronology*. New York: Academic. 267 pp.

Harkness, D. D., Burleigh, R. 1974. Possible carbon-14 enrichment in high altitude wood. *Archaeometry* 16:121–27

Hays, J. D., Imbrie, J., Shackleton, N. J. 1976. Variations in the Earth's orbit: Pacemaker of the Ice Ages. *Science* 194: 1121–32

Houtermans, J. C. 1966. On the quantitative relationships between geophysical parameters and the natural C 14 inventory. *Z. Phys.* 193:1–12

Houtermans, J. C. 1971. *Geophysical interpretations of bristlecone pine radiocarbon measurements using a method of Fourier analysis for unequally-spaced data*. PhD thesis. Univ. Berne, Switzerland.

Houtermans, J. C., Suess, H. E., Oeschger, H. 1973. Reservoir models and production rate variations of natural radiocarbon. *J. Geophys. Res.* 78:1897–1908

Huber, B. 1970. Dendrochronology of central Europe. See Olsson 1970a, pp. 233–35

Jansen, H. S. 1962. Comparison between ring-dates and C 14-dates in a New Zealand kauri tree. *N.Z. J. Sci.* 5:74–84

Jansen, H. S. 1970. Secular variations of radiocarbon in New Zealand and Australian trees. See Olsson 1970a, pp. 261–74

Kanwisher, J. 1963. On the exchange of gases between the atmosphere and the sea. *Deep Sea Res.* 10:195

Karlén, I., Olsson, I. U., Kallberg, P., Kilicci, S. 1964. Absolute determination of the activity of two C^{14} dating standards. *Ark. Geofys.* 4:465–71

Keeling, D. C. 1973. The carbon dioxide cycle: reservoir models to depict the exchange of atmospheric carbon dioxide with oceans and land plants. In *Chemistry of the Lower Atmosphere*, ed. S. I. Rasool, pp. 251–329. New York: Plenum

Kigoshi, K., Hasegawa, H. 1966. Secular variations of atmospheric radiocarbon concentration and its dependence on geomagnetism. *J. Geophys. Res.* 71:1065–71

Kigoshi, K., Kobayashi, H. 1966. Gakushuin natural radiocarbon measurements V. *Radiocarbon* 8:54–73

Kondratyev, K. Ya., Nikolsky, G. A., Murcray, D. G., Kosters, J. J., Gast, P. R. 1971. The solar constant from data of balloon investigations in the USSR and the USA. In *Space Research*, 11:695–703. Berlin: Akademic

Konstantinov, B. P., Kocharov, G. E. 1965. Astrophysical phenomena and radio-carbon. *Dokl. Akad. Nauk. SSSR* 165:63–4

Konstantinov, B. P., Kocharov, G. E. 1967. Astrophysical phenomena and radio-carbon. *A. F. Loffe Physico-Tech. Inst.*, *Prepr. 064*, 43 pp.

Labeyrie, J., Delibrias, G., Duplessy, J. C. 1970. The possible origin of natural carbon radioactivity fluctuations in the past. See Olsson 1970a, pp. 539–46

Lal, D. 1965. Some aspects of astrophysical studies based on observations of isotopic changes. *Proc. Intl. Conf. Cosmic Rays, London, England*, pp. 81–91

Lal, D. 1974. Long term variations in the cosmic ray flux. *Philos. Trans. R. Soc. London Ser. A* 277:395–411

LaMarche, V. C., Jr., Harlan, T. P. 1973. Accuracy of tree-ring dating of bristlecone pine for calibration of the radiocarbon time scale. *J. Geophys. Res.* 78:8849–58

Lamb, H. H. 1965. The early Medieval Warm Epoch and its sequel. *Paleogeogr., Paleoclimatol., Paleoecol.* 1:13–37

Lanzerotti, L. J. 1977. Solar and galactic energetic particles. See Damon, 1977b, pp. 383–403

Lerman, J. C. 1970. Discussion of causes of secular variations. See Olsson 1970a, pp. 609–10

Lerman, J. C. 1972. Carbon 14 dating: origin and correction of isotope fractionation errors in terrestrial living matter. See Rafter & Grant-Taylor 1972, pp. H16–H28

Lerman, J. C. 1974a. Isotope "paleothermometers" on continental matter: Assessment. In *Les Méthodes Quantitatives d'Étude des Variations du Climat au Cours du Pleistocene*, pp. 163–81. Paris: Int. CNRS, no. 219

Lerman, J. C. 1974b. *Les isotopes du carbone: variation de leur abondance naturelle, application aux corrections de datations radiocarbone, a l'étude du metabolisme vegetal et aux paleoclimats*. PhD dissertation, Univ. Paris, and also Note—CEA. In press

Lerman, J. C., Mook, W. G., Vogel, J. C. 1967. Effect of the Tunguska Meteor and sunspots on radiocarbon in tree rings. *Nature* 216: 990–91

Lerman, J. C., Mook, W. G., Vogel, J. C. 1970. C 14 in tree rings from different localities. See Olsson 1970a, pp. 275–301

Lerman, J. C., Mook, W. G., Vogel, J. C., de Waard, H. 1969. C 14 in Patagonian tree rings. *Science* 165:1123–25

Libby, L. M., Lukens, H. R. 1973. Production of radiocarbon in tree rings by lightning bolts. *J. Geophys. Res.* 78:5902

Libby, L. M., Pandolfi, L. J. 1976. Isotopic

tree thermometers: Correlation with radiocarbon. *J. Geophys. Res.* 81:6377–81

Libby, W. F. 1955. *Radiocarbon Dating.* Chicago: Univ. Chicago Press. 175 pp. 2nd ed.

Lingenfelter, R. E., Ramaty, R. 1970. Astrophysical and geophysical variations in C 14 production. See Olsson 1970a, pp. 513–37

Long, A., Arnold, L. D., Damon, P. E., Ferguson, C. W., Lerman, J. C., Wilson, T. A. 1978. Radial translocation of carbon in bristlecone pine. *Proc. 9th Int. Radiocarbon Conf.,* Los Angeles, San Diego. In press

Marshall, L. 1966. Non-anti-matter nature of the Tunguska Meteor. *Nature* 212: 1226–27

McKerrell, H. 1975. Correction procedures for C-14 dates. See Fleming 1975, pp. 47–100

Michael, H. N., Ralph, E. K. 1972. Discussion of radiocarbon dates obtained from precisely dated *Sequoia* and Bristlecone Pine samples. See Rafter & Grant-Taylor 1972, pp. 28–43

Mielke, J. E., Long, A. 1969. Smithsonian Institution radiocarbon measurements V. *Radiocarbon* 11:163–82

Moljk, A., Drever, R. W. P., Curran, S. C. 1957. The background of counters and radiocarbon dating. *Proc. R. Soc. London* 239:433–45

Muller, R. A. 1977. Radioisotope dating with a cyclotron. *Science* 196:489–94

Münnich, K. O. 1957. Heidelberg natural radiocarbon measurements I. *Science* 126: 194–99

Nelson, D. E., Korteling, R. G., Stott, W. R. 1977. Carbon-14: Direct determination at natural concentrations. *Science* 198:507–8

Neustupný, E. 1970a. The accuracy of radiocarbon dating. See Olsson 1970a, pp. 23–34

Neustupný, E. 1970b. Radiocarbon chronology of central Europe from c. 6450 B.P. to c. 3750 B.P. See Olsson 1970a, pp. 105–8

Neustupný, E. 1970c. A new epoch in radiocarbon dating. *Antiquity* 44:38–45

Nydal, R. 1968. Further investigation on the transfer of radiocarbon in nature. *J. Geophys. Res.* 73:3617–35

Oeschger, H., Houtermans, J., Loosli, H., Wahlen, M. 1970. The constancy of cosmic radiation from isotope studies in meteorites and on the Earth. See Olsson 1970a, pp. 471–500

Oeschger, H., Siegenthaler, U., Schotterer, U., Gugelmann, A. 1975. A box diffusion model to study the carbon dioxide exchange in nature. *Tellus* 27:168–92

Oeschger, H., Wahlen, M. 1975. Low level counting techniques. *Ann. Rev. Nucl. Sci.* 25:423–63

Olsson, I. U. 1968. Modern aspects of radiocarbon datings. *Earth-Sci. Reviews* 4:203–18

Olsson, I. U. 1970a. ed. *Radiocarbon Variations and Absolute Chronology.* Proc. XII Nobel Symp., New York: Wiley. 652 pp.

Olsson, I. U. 1970b. Explanation of Plate IV. See Olsson 1970a, pp. 625–26

Olsson, I. U. 1974. The eighth international conference on radiocarbon dating. *Geol. Foeren. Stockholm Foerh.* 96:37–44

Olsson, I. U., El-Gammal, S., Goksu, Y. 1969. Uppsala natural radiocarbon measurements IX. *Radiocarbon* 11:515–44

Olsson, I. U., Klasson, M., Abd-el-Mageed, A. 1972. Uppsala natural radiocarbon measurements XI. *Radiocarbon* 14:247–71

Olsson, I. U., Osadebe, F. A. N. 1974. Carbon isotope variations and fractionation corrections in ^{14}C dating. *Boreas* 3: 139–46

Pauling, L. 1958. Genetic and somatic effects of carbon-14. *Science* 128:1183–86

Pearson, G. W., Pilcher, J. R., Baillie, M. G. L., Hillam, J. 1977. Absolute radiocarbon dating using a low altitude European tree-ring calibration. *Nature* 270:25–28

Provasoli, L. 1963. Organic regulation of phytoplankton fertility. In *The Sea,* ed. M. N. Hill, 2:165–219. New York: Interscience

Rafter, T. A., Fergusson, G. J. 1957. "Atomic bomb effect"—recent increase of carbon-14 content of the atmosphere and biosphere. *Science* 126:557–58

Rafter, T. A., Grant-Taylor, T., eds. 1972. *Proc. 8th Int. Radiocarbon Dating Conf.* Wellington, New Zealand: R. Soc. N.Z., 723 pp. 2 Vol.

Rafter, T. A., O'Brien, B. J. 1970. Exchange rates between the atmosphere and the ocean as shown by recent C 14 measurements in the South Pacific. See Olsson 1970a, pp. 355–78

Rafter, T. A., O'Brien, B. J. 1972. ^{14}C measurements in the atmosphere and in the South Pacific Ocean. —a recalculation of the exchange rates between the atmosphere and the ocean. See Rafter & Grant-Taylor 1972, pp. C17–C42.

Ralph, E. K., Michael, H. N. 1970. MASCA radiocarbon dates for Sequoia and bristlecone-pine samples. See Olsson 1970a, pp. 619–24

Ralph, E. K., Michael, H. N., Han, M. C.

1973. Radiocarbon dates and reality. *MASCA Newslett.* 9:1–20

Ralph, E. K., Stuckenrath, R. 1960. Carbon-14 measurements of known age samples. *Nature* 188:185–87

Renfrew, C. 1973. *Before Civilization.* New York: Knopf. 292 pp.

Ronov, A. B., Yaroshevsky, A. A. 1969. Chemical composition of the earth's crust. In *The Earth's Crust and Upper Mantle, Geophys. Monogr. 13,* ed. P. J. Hart. Washington, D. C.: Amer. Geophys. Union

Sakharov, A. D. 1958. Radiocarbon from nuclear tests and "non-threshold" effects of atomic radiations. *At. Energ.* 5:576–80. Transl. in *At. Energy (USSR)* 4:757–62 (from Russian)

Schnitzer, R., Aberth, W. H., Brown, H. L., Anbar, M. 1974. Mass spectrometric carbon dating technique. *Proc. 22nd Ann. Conf. Mass Spectrom. and Allied Top.,* p. 64

Schnitzer, R., Anbar, M. 1976. Scope and limitations of mass spectrometric determination of carbon-14: *Proc. 24th Ann. Conf. Mass Spectrom. and Allied Top.,* p. 361

Schove, D. J. 1955. The sunspot cycle 649 B.C. to 2000 A.D. *J. Geophys. Res.* 60:127–45

Schulman, E. 1956. *Dendroclimatic Changes in Semiarid America.* Tucson: Univ. Arizona Press. 142 pp.

Schwarcz, H. P. 1969. Isotopes in nature. In *Handbook of Geochemistry,* ed. K. H. Wedepohl, II-1:6-B-1–6-B-16. New York: Springer

Sellers, W. D. 1974. A reassessment of the effect of CO_2 variations on a simple global climatic model. *J. Appl. Meteorol.* 13:831–33

Skopintsev, B. A. 1950. Organic matter in natural waters. *Tr. Geol. Inst. Akad. Nauk SSSR* 17:29

Smith, E. v. P., Gottlieb, D. M. 1975. Solar flux and its variations. In *Possible Relationships between Solar Activity and Meteorological Phenomena, NASA Publ. No. SO-377,* ed. W. R. Bandeen, S. P. Maran, pp. 97–118

Smith, P. J. 1967a. The intensity of the ancient geomagnetic field: a review and analysis. *Geophys. J. R. Astron. Soc.* 12:321–62

Smith, P. J. 1967b. Ancient geomagnetic field intensities—I. Historic and archaeological data, sets H1–H9. *Geophys. J. R. Astron. Soc.* 13:417–19

Smith, P. J. 1967c. Ancient geomagnetic field intensities—II. Geological data: sets G1–

G21; historic and archaeological data: sets H10–H13. *Geophys. J. R. Astron. Soc.* 13:483–86

Sternberg, R. S., Damon, P. E. 1976. Sensitivity of radiocarbon fluctuations and inventory to geomagnetic and reservoir parameters. See Gulliksen & Nydal 1977

Stuiver, M. 1961. Variations in radiocarbon concentration and sunspot activity. *J. Geophys. Res.* 66:273–76

Stuiver, M. 1965. Carbon-14 content of 18th-and 19th-century wood, variations correlated with sunspot activity. *Science* 149:533–35

Stuiver, M. 1969. Yale natural radiocarbon measurements IX. *Radiocarbon* 2:545–658

Stuiver, M. 1971. Evidence for the variation of atmospheric ^{14}C content in the late Quaternary. In *The Late Cenozoic Glacial Ages,* ed. K. K. Turekian, pp. 69–70. New Haven: Yale Univ. Press

Stuiver, M. 1974. Natural radiocarbon in the 19th century (abstract). *Proc. Geol. Soc. Amer. Meet., Boulder, Colorado.*

Stuiver, M., Robinson, S. W. 1974. University of Washington Geosecs North Atlantic carbon-14 results. *Earth Planet. Sci. Lett.* 23:87–90

Stuiver, M., Suess, H. E. 1966. On the relationship between radiocarbon dates and true ages. *Radiocarbon* 8:534–40

Suess, H. E. 1955. Radiocarbon concentration in modern wood. *Science* 122:415–17

Suess, H. E. 1961. Secular changes in the concentration of atmospheric radiocarbon. In *Proc. Highland Park, Ill., Conf. NAS-NRC Publ. 845,* pp. 90–95

Suess, H. E. 1965. Secular variations of the cosmic-ray-produced carbon-14 in the atmosphere and their interpretations. *J. Geophys. Res.* 70:5937–52

Suess, H. E. 1967. Bristlecone pine calibration of the radiocarbon time scale from 4100 B.C. to 1500 B.C. In *Radioactive Dating and Methods of Low-Level Counting,* pp. 143–51. Vienna: IAEA

Suess, H. E. 1968. Climatic changes, solar activity and the cosmic-ray production rate of natural radiocarbon. *Meteorol. Monogr.* 8:146–50

Suess, H. E. 1970a. Bristlecone pine calibration of the radiocarbon timescale 5200 B.C. to the present. See Olsson 1970a, pp. 303–11

Suess, H. E. 1970b. The three causes of the secular C 14 fluctuations, their amplitudes and time constants. See Olsson 1970a, pp. 595–606

Suess, E. H. 1978. The carbon-14 level during the fourth and second half of the fifth millennium B.C. and the carbon-14 calibration curve. See Gulliksen & Nydal, 1978. In press

Suess, H. See Houtermans 1971

Switsur, V. R. 1973. The radiocarbon calendar recalibrated. *Antiquity* 47:131–37

Tamers, M. A. 1969. Radiocarbon dating of Recent events. *Atompraxis* 15:1–6

Tauber, H. 1958. Difficulties in the application of C 14 results in archaeology. *Archaeol. Austriaca* 24:59–69

Verniani, F. 1966. The total mass of the earth's atmosphere. *J. Geophys. Res.* 71:385–92

Vinogradov, A. P., Devirts, A. L., Dobkina, E. I. 1966. Concentration of C^{14} in the atmosphere at the time of the Tunguska catastrophe and antimatter. *Dokl. Akad.*

Nauk. 168:900–3 (Transl. in *Geochemistry,* pp. 185–88)

Waldmeier, M. 1961. *The Sunspot Activity in the Years 1610–1960.* Zurich: Schulthess

Watkins, T. 1975. ed. *Radiocarbon: Calibration and Prehistory.* Edinburgh: Edinburgh Univ. Press. 147 pp.

Wendland, W. M., Donley, D. L. 1971. Radiocarbon-calendar age relationship. *Earth Planet. Sci. Lett.* 2:135–39

Willis, E. H., Tauber, H., Münnich, K. O. 1960. Variations in the atmospheric radiocarbon concentration over the past 1300 years. *Radiocarbon* 3:1–4

Yang, Albert In Che, Fairhall, A. W. 1972. Variations of natural radiocarbon during the last 11 millenia and geophysical mechanisms for producing them. See Rafter & Grant-Taylor 1972, p. A44–A57

Ann. Rev. Earth Planet. Sci. 1978. 6: 495–523

SYNOPTIC EDDIES IN THE OCEAN

✲10103

M. N. Koshlyakov and A. S. Monin

P. P. Shirshov Institute of Oceanology, USSR Academy of Sciences, Moscow, USSR

1 DEFINITIONS

Synoptic oceanic eddies are defined as nonstationary eddylike disturbances of the oceanic circulation with a horizontal scale of the order of the Rossby internal scale $R = \bar{N}hf^{-1}$, where f is the Coriolis parameter, h the thickness of the baroclinic layer in the ocean, and $\bar{N}$ the average over the baroclinic layer Väisälä frequency; in the internal parts of the anticyclonic subtropical gyres R is of the order of 50 km. The use of the term "synoptic" instead of the term "mesoscale" emphasizes the physical analogy between the phenomenon under consideration and the synoptic eddies in the atmosphere (cyclones and anticyclones). This analogy can be made on the basis of their similar physical natures (Rossby waves, the determining role for which is played by the Coriolis parameter), the suggested prevailing mechanism of generation (baroclinic instability of large-scale currents), the corresponding horizontal scales (Rossby atmospheric and oceanic scales) and in the quasi-geostrophic character of the motion.

The experimental data obtained so far make it possible to subdivide the synoptic oceanic eddies into two classes:

1. Frontal eddies that are produced by the cutoff of meanders from such frontal currents as the Gulf Stream and the Kuroshio.
2. Open-ocean eddies that are, as it is understood at the present time, quasi-horizontal and quasi-geostrophic waves of the synoptic scale.

This review deals primarily with the open-ocean eddies discovered and experimentally studied during the past decade. The frontal eddies are considered mainly to elucidate common features and differences between them and the open-ocean eddies. The theoretical part of the review is subordinate and deals primarily with the results directly related to experiment.

2 OPEN-OCEAN SYNOPTIC EDDIES

2.1 First Serious Indications

The most fundamental contribution to the studies of synoptic eddies of the open ocean has been made by the Soviet POLYGON-70 experiment (Brekhovskikh et al

495

0084-6597/78/0515-0495$01.00

1971) and the American–British Mid-Ocean Dynamics Experiment (MODE-1) (US POLYMODE Organizing Committee, 1976). However, these were a logical continuation of the studies of nonstationary ocean currents carried out earlier by oceanographers of different countries. For instance, POLYGON-70 was one in a series of the Soviet experiments specially aimed at studying variability of the sea and ocean currents on the basis of long-term measurements of currents at moorings. These "polygon" studies were initiated by V. B. Stockmann, who made a series of

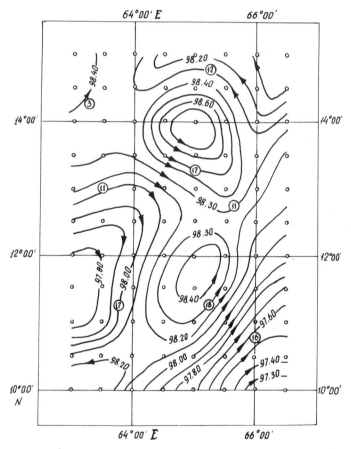

Figure 1   Geostrophic currents at a depth of 150 m from the hydrographic survey in the POLYGON-67 area on 20 March to 6 April, 1967 [after Koshlyakov, Galerkin & Truong Din Hien (1970)]. Numbers near the stream lines are the values $Q \equiv -\int_{z_0}^{z} \tilde{\rho}(x, y, z)\, dz$ in units corresponding to density in $(\sigma_t - 20)$, and depth in hundreds of meters; $z_0 = 1500$ m is the depth of the surface of no motion; $\tilde{\rho}(x, y, z)$ is the density distribution smoothed in the horizontal plane for eliminating random "noise." Numbers in circles designate the velocity in cm sec^{-1}. Dots show the array of the bathometric stations.

long-term measurements of currents in the Caspian Sea as early as 1935 (Stockmann & Ivanovsky 1937). (By "polygon" we mean long-term measurements concentrated in a limited ocean area.) In 1956 a polygon was deployed in the Black Sea (Ozmidov 1962), in 1958 in the North Atlantic (Ozmidov & Yampolsky 1965), and in 1967 in the Arabian Sea (Stockmann et al 1969).

Of the experiments listed above, the Arabian Sea polygon (POLYGON-67) was the first during which synoptic oceanic eddies were detected, though mainly by an indirect method (Figure 1). A repeated survey of the same region revealed that the synoptic-scale currents were primarily nonstationary. The measurement of currents at moorings during POLYGON-67 was not extensive enough to elucidate directly the spatial structure of the current field. However, a rather good coincidence between the measured and the calculated currents was obtained (Koshlyakov, Grachev & Truong Din Hien 1972).

Now one can say with certainty that the density fluctuations shown in Figure 2 were caused by the passing of baroclinic synoptic eddies. The prevailing period of the fluctuations has proved to be 95 days (Yasui 1961). Very interesting is the intermittence of the fluctuations—the eddies were obviously spreading by separate "packets," the variability period of the amplitude fluctuation being approximately a year and a half.

Results of the measurements of deep currents carried out with the Swallow neutrally buoyant floats in the area to the west of Portugal in 1958 (Swallow & Hamon 1960) and in the vicinity of Bermuda in 1959–1960 (Crease 1962, Swallow 1971) made a great impact among oceanographers. The Bermuda measurements, the results of which appeared to be particularly important, were made at depths of 2 and 4 km. Despite the assumption of a weak quasi-stationary water transport in the meridional direction, findings were made of nonstationary currents with velocities of about 40 cm sec^{-1}(!) at a depth of 4 km. The measurements were not intensive

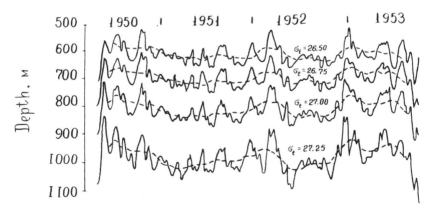

Figure 2 Low-frequency fluctuations of isopycnic surface depths in 1950–1953 from the measurements at the "Tango" station (24°N, 135°E), after Yasui (1961). Dashed curves denote the result of smoothing with a period of 100 days.

enough to be able to reveal the structure of the current; nevertheless a rough estimate was made of the lateral and time scales of the velocity field, which turned out to be 100 km and 50 days.

Wunsch (1972) has calculated the temperature fluctuation spectrum at a depth of 500–600 m from the hydrographic measurements that were made at H. Stommel's initiative at a point 25 km to the east-southeast of Bermuda every half month from 1954 to 1969. The obtained spectrum peak in the period of 100 days is evidently related (again as we now understand it) to baroclinic synoptic eddies.

Having applied spectral analysis to the two-year (1965–1967) measurements of currents at several depths at an ocean point immediately north of the Gulf Stream (39°N, 70°W), Thompson (1971) has obtained pronounced energy spectra maxima in a period of about 40 days.

The observations described above were not the only ones (although from our point of view, they were the most important) that by the late 1960s and the early 1970s showed the existence in the ocean depths of strong nonstationary long-period motions. However, there were a number of unanswered cardinal questions such as:

1. Are these currents typical for the ocean? Do they exist in the really "open" ocean at a great distance from the frontal currents like the Gulf Stream and the Kuroshio?
2. What is the nature of the currents? Is this turbulence or waves, and if waves, what are they—plane or two-dimensional? Do they fill up the space continuously or do they represent individual disturbances?
3. What are the scales of the current field? How are the spatial and the time scales interrelated? Does this interrelation conform to the known theoretical models of nonstationary oceanic motions?
4. What is the energy of these currents compared to the energy of other components of the ocean motion spectrum? What is the mechanism of their generation?

It became obvious that a new experiment would be necessary to answer these questions, though maybe in a partial or preliminary form, through direct current measurements. POLYGON-70 was such an experiment.

2.2 POLYGON-70

The POLYGON-70 experiment was deployed in spring–summer, 1970, in the tropical zone of the North Atlantic, in the eastern part of the North Equatorial Current (Brekhovskikh et al 1971). The main pycnocline centered at 100–150 m depths is very sharp here. According to the computations of currents from the given density field (Yenikeyev & Koshlyakov 1973), the large-scale geostrophic current in the POLYGON-70 area is directed westward-southwestward in the upper layer of the ocean and has a velocity of 3 cm sec^{-1}; at depths below 300 m it is replaced by a deep countercurrent the velocity of which does not exceed 1 cm sec^{-1}.

The observations during POLYGON-70 included mainly measurements of currents at 17 moorings spaced along the arms of the rectangular cross with a center at 16°30′ N, 33°30′ W (Figures 3, 4); the length of each arm was 100 km, and the measurements were made at 10 depths from 25 to 1500 m. This system of stations was maintained continuously from late February to early September, 1970.

The spectral analysis of the time series of velocity components obtained during the POLYGON-70 measurements has shown a deep energy density minimum in the period of 3–4 days (Vasilenko, Mirabel & Ozmidov 1976). This energy density minimum can serve as a basis for the low-pass filtration of the above-mentioned time series, which was made with an effective period of 3.5 days for all the points and depths of the measurements. Some of the distributions of the synoptic-scale currents

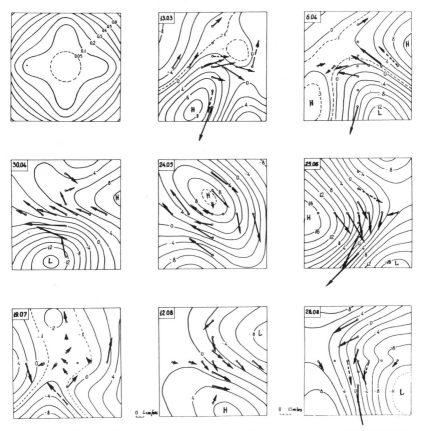

Figure 3 Evolution of the synoptic-scale currents at 300 m depth from the POLYGON-70 data, after Grachev & Koshlyakov (1977). Dates are shown in the upper left-hand corners of the figures. Arrows stand for the low-frequency components of velocity vectors at observational points; dotted arrows have been obtained by interpolation in depth or time. The absence of arrow at a point of observations means the absence of measurements. Curves are stream lines calculated with the aid of objective analysis; numbers near the curves show the stream function values in 10^7 cm^2 sec^{-1}; H is the high pressure, L the low pressure. Distance and velocity scales are given below. The square is centered at 16°30'N, 33°30'W; the square side is 280 km. In the upper left-hand corner the distribution of the mean error of stream function interpolation is given at the presence of the initial data for all the thirteen points of measurements used.

obtained in this experiment are presented in Figures 3 and 4. The objective analysis was performed by the method of optimum interpolation of the homogeneous and isotropic random vector field according to Gandin (1964). The analysis of the synoptic patterns for different depths and dates, including those shown in Figures 3 and 4, leads to the following main conclusions (Koshlyakov & Grachev 1973, Koshlyakov 1978):

1. In the POLYGON-70 area several cyclonic and anticyclonic eddylike velocity disturbances were recorded, of which one anticyclonic eddy with the center in the vicinity of the polygon center in the second half of May was measured particularly well.

2. "Close packing" of the eddies obviously prevailed. Weak current periods in the polygon center are interpreted as the periods of passing through the polygon area of saddle regions between four eddies.

Figure 4   Evolution of the synoptic-scale currents at 1000 m depth from the POLYGON-70 data, after Grachev & Koshlyakov (1977).

3. The lateral scale of the eddies (distance from the eddy center to the maximum velocity point) was very stable, decreasing from 110–120 km at a depth of 300 m to 100 km at a depth of 1000 m.
4. The eddies moved westward (with a small southward component) at an average velocity of 5–6 cm sec^{-1}. This value was particularly stable for the main anti-cyclone.
5. The slope of the axis of the main anticyclone in the direction nearly opposite to that of its drift was definitely recorded (Figures 3 and 4). This slope resulted in a 60-km shift between the positions of the eddy center at depths of 300 and 600 m in the second half of May, which at a 440-km wave length corresponds to 50° phase shifts of velocity fluctuations.
6. On the average, current velocity in the eddy field was of the order of 10 cm sec^{-1} at depths of 200–1000 m, but at some points it was instantaneously as great as 25 cm sec^{-1} at depths of 200–300 m, 35 cm sec^{-1} at depths of 400–600 m, 20 cm sec^{-1} at depths of 1000 m, and 10 cm sec^{-1} at depths of 1500 m. A sufficiently reliable indication of velocity increase was obtained in the rear part of the main anticyclone during May–June from 10 to nearly 17 cm sec^{-1} at a depth of 300 m and an even greater increase at a depth of 600 m.

A physical interpretation of the POLYGON-70 data in the light of the Rossby wave theory (see Section 5) was presented in a number of papers (Koshlyakov & Grachev 1973, Koshlyakov 1973, McWilliams & Robinson 1974, Fomin & Yampolsky 1977, Brekhovskikh et al 1978). This interpretation indicates that from the theoretical point of view during POLYGON-70 thorough direct measurements of baroclinic Rossby waves were made in the open ocean for the first time. It will be shown in Section 5 that the properties of the main eddy of POLYGON-70 described under items 5 and 6 above are indicative of the development of this eddy due to baroclinic instability of the large-scale current.

The vertical structure of the synoptic currents in the POLYGON-70 area was studied by Vasilenko & Mirabel (1977) through the expansion of the current measurements in the system of vertical natural orthogonal functions (Obukhov 1960). It was established that the first three modes of this expansion practically covered the vertical variability of the currents.

2.3 MODE-1

MODE-1 (Mid-Ocean Dynamics Experiment) was the second experiment specially aimed at studying synoptic-scale ocean currents. Its intensive phase was carried out by oceanographers from the United States and the United Kingdom in March–July 1973 in the Sargasso Sea area with the center at 28° N, 69°40′ W, and a radius of about 200 km. During MODE-1 a wide complex of oceanographic observational methods and techniques was used to make the following main investigations: (a) current and temperature measurements at more than 20 moorings in the layer from 500 m to the ocean bottom, with 4 depths of current measurements and 7 depths of temperature measurements, on the average, at each mooring; (b) measurements of currents with the SOFAR floats at a depth of 1500 m — altogether about 25 launchings of the floats for a period of 1–4 months each during the intensive phase; (c) 8 density surveys of the area mainly down to the ocean bottom. All these measurements have yielded maps of synoptic eddies for different depths and dates (for instance, McWilliams 1976a, US POLYMODE Organizing Committee 1976). Examples of these maps are given in Figure 5.

The principal result of MODE-1, as well as that of POLYGON-70, is the discovery of several close packed synoptic eddies. Particularly thorough measurements were

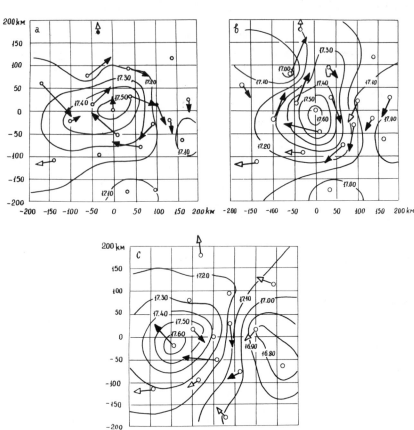

Figure 5   Evolution of the synoptic components of the current and temperature fields at 420 m depth from the MODE-1 data (POLYMODE Organizing Committee, 1976). (*a*) 15–18 April; (*b*) 9–12 May; (*c*) 2–5 June, 1973. The square is centered at 28°N, 69°40'W. Circles denote the mooring array. Arrows stand for velocity vectors obtained through the low-pass filtration of velocity time series with a period of 4 days; nonblackened arrows are not quite reliable. The velocity scale is given below. Curves are temperature in °C isolines constructed by the optimum interpolation method after Gandin from the measurements at moorings.

made of one anticyclonic eddy (Figure 5), which by a number of its parameters much resembled the main anticyclone of POLYGON-70. The lateral scale of the MODE-1 eddies was 90–95 km, somewhat less than that of POLYGON-70 eddies, which is explained by a smaller local value of the Rossby internal scale. The main eddy of MODE-1 drifted westward at an average velocity of 2 cm sec^{-1}, much more slowly than did the POLYGON-70 eddies because of a practical absence in the MODE-1 area of mean large-scale currents. Current velocity in the MODE-1 eddy field turned out to be somewhat higher than that in the POLYGON-70 area. The temperature measurements at the MODE-1 central mooring showed that the main MODE-1 anticyclone was spreading to the very ocean bottom (US POLYMODE Organizing Committee 1976). The interpretation of the MODE-1 data in the light of the Rossby wave theory shows that the vertical structure and the westward drift of the MODE-1 eddies are well simulated by a combination of two barotropic and two baroclinic (of the first mode) Rossby waves (McWilliams & Flierl 1976).

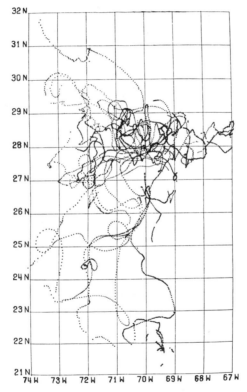

Figure 6 Trajectories of SOFAR floats at 1500 m depth in the MODE-1 area from October, 1972, through December, 1974 (Rossby, Voorhis & Webb 1975). Almost all the floats were launched in a circle with a 100 km radius and the center at 28°N, 69°40′W. The floats were in operation mainly from a month to two years. Dots show float positions every other day.

The SOFAR float measurements (Figure 6) were made at a depth where the barotropic component of the synoptic-scale currents was prevalent (for instance, McWilliams & Flierl 1976). Apart from orbital motions of the floats in the eddy field, Figure 6 attracts attention by the prevalence of a westward transport of the floats (i.e. in the direction of the phase velocity vector of the eddies) compared to an eastward transport and the dispersion of the floats primarily in the meridional direction. All this is indicative of an appreciable part of the advective (turbulent) form of motions in the eddy field; the corresponding effective value of the horizontal turbulent diffusion coefficient has turned out to be $8 \times 10^6 \, cm^2 \, sec^{-1}$ (Freeland, Rhines & Rossby 1975).

2.4   Other Observations

The great amounts of experimental data collected so far demonstrate the occurrence of synoptic eddies in different areas of the World Ocean. In many cases eddies are traced among the old data (mainly among the density survey results) that were obtained by the 1970s but did not receive the attention due them until POLYGON-70 and MODE-1 experiments. For instance, as far back as 1958 the density surveys and direct current measurements from aboard a drifting ship definitely revealed a typical (100-km scale) anticyclonic synoptic eddy in the Pacific Ocean north of New Guinea (Burkov & Ovchinnikov 1960). Current velocity in the eddy field was as great as $75 \, cm \, sec^{-1}$ at 100–200 m depths. The eddy found in the southern part of the Equatorial Counter Current moved to the east at a velocity of nearly $15 \, cm \, sec^{-1}$.

The density survey and the surface current measurements made with the electromagnetic sensor in summer, 1964, revealed a very strong (with the surface current velocity of $1 \, m \, sec^{-1}$) anticyclonic eddy at 35° N off the eastern coast of Australia (Hamon 1965). The subsequent observations (Andrews & Scully-Power 1976) confirmed the quasi-constant occurrence of anticyclonic eddies in this part of the East Australian Current.

Traces of the baroclinic synoptic eddies were definitely established on a system of meridional and zonal sections through the North Equatorial Current in the vicinity of the Hawaiian Islands. These sections were repeated 16 times with an interval of one month in 1964–1965 (Wyrtki 1967, Bernstein & White 1974). According to these data, the eddies apparently drifted westward at a mean velocity of about $4.5 \, cm \, sec^{-1}$, approximately 3 times as fast as did the westward large-scale geostrophic current in the upper ocean. There are indications that the amplitude of the eddies grew during the periods of amplification of the above-mentioned zonal current (Seckel 1975).

Sound velocity maps constructed from the surveys of 1966 and 1969 (Beckerle & La Casce 1973) show well-pronounced synoptic eddies at a depth of 800 m in the southwestern part of the Sargasso Sea. Disturbances of the oceanographic fields associated with the synoptic eddies were observed in the eastern part of the near-equatorial Pacific (White 1973), in the area southwest of the southern extremity of Africa (Duncan 1968), and in the central part of the southern half of the Pacific (Patzert & Bernstein 1976). In the spectra of temperature fluctuations at a depth of 250 m based on the 1966–1969 measurements from aboard weather ships to the south and to the north of the central part of the North Atlantic Current, there are clearly defined peaks in the periods of about 100 days that can naturally be related to baroclinic synoptic eddies (Gill 1975). Note, however, the absence of such peaks in the similar spectra in the vicinity of the Labrador, the East Greenland, the Norwegian, and the Irminger currents (Gill 1975).

The foregoing discussion dealt only with eddies of the synoptic scale occurring over the entire ocean depth and dominating in the main thermocline, immediately above the latter and in the ocean depths. The observations, however, reveal the presence in the upper ocean (the "seasonal" thermocline and the near-surface homogeneous layer) of eddies of smaller sizes – with a lateral scale of 5–50 km. Such eddies were reported by POLYGON-70 (Kort, Byshev & Tarasenko 1974) and MODE-1 (for instance, Voorhis, Schroeder & Leetmaa 1976), and were found in the California Current [McEwen 1948(!), Reid, Schwartzlose & Brown 1963], in the northern (Swallow 1971) and the southwestern (Beckerle 1972) parts of the Sargasso Sea, under the ice in the Arctic Basin (Hunkins 1974) and in other parts of the World Ocean.

Even the incomplete review presented above indicates convincingly that the phenomenon of synoptic eddy formation is typical for the World Ocean. The general tendency is such that this phenomenon is particularly clear-cut in the regions of the most pronounced large-scale currents. This conclusion is obviously related to the problem of genesis of synoptic eddies; we return to this question after a brief consideration of frontal eddies.

3 FRONTAL SYNOPTIC EDDIES

As numerous observations show (for instance, Hansen 1970), the Gulf Stream over its path from Cape Hatteras to at least the Grand Banks represents an unstable jet current characterized by the development of horizontal waves 300–400 km in length. The transition of these waves to so-called meanders, and the subsequent cutoff of the meanders and their transformation into cold cyclonic eddies to the right and warm anticyclonic eddies to the left of the Gulf Stream is also well known. The classical example of the cutoff of a Gulf Stream cyclonic meander and its transformation into a cold eddy is given by Fuglister & Worthington (1951; see also Monin, Kamenkovich & Kort 1977), who described the results of several quasi-synchronous hydrographic surveys of the Gulf Stream made in the area between 56° and 65° W in April, 1950. The rotation velocity in the eddy field near the ocean surface immediately after the cutoff was 3 cm sec^{-1}; it decreased with depth due to the geostrophic effect.

The formation of warm anticyclonic eddies north of the Gulf Stream is almost analogous. According to Fuglister (1972), five or somewhat more pairs of the Gulf Stream cyclones and anticyclones are formed per year in the path from Cape Hatteras to the Grand Banks.

The comparison of the oceanographic and the meteorological data shows that the formation of the Gulf Stream cyclones and anticyclones is qualitatively similar to the process of the cutoff of the meanders and their transformation into cyclonic and anticyclonic eddies in the western stream in the upper atmospheric layers (for

instance, Palmen & Newton 1969). It may be suggested that this fact is indicative of
the unity of the formation mechanism of the atmospheric and the oceanic eddies,
which apparently consists of baroclinic instability of the atmospheric and the oceanic
zonal currents.

The formation of cold cyclonic (for instance, Masuzawa 1957) and warm anticyclonic (for instance, Kawai 1972) eddies as a result of the cutoff of the Kuroshio
meanders in the area immediately east of Japan is analogous, for the most part, to the
eddy formation in the Gulf Stream field. The size, amplitude, and duration of an
anticyclonic eddy, shown in Figure 7, are amazing. Surface current velocity in the

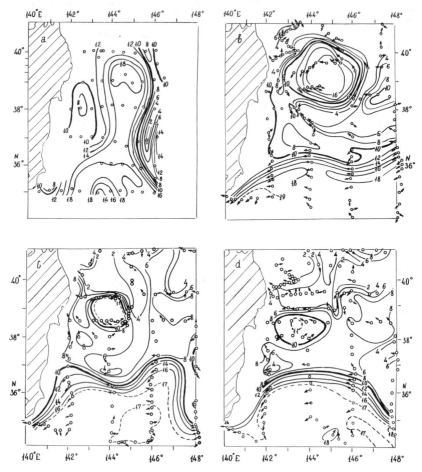

Figure 7 Temperature (°C) distributions at 200 m depth and current directions at the
ocean surface east of the Island of Honshu (Kawai 1972): (*a*) 1–21 July, 1960; (*b*) 12–21
August, 1960; (*c*) 15 February–14 March, 1961; (*d*) 3–27 August, 1961. Circles show the
bathometric station array; current direction was measured with an electromagnetic sensor.

eddy field immediately after its formation was as great as 2 m sec^{-1}. It should be emphasized, however, that, unlike the Gulf Stream frontal eddies and the Kuroshio cyclones, the Kuroshio anticyclones after their formation are not of a solitary eddylike type drifting in a surrounding quiet ocean. On the contrary, the vast area of the Pacific Ocean east of Japan to nearly 180° longitude represents the most hydrologically and hydrodynamically complicated region in the world. It is filled with pulsating streams and branches of the Kuroshio, the Oyashio, and the North Pacific Current, as well as moving and continuously deforming eddies of both signs that are formed mainly from the cutoff of the meanders of the above-mentioned currents but are partially due to a simple recirculation effect at the internal sides of the wave-like bends of their streams. The results of the studies of the structure and dynamics of the above-mentioned ocean area are described in the papers by Masuzawa (1955), Bubnov (1960), Ichiye (1956), Koshlyakov (1961), Bulgakov (1967), Barkley (1968), Kawai (1972), Byshev, Grachev & Ivanov (1976), Pavlychev (1975), Kitano (1975), Bernstein & White (1977), Wilson & Dugan (1977), and other investigators. A definite dynamical analogy between this part of the Pacific Ocean and the comparable region of the North Atlantic and the Labrador currents east of the Grand Banks is quite possible.

Let us consider in brief the properties of the frontal (mainly cyclonic) eddies of the Gulf Stream. A radial cross section of the typical Gulf Stream cyclone ("ring") is seen in the western (at 60° W) part of the section in Figure 8. A solitary character of the ring is well pronounced. The young Gulf Stream rings are characterized by the following parameters (Fuglister 1963, 1972, Barrett 1971, Parker 1971, Richardson, Strong & Knauss 1973, Cheney et al 1976). The diameter of the ring, defined as the mean width of the cold anomaly region, is nearly 200 km. The horizontal temperature difference in the main thermocline may be as great as 10–12°C, which corresponds to a difference of 600–700 m in heights of the isothermic surfaces! The velocity of rotation in the upper part of the ring is often 2 m sec^{-1} and even more. The rings penetrate the ocean as deep as at least 3000 m and possibly to the bottom (for instance, Fuglister 1963). The young rings "leave signatures" at the ocean surface in the form of a low-temperature patch that makes possible their detection from satellites (for instance, Vukovich 1976). As the ring is aging, mainly in the process of by-layer turbulent mixing of its water with the surrounding water of the Sargasso Sea (for instance, Lambert 1974), a slow decrease of its diameter, horizontal temperature difference, and rotation velocity becomes obvious (Fuglister 1972; Cheney & Richardson 1976). Some months after the ring's formation the temperature signal at the ocean surface disappears (usually due to direct atmospheric influence).

Having been formed, the Gulf Stream rings drift, as a rule, westward and southwestward in the Sargasso Sea at a mean velocity of about 3 cm sec^{-1} (Parker 1971; Lai & Richardson 1977). It may be supposed that this drift is caused both by the internal dynamics of the rings (Warren 1967, Flierl 1977), and by the effect of a large-scale current directed, apparently, over the entire ocean column westward and southwestward in the Sargasso Sea east and south of the Gulf Stream (Keondjian 1972, Worthington 1977, Schmitz 1977). The cardinal question is whether the drift of the rings is principally of advective or of wave character. Observations of the

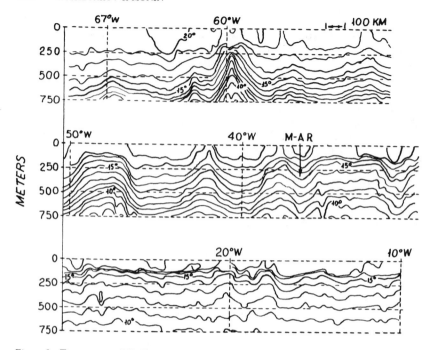

Figure 8 Temperature (°C) distribution on the section along 34°30′N in the Atlantic Ocean on 22 January to 2 February, 1975, after Seaver (US POLYMODE Organizing Committee, 1976). Distance between the bathythermograph stations was 20 km. M-AR denote the axis position of the Mid-Atlantic Ridge.

physical and chemical properties of the rings (Lambert 1974) and the measurements with the neutrally buoyant floats (Cheney et al 1976) have shown that the upper portion of the rings. to depths of approximately 700–1000 m, is occupied mainly by the water moving together with the ring; this is principally slope water originating from the Atlantic region north of the Gulf Stream. On the contrary, in the ocean depths the "wave" form of motion evidently prevails and the trajectories of the particles cross the ring area (Cheney et al 1976).

The Gulf Stream rings disappear either by being absorbed by the Florida Current or by completely decaying in the Sargasso Sea (Parker 1971, Lai & Richardson 1977). The average life duration of a ring is estimated as two to three years (Lai & Richardson 1977). Combining this estimation with the estimation of the number of rings formed per year (5 or somewhat more, Fuglister 1972), we deduce that about 15 rings should be observed at a time in the Sargasso Sea. The latter conclusion is in good agreement with the experimental result represented in Figure 9—altogether 11 rings were seen at a time, although it is possible that 1–2 rings in the eastern or the southern parts of the area were omitted. It should be noted that in the region of their greatest occurrence (Figure 9) the rings occupy about one third of the area of the corresponding part of the ocean! It is obvious that together with the warm Gulf Stream anticyclones the rings should act a rather significant part in heat exchange between the subtropical and the subpolar zones of the ocean.

The warm Gulf Stream anticyclones, unlike its cold cyclones, are characterized by noticeably smaller horizontal and vertical scales and a smaller amplitude (energy) (for instance, Sounders 1971). After their formation they drift at a velocity of about 5 cm sec^{-1} westward and southwestward in the region north of the Gulf Stream and decay by coalescing with the Gulf Stream in the vicinity of Cape Hatteras (Lai & Richardson 1977). Their life duration averages half a year so that in the slope water region nearly three Gulf Stream anticyclones can be recorded at a time.

The general features of the frontal eddies and of the open-ocean eddies described above, which make it possible to combine them into a single type of the oceanic synoptic eddies, are obvious—they are the same as the general features of the oceanic and the atmospheric eddies pointed out in the introductory section. It follows from the foregoing, however, that the frontal eddies have some specific features distinguishing them, say, from the POLYGON-70 or MODE-1 eddies:

1. Frontal eddies are formed in the regions of the frontal stream oceanic currents through the cutoff of the meanders of these currents; for this reason, frontal eddies immediately after their formation contain inside them water that has a different origin from the surrounding water.

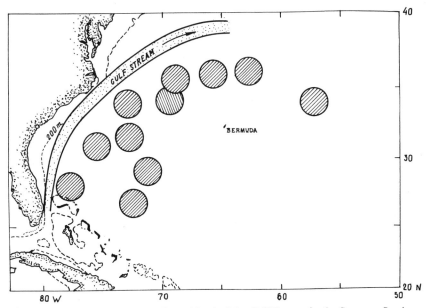

Figure 9   Distribution of cold cyclones (rings) of the Gulf Stream in the Sargasso Sea in November, 1971 (Lai & Richardson 1977). Ring positions were determined by bathy-thermograph measurements. Unlike the figure in the original paper, the present figure gives the natural size of the rings, i.e. 200 km in diameter.

2. Frontal eddies are quasi-solitary formations (this conclusion, strictly speaking, does not relate to the Kuroshio anticyclones). On the one hand, from the frontal current only anticyclones are formed; on the other hand, only cyclones.

3. Advective form of motion is of great significance in the drift of frontal eddies; at any rate, this form of motion prevails in the upper parts of these eddies.

4. Frontal eddies are exceptionally concentrated formations. Their specific kinetic energy is two orders of magnitude higher than the specific kinetic energy of the typical open-ocean eddies.

However, it would be naïve to suppose that the division of the oceanic eddies into the two types settles the problem and that in the ocean there are no formations intermediate in their features between, say, the Gulf Stream rings and the eddies of the POLYGON-70 type. It follows from the foregoing that the eddies of the Kuroshio-Oyashio region are in a definite sense such intermediate formations. Another example of this type of formation is represented by the central part of the section in Figure 8. Whereas a typical Gulf Stream ring is seen in the western part of the section, and a continuous symmetrical field of wave disturbances similar to those recorded during POLYGON-70 and MODE-1 in the eastern part (for instance, in the area of 20° W), the entire middle part of the section is occupied by the intermediate-type disturbances—they are not solitary formations of the ring type nor a symmetrical field of the rises and falls of the isothermic surfaces; the regions of the rises corresponding to the cyclonic sign of the eddy are more localized and more clearly pronounced in the middle part of the section than in the regions of isothermic falls. Note that the growth of the degree of "loneliness" of the disturbances from the east to the west in Figure 8 follows the increase of their energy; as will be seen from Section 5, this finds explanation in the theory of oceanic synoptic eddies.

Weakened cyclonic eddies in the Atlantic Ocean south of the North Atlantic Current and east of 55° W (to which, in particular, the isothermic rise regions in the middle part of the section in Figure 8 can be related) were called by the U.S. oceanographers "big babies" (McCartney, Worthington & Schmitz 1977). It is quite possible that these eddies are formed through the cutoff of the meanders of separate branches into which the Gulf Stream divides on passing the Grand Banks traverse.

4 ESTIMATES OF ENERGY

Two important facts should be noted. First, if the regions of the stable jetlike western and equatorial currents are excluded, the mean density of the available potential energy of the large-scale oceanic currents [i.e. the store P_m of potential energy related to the deviation of isopycnic surfaces from the horizontal position (see Section 5)] exceeds the average density of kinetic energy K_m by approximately 1000 times (Stommel 1966, Gill, Green & Simmons 1974, Vulis & Monin 1975). Second, the mean densities of the available potential P_e and kinetic K_e energy of synoptic oceanic eddies are values of the same order of magnitude (for instance, Kamenkovich & Reznik 1978). The difference between the properties of large-scale currents and eddies is easy to understand if it is taken into consideration that P is proportional to the mean square of depth of disturbances of isopycnic surfaces, and K, because of the geostrophic relationship, is proportional to the mean square of the slopes of isopycnic surfaces (Gill, Green & Simmons 1974, Kamenkovıch & Reznik 1978); thus the P/K ratio is defined by the horizontal scale of the current field.

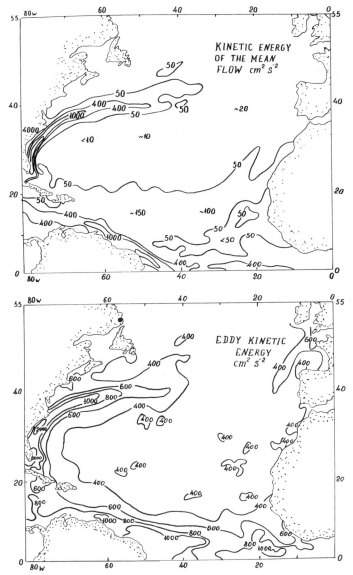

Figure 10   Distributions of specific kinetic energy of the mean currents and mean specific kinetic energy of the synoptic eddies in the near-surface layer of the North Atlantic (Wyrtki, Magaard & Hager 1976). Energy density in cm² sec⁻².

The maps in Figure 10 are constructed from the navigational data; energy deviation of the surface current from the climatic mean was interpreted as a synoptic disturbance. Figure 10 has two basic peculiarities: (a) general similarity of K_m and $\overline{K}_e$ distributions (the dash means averaging in time); (b) essential (on the average, by an order of magnitude) excess of $\overline{K}_e$ over K_m, except for the Gulf Stream region to 60° W. Rather interesting is the high degree of occurrence of synoptic disturbances in the near-equatorial region of the ocean.

The distribution of $\overline{P}_e$ in the core of the main thermocline in the North Atlantic was studied by Dantzler (1977) from numerous bathythermograph measurements. The results show a very strong maximum in the Gulf Stream region, weak maxima to the southeast of the Azores and in the North Equatorial Current (POLYGON-70), and a distinctly pronounced minimum along approximately 25° N, i.e. in the zone of a practical absence of large-scale currents. Analyzing the temperature sections from Fuglister's (1960) Atlas, Gill, Green & Simmons (1974) have obtained $P_m/\overline{P}_e \approx 10$ for the main thermocline in the central part of the major anticyclonic gyre of the North Atlantic. This estimate is in good agreement with the POLYGON-70 and MODE-1 results giving $\overline{K}_e/K_m \approx 100$ for depths corresponding to the thermocline. The analysis of the long-time current measurements at depths of 2000 to 5000 m in the Gulf Stream area between 70° and 50° W and south of the latter has shown a rapid decrease of $\overline{K}_e$ southward from the Gulf Stream (Schmitz 1977). The $\overline{K}_e/K_m$ ratio at these depths at 34° N appeared to be of the order of 10.

All these results are alike in two points:

1. Eddies are stronger where large-scale currents are stronger.
2. Kinetic (and approximately equal available potential) energy of eddies is essentially higher, as a rule, than kinetic energy of large-scale currents and at the same time much lower than available potential energy of the latter.

These conclusions give very strong experimental evidence in favour of the theory of eddy generation through baroclinic instability of the large-scale oceanic currents.

5   ELEMENTS OF THE THEORY

5.1   *Some General Relationships*

The contents of this section and the next two sections are based largely on the chapter "Rossby waves" written by Kamenkovich & Reznik (1978) for the book "Ocean Hydrodynamics." We do not refer specially to the paper by Kamenkovich & Reznik in these sections, but ask the reader to bear the above-mentioned circumstance in mind.

Taking into account the experimentally established (see above) spatial and time scales of synoptic eddies, we can write the equations for the local dynamics of the eddies in a definite ocean region (except for a narrow equatorial zone) in the following form. Horizontal velocities u, v are expressed in terms of pressure perturbation p' by geostrophic formulae $u = -\rho_0^{-1} f_0^{-1}(\partial p'/\partial y)$, $v = \rho_0^{-1} f_0^{-1}(\partial p'/\partial x)$, and density perturbation ρ' by the hydrostatic equation $\rho' = -g^{-1}(\partial p'/\partial z)$. Equations for the vertical component of the absolute vorticity and for entropy are reduced to the form:

$$\frac{1}{\rho_0 f_0}\left(\frac{\partial}{\partial t} + u\frac{\partial}{\partial x} + v\frac{\partial}{\partial y}\right)\Delta p' + \beta_0 v - f_0\frac{\partial w}{\partial z} = \frac{1}{\rho_0^2}\frac{\partial(\rho',p')}{\partial(x,y)}; \tag{1}$$

$$\left(\frac{\partial}{\partial t} + u\frac{\partial}{\partial x} + v\frac{\partial}{\partial y}\right)\rho' - \frac{\rho_0}{g}N^2(z)\cdot w = 0. \tag{2}$$

These equations are written in the local Cartesian coordinates: The x-axis is directed to the east, the y-axis to the north, and the z-axis vertically upwards. The origin of coordinates lies at an undisturbed ocean surface in the center of the region under consideration, $\Delta \equiv \partial^2/\partial x^2 + \partial^2/\partial y^2$; t is the time, w is the vertical velocity, and g is the acceleration of gravity; $f_0 \equiv 2\Omega\sin\phi_0$ and $\beta_0 \equiv 2\Omega a^{-1}\cos\phi_0$ are the Coriolis parameter and its latitudinal variation in the center of the region, where Ω is the angular velocity of the Earth's rotation, a its radius, and ϕ_0 the geographic latitude; ρ_0 is the mean water density in the ocean,

$$N^2(z) = \left[-g\rho_0^{-1}(d\rho_s/dz) - g^2c^{-2}(z)\right]$$

is the square of Väisäla frequency, and $\rho_s(z)$ and $c(z)$ are the vertical distributions of density and sound velocity in the undisturbed state. The first and the second terms in the left-hand part of (1) describe local change and advection of the relative vorticity and advection of the "planetary vorticity," i.e. changes of vorticity of a meridionally moving particle due to variation of the Coriolis force; the third term describes the effect of vortex stretching as a result of the horizontal current divergence in the Coriolis force field. The right-hand part of the equation describes the vorticity change of a particle due to the baroclinity effect (Bjerknes solenoids).

With the use of the geostrophic and hydrostatic equations and (2), Equation (1) can be transformed to contain one variable p':

$$\left(\frac{\partial}{\partial t} - \frac{1}{\rho_0 f_0}\frac{\partial p'}{\partial y}\frac{\partial}{\partial x} + \frac{1}{\rho_0 f_0}\frac{\partial p'}{\partial x}\frac{\partial}{\partial y}\right)\mathscr{L}p' + \frac{\beta_0}{\rho_0 f_0}\frac{\partial p'}{\partial x} = \frac{1}{g\rho_0^2}\frac{\partial\left(p', \frac{\partial p'}{\partial z}\right)}{\partial(x,y)};$$

$$\mathscr{L}p' \equiv \frac{1}{\rho_0 f_0}\Delta p' + f_0\frac{\partial}{\partial z}\left[\frac{1}{\rho_0 N^2(z)}\frac{\partial p'}{\partial z}\right]. \tag{1a}$$

Equation (1a) is called the vorticity equation in a quasi-geostrophic approximation. Its analog was used for the first time by Obukhov (1949) and Charney (1949) in a principal scheme of a short-term weather forecast; later the equation was studied by many authors (for instance, Charney & Stern 1962, Phillips 1963, Pedlosky 1964a,b, Monin 1969, Kamenkovich 1973, Monin, Kamenkovich & Kort 1977).

Under usual boundary conditions at the boundaries of an ocean region it is not difficult to derive the following integral equation of energy

$$\int_V \rho_0\frac{u^2+v^2}{2}dV + \int_V \frac{g^2}{\rho_0 N^2}\frac{\rho'^2}{2}dV + \int_\Sigma g\rho_0\frac{\zeta^2}{2}d\Sigma = \text{const}, \tag{3}$$

where Σ and V are the area of the horizontal section of the region and its volume and $\zeta(x, y, t)$ is the ocean surface disturbance. The first term in the left-hand part of (3) represents integral kinetic energy of synoptic eddies, and the sum of the two other terms is integral available potential energy of eddies [value $(g^2/N^2)(\rho'^2/2)$ is the volume density of available potential energy].

The estimation of the terms of the vorticity equation (1a) was made by McWilliams (1976b) from the current measurements with the floats and the density surveys during MODE-1. It has turned out that during short time intervals (about 10 days) the local changes of relative vorticity were caused mainly by its advection. However, if one passes to the vorticity equation averaged by the time period of the order of 60 days, vorticity balance changes essentially—the linear terms having the same order become the leading ones in Equation (1a); it is this balance that determines the mean drift of eddies to the west. Similar results were obtained from the analysis of the measurements of current during POLYGON-70 (Brekhovskikh et al 1978). The analysis of the heat balance equation similar to Equation (2) but with the replacement of density ρ' by heat quantity $c\rho_0 T'$, where T is the temperature and c the heat capacity, was made by Bryden (1976) from the temperature and current measurements in the 700- to 2000-m layer at the MODE-1 moorings. The analysis produced the same order of all the terms of the equation and revealed the importance of advective effects in the eddy field. The density surveys carried out during MODE-1 showed a significant amount of horizontal advection in heat balance in the near-surface layer of the ocean as well (Voorhis, Schroeder & Leetmaa 1976). A similar conclusion can be drawn from the density survey data obtained during POLYGON-67 (Koshlyakov, Galerkin & Truong Din Hien 1970).

5.2   Baroclinic Instability of the Large-Scale Currents

As stated above, the hypothesis of synoptic eddy formation through baroclinic instability of the large-scale oceanic currents is the leading one at the present time. Consider for simplicity the case of a zonal large-scale geostrophic current $U(y, z)$ in an infinite channel limited by vertical zonal walls, a horizontal bottom, and a "solid lid" at the fluid surface. By relating Equations (1)–(2) to the sum of the zonal current and synoptic eddies, by considering such eddy-associated perturbations that are first, periodic in x with the period X and second, small in amplitude, and consequently by neglecting the nonlinear terms in Equations (1) and (2), we obtain the following equation describing energy exchange between the large-scale current and the eddies:

$$\frac{\partial}{\partial t} \int_V \left(\frac{u^2 + v^2}{2} + \frac{g^2}{N^2 \rho_0^2} \frac{\rho'^2}{2} \right) dV = \int_V (-uv) \frac{\partial U}{\partial y} dV + \int_V \left(-\frac{g^2}{N^2 \rho_0^2} v\rho' \frac{\partial \bar{\rho}}{\partial y} \right) dV. \quad (4)$$

The first term in the right-hand part of (4) describes the energy change of the eddies due to barotropic instability of the mean current, the second, the energy change due to its baroclinic instability; in the latter case the eddy energy source lies in available potential energy of the mean current. The estimates show that, except for the regions of the jetlike currents of the type of the Florida Current, the ocean is characterized by a decisive (on the average, by an order of magnitude) prevalence of the second term of the right-hand part (4) over the first one.

In the paper of Eady (1949) on baroclinic instability of the zonal current with a linear vertical velocity shift it is shown that as the shift exceeds a certain critical value, the current becomes unstable and the maximum rate of growth of perturbations is realized on a scale close to the Rossby internal scale (see Section 1). This fundamental result can account for the prevailing horizontal scale of the synoptic eddies.

Baroclinic instability of the oceanic currents proper was the subject of a large number of the theoretical studies (for instance, Pedlosky 1964a, 1964b, 1975a, 1975b, Tareev 1965, Schulman 1967, Robinson & McWilliams 1973, Gill, Green & Simmons 1974, Orlansky & Cox 1973). The most important result of these studies lies in the conclusion that under some additional conditions quite realizable in the ocean the vertical velocity shift of only several centimeters per second can be sufficient for instability of the large-scale current and the generation of eddies.

Let us deduce an important property of the unstable synoptic disturbances of the current field. Assume for simplicity $U = U(z)$ and thus remove the first term from the right-hand part (4). Pressure perturbation is given as a plane wave

$$p' = \Phi(z) \cos \left[kx - kc_r t + \theta(z) \right] \exp (kc_i t)$$

(k is the wave number, c_r the phase velocity, $\Phi(z)$ the amplitude, $\theta(z)$ the initial phase, c_i the amplitude change rate). By substituting the corresponding geostrophic velocity perturbations v and hydrostatic density perturbations ρ' in the right-hand part (4) we obtain the relationship

$$\frac{\partial}{\partial t} \int_V (K_e + P_e) dV = - \frac{k}{2f_0} e^{2kc_i t} \int_V \Phi^2(z) \frac{d\theta}{dz} \left(\frac{\partial \bar{\rho}}{\partial y} \middle/ \frac{d\rho_s}{dz} \right) dV. \tag{5}$$

It follows from (5) that the necessary condition for the existence of the baroclinically unstable disturbances is a disturbance phase shift with depth, the phase shift sign being determined by the amplification of damping of the disturbance. By comparing Equation (5) with the POLYGON-70 results described under Section 2.2 one becomes easily convinced that both by the signs and by the orders of magnitudes the inclination of the axis of the POLYGON-70 main anticyclone and the rate of its energy change were interrelated just in the way indicated by theory for the case of eddy intensification due to baroclinic instability of the zonal large-scale current (Koshlyakov & Yenikeyev 1977).

5.3   Wave Interpretation

Eliminating from (1a) the nonlinear terms and assuming the ocean bottom horizontal we find the solution of a linearized vorticity equation under usual boundary conditions in the form

$$p' = \sum_{n=0}^{\infty} p_n(x, y, t) P_n(z),$$

where functions $p_n(x, y, t)$ satisfy equations

$$\left(\Delta - \frac{f_0^2}{gh_n} \right) \frac{\partial p_n}{\partial t} + \beta_0 \frac{\partial p_n}{\partial x} = 0, \qquad n = 0, 1, \ldots, \tag{6}$$

and h_n and $P_n(z)$ are eigenvalues and eigenfunctions of the problem

$$\frac{d}{dz}\left(\frac{1}{\rho_0 N^2}\frac{dP_n}{dz}\right) + \frac{1}{gh_n\rho_0}P_n = 0;$$ (7)

$$\frac{g}{N^2}\frac{dP_n}{dz} + P_n = 0 \quad \text{at} \quad z = 0; \qquad \frac{dP_n}{dz} = 0 \quad \text{at} \quad z = -H.$$ (8)

It can be shown that $P_0(z)$ does not practically depend on z (barotropic mode of pressure perturbations), and $P_n(z)$, $n = 1, 2, \ldots$, have n zeroes within the interval $(-H, 0)$ (baroclinic modes). The values h_n are usually called equivalent depths; at $N = 2 \times 10^{-3}$ sec^{-1} and $H = 4$ km, $h_0 = 4$ km, h_1 is of the order of 70 cm, and h_2 about 15 cm, etc. If $p_n(x, y, t)$ is given in the form of plane harmonic waves, the following dispersion relationships are easily obtained:

$$\omega_n = -\frac{\beta_0 m_1}{m_1^2 + m_2^2 + f_0^2/gh_n}, \qquad n = 0, 1, 2, \ldots,$$ (9)

where m_1 and m_2 are the wave vector components along the axes x and y and ω is the angular frequency. The solution of Equation (6) requires the statement of the initial and the boundary (in x and y) conditions for functions p_n. In case of an ocean infinite in x and y the problem is easily solved with the aid of the Fourier transform. If at $t = 0$

$$p_n(\mathbf{x}, 0) = \int\int_{-\infty}^{\infty} \mathscr{P}(\mathbf{k}) \exp(i\mathbf{k}\mathbf{x})\, d\mathbf{k}, \qquad \mathbf{k} = (m_1, m_2), \qquad \mathbf{x} = (x, y);$$ (10)

then at $t > 0$ we shall have

$$p_n(\mathbf{x}, t) = \int\int_{-\infty}^{\infty} \mathscr{P}(\mathbf{k}) \exp\{i[\mathbf{k}\mathbf{x} - \omega_n(\mathbf{k})t]\}\, d\mathbf{k},$$ (11)

where the dependence of ω_n on $\mathbf{k}$ is determined by relationships (9).

The relationships (9) describing the local dynamics of the so-called Rossby waves (Rossby et al 1939) are of a fundamental importance for the studies of synoptic eddies. Since $\omega > 0$ we have $m_1 < 0$, and thus the free Rossby waves must move with a westward component that is corroborated by the measurements of synoptic eddies in the ocean regions with weak mean currents. Considering further for simplicity a wave moving in a purely zonal direction ($m_2 = 0$), assume, according to the observations, $m_1 = R^{-1}$ and take into account the relationship $R = (gh_1)^{1/2}f_0^{-1}$ relating the Rossby internal scale to the equivalent depth of the first baroclinic mode. Then for the phase velocity $c = \omega/m_1$ of the first baroclinic mode we shall obtain the relationship $c = -\beta_0 R^2/2$. At $\beta_0 = 2 \times 10^{-13}$ cm^{-1} sec^{-1} and $R = 5 \times 10^6$ cm we have $|c| = 2.5$ cm sec^{-1}, which is also in good agreement with the measurements of synoptic eddies, for instance during MODE-1. This fact alone indicates the prevalence of the first baroclinic mode in the synoptic disturbances of the oceanic currents. The latter conclusion is corroborated also by the studies of the vertical structure of the observed synoptic eddies (Vasilenko & Mirabel 1977, McWilliams & Flierl 1976).

The ocean model of two density layers ρ_1 and ρ_2 makes it possible in the framework of the Rossby wave linear theory to take into account approximately the influence of the large-scale current and the ocean bottom slope on the propagation of stable waves (Robinson & McWilliams 1973). For the case of the small thickness $h^{(1)}$ of the upper layer compared to the thickness $h^{(2)}$ of the lower layer, the presence of the mean current (velocity components U and V) only in the upper layer, and the direction of wave propagation not very close to the meridional direction, the dispersion relationships for the barotropic (ζ) and the baroclinic (ρ) Rossby waves can be written as

$$\omega_\zeta = \frac{1}{m_1^2 + m_2^2}\left[-\beta_0 m_1 + \frac{f_0}{h^{(2)}}(m_2 H_x - m_1 H_y) \right];$$ (12)

$$\omega_\rho = \frac{1}{m_1^2 + m_2^2 + \rho_0 f_0^2 (\Delta\rho \cdot g \cdot h^{(1)})^{-1}}\left[-\beta_0 m_1 + (m_1^2 + m_2^2)(m_1 U + m_2 V) \right],$$ (13)

where H_x and H_y are the zonal and the meridional slopes of the bottom, $\Delta\rho \equiv \rho_2 - \rho_1$. Summing two baroclinic waves propagating at an angle to one another, McWilliams & Robinson (1974) succeeded in the approximation of a westward drift of the main POLYGON-70 eddy which indicates, once again, the fitness in the first approximation of the Rossby wave linear models for the description of the mean drift of synoptic eddies in the ocean regions with a moderate eddy energy. Taking into account that $R = (\Delta\rho \cdot \rho_0^{-1} g h^{(1)})^{1/2} f_0^{-1}$, at $m_1^2 = R^{-2}$ and $m_2 = V = 0$, we can infer from (15) that $c_\rho = (-\beta_0 R^2 + U)/2$.

Bottom relief effects on the propagation of the Rossby waves in a barotropic ocean were studied by Rhines (1969a,b), Rhines & Bretherton (1973), Odulo (1975a,b), and Volosov (1976). The Rossby waves in a closed barotropic ocean were considered by Longuet–Higgins (1964, 1965), Pedlosky (1965), and Rhines & Bretherton (1973), and in a baroclinic ocean by Rattray & Charnell (1966). The induced Rossby waves generated by wind oscillations above the ocean were studied most thoroughly by Kamenkovich & Reznik (1978). The main conclusion is that the wind oscillations of the spatial and time scales of about 2000 km and one day generate mainly barotropic fluctuations of currents in the ocean. It follows from this that the direct wind effect cannot be considered as some essential factor of the generation of the synoptic oceanic eddies.

The development of the theory of the nonlinear Rossby waves is very difficult and therefore until recently it was made principally for the case of a barotropic ocean. In the latter case and in the assumption that the ocean bottom is horizontal, the vorticity equation (1a) can be rewritten (with the replacement of p' by stream function perturbation $\psi = p'/\rho_0 f_0$) in the form

$$\left(\Delta - \frac{f_0^2}{gH} \right)\frac{\partial\psi}{\partial t} + \frac{\partial(\psi, \Delta\psi)}{\partial(x, y)} + \beta_0 \frac{\partial\psi}{\partial x} = 0.$$ (14)

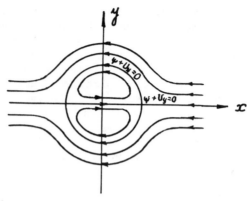

Figure 11 Schemes of stream lines in the field of a two-dimensional solitary barotropic Rossby wave in the coordinate system moving together with the wave, after Larichev & Reznik (1976a).

Of particular interest are the solutions of Equation (14) in the form of the two-dimensional solitary Rossby waves (solitons) (Larichev & Reznik 1976a,b) in the field of which stream function perturbations (Figure 11) attenuate exponentially in all the directions from the wave center. As can be seen from Figure 11, the central part of the soliton is occupied by the water moving with the wave (eddy). The physics of the soliton is such that it can move (relative to a constant zonal current) either to the east or to the west but in the latter case at a velocity necessarily exceeding the maximum possible velocity of the periodic barotropic Rossby waves $\beta_0 g H / f_0^2$ [see Equation (9)]. Mention should also be made of the Gulf Stream ring baroclinic model constructed by Flierl (1977), which gives a very good accordance with the experimental data.

5.4 *Numerical Models*

The greatest success in the theoretical study of the oceanic synoptic currents has so far been made through their numerical simulation on the fast-acting computers. These models can be divided into two main classes. 1. Local models that describe the dynamics of the eddy field in some limited ocean region (for instance, Rhines 1975, 1977, Bretherton & Haidvogel 1976, Owens & Bretherton 1977). 2. Eddy resolving models of the general oceanic circulation (for instance, Holland & Lin 1975, Robinson et al 1977. Semtner & Mintz 1977, Sarkisyan, Seidov & Semenov 1978). The models of the first kind describe nonlinear interaction between eddies and the response of the eddy field to the change of the "external" parameters of the ocean, such as bottom relief or stratification. The use of this approach has made it possible to obtain the eddy field that most resembles by its properties the one measured during MODE-1 (Owens & Bretherton 1977). A very important study was carried out by Rhines (1975), who has shown that at typical energy values the synoptic eddies in the ocean combine the properties of two-dimensional quasi-geostrophic turbulence and the classical Rossby waves. Rhines has obtained an especially interesting and original result, according to which the eddy field preassigned in the form of chaotic mesoscale two-dimensional turbulence under the directive action of the Earth's rotation and sphericity and due to internal interactions between the eddies is transformed into quasi-ordered nonlinear Rossby waves characterized by some prevailing spatial and time scales. The latter ones are determined by the mean energy of the eddies and stratification parameters and bottom relief, and even at a relatively high energy of the eddies they are interrelated as it is approximately indicated by the linear theory.

The models of the second class are aimed at solving the problems of the generation of synoptic eddies in the ocean, their transport, dissipation, and interaction with the large-scale oceanic circulation. The very fact of simulating eddies in such models is a very important achievement of theoretical oceanography. The most advanced models (for instance, Semtner & Mintz 1977) give the following directions of energy exchange. In the ocean regions adjacent to intensive currents of the Gulf Stream type, eddies are formed due to barotropic-baroclinic instability of the large-scale currents; in the rest of the ocean the baroclinic effect is decisively dominant. In their turn, the eddies expend energy mainly for bottom friction and for horizontal turbulent viscosity. The transition of eddy energy to the kinetic energy of the mean circulation in the deep ocean is noticeable; this is the effect of the so-called "negative viscosity" similar to that observed in the atmosphere (for instance, Oort & Piexoto 1974). The calculations show that the synoptic eddies in the ocean account for an appreciable meridional heat transport; the corresponding coefficient of horizontal turbulent heat conductivity is of the order of 10^7 cm^2 sec^{-1}.

ACKNOWLEDGMENTS

The authors are grateful to V. M. Kamenkovich and G. M. Reznik for their kind permission to use the manuscript of their paper "Rossby waves" and to E. A. Tsvetkova for the translation of the present review from Russian into English.

Literature Cited

Andrews, J. C., Scully-Power, P. 1976. The structure of an East Australian Current anticyclonic eddy. *J. Phys. Oceanogr.* 6:756–65

Barkley, R. A. 1968. The Kuroshio-Oyashio front as a compound vortex street. *J. Mar. Res.* 26:83–104

Barrett, J. R. 1971. Available potential energy of Gulf Stream rings. *Deep-Sea Res.* 18:1221–31

Beckerle, J. C. 1972. Prediction of mid-oceanic frontal passage confirmed in near-surface current measurements. *J. Geophys. Res.* 77:1637–46

Beckerle, J. C., La Casce, E. O. 1973. Eddy patterns from horizontal sound velocity variations in the main thermocline between Bermuda and Bahamas. *Deep-Sea Res.* 20:673–75

Bernstein, R. L., White, W. B. 1974. Time and length scales of baroclinic eddies in the Central North Pacific Ocean. *J. Phys. Oceanogr.* 4:613–24

Bernstein, R. L., White, W. B. 1977. Zonal variability in the distribution of eddy energy in the mid-latitude North Pacific Ocean. *J. Phys. Oceanogr.* 7:123–26

Brekhovskikh, L. M., Fedorov, K. N., Fomin, L. M., Koshlyakov, M. N., Yampolsky, A. D. 1971. Large-scale multi-buoy experiment in the Tropical Atlantic. *Deep-Sea Res.* 18:1189–1206

Brekhovskikh, L. M., Grachev, Yu. M., Koshlyakov, M. N., Fomin, L. M. 1978. Some results of the studies of synoptic eddies in the ocean. *Meteorol. Gidrol.* No. 2, pp. 5–14

Bretherton, F. P., Haidvogel, D. B. 1976.

Two-dimensional turbulence above topography. *J. Fluid Mech.* 78:129–54

Bryden, H. 1976. Horizontal advection of temperature for low frequency motions. *Deep-Sea Res.* 23:1165–74

Bubnov, V. A. 1960. On water dynamics of the Kuroshio-Oyashio frontal zone. *Tr. Gidrofiz. Inst.* 22:15–26

Bulgakov, N. P. 1967. Principal features of the structure and position of the sub-Arctic front in the North-West Pacific. *Okeanologiya* 7:879–88

Burkov, V. A., Ovchinnikov, I. M. 1960. Studies of the equatorial currents north of New Guinea. *Trudy Inst. Okeanol. Akad. Nauk SSSR* 40:121–134

Byshev, V.I., Grachev, Yu. M.. Ivanov, Yu. A. 1976. Studies of the velocity field mesostructure in the North Pacific Current. *Okeanologiya* 16:216–21

Charney, J. G. 1949. On a physical basis for numerical prediction of large-scale motions in the atmosphere. *J. Atmos. Sci.* 6:371–85

Charney, J. G., Stern, M. H. 1962. On the stability of internal baroclinic jets in a rotating atmosphere. *J. Atmos. Sci.* 19:159–72

Cheney, R. E., Richardson, P. L. 1976. Observed decay of a cyclonic Gulf Stream ring. *Deep-Sea Res.* 23:143–55

Cheney, R. E., Gemmill, W. H., Shank, M. K., Richardson, P. L., Webb, D. 1976. Tracking a Gulf Stream ring with SOFAR floats. *J. Phys. Oceanogr.* 6:741–49

Crease, J. 1962. Velocity measurements in the deep water of the western North Atlantic, Summary. *J. Geophys. Res.* 67:3173–76

Dantzler, H. 1977. Potential energy maxima in the tropical and subtropical North Atlantic. *J. Phys. Oceanogr.* In press

Duncan, C. P. 1968. An eddy in the subtropical convergence south-west of South Africa. *J. Geophys. Res.* 73:531–34

Eady, E. T. 1949. Long waves and cyclonic waves. *Tellus* 1:33–52

Flierl, G. 1977. The application of linear quasigeostrophic dynamics to Gulf Stream rings. *J. Phys. Oceanogr.* 7:365–79

Fomin, L. M., Yampolsky, A. D. 1977. Local kinematics of synoptic eddylike disturbances in the velocity field of the oceanic currents. *Dokl. Akad. Nauk SSSR* 232:50–3

Freeland, H., Rhines, P., Rossby, T. 1975. Statistical observations of the trajectories of neutrally buoyant floats in the North Atlantic. *J. Mar. Res.* 33:383–404

Fuglister, F. C. 1960. *Atlantic Ocean Atlas of Temperature and Salinity Profiles and Data from the International Geophysical Year of 1957–1958.* Woods Hole, Mass. Woods Hole Oceanogr. Inst. 209 pp.

Fuglister, F. C. 1963. Gulf Stream '60. In *Progress in Oceanography*, ed. M. Sears. 1:265–385. Oxford: Pergamon Press

Fuglister, F. C. 1972. Cyclonic rings formed by the Gulf Stream, 1965–66. In *Studies in Physical Oceanography*, ed. A. Gordon, 1:137–67. New York, London, Paris: Gordon & Breach

Fuglister, F. C., Worthington, L. V. 1951. Some results of a multiple ship survey of the Gulf Stream. *Tellus* 3:1–14

Gandin, L. S. 1964. Optimum interpolation of vector fields. *Trudy Gl. Geofiz. Obs.* 165:47–59

Gill, A. E. 1975. Evidence for mid-ocean eddies in weather ship records. *Deep-Sea Res.* 22:647–52

Gill, A. E., Green, J. S. A., Simmons, A. J. 1974. Energy partition in the large-scale ocean circulation and the production of mid-ocean eddies. *Deep-Sea Res.* 21:499–528

Grachev, Yu. M., Koshlyakov, M. N. 1977. Objective analysis of synoptic eddies in POLYGON-70. *POLYMODE News*, No. 23, unpublished document. Woods Hole, Mass. Woods Hole Oceanogr. Inst.

Hamon, B. V. 1965. The East Australian Current, 1960–64. *Deep-Sea Res.* 12:899–921

Hansen, D. 1970. Gulf Stream meanders between Cape Hatteras and the Grand Banks. *Deep-Sea Res.* 17:495–511

Holland, W. R., Lin, L. B. 1975. On the generation of mesoscale eddies and their contribution to the oceanic general circulation. *J. Phys. Oceanogr.* 5:642–69

Hunkins, K. L. 1974. Subsurface eddies in the Arctic Ocean. *Deep-Sea Res.* 21:1017–33

Ichiye, T. 1956. On the behaviour of the vortex in the Polar Front Region. *Oceanogr. Mag.* 7:115–32

Kamenkovich, V. M. 1973. *Fundamentals of the ocean dynamics.* Leningrad: Gidrometeoizdat. 238 pp.

Kamenkovich, V. M., Reznik, G. M. 1978. Rossby waves. In *Ocean Physics*, Vol. 2: *Ocean Hydrodynamics*, ed. V. M. Kamenkovich, A. S. Monin, pp. 300–58. Moscow: Nauka

Kawai, H. 1972. Hydrography of the Kuroshio extension. In *Kuroshio: Physical Aspects of the Japan Current*, ed. H. Stommel, K. Yoshida, pp. 235–52. Seattle: Univ. Washington Press

Keondjian, V. P. 1972. Diagnostic calculations of current velocities at 16 depths in the North Atlantic. *Izv. Akad. Nauk SSSR, Fiz. Atmos. Okeana* 8:1297–307

Kitano, K. 1975. Some properties of the warm eddies generated in the confluence zone of the Kuroshio and Oyashio currents. *J. Phys. Oceanogr.* 5:245–52

Kort, V. G., Byshev, V. I., Tarasenko, V. M. 1974. Synoptic variability of currents on the Atlantic polygon. In *The Atlantic Hydrophysical Polygon-70*, ed. V. G. Kort, V. S. Samoilenko, pp. 181–88. Moscow: Nauka

Koshlyakov, M. N. 1961. On water dynamics in the northwestern Pacific Ocean. *Trudy Inst. Okeanol. Akad. Nauk SSSR* 38:31–55

Koshlyakov, M. N. 1973. Results of the observations on the Atlantic polygon in 1970 in the light of some free Rossby wave models. *Okeanologiya* 13:760–67

Koshlyakov, M. N. 1978. Synoptic eddies in the ocean. In *Ocean Physics*, Vol. 1: *Ocean Hydrophysics*, ed. V. M. Kamenkovich, A. S. Monin, pp. 62–84. Moscow: Nauka.

Koshlyakov, M. N., Galerkin, L. I., Truong Din Hien. 1970. On the mesostructure of the open-ocean geostrophic currents. *Okeanologiya* 10:805–14

Koshlyakov, M. N., Grachev, Yu. M., Truong Din Hien. 1972. On the methods of studying the quasi-stationary oceanic currents. *Okeanologiya* 12:728–34

Kosklyakov, M. N., Grachev, Yu. M. 1973. Meso-scale currents at a hydrophysical polygon in the tropical Atlantic. *Deep-Sea Res.* 20:507–26

Koshlyakov, M. N., Yenikeyev, V. Kh. 1977. Synoptic-statistical analysis of the current field in POLYGON-70. *POLYMODE News*, No. 23, unpublished document. Woods Hole, Mass.: Woods Hole Oceanogr. Inst.

Lai, D. Y., Richardson, P. L. 1977. Distribution and movement of Gulf Stream rings. *J Phys. Oceanogr.* 7, No. 5:670–83

Lambert, R. B. 1974. Small-scale dissolved oxygen variations and the dynamics of Gulf Stream eddies. *Deep-Sea Res.* 21:529–46

Larichev, V. D., Reznik, G. M. 1976a. On two-dimensional solitary Rossby waves. *Dokl. Akad. Nauk SSSR* 231:1077–79

Larichev, V. D., Reznik, G. M. 1976b. High nonlinear two-dimensional solitary Rossby waves. *Okeanologiya* 16:961–67

Longuet-Higgins, M. S. 1964. Planetary waves on a rotating sphere. *Proc. R. Soc. London, Ser. A* 279:446–73

Longuet-Higgins, M. S. 1965. Planetary waves on a rotating sphere II. *Proc. R. Soc. London, Ser. A* 284:40–68

Masuzawa, J. 1955. An outline of the Kuroshio in the Eastern Sea of Japan

(Currents and water masses of the Kuroshio System, IV). *Oceanogr. Mag.* 7:29–47

Masuzawa, J. 1957. An example of cold eddies south of the Kuroshio. *Rec. Oceanogr. Works Jpn* 3:1–7

McCartney, M. S., Worthington, L. V., Schmitz, W. J. 1977. Large cyclonic rings from the northeast Sargasso Sea. *J. Geophys. Res.* In press

McEwen, G. F. 1948. The dynamics of large horizontal eddies (axes vertical) in the ocean off Southern California. *J. Mar. Res.* 7:188–216

McWilliams, J. C. 1976a. Maps from the Mid-Ocean Dynamics Experiment: Part I. Geostrophic streamfunction. *J. Phys. Oceanogr.* 6:810–27

McWilliams, J. C. 1976b. Maps from the Mid-Ocean Dynamics Experiment: Part II. Potential vorticity and its conservation. *J. Phys. Oceanogr.* 6:828–46

McWilliams, J. C., Flierl, G. 1976. Optimal quasi-geostrophic wave analyses of MODE Array Data. *Deep-Sea Res.* 23:285–300

McWilliams, J. C., Robinson, A. R. 1974. A wave analysis of the POLYGON array in the Tropical Atlantic. *Deep-Sea Res.* 21:359–68

Monin, A. S. 1969. *Weather Forecasting as a Problem in Physics.* Moscow: Nauka. 183 pp.

Monin, A. S., Kamenkovich, V. M., Kort, V. G. 1977. *Variability of the Oceans.* New York: Wiley (Russian edition, Gidrometeoizdat, 1974)

Obukhov, A. M. 1949. On the geostrophic wind. *Dokl. Akad. Nauk SSSR* 13:281–306

Obukhov, A. M. 1960. On statistically orthogonal expansions of empirical functions. *Izv. Akad. Nauk SSSR, Ser. Geofiz.* 3:432–40

Odulo, A. V. 1975a. Propagation of long waves in a rotating basin of varying depth. *Okeanologiya* 15:18–24

Odulo, A. V. 1975b. Propagation of long waves in an infinite ocean of varying depth. *Okeanologiya* 15:781–85

Oort, A. H., Peixoto, J. P. 1974. The annual cycle of the energetics of the atmosphere on a planetary scale. *J. Geophys. Res.* 79:2705–19

Orlansky, I., Cox, M. D. 1973. Baroclinic instability in ocean currents. *Geophys. Fluid Dyn.* 4:297–332

Owens, W. B., Bretherton, F. P. 1977. A numerical study of mid-ocean mesoscale eddies. *Deep-Sea Res.* In press

Ozmidov, R. V. 1962. Statistical characteristics of horizontal macroturbulence in the

Black Sea. *Trudy Inst. Okeanol. Akad. Nauk SSSR* 60:114–29

Ozmidov, R. V., Yampolsky, A. D. 1965. Some statistical characteristics of velocity and density variations in the ocean. *Izv. Akad. Nauk SSSR, Fiz. Atmos. Okeana* 1:615–22

Palmen, E., Newton, C. W. 1969. Atmospheric circulation systems. Their structure and physical interpretation. New York & London: Academic

Parker, C. E. 1971. Gulf Stream rings in the Sargasso Sea. *Deep-Sea Res.* 18:981–94

Patzert, W. C., Bernstein, R. L. 1976. Eddy structure in the central South Pacific. *J. Phys. Oceanogr.* 6:392–94

Pavlychev, V. P. 1975. Water regime and the sub-Arctic front position in the northwestern Pacific. *Izv. Tikhookean. Nauchno-Issled. Inst. Rybn. Khoz. Okeanogr.* 96:3–18

Pedlosky, J. 1964a. The stability of currents in the atmosphere and the ocean. Part I. *J. Atmos. Sci.* 21:201–19

Pedlosky, J. 1964b. The stability of currents in the atmosphere and the ocean. Part II. *J. Atmos. Sci.* 21:342–53

Pedlosky, J. 1965. A study of the time-dependent ocean circulation. *J. Atmos. Sci.* 22:267–72

Pedlosky, J. 1975a. On secondary baroclinic instability and the meridional scale of motion in the ocean. *J. Phys. Oceanogr.* 5:603–7

Pedlosky, J. 1975b. On the amplitude of baroclinic wave triads and meso-scale motion in the ocean. *J. Phys. Oceanogr.* 5:608–14

Phillips, N. A. 1963. Geostrophic motion. *Rev. Geophys.* 1:123–76

Rattray, M., Charnell, R. L. 1966. Quasi-geostrophic free oscillations in enclosed basins. *J. Mar. Res.* 24:82–102

Reid, J. L., Schwartzlose, R. A., Brown, D. M. 1963. Direct measurements of a small surface eddy off northern Baja California. *J. Mar. Res.* 21:205–18

Richardson, P. L., Strong, A. E., Knauss, J. A. 1973. Gulf Stream eddies: Recent observation in the western Sargasso Sea. *J. Phys. Oceanogr.* 3:297–301

Rhines, P. B. 1969a. Slow oscillations in an ocean of varying depth. Part I. Abrupt topography. *J. Fluid Mech.* 1:161–89

Rhines, P. B. 1969b. Slow oscillations in an ocean of varying depth. Part II. Islands and seamounts. *J. Fluid Mech.* 31:191–205

Rhines, P. B. 1975. Waves and turbulence on a β-plane. *J. Fluid Mech.* 69:417–43

Rhines, P. B. 1977. The dynamics of unsteady currents. In *The Sea*, ed. E. D. Goldberg, I. N. McCane, J. J. O'Brien, J. H. Steele, 6:189–318. New York: Wiley

Rhines, P. B., Bretherton, F. 1973. Topographic Rossby waves in a rough-bottomed ocean. *J. Fluid Mech.* 61:583–607

Robinson, A., McWilliams, J. 1973. The baroclinic instability of the open ocean. *J. Phys. Oceanogr.* 4:281–94

Robinson, A. R., Harrison, D. E., Mintz, Y., Semtner, A. J. 1977. Eddies and the general circulation of an idealized oceanic gyre: a wind and thermally driven primitive equation numerical experiment. *J. Phys. Oceanogr.* 7:182–207

Rossby, C. G., et al. 1939. Relation between variations in the intensity of the zonal circulation of the atmosphere and the displacement of the semi-permanent centers of action. *J. Mar. Res.* 2:38–55

Rossby, T., Voorhis, A., Webb, D. 1975. A quasi-Lagrangian study of mid-ocean variability using long range SOFAR floats. *J. Mar. Res.* 33:355–82

Sarkisyan, A. S., Seidov, D. G., Semenov, E. V. 1978. A numerical model of the synoptic-scale ocean currents. *Okeanologiya.* 18, No. 1:5–10

Schmitz, W. J. 1977. On the deep general circulation in the western North Atlantic. *J. Mar. Res.* 35:21–28

Schulman, E. E. 1967. The baroclinic instability of a mid-ocean circulation. *Tellus* 19:292–305

Seckel, G. 1975. Seasonal variability and parameterization of the Pacific North Equatorial current. *Deep-Sea Res.* 22:379–401

Semtner, A. J., Mintz, Y. 1977. Numerical simulation of the Gulf Stream and mid-ocean eddies. *J. Phys. Oceanogr.* 7:208–30

Sounders, P. M. 1971. Anticyclonic eddies formed from shoreward meanders of the Gulf Stream. *Deep-Sea Res.* 18:1207–19

Stockmann, V. B., Ivanovsky, I. I. 1937. Results of the stationary studies of currents off the western coast of the middle Caspian Sea. *Meteorol. Gidrol.* 3–5:154–60

Stockmann, V. B., Koshlyakov, M. N., Ozmidov, R. V., Fomin, L. M., Yampolsky, A. D. 1969. Long-time measurements of the spatial and time variability of the physical fields on oceanic polygons as a new stage in ocean studies. *Dokl. Akad. Nauk SSSR* 186:1070–73

Stommel, H. 1966. *The Gulf Stream. A Physical and Dynamical Description.* Univ. of Calif. Press. Cambridge Univ. Press. 248 pp. 2nd ed.

Swallow, J. C. 1971. The "Aries" current

measurements in the western North Atlantic. *Philos. Trans. R. Soc. London* 270:451–60

Swallow, J. C., Hamon, B. V. 1960. Some measurements of deep currents in the eastern North Atlantic. *Deep-Sea Res.* 6:155–68

Tareev, B. A. 1965. Unstable Rossby waves and nonstationary character of the ocean currents. *Izv. Akad. Nauk SSSR, Fiz. Atmos. Okeana* 1:426–38

Thompson, R. 1971. Topographic Rossby waves at a site north of Gulf Stream. *Deep-Sea Res.* 18:1–19

US POLYMODE Organizing Committee. 1976. *U.S. POLYMODE Program and Plan.* Cambridge: Mass. Inst. Technol. 98 pp.

Vasilenko, V. M., Mirabel, A. P. 1977. On the vertical structure of the ocean currents in different frequency ranges. *Izv. Akad. Nauk SSSR, Fiz. Atmos. Okeana* 13:328–31

Vasilenko, V. M., Mirabel, A. P., Ozmidov, R. V. 1976. On the spectra of current velocity and turbulent viscosity coefficient in the Atlantic Ocean. *Okeanologiya* 16:55–60

Volosov, V. M. 1976. Nonlinear topographic Rossby waves. *Okeanologiya* 16:389–96

Voorhis, A. D., Schroeder, E. H., Leetmaa, A. 1976. The influence of deep mesoscale eddies on sea surface temperature in the North Atlantic subtropical convergence. *J. Phys. Oceanogr.* 6:953–61

Vukovich, F. M. 1976. An investigation of a cold eddy on the eastern side of the Gulf Stream using NOAA 2 and NOAA

3 satellite data and ship data. *J. Phys. Oceanogr.* 6:605–12

Vulis, I. A., Monin, A. S. 1975. On available potential energy in the ocean. *Dokl. Akad. Nauk SSSR* 221:597–600

Warren, B. A. 1967. Notes on translatory movement of rings of current in the Sargasso Sea. *Deep-Sea Res.* 14:505–24

White, W. 1973. An oceanic wake in the Equatorial Undercurrent downstream from the Galapagos archipelago. *J. Phys. Oceanogr.* 3:156–61

Wilson, W. S., Dugan, J. P. 1977. Mesoscale thermal variability in the vicinity of the Kuroshio Extension. *J. Phys. Oceanogr.* In press

Worthington, L. V. 1977. *On the North Atlantic Circulation.* Johns Hopkins Oceanogr. Stud. Baltimore: Johns Hopkins Press

Wunsch, C. 1972. The spectrum from two years to two minutes of temperature fluctuations in the main thermocline at Bermuda. *Deep-Sea Res.* 19:577–94

Wyrtki, K. 1967. The spectrum of ocean turbulence over distance between 40 and 1000 kilometers. *Dtsch. Hydrogr. Z.* 20:176–86

Wyrtki, K., Magaard, L., Hager, J. 1976. Eddy energy in the oceans. *J. Geophys. Res.* 81:2641–46

Yasui, M. 1961. Internal waves in the open ocean (an example of internal waves progressing along the oceanic frontal zone). *Oceanogr. Mag.* 12:157

Yenikeyev, V. Kh., Koshlyakov, M. N. 1973. Geostrophic currents of the tropical Atlantic. *Okeanologiya* 13:947–62

AUTHOR INDEX

CUMULATIVE INDEXES

CONTRIBUTING AUTHORS VOLUMES 2 - 6

CHAPTER TITLES VOLUMES 2 - 6